Sustainable Food Supply Chains

Sustainable Food Supply Chains

Planning, Design, and Control through Interdisciplinary Methodologies

Edited by

Riccardo Accorsi
Alma Mater Studiorum—University of Bologna, Bologna, Italy

Riccardo Manzini
Alma Mater Studiorum—University of Bologna, Bologna, Italy

Academic Press is an imprint of Elsevier
125 London Wall, London EC2Y 5AS, United Kingdom
525 B Street, Suite 1650, San Diego, CA 92101, United States
50 Hampshire Street, 5th Floor, Cambridge, MA 02139, United States
The Boulevard, Langford Lane, Kidlington, Oxford OX5 1GB, United Kingdom

Library of Congress Cataloging-in-Publication Data
A catalog record for this book is available from the Library of Congress

British Library Cataloguing-in-Publication Data
A catalogue record for this book is available from the British Library

ISBN 978-0-12-813411-5

For information on all Academic Press publications
visit our website at https://www.elsevier.com/books-and-journals

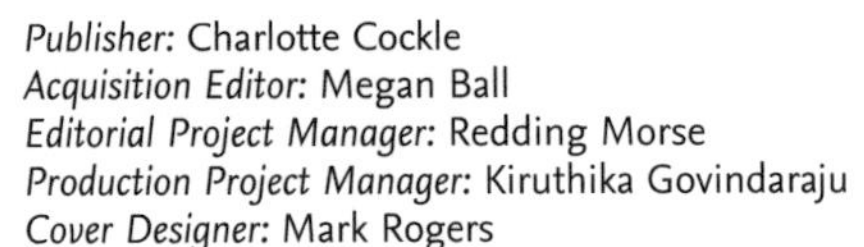

Publisher: Charlotte Cockle
Acquisition Editor: Megan Ball
Editorial Project Manager: Redding Morse
Production Project Manager: Kiruthika Govindaraju
Cover Designer: Mark Rogers

Typeset by SPi Global, India

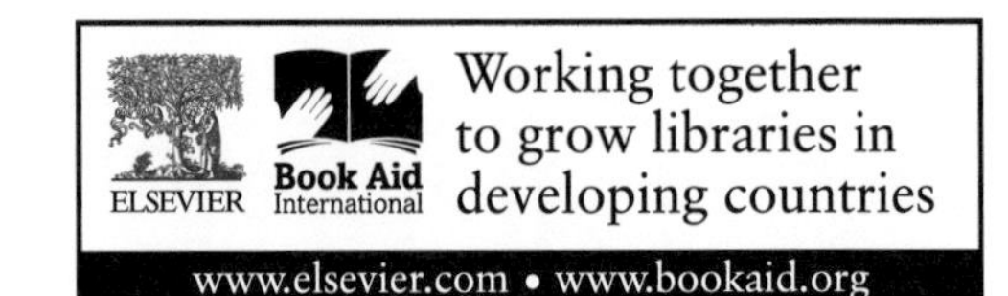

Contents

14. Quality assessment of temperature-sensitive high-value food products: An application to Italian fine chocolate distribution

Riccardo Manzini, Riccardo Accorsi, Marco Bortolini and Andrea Gallo

15. Models and technologies for the enhancement of transparency and visibility in food supply chains

Fredrik Nilsson, Malin Göransson and Klara Båth

16. Forecasting for food demand

Fotios Petropoulos and Shawn Carver

17. Food systems sustainability: The complex challenge of food loss and waste

Matteo Vittuari, Fabio De Menna, Laura García-Herrero, Marco Pagani, Laura Brenes-Peralta and Andrea Segrè

Contributors

Riccardo Accorsi (1, 61, 83, 115, 131, 145, 201, 277, 293, 305, 339, 351), Department of Industrial Engineering, Alma Mater Studiorum—University of Bologna, Bologna, Italy

Omar Ahumada (317), Autonomous University of Occident, Culiacan, Mexico

Renzo Akkerman (105), Operations Research and Logistics Group, Wageningen University, Wageningen, The Netherlands

Giulia Baruffaldi (115, 131, 293), Department of Industrial Engineering, Alma Mater Studiorum—University of Bologna, Bologna, Italy

Klara Båth (219), Microbiology and Hygiene, RISE Research Institutes of Sweden, Gothenburg, Sweden

Behzad Behdani (167), Operations Research and Logistics, Wageningen University, Wageningen, The Netherlands

Rajeev Bhat (23), ERA Chair for Food (By-)Products Valorization Technologies, Estonian University of Life Sciences, Tartu, Estonia

Jacqueline M. Bloemhof (159, 167), Operations Research and Logistics, Wageningen University, Wageningen, The Netherlands

Marco Bortolini (201, 305, 339), Department of Industrial Engineering, Alma Mater Studiorum—University of Bologna, Bologna, Italy

Laura Brenes-Peralta (249), Department of Agricultural and Food Sciences, Alma Mater Studiorum—University of Bologna, Bologna, Italy

Shawn Carver (237), Fiddlehead Technology Inc., Moncton, NB, Canada

Fabio De Menna (249), Department of Agricultural and Food Sciences, Alma Mater Studiorum—University of Bologna, Bologna, Italy

Ferruh Erdogdu (83), Department of Food Engineering, Ankara University, Ankara, Turkey

Yun Fan (167), Operations Research and Logistics, Wageningen University, Wageningen, The Netherlands

Emilio Ferrari (1), Department of Industrial Engineering, Alma Mater Studiorum—University of Bologna, Bologna, Italy

Andrea Gallo (201, 339), Department of Industrial Engineering, Alma Mater Studiorum—University of Bologna, Bologna, Italy

Mauro Gamberi (305), Department of Industrial Engineering, Alma Mater Studiorum—University of Bologna, Bologna, Italy

Federica Garbellini (351), Department of Industrial Engineering, Alma Mater Studiorum—University of Bologna, Bologna; CAMST—La Ristorazione Italiana Soc. Coop. A r.l., Villanova di Castenaso, Italy

Rodolfo García-Flores (261), CSIRO Data61, Melbourne, VIC, Australia

Laura García-Herrero (249), Department of Agricultural and Food Sciences, Alma Mater Studiorum—University of Bologna, Bologna, Italy

Francesca Giavolucci (351), Department of Industrial Engineering, Alma Mater Studiorum—University of Bologna, Bologna, Italy

Zhané Goff (145), Department of Industrial and Systems Engineering, Northern Illinois University, DeKalb, IL, United States

Malin Göransson (219), Packaging Logistics, Department of Design Sciences, Lund University, Lund, Sweden

Christian James (185), Food Refrigeration and Process Engineering Research Centre (FRPERC), Grimsby Institute, Grimsby, United Kingdom

Ivi Jõudu (23), ERA Chair for Food (By-)Products Valorization Technologies, Estonian University of Life Sciences, Tartu, Estonia

Pablo Juliano (261), CSIRO Agriculture and Food, Melbourne, VIC, Australia

Argyris Kanellopoulos (159), Operations Research and Logistics, Wageningen University, Wageningen, The Netherlands

Sara Limbo (49), Department of Food, Environmental and Nutritional Sciences—DeFENS, Università degli Studi di Milano, Milano, Italy

Riccardo Manzini (1, 115, 131, 201, 293, 351), Department of Industrial Engineering, Alma Mater Studiorum—University of Bologna, Bologna, Italy

Holger Meinke (39), Tasmanian Institute of Agriculture, University of Tasmania, Hobart, TAS, Australia

Christine Nguyen (145), Department of Industrial and Systems Engineering, Northern Illinois University, DeKalb, IL, United States

Fredrik Nilsson (219, 293), Packaging Logistics, Department of Design Sciences, Lund University, Lund, Sweden

Marco Pagani (249), Department of Agricultural and Food Sciences, Alma Mater Studiorum—University of Bologna, Bologna, Italy

Stefano Penazzi (277), Department of Industrial Engineering, Alma Mater Studiorum—University of Bologna, Bologna, Italy; The Logistics Institute, University of Hull, Hull, United Kingdom

Karolina Petkovic (261), CSIRO Manufacturing, Clayton South, VIC, Australia

Fotios Petropoulos (237), School of Management, University of Bath, Bath, United Kingdom

Luciano Piergiovanni (49), Department of Food, Environmental and Nutritional Sciences—DeFENS, Università degli Studi di Milano, Milano, Italy

Francesco Pilati (131, 305), Department of Industrial Engineering, Alma Mater Studiorum—University of Bologna, Bologna, Italy

Daniele Santi (131), Department of Industrial Engineering, Alma Mater Studiorum—University of Bologna, Bologna, Italy

Fabrizio Sarghini (83), Department of Agriculture, University of Naples Federico II, Naples, Italy

Andrea Segrè (249), Department of Agricultural and Food Sciences, Alma Mater Studiorum—University of Bologna, Bologna, Italy

Helena M. Stellingwerf (159), Operations Research and Logistics, Wageningen University, Wageningen, The Netherlands

Alessandro Tufano (115, 351), Department of Industrial Engineering, Alma Mater Studiorum—University of Bologna, Bologna, Italy

J. Rene Villalobos (317), International Logistics and Productivity Improvement Laboratory, Arizona State University, Tempe, AZ, United States

Matteo Vittuari (249), Department of Agricultural and Food Sciences, Alma Mater Studiorum—University of Bologna, Bologna, Italy

Luca Volpe (293), Department of Industrial Engineering, Alma Mater Studiorum—University of Bologna, Bologna, Italy

Foreword

The sustainable production and distribution of food is one of the most crucial problems, if not the definitive problem, facing the world today. As our population grows and standards of living rise, humanity is developing an appetite for foodstuffs once enjoyed by only a subset of wealthy consumers in developed nations. Supply chains for all products are becoming increasingly far-flung and complex, but those associated with food must support the insatiable demand of consumers ordering from a huge, diverse menu no longer limited by locale and season. Seemingly insignificant or benign choices may lead to unintended consequences, especially with respect to environmental and social issues: who can forget how consumers in North America and Europe developing a taste for quinoa reportedly led to shortages and price increases in Peru and Bolivia, effectively removing what was a local nutritional staple from the plates of Andean farmers?

Perhaps we can get by with fewer clothes in our closets or shop for apparel at secondhand stores, but the mantra of "reduce, reuse, and recycle" is less applicable to nutrition. A systematic approach to designing, implementing, and monitoring food supply chains (FSCs) is required. Bologna is considered the culinary capital of Italy, a country revered worldwide for its edible traditions, so it is appropriate that the University of Bologna is home to the Food Supply Chain Center, which takes such an approach. Through their work at the Center and their numerous publications and workshops, Dr. Accorsi and Dr. Manzini have devoted much of their professional lives to studying food supply chain systems, addressing the important questions in a cross-disciplinary and inclusive manner. As documented in a myriad of case studies, their industry partnerships have provided quantifiable benefits to companies and cooperatives within the food sector, while their active recruitment and advising of students is establishing a new generation of FSC researchers.

This book represents their latest effort in the field: they have compiled and edited two dozen chapters, starting with the design of agricultural systems and finishing with the decision support tools for specific food subsectors. Contributions from experts span a multitude of disciplines ranging from agricultural science to economics. These collected works are curated and ordered to inform academics and practitioners alike as to how to ask the right questions, use appropriate models to make use of available data, and translate results into policies and other relevant decisions.

Susan Cholette

Decision Sciences, San Francisco State University, College of Business, San Francisco, CA, United States

Preface

The idea behind this book has specific roots: Milan, July of 2015. The Universal Exposition is hosted by the Italian economic capital, in the north of what is perhaps the most renowned country worldwide for food traditions and industry. Over a 6-month period, more than 140 participating countries are called to show the best of their technology and best practices in order to offer a concrete answer to a vital question: how to guarantee healthy, safe, quality, and affordable food for everyone, while respecting the planet and its equilibrium. Economic development, agriculture, energy, and sustainability issues combine to title the Exposition's motto: "Feeding the Planet, Energy for Life."

The governments, citizens, and industries of the entire planet look at the Exposition as a great learning and sharing opportunity, leading each of us to consider ourselves part of a community and to face the hard challenges ahead, as Humanity. Food and water wars, food-induced migrations, environment and natural resources protections, pollutions, intensive agriculture, the debate between organic versus GMO products, and land-use control are just a few examples.

The central role of food in the way we conceive, plan, and design our society is the main finding of the Exposition and this encourages new choices for citizens, students, scholars and researchers, practitioners, and policymakers as well. Beyond these common hopes, some other personal and private pictures and memories arise from that event.

My colleagues and I had the chance to present at the Exposition some parts of the research we developed at the Food Supply Chain Center at the University of Bologna (http://foodsupplychain.din.unibo.it/), which focuses on the creation of decision-support tools that aid the design of sustainable food supply chain (FSC) operations from an industrial engineering perspective. In the era of the Internet of Things, virtualization of physical processes and gamification of reality, we developed a user-friendly ICT platform integrating the entities active in common food supply ecosystems and virtualizing the flows of food from the agricultural phases, through processing and transformation, to the consumer market. This tool (Accorsi et al., 2017) aims at enabling the users/planners to experience the effect of a decision on a given link of the network to the whole food ecosystem. Such a method is called *gaming*, because the users/planners play the levers of the network and experience the impact of their choices on the overall sustainability of the FSC as a whole ecosystem. Waste generation, GHG emissions generated by food production and transportation, water and energy use, and labor involvement at each stage are accounted for by the tool as a result of the decision of the planner, and results are shown through an easy and colorful dashboard.

Milan, Expo 2015. A shy child, encouraged by his mum, played the *FSC Game* with us, simulating and measuring the impacts of his decisions on the economic, environmental, and social sustainability of a food ecosystem.

That's what I most carried with me from the Universal Exposition. Not the shining exhibition stands, the healthy and delicious foods we tasted, the spectacular dances or fireworks, the solemn speeches from politicians. But an enthusiastic child playing as the "planner of tomorrow" at the global challenges that humanity will have to tackle: feeding the planet in a sustainable manner.

To learn from this lesson and to educate new classes of planners, new multidisciplinary methodologies and systemic approaches for the planning and design of sustainable FSCs must be introduced and formulated.

This book is intended to collect and link along a *fil-rouge* some models, tools, and approaches coming from different disciplines (i.e., agricultural science, food science, industrial and manufacturing engineering, operations research, economics, packaging science, transportation engineering) together involved in the design, planning, and control of FSCs and food ecosystems.

Starting from the agricultural systems and following the physical flow and cycle of food along processing, packing, storage, transportation, and consumption, a modeling approach is adopted at each stage in order to formulate practical taxonomy frameworks and quantitative models supporting a planner in design, planning, and control food systems and operations.

As a result, this book offers a wide overview of the issues affecting FSCs and seeks to provide practical and effective tools for their management. Given such an ample environment, the realization of this project would have been impossible for us alone to complete. We thus involved an enthusiastic group of selected experts, whose scientific contributions shaped most of the recent literature in the field. As editors, we spent most of our efforts in filling the gaps among the topics, connecting the contents, and reducing the overlap among chapters.

This book is mainly for master students and young researchers belonging to different disciplines (e.g., agriculture, engineering, economics, computer science, or mathematics faculties) but also for managers and practitioners of the food industry and policymakers called to rule the sector, or anyone who wants to better understand the decision drivers behind the design, planning, and control of sustainable FSCs. We do not expect to have covered all the issues and aspects affecting the agro-food ecosystems, but our efforts went in that direction, and future volumes might cover the still unhandled topics.

I warmly thank all the people who contributed to this project, which has heavily occupied me over the last 2 years: the coauthors and chapter contributors, the colleagues of the Food and Wine Supply Chain Council (FWSCC), the closest colleagues and friends of the Department of Industrial Engineering at University of Bologna, the companies involved in most of the research projects we hereby disseminated, the officers of the publishing house, and the scholars and researchers who have contributed in the last decade to the state-of-the-art of the field that has inspired and motivated us every single day.

Lastly, I am grateful to my family for all their support, and particularly to my wife, Debora, for her patience for the time I stole from us. I dedicate this book to them with much love.

Riccardo Accorsi

Reference

Accorsi, R., Bortolini, M., Baruffaldi, G., Pilati, F., Ferrari, E., 2017. Internet-of-things paradigm in food supply chains control and management. Proc. Manuf. 11, 889–895.

Sustainable food supply chain: Planning, design, and control through interdisciplinary methodologies

Riccardo Accorsi and Riccardo Manzini
Department of Industrial Engineering, Alma Mater Studiorum—University of Bologna, Bologna, Italy

The agro-food industry is a central and leading sector of the world economy (Manzini and Accorsi, 2013). In such a sector the business opportunities are wide, as are the challenges to be faced and the environmental and social externalities to measure and manage. By 2050, food demand will be doubled, because of further growth of the world population and larger consumption of animal food. As a consequence, food supply chains (FSCs) are expanding accordingly and products are shipped across countries involving several actors, responsible for processing, storage, and transportation operations, following a trend expected to accelerate in the future. The intensification of FSC processes requires investigation as to how sustainable these are, not just for practitioners and companies, but also for the environment and society as a whole (Accorsi, 2019).

Although there is an increasing focus of governments and authorities on the food sector, the design of sustainable FSCs is indeed far from a reality (Notarnicola et al., 2017). Many concerns still affect the well-known three pillars of sustainability in the food sector. The greenhouse gas (GHG) emissions from food and beverage production and transportation and their effects on climate change (Sala et al., 2017), the exploitation of natural resources like water and soil (Kummu et al., 2012), the competition for land between food and biofuels (Rathmann et al., 2010; Fischer et al., 2010a, b; Cobuloglu and Büyüktahtakın, 2015), the intensive use of genetically modified organisms (GMOs), pesticides, and chemical fertilizers vs. food organic models (Gerdes et al., 2012; Roy et al., 2009; Boye and Arcand, 2013; McLaughlin and Kinzelback, 2015), and the management of food waste (Garrone et al., 2014; Lebersorger and Schneider, 2014; Aiello et al., 2014) are just some of the major environmental issues affecting food ecosystems. From an economic perspective, volatile and unaffordable prices, bottlenecks and operational inefficiencies, and food losses, as well as food contamination and product counterfeiting, influence the profitability of food companies and the benefits for the actors and the stakeholders throughout the supply chain. Furthermore, food insecurity in less-developed countries and the impact that consumers' habits in the richest countries have on their own health results evidently in social issues, whose obesity and food desert (Thomas, 2010; Sadler et al., 2016) are claimed effects.

Although these issues independently affect the environmental, economic, and social dimensions of sustainability, they mutually arise from the lack of awareness of what role the operations play throughout the FSCs, of the way these are implemented and how interdependent they are, and, lastly, which costs, impacts, and externalities result from each single stage and process from farm to table.

While a large body of literature independently addresses the sustainability of several food supply chain processes and operations (Zhu et al., 2018; Accorsi et al., 2018a), less attention is given to interdisciplinary methodologies attempting the economic, environmental, and social sustainability of the whole food ecosystem.

Arising from the assumption that larger and more complex food supply chains require advanced skills and methods able to embrace their multidisciplinary nature, this book collects models, taxonomy frameworks, quantitative methods, and tools from several disciplines within the extended field of FSC ecosystems to address development and sustainability goals.

This book, intended mainly for master students and young researchers, food industry managers, and policymakers, is organized upon a *fil-rouge* that follows the physical flow of food from the growers to the consumers. Thus, it involves the different stages of the FSC and treats their main issues through a set of practical models and quantitative tools that support decision making. The book is organized in three parts: Part A introduces the issues affecting the sustainability of FSCs and defines the boundaries of a food ecosystem; the chapters of Part B follow the cycle of food from farm to fork and provide models, methods, and tools for the planning and management of such stages; Part C focuses on some significant and diffuse FSCs whose impacts have to be properly managed: fruits and vegetables, catering, and the meat industries. The outline of this book is summarized in Table 1.

Where Table 1 summarizes the issues, problems, and supply chain phase focused on within each chapter and classifies the disciplines involved, Fig. 1 provides a schematic map of the contents. The aim of this map is to exemplify the physical flows of food from farms/crops to the consumers' tables, and to show the number of actors involved, the network of facilities visited, and the set of entities contributing somehow to the food ecosystem. The steps and entities of the FSC are exemplified in a schematic way as they appear chapter by chapter when the associated decision-support models are introduced. The authors encourage the reader to look at Fig. 1 as a battleship board and to see the references to its entities accordingly (e.g., *G1* – a port).

Chapter 1 defines the elements composing a food ecosystem that goes from the rural areas to the demand points. By doing this, it also formalizes strategic planning models intended for the location of sustainable FSC uses, that is, crops (*D2*),

TABLE 1 The book outline

Part	Chapter	Goals	Involved disciplines
Part A	Chapter 1	Gives a systemic definition of an inclusive Food Ecosystem from the growers/farmers to the consumers and formalizes strategic models to design sustainable FSCs involving land-use allocation and network infrastructure planning.	Land-use planning Operations management
	Chapter 2	Summarizes the main issues affecting modern FSCs and provides a taxonomy framework to understand the challenges to be faced.	Agriculture science Food science
Part B	Chapter 3	Provides a modeling framework for the main decision problems affecting agricultural systems, their design, planning, and coordination.	Agriculture science Land-use planning Operations management
	Chapter 4	Introduces shelf-life models for the management and control of food processing and distribution processes.	Food science Food engineering
	Chapter 5	Discusses the role of packaging hierarchy in food product-package systems and provides a support-design top-down procedure for multidisciplinary and sustainable food packaging solutions.	Industrial engineering Packaging science
	Chapter 6	Explores the role of advanced computerized applications, mathematical models, and virtualization in the design of innovative and sustainable food primary package solutions.	Packaging science Mechanical engineering Food engineering
	Chapter 7	Focusing on food industry, provides production planning and scheduling models to increase efficiency of food processing and to reduce losses.	Operations management
	Chapter 8	Focusing on food catering industry, provides ready-to-practice models to aid the design of production resources in a food job-shop system.	Operations management
	Chapter 9	Overviews what happens along storage operations, and illustrates models and tools to manage the warehousing activities of perishable products.	Food science Operations management
	Chapter 10	Introduces decision-support models for the design of food distribution networks, and vehicle routing and delivery dispatching.	Operations management
	Chapter 11	Focuses on the impact of vehicle routing in cold-chains with the attempt of improving the sustainability of food deliveries.	Operations management
	Chapter 12	Overviews the main food transportation systems and technologies with practical examples of the impact of temperatures.	Operations management Transportation science
	Chapter 13	Provides a taxonomy framework for the food conservation, refrigeration, and transportation technologies.	Packaging science Mechanical engineering Transportation science
	Chapter 14	Illustrates how to use climate-controlled chambers to simulate the impact of environmental stresses, occurring along storage and distribution, on the quality of food products.	Food science Packaging science Operations management

TABLE 1 The book outline—cont'd

Part	Chapter	Goals	Involved disciplines
	Chapter 15	Exemplifies the adoption of information and communication technology (ICT) tools to improve the transparency of FSC operations and the safety and quality of delivered food.	Food science Packaging science Operations management
	Chapter 16	Provides forecasting models to manage food demand properly and avoid losses.	Operations management Economics
	Chapter 17	Introduces the topic of food losses and waste with a taxonomy framework that covers new practices to improve sustainability of FSCs.	Economics Policy
	Chapter 18	Provides decision-support models for the management and reduction of food losses along FSCs.	Operations management
	Chapter 19	Deals with the logistics of food in urban systems and suggests how to connect rural areas and consumers in a sustainable manner.	Land-use planning Agriculture science Energy management
	Chapter 20	Focuses on the logistics of food packaging and uses computerized tools to support the design of reusable packaging networks.	Packaging science Operations management
	Chapter 21	Sustains the penetration of renewable energies to power FSC facilities and operations by providing a strategic decision-support model for an integrated smart-grid and FSC network.	Land-use planning Energy management Operations management
Part C	Chapter 22	Presents planning tools for the control and coordination of the supply chain of seasonal products such as fresh fruits and vegetables.	Agriculture science Operations management
	Chapter 23	Focuses on the sustainability of the meat industry and presents a tactical decision-support model for the management of waste and the valorization of by-product.	Operations management
	Chapter 24	Illustrates the main issues of the food service industry, and provides a support-decision method for the scheduling and planning of meals production and delivery.	Operations management

processing facilities (*J2*), warehousing facilities (*L1*), renewable power plants (*H1*), carbon plantings (*I1*), and the allocation of distribution flows in between with the purpose of mitigating the externalities of the FSC. It is worth noting how renewables and carbon plantings respond to the environmental externalities of FSCs as they enable a reduction in the impacts and mitigate the associated carbon emissions (Accorsi et al., 2016).

While Chapter 2 overviews the issues and challenges affecting the global FSC, Chapter 3 focuses on the main planning problems affecting the agricultural systems (*B5-E6*).

Chapter 4 deals with models for the control shelf life and safety decay of food products along distribution processes, while Chapters 5 and 6 focus on the role of packaging and explore approaches and methodology for the design of sustainable food packaging solutions (*L1*, *3-O1*).

In Chapter 7 the management and scheduling of food processing activities in the production facility (*J2*) are formalized through an operational mathematical model. Chapter 8 deals with typical production systems (*Q4*), known as food job-shop (Tufano et al., 2018), designed for the food service industry that serve restaurants, canteens, hospitals, and schools. This chapter introduces quantitative methods to establish the adequate number of processing resources within the facility layout.

Chapter 9 concerns the warehousing activities (*M2*) of perishable products and provides decision-support models and tools for safe management (see also Accorsi et al., 2018b).

In Chapter 10, food distribution planning models (*J2-N2*) are formalized in terms of vehicle routing and network design problems.

Chapter 11 (*P2*) exemplifies the impact of vehicle routing models on the sustainability of the cold-chain.

Chapters 12 and 13 overview (e.g., *L1*, *F5*, *M3*) transportation modes and conservation systems used for the storage and distribution of food products.

Chapter 14 illustrates the application of the climate-controlled chamber to simulate the impact of environmental stresses experienced along storage and transportation (*L1*, *L3*) on food and package quality (Manzini et al., 2017).

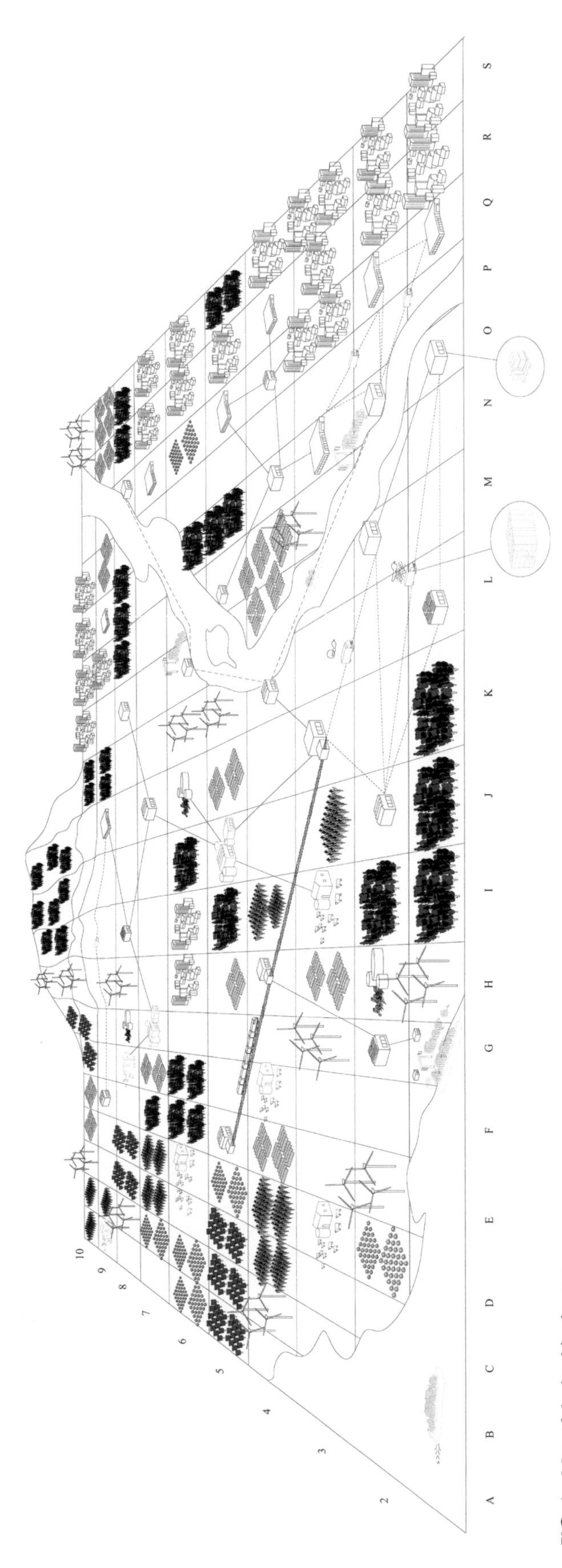

FIG. 1 Map of the book's chapters.

In Chapter 15 smart labels and other information technology tools (*L1*, *O1*) are adopted to improve the transparency of the FSC and to prevent food losses.

Chapter 16 looks at the retailer (*Q1-Q2*) and consumers (*R1*) stage by providing forecasting models intended for seasonal products like food.

The purpose of Chapter 17 is to discuss the drivers of food losses and waste and identify strategies, policy, and behaviors to prevent them.

To further support the minimization of losses and discarded food (*H2*), Chapter 18 provides tactical operations management models to aid the distribution of fresh products from crops (*D5*) to consumers (*H6*).

Chapter 19 is intended to design the sustainable configuration of a rural-urban ecosystem (*O6-S10*), including renewable fields (*R9*) and green areas (*N8*), where food is produced in crops and delivered to the city dwellers by minimizing the overall carbon emissions generated by transportation and mobility.

Chapter 20 follows the reverse flow of reusable packaging (*O1*) from the consumers to the growers and suggests the use of intermodality to reduce the carbon emissions from transportation.

Chapter 21 provides strategic network design models to foster the adoption of renewables (*G3-H3*) for powering FSC processes. The last chapters focus on specific supply chains.

Chapter 22 formalizes mathematical models to better coordinate agricultural and logistics operations for fresh fruits and vegetables (*B5-E5*).

Chapter 23 deals with the valorization of by-products from the meat industry (*J5*), and Chapter 24 discusses the layout design in job-shop facilities and the control of recipe-driven production activities (*Q4*) for the food catering industry.

Despite the works collected in this book, many cells in the map of Fig. 1 still need to be filled with new entities, processes, and related support-decision methods, belonging to the wide food ecosystem. In the end, we leave willing and creative readers hopefully inspired by these pages to build upon this map and provide suggestions and topics for future volumes and editions.

References

Accorsi, R., 2019. Planning sustainable food supply chains to meet growing demands. In: Encyclopedia of Food Security and Sustainability. Reference Module in Food Sciences. vol. 3. Elsevier, pp. 45–53. https://doi.org/10.1016/B978-0-08-100596-5.22028-3.

Accorsi, R., Cholette, S., Manzini, R., Tufano, A., 2018a. A hierarchical data architecture for sustainable food supply chain management and planning. J. Clean. Prod. 203, 1039–1054.

Accorsi, R., Baruffaldi, G., Manzini, R., 2018b. Picking efficiency and stock safety: a bi-objective storage assignment policy for temperature-sensitive products. Comput. Ind. Eng. 115, 240–252.

Accorsi, R., Cholette, S., Manzini, R., Pini, C., Penazzi, S., 2016. The land-network problem: ecosystem carbon balance in planning sustainable agro-food supply chains. J. Clean. Prod. 112 (1), 158–171.

Aiello, G., Enea, M., Muriana, C., 2014. Economic benefits from food recovery at the retail stage: an application to Italian food chains. Waste Manag. 34, 1306–1316.

Boye, J.I., Arcand, Y., 2013. Current trends in green technologies in food production and processing. Food Eng. Rev. 5 (1), 1–17.

Cobuloglu, H.I., Büyüktahtakın, I.E., 2015. Food vs. biofuel: an optimization approach to the spatio-temporal analysis of land-use competition and environmental impacts. Appl. Energy 140, 418–434.

Fischer, G., Prieler, S., van Velthuizen, H., Lensink, S.M., Londo, M., de Witc, M., 2010b. Biofuel production potentials in Europe: sustainable use of cultivated land and pastures. Part I: land productivity potentials. Biomass Bioenergy 34, 159–172.

Fischer, G., Prieler, S., van Velthuizen, H., Berndes, G., Faaij, A., Londo, M., de Witc, M., 2010a. Biofuel production potentials in Europe: sustainable use of cultivated land and pastures, Part II: land use scenarios. Biomass Bioenergy 34, 173–187.

Garrone, P., Melacini, M., Perego, A., 2014. Opening the black box of food waste reduction. Food Policy 46, 129–139.

Gerdes, L., Busch, U., Pecoraro, S., 2012. GMOfinder—a GMO screening database. Food Anal. Methods 5, 1368–1376.

Kummu, M., de Moel, H., Porkka, M., Siebert, S., Varis, O., Ward, P.J., 2012. Lost food, wasted resources: global food supply chain losses and their impacts on freshwater, cropland, and fertilizer use. Sci. Total Environ. 438, 477–489.

Lebersorger, S., Schneider, F., 2014. Food loss rates at the food retail, influencing factors and reasons as a basis for waste prevention measures. Waste Manag. 34, 1911–1919.

Manzini, R., Accorsi, R., 2013. The new conceptual framework for food supply chain assessment. J. Food Eng. 115 (2), 251–263.

Manzini, R., Accorsi, R., Piana, F., Regattieri, A., 2017. Accelerated life testing for packaging decisions in the edible oils distribution. Food Packaging Shelf Life 12, 114–127.

McLaughlin, D., Kinzelback, W., 2015. Food security and sustainable resource management. Water Resour. Res. 51, 4966–4985.

Notarnicola, B., Tassielli, G., Renzulli, P.A., Castellani, V., Sala, S., 2017. Environmental impacts of food consumption in Europe. J. Clean. Prod. 140 (2), 753–765.

Rathmann, R., Szklo, A., Schaeffer, R., 2010. Land use competition for production of food and liquid biofuels: an analysis of the arguments in the current debate. Renew. Energy 35, 14–22.

Roy, P., Nei, D., Orikasa, T., Xu, Q., Okadome, H., Nakamura, N., Shiina, T., 2009. A review of life cycle assessment (LCA) on some food products. J. Food Eng. 90, 1–10.

Sadler, R.C., Gilliland, J.A., Arku, G., 2016. Theoretical issues in the 'food desert' debate and ways forward. GeoJournal 81 (3), 443–455.

Sala, S., Assumpció, A., McLaren, S.J., Notarnicola, B., Saouter, E., Sonesson, U., 2017. In quest of reducing the environmental impacts of food production and consumption. J. Clean. Prod. 140 (2), 387–398.

Thomas, B.J., 2010. Food deserts and the sociology of space: distance to food retailers and food insecurity in an Urban American neighborhood. World Academy of Science. Eng. Technol. 43, 19–28.

Tufano, A., Accorsi, R., Garbellini, F., Manzini, R., 2018. Plant design and control in food service industry. A multidisciplinary decision-support system. Comput. Ind. 103, 72–85.

Zhu, Z., Chu, F., Dolgui, A., Chu, C., Zhou, W., Piramuthu, S., 2018. Recent advances and opportunities in sustainable food supply chain: a model-oriented review. Int. J. Prod. Res. 56 (17), 5700–5722.

Chapter 1

Modeling inclusive food supply chains toward sustainable ecosystem planning

Riccardo Accorsi, Emilio Ferrari and Riccardo Manzini

Department of Industrial Engineering, Alma Mater Studiorum—University of Bologna, Bologna, Italy

Abstract

The increasing global food demand is forcing the food industry to identify new effective strategies for production and distribution, as was already experienced in the 1960s and 1970s with the birth of green chemistry and the explosion of fossil-fueled agriculture.

The globalization of the food trade has bridged the barriers between production and consumption without respecting the balance of natural resources, resulting in enlarged gaps between developed and developing countries. In the food sector many issues affect the three dimensions of sustainability: economic, environmental, and social. These include land-use change and deforestation to widen farms, pastures, and biofuel cultivations, land grabbing to establish intensive agriculture, clean water consumption, soil and air pollution, lack of supply chain infrastructures, volatile prices, and climate change. Together these issues make current food supply chains (FSCs) unsustainable over the long term and challenge the future of the food industries and society as well. Such issues also reveal the lack of connections and coordination among actors involved in FSCs and open debate about the neglected role of the physical and logistic/distribution infrastructures in addressing long-term sustainability targets.

This chapter illustrates a hierarchical framework aimed at modeling production and distribution food ecosystems through a set of interdisciplinary parameters and decision variables. The decision levers identified in this framework describe how the food ecosystem behaves according to an input-output flow analysis. The framework formulates a set of planning decision problems via mixed linear programming. The definition of the inclusive food ecosystem inspires collecting multidisciplinary parameters that impose collaboration between decision-makers. Furthermore, as part of the ecosystem, logistics and distribution processes are involved in the planning issues beyond the common perception that sees an FSC as a sequence of independent stages.

1 Introduction

Food systems have played a central role in the progress of society. Early, crops and farms remained close to the village, as the village dwellers worked the farms and their crops were the main source of sustenance. Agriculture influenced people's well-being on the basis of the natural resources available (e.g., water energy, water, fertile soil) and favorable climate conditions (e.g., regular seasons, mild weather). In the past, people migrated toward fertile ecosystems for living and growing their food (Steel, 2008; Leduc and Van Kann, 2013).

The industrial and technological revolution of the last two centuries deeply changed our economy and the agriculture sector as well. Food production and distribution processes have gradually evolved, powered by cheap energy sources and fossil fuels. Crops and farms enhanced their yields due to the advent of green chemistry, and new storage and transportation systems made food cheaper, safer, and more affordable (Jones et al., 2016; Stephens et al., 2017). People stopped migrating toward fertile ecosystems and established new patterns for food production and distribution on a large scale. Namely, people stopped moving and food began to travel.

Nowadays, global trade has made such patterns undeniable. Food is shipped across countries through processing, storage, and distribution operations that entail multiple actors and logistics networks. FSC networks require advanced management skills and new decision-support models and tools able to embrace their multidisciplinary nature.

Most of the distribution activities worldwide deal with food and energy (O'Neill et al., 2010). The global trade allows safer and more affordable food to be provided to many consumers, despite the season and area of consumption (Govindan, 2018). Matching supply and demand is considered a target of the FSC (Apaiah et al., 2005). However, food insecurity still affects about 15% of the world population (FAO, IFAD, and WFP, 2012; FAO, 2013), while the long-term economic, environmental, and social sustainability of the food sector are far from being adequately addressed.

Sustainable Food Supply Chains. https://doi.org/10.1016/B978-0-12-813411-5.00001-6

In the last decade, global food production has grown exponentially to satisfy new markets like India and China. By 2050, the global food demand will be doubled, because of further increases in the world population and larger consumption of animal food (Koning and Ittersum, 2009) resulting from changing food habits. In developed countries, food demand is pushed by increasingly more sophisticated consumer expectations. As a consequence, FSCs are expanding accordingly with the trend expected to accelerate in the future (Ahumada and Villalobos, 2009; World Bank, 2013). The intensification of FSC processes causes policymakers and practitioners to question whether or not these are sustainable for the environment, not just for the stakeholders and companies of the sector.

Scholars and policymakers are debating the impact that food systems have on the environment and will have in the future (FAO, 2012b). So far, only partial answers are available. One answer is based on the expected continuing of technological advances to increase crop yields further, theoretically expanding them by 80% (Penning de Vries et al., 1995). Recent literature, however, suspects that such an estimate was overly optimistic, as it does not account for the growth of urban areas and the necessity of wild areas to protect biodiversity and areas devoted to energy production (i.e., biofuels). Koning and Ittersum (2009) prudently reduce this value to 40%, concluding that although we have enough soil to feed the planet, the impacts and the externalities need to be quantified and assessed.

The globalization of the food sector changed the metabolism between society and the environment (Lowe et al., 2008; Raspor, 2008; Notarnicola et al., 2012; Amate and Gonzalez de Molina, 2013). From a natural source of sustenance, agriculture became a resource-intensive industry. Food operations involve production, packaging, storage, and logistics, as for any other product. The role of nonagricultural activities throughout the FSC thus increases and several industrial processes develop in between growers and consumers. These include consolidating raw materials, food processing, packaging, conservation, distribution, and, lastly, the treatment of waste and losses through recycling or valorization chains. The physical distance between growing and consuming areas is proportional to the number of stages and actors involved along the supply chain.

Such distribution networks need innovative planning and management methods and tools that enforce the sustainability of the food ecosystem as a whole. These methods must identify sustainable practices and guidelines for:

1. the *fair exploitation* of natural *resources*,
2. the *reduction* of the *human-induced impacts* on the environment, and
3. the *incorporation* of economic, environmental, and social *externalities* in the agro-food industry decisions.

Use of natural resources, as an example, must not exceed the regeneration rate (Leduc and Van Kann, 2013; FAO, 2012a). Such practice is encouraged even more when producing and consuming areas are different. The consumers typically are not aware of operations beyond the supply of food and the environmental and social impacts these operations provoke. It is not uncommon, indeed, that the natural ecosystems of less-developed countries are overstressed to provide cheap food to the richest countries (Tukker et al., 2008). Globalization strengthens the connections among geographical areas but affects the overall sustainability of global ecosystems (Amate and Gonzalez de Molina, 2013).

Therefore, the question to answer is no longer how to feed the planet, but how do it in a sustainable manner. According to this philosophy, FSCs need to become not just more efficient and affordable, but even more sustainable and resilient. Despite the implementation of rules, laws, and international agreements that promote environmentally friendly practice in any industrial sector (Nilsson, 2004; Massoud et al., 2010), the sustainability of FSCs is still an ambitious goal (Partidario et al., 2007). Many metrics and indicators confirm this fact. Whether the ratio of the energy consumed by FSC operations to the nutritional value of supplied food is a metric of sustainability (Amate and Gonzalez de Molina, 2013), recent literature confirms that this parameter has grown exponentially in the last decades (Conforti and Giampietro, 1997; Gallo et al., 2017).

Possible solutions lie in the identification of the best compromise between short- and long-term objectives bridging together decisional perspectives and targets (Spiertz, 2010). The management of the environmental impacts in the planning of FSCs requires first their quantification and assessment. Some of these are measured through greenhouse gas (GHG) emissions, soil degradation, energy consumption, eutrophication, and climate change. The contribution of each actor and stage of an FSC to these impacts should be allocated and quantified. To do this, a new inclusive definition of the boundaries of the food ecosystems is introduced. Fig. 1 conceptualizes the agro-food ecosystem extended beyond the spatial boundaries, and including agriculture, processing, and packaging, storage and transportation, and some mitigation strategies.

According to this model, widely discussed by Accorsi et al. (2016a), an FSC is assumed to be a distributed ecosystem aimed at providing food to consumers while optimizing the available resources and the environmental externalities. The long-term sustainability of this ecosystem is crucial to meet the goals of the Kyoto Protocol (Almer and Winkler, 2017), but requires joint and integrated multidisciplinary tools able to account for and incorporate economic, technological, organizational, social, and environmental aspects into the strategic planning and design of sustainable FSCs.

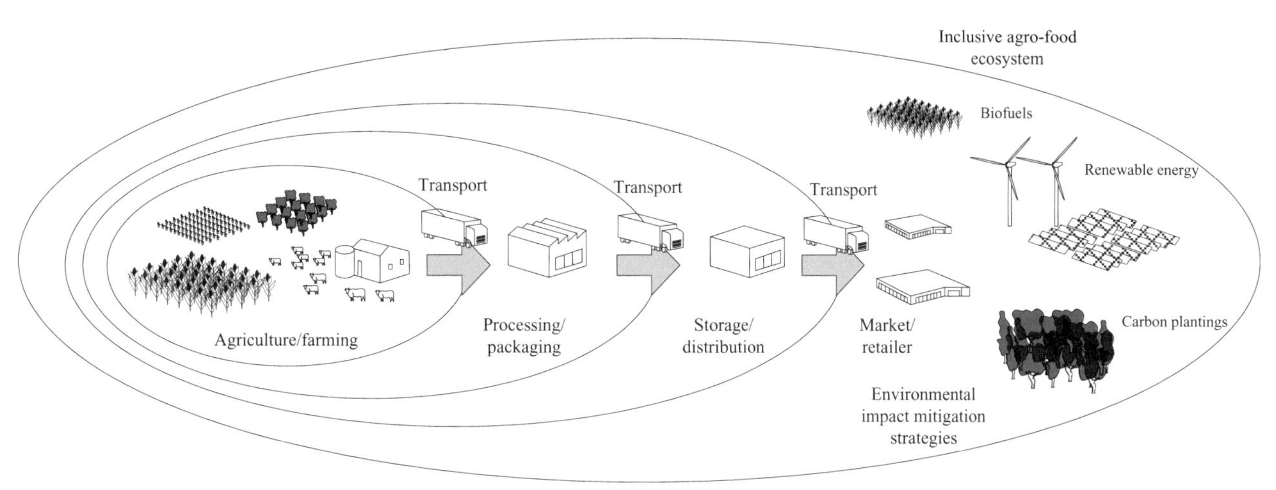

FIG. 1 Inclusive food supply chain ecosystem.

1.1 Scope of this chapter

The growth of global food demand elicits debate on how to reconcile economic growth of the sector with environmental sustainability. New planning approaches should include both agriculture and supply chain operations. Economic, environmental and social targets should drive a revolutionary change in the management and planning of food systems. New planning approaches aimed to design sustainable inclusive ecosystems should be formulated and diffused in practice. These methods may build upon existing techniques, that is, life cycle assessment (LCA), which assess the impacts of the food life-cycle from crop to consumer (Accorsi et al., 2013). Decision making in food systems requires strengthening the interaction between scientific disciplines (i.e., agriculture science, food science, biology, industrial, environmental, and civil engineering, computer science, operations research, geography, business and management science) and adopting new quantitative decision-support tools, methodologies, and approaches still lacking (McCown, 2002a, b).

This chapter illustrates a framework that collects strategic planning problems intended to design an FSC assumed as a closed ecosystem. The scope is to illustrate new ways to perceive a food system and to aid its design. The framework is organized through a top-down methodology, whose steps are summarized as follows:

1. First, it aims at drawing the food system in a quantitative manner through a set of multidisciplinary parameters that describe the behavior of the system and quantify the available resources and the impacts resulting from their exploitation.
2. Then, it defines a set of planning problems that can be formulated through tailored decision variables.
3. Lastly, it allows the formulation of models that quantify benefits and externalities of each choice and incorporate both in the planning process. It is worth noting that the scale of the planning problem is no longer local, but it overcomes the spatial boundaries according to the inclusive perspective illustrated in Fig. 1.

While agriculture is spatially confined, the main processes of FSCs are distributed over a wide geography. The planning problems adopted in agriculture deal with the land-use allocation, the scheduling of sowing, cropping and harvesting, and the management of production resources (e.g., labor, machinery). The other FSC processes require the procurement of packaging materials, the location of facilities, the production planning, the inventory management of perishable products, the design and control of storage systems, and the strategic-tactical planning of logistics.

The integration of all FSC stages reflects on the integration of decision problems that are currently solved independently. These problems are formulated via linear and mixed integer programming as commonly adopted in operations research. However, the planning problems can be approached using other quantitative techniques (e.g., simulation, data mining and analytics, machine learning, empirical methods, and analytical methods).

The framework also sustains the fulfillment of new targets (e.g., GHG emission minimization), in comparison with the traditional maximization of profit and yields. Furthermore, it looks at the FSC as a whole and avoids single-stage (partial) optimization (Lowe and Preckel, 2004; Lucas and Chhajed, 2004; see Ahumada and Villalobos, 2009 for an extended survey). The proposed framework indeed seeks for minimizing the externalities from crop to consumer, while guaranteeing the fulfillment of the food demand.

The remainder of this chapter is organized as follows. Section 2 describes the methodology of analysis. After the illustration of the food ecosystem parameters, examples of planning problem formulation are given. These examples focus on the minimization of the CO_2 emissions resulting from food production and distribution. The discussion of the potential application of the framework and the expected benefits is treated in Section 3. Section 4 concludes the chapter with a proposal for future developments.

2 A framework to model food supply chains

The intensive way that agriculture is developed compels careful exploitation of the natural resources of energy, soil, and water and the reduction of the impacts and externalities. At the actual current rate, the exploitation of natural resources is no longer sustainable. Furthermore, the externalities of FSCs are not quantified, as a broad and deep visibility into such operations is lacking. Typically, food companies prefer areas where production conditions, for example, mild climate and cheap labor, are more favorable, even though it might lead to overstressing some geographic areas (Meinke et al., 2009).

The supply of quality, affordable, and environmentally friendly food is still a hotspot for both practitioners and scholars (Spiertz, 2010; Brussard et al., 2010). There is an urgent need for integrated and multidisciplinary approaches able to shed light on externalities generated by food systems and to support their minimization through quantitative approaches (Luning and Marcelis, 2009). This chapter proposes a framework for the design of sustainable FSCs that:

(1) supports the conceptual modeling of an inclusive food ecosystem;
(2) focuses on the minimization of the externalities;
(3) merges the planning problems of agriculture, production, and distribution;
(4) fosters a multidisciplinarian collection of data to understand the food ecosystem as a whole;
(5) sustains the design of cleaner agricultural, production, and distribution processes.

This framework integrates the design of the supply chain network, that is, food consolidation, processing, and storage facilities, with the allocation of land uses, that is, functions of an area.

Literature provides several FSC design models merging agriculture and logistics operations (Ahumada and Villalobos, 2009). These models use optimization to aid planning of a specific stage/process, rather than to adopt a broad perspective, mainly because of modeling and computational complexity and lack of large real-world datasets. Conversely, this framework leads to the collection and manipulation of a wide set of parameters belonging to agriculture, production, and distribution fields in agreement with a comprehensive and referenced data architecture (Accorsi et al., 2018c).

In order to lead the decision maker throughout data collection and the formulation of the planning problem, a multistep methodology is set as follows:

1. Geography mapping;
2. Agro-climate data collection;
3. Setting spatial functions;
4. Analyzing potentials;
5. Modeling agro-food parks;
6. Modeling agro-food ecosystems.

Descriptions of each step are found in the following sections. The framework leads the decision maker across the strategic planning of an FSC, attempting to reach environmental sustainability of the operations and the minimization of externalities, while fulfilling the expected food demand.

2.1 Geography mapping

The first step defines the geographical boundaries of the analysis. The planner identifies on the map areas to be candidates for the allocation of crops/orchards/farms or other FSC services. The scale of the areas (e.g., 1 ha, 10 ha) depends on the accuracy of the spatial grid and on the availability of spatial data. For example, the meteorological stations spread over a national scale represent robust sources of weather and climate spatial data, but soil properties change frequently over the same grid. Deciding the proper spatial unit to adopt is indeed crucial. This should be reasonably homogeneous in its properties (e.g., soil thickness, structure, weather), but not too small, in order to avoid redundant instances and computational troubles (Ostendorf, 2011).

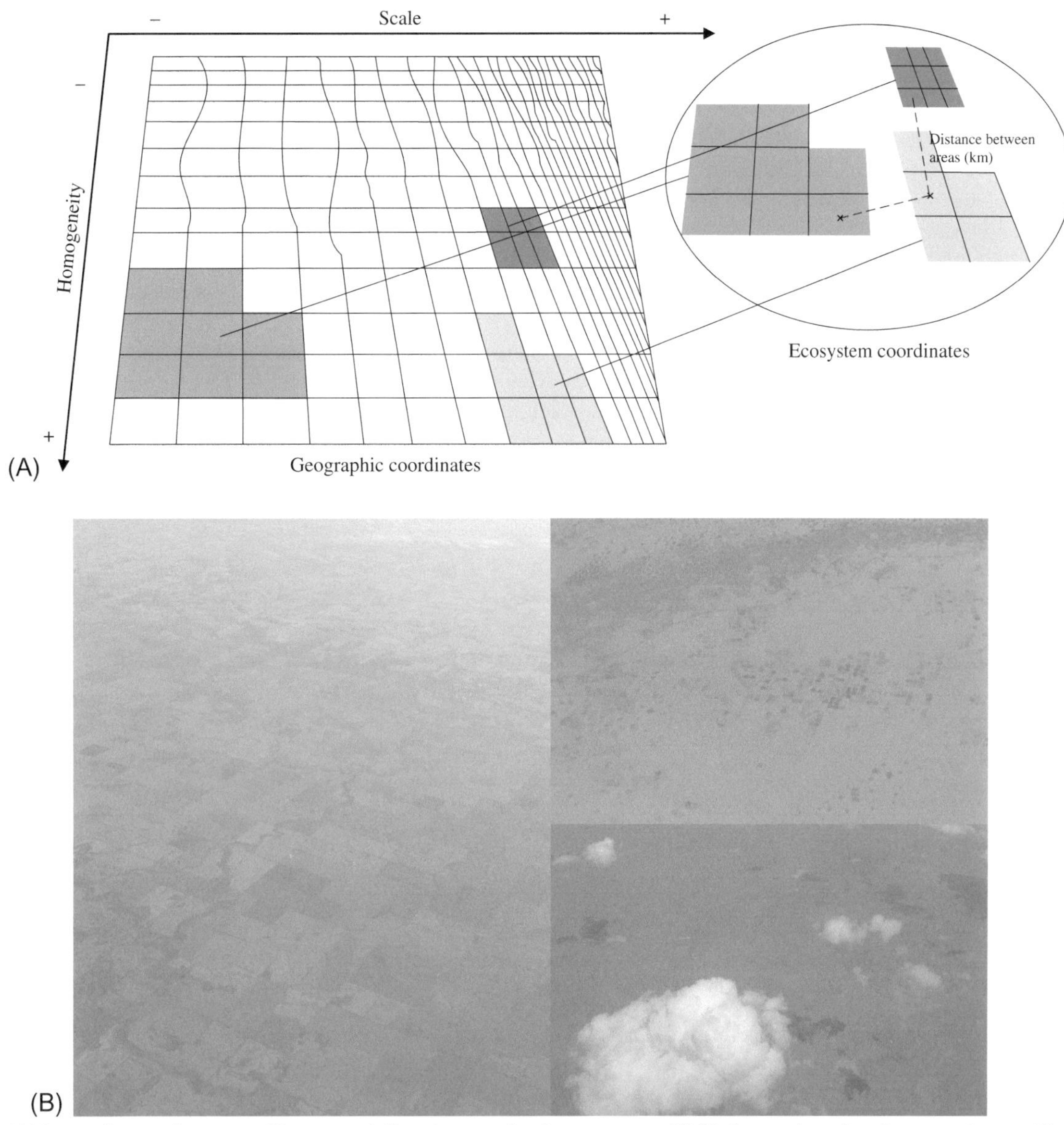

FIG. 2 (A) Levers for mapping geographic areas and allocating to a planning ecosystem; (B) Bird's-eye view of rural areas as they would appear to a planner: from Oregon's skies (left); over Niger (right above); somewhere above Central Africa's skies (right below).

Fig. 2A illustrates the two levers for the selection of the planning domain. The scale affects the size of the spatial unit, whilst the homogeneity indicates the level of similarity between the shape of the selected areas. Areas with different scale and homogeneity can be selected, but need to be normalized into a new system of coordinates, namely *ecosystem coordinates*. In this new reference system the spatial unit is defined in agreement with availability of spatial data and routing distance among each unit calculated. This allows considering areas spread over a large geography within a planning ecosystem that break the physical barriers of distance. The areas involved at this step are typically cleared (i.e., without both building and forests) because they will be allocated to new land uses. Nevertheless, the homogeneity of the spatial units may vary with mountains, rivers, and other natural barriers.

2.2 Agro-climate data collection

The second step is the collection of spatial agro-climate data that describes the characteristics of the selected areas. The planning of the agro-food ecosystem entails the assessment of the benefits resulting from land-use allocation to the ecosystem areas (Glen, 1987; Lucas and Chhajed, 2004).

In dealing with agricultural systems, the literature widely discusses the strategies to maximize the crop yield, optimize the use of water, and minimize the environmental impacts (Pacini et al., 2004; Ahumada and Villalobos, 2009; Bryan et al., 2011). Other studies provide prediction models for crop yields. The crop yield is affected by two interdependent environmental drivers: the soil properties and the weather conditions (Mathe-Gaspar et al., 2005). Agriculture, indeed, is the most climate-sensitive sector, with outdoor production sensibly affected by temperature and precipitation (Coehlo and Costa, 2010). Soil properties specify the ability to accept, retain, and transmit water to plants (Van Ittersum et al., 2003; Jones et al., 2003). Climate and soil data series are used to formulate empirical models as well as to fuel simulation models (Soltani and Hoogenboom, 2007) for crop yield prediction. These models typically require the value of parameters like temperature, rainfall, air humidity, wind speed, solar radiation, soil texture and slope gradient, moisture characteristics in the root, and others (Schultink et al., 1987), whose collection is often infeasible without remote sensing tools; in addition, accessible datasets are available only for some countries.

Notwithstanding poor data availability, better understanding as to which parameters are of interest and for which purpose is one of the aims of this framework. Fig. 3 exemplifies such parameters intended, first, to predict the crop yields of the ecosystem's areas, and then to assess the potential of other uses of land like renewable energy production (e.g., solar and wind energy) of FSC services. Through the definition of the main parameter, the rationale of Fig. 3 anticipates how this framework supports synergies between FSC services, involving agriculture and distribution facilities but also renewables and carbon plantings as planning levers to achieve the environmental sustainability of the ecosystem.

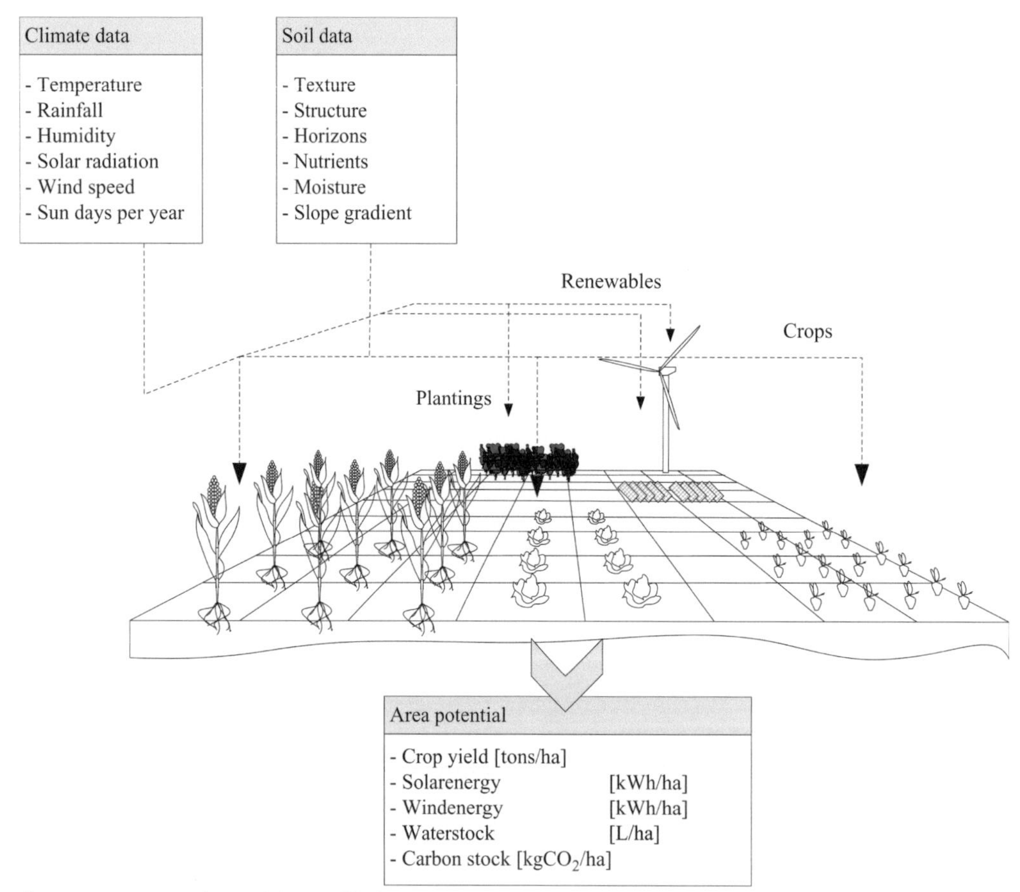

FIG. 3 Agro-climate parameters and potentials quantified per area.

2.3 Set spatial functions

This step aims at defining the alternative spatial functions, or land-uses, allocated to the areas of the food ecosystem. The land-use allocation problem may consider ample opportunities, in agreement with the planning purpose (Yewlett, 2001; Witlox, 2005). The target of the planner gives a first input about the different land uses to consider: different crops and varieties, the supply chain services, renewables, etc. The demand to fulfill influences the solution of the land-use allocation problem. For example, the presence of localized or distributed demand determines the importance of supply chain services in comparison with crops and agricultural uses. The importance of handling the land-use problem along with the planning of sustainable FSCs is highlighted by the literature, which identifies land-use change and transportation as the most significant sources of air pollution (Desjardins et al., 2007; Ovando and Caparros, 2009).

The sustainable food ecosystem is built in this framework upon a set of land uses that are somehow influenced by weather and soil. These four spatial functions are classified as follows:

1. *Production use*. This includes the spatial functions for primary food production, which refers to agriculture and farming (i.e., crops, farms, raw food preprocessing).
2. *Resource harvest use*. This set provides the resources needed for production and distribution processes (i.e., renewable energy, water source).
3. *Mitigation use*. This spatial function aims at mitigating the externalities generated within the food ecosystem. Examples of environmental externalities are GHGs, while possible land-intensive mitigation strategies can be plantings to sink carbon, that is, afforestation processes.
4. *Supply chain service*. This set implements supply chain services provided through a network of production/processing facilities (treated by Penazzi et al., 2017; Tufano et al., 2018), storage facilities (discussed in Accorsi et al., 2017e), and packaging facilities or other logistic nodes.

By including these spatial functions into a closed (but distributed) ecosystem (Fig. 4), the framework investigates the interaction among rural and green areas, renewable sources, and supply chain infrastructures and their contribution to generate environmental impacts. The resulting land-use allocation problem involves further aspects of the FSC other than just agriculture and is intended to consider the land used as a precious resource to manage carefully. Nevertheless, it is worth stating that, although we limit the analysis to those processes requiring land, other environmental stressors can exist in an FSC, for example, fertilizers, farming sewage, and water use (Tilman et al., 2002; Sheerr and McNeely, 2008; Spiertz, 2010; Walters and Hansen, 2012; Notarnicola et al., 2012).

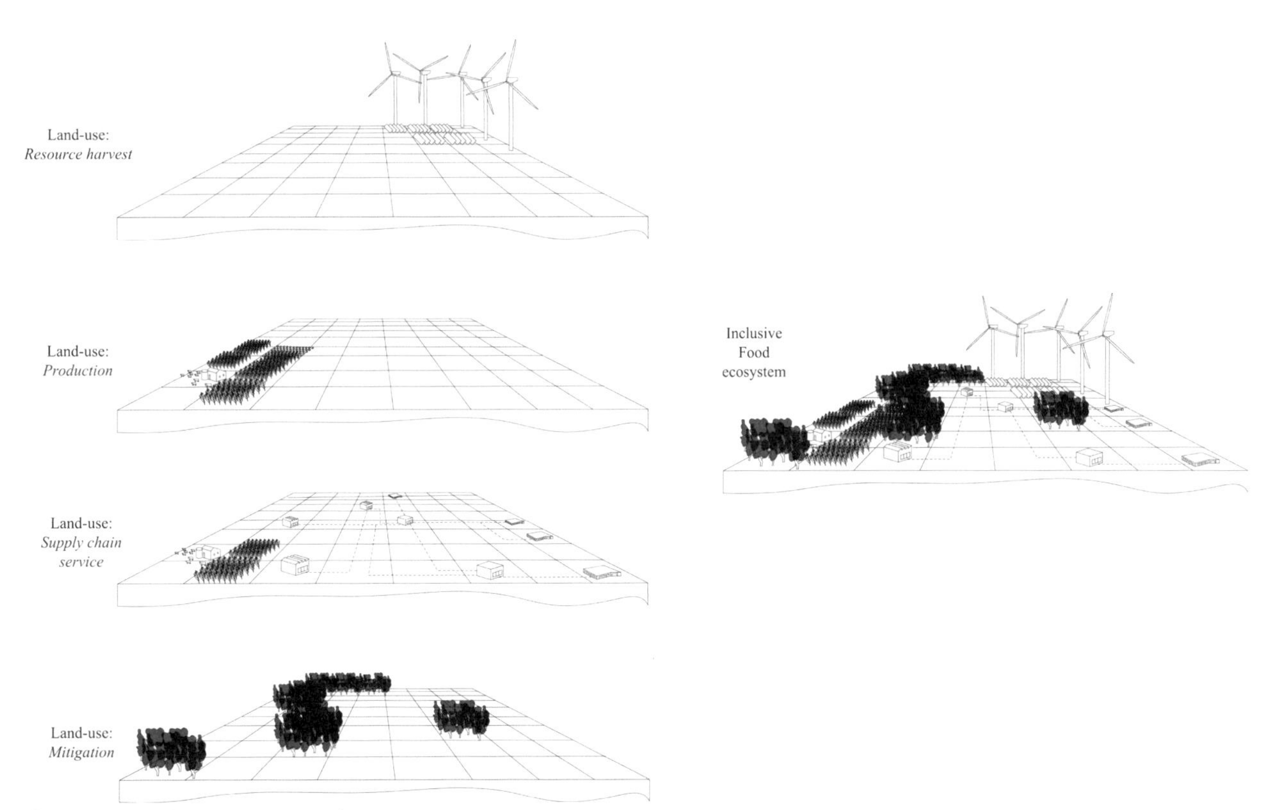

FIG. 4 Inclusive food ecosystem services.

2.3.1 Production

Dealing with *production* use, two opposite patterns are diffused in practice: intensive vs. organic production. Intensive production adopts efficient machineries (Cascini et al., 2013), chemical nutrients and fertilizers, enhancing food standardization through biotechnology and genetically modified organisms (GMOs) (Brussard et al., 2010). Conversely, organic production exploits biodiversity and organic fertilizers to supply safer and more sustainable products.

Given a crop, these two patterns can be formulated as alternative land uses, with different yields but generating different impacts. The land-use allocation aided by this framework will identify the trade-off between intensive and organic production by designing sustainable agro-food production parks.

2.3.2 Mitigation

The *mitigation* functions combine grazing and green areas (i.e., forest, carbon plantings) responsible for mitigating and containing the environmental externalities resulting from primary production and from supply chain services. Although both forest and crops sink carbon emissions, forests require less water and represent the cheapest solution to mitigate global warming (Strengers et al., 2008; West et al., 2010; Seguin, 2011; Ackerman and Stanton, 2013). Lastly, the mitigation use of land generates even revenue in the presence of cap-and-trade or carbon tax regimes.

2.3.3 Resource harvest

Resource harvest functions use areas to collect energy from renewables (i.e., solar, wind energy, and biofuels). Such energy is used to cover the power load required by primary production and other supply chain services (i.e., harvesting, irrigation, processing, packaging, storage, transportation) throughout the food ecosystem.

2.3.4 Supply chain service

Supply chain service is obtained through the allocation of processing, packaging, and storage facilities that allow conveying food from growers to retailers.

2.4 Analyze potentials

The allocation of such land uses within this inclusive ecosystem enforces interaction between different spatial functions bridging the gap between cost effectiveness and sustainability of the FSC. To assess the best configuration of land use of the food ecosystem, the potentials of areas need to be quantified.

2.4.1 Crop yield

Dealing with *production* use, the application of crop models has a long tradition in the agronomy and agro-ecological literature (Multsch et al., 2011). The research school at Wageningen University widely contributed to develop yield forecasting models studied and used for years (Bouman et al., 1996). Such models find application also in land-use planning on local and regional scales (Van Ittersum et al., 2003). A crop model predicts the yield and quantifies the benefit of the land-use allocation problem, which can be supported by quantitative mathematical formulations.

Most of the crop models assume simplified conditions such as homogeneous soil type and stable weather, and they trust the accuracy of spatial data. A hierarchy of growing factors is stated by the literature (Van Ittersum and Rabbinge, 1997). This includes defining factors (e.g., carbon dioxide, solar radiation, temperature, crop physiology, phenology, and architecture), which determine the nominal production associated with standard conditions; some limiting factors (e.g., water, nutrients), which determine the effective production with available resources; and some reducing factors (e.g., weeds, pests, diseases, pollutants), which affect the expected yield. For a comprehensive taxonomy of the most adopted crop models, the authors reference Jones et al. (2003).

2.4.2 Renewable potential

Dealing with the *resource harvest* use, we consider the energy collected by an area assuming a cover of renewable energy plants. The renewable potential is, for example, the electricity from photovoltaic systems (PVs) or wind plant (Savino et al., 2017). The characteristics of an area (i.e., latitude, longitude, altitude, slope gradient) and its climate (i.e., solar irradiation, wind speed) determine the energy potential, while the characteristics of the soil constrain the installation of solar and wind plants. Renewables can power several FSC processes (i.e., farming, processing, storage) in a sustainable manner, but could reduce land devoted to crops. This trade-off is a hotspot of current debate on biofuels (Koning and Ittersum, 2009; Fischer et al., 2010). The framework addresses such debate by controlling the allocation of land use involving agricultural and energy systems into a unique problem.

2.4.3 Mitigation potential

An example of mitigation potential is the amount of CO_2 an area is able to absorb. The literature widely discusses the strategies to mitigate GHG emissions through the adequate use of land, like carbon plantings (Smith et al., 2007; Ovando and Caparros, 2009). Although without incentives growers do not implement such strategies, some hidden benefits result from these practices. These include carbon sequestration in the soil to increase yields; the protection of biodiversity, which reduces the dependence on pesticides; and the increase of organic nutrients in soil and water.

2.4.4 Supply chain service

The *supply chain* service (SCS) is provided by the processing, packaging, and storage facility that allow the transformation and distribution of food. The establishment of these facilities is not typically driven by the climate parameters. The design of a supply chain network and the planning of the operations is led by cost targets and infrastructural constraints that do not involve climate aspects (Accorsi et al., 2018b). Nevertheless, the adoption of climate parameters is recommended to reduce energy consumption in warehouses (Accorsi et al., 2017d, 2018a) and distribution systems (Gallo et al., 2017) and along delivery operations (Accorsi et al., 2017a). Areas with favorable weather, for example, or partially exposed to solar radiation can be effective candidates for warehouses that require refrigerated rooms for perishable products.

While still not handled in planning, the weather conditions also affect the quality and safety of delivered food and packages, as stated in recent literature from the authors of this volume (Valli et al., 2013; Manzini and Accorsi, 2013; Accorsi et al., 2014, 2016b; Ayyad et al., 2017; Manzini et al., 2017). Therefore, new planning patterns are expected to assess the potential of an area in terms of supply chain service. For instance, the SCS potential of a spatial unit decreases as its distance from existing logistic infrastructures (i.e., ports, distribution centers, rail stations) increases (Accorsi et al., 2015b); moreover, the SCS potential should decrease when the climate conditions affect the conservation of products or the cost for refrigeration (i.e., cold chains in areas with warm and humid climate).

This framework uses the quantified potentials to fuel planning models that solve land-use and network-design problems able to fulfill a given demand while minimizing the impacts of the food ecosystem. Different potentials refer to different metrics and influence the target to optimize. For example, the equivalent carbon footprint, that is, CO_2eq., which normalizes the GHG emissions generated along the life cycle of a process/product, can be used for the environmental externalities of a food ecosystem (Wright et al., 2011). In order to minimize the carbon footprint, mitigation uses such as renewables, which reduce the need for fossil-fueled energy, and carbon plantings, which sink carbon emissions, can be considered, rather than locating organic vs. intensive crops. The collected potentials, varying area by area (Fig. 5), will determine the optimal configuration of the ecosystem that minimizes costs, together with impacts and used soil.

It is worth noting that the number of input parameters increases with the observed geography. The management and manipulation of this data can be supported by a relational database that stores climate, soil, and other supply chain records into a geographic information system (GIS). Primary data propels predicting models, for example, crop yield model, in order to calculate secondary data as the areas' potentials (Witlox, 2005; Jones et al., 2003; Doralswamy et al., 2003, Soltani and Hoogenboom, 2007).

2.5 Modeling sustainable agro-food parks

So far, the steps of this framework lead the planner through the definition of the geographical boundaries of the food ecosystem, the collection of spatial data (some referenced sources in FAO, 1976, 1997, 2000), and the assessment of the potentials benefits resulting from the allocation of spatial functions.

This step aims at assessing the environmental impacts of alternative agricultural systems. These are built through the combination of land uses for primary production, for example, crops, farms. However, other ecosystem services as resource harvest and mitigation can be involved to minimize the carbon footprint of agro-food park.

It is worth noting that the problem of allocating the spatial functions to fulfill a given demand while minimizing the carbon footprint of food production assumes the shape of a typical optimization problem (1).

$$\min c^T \cdot x \tag{1}$$

$$\begin{aligned} \text{s.t.}\quad A \cdot x &\leq b \\ C \cdot x &\geq d \\ x &\geq 0 \end{aligned}$$

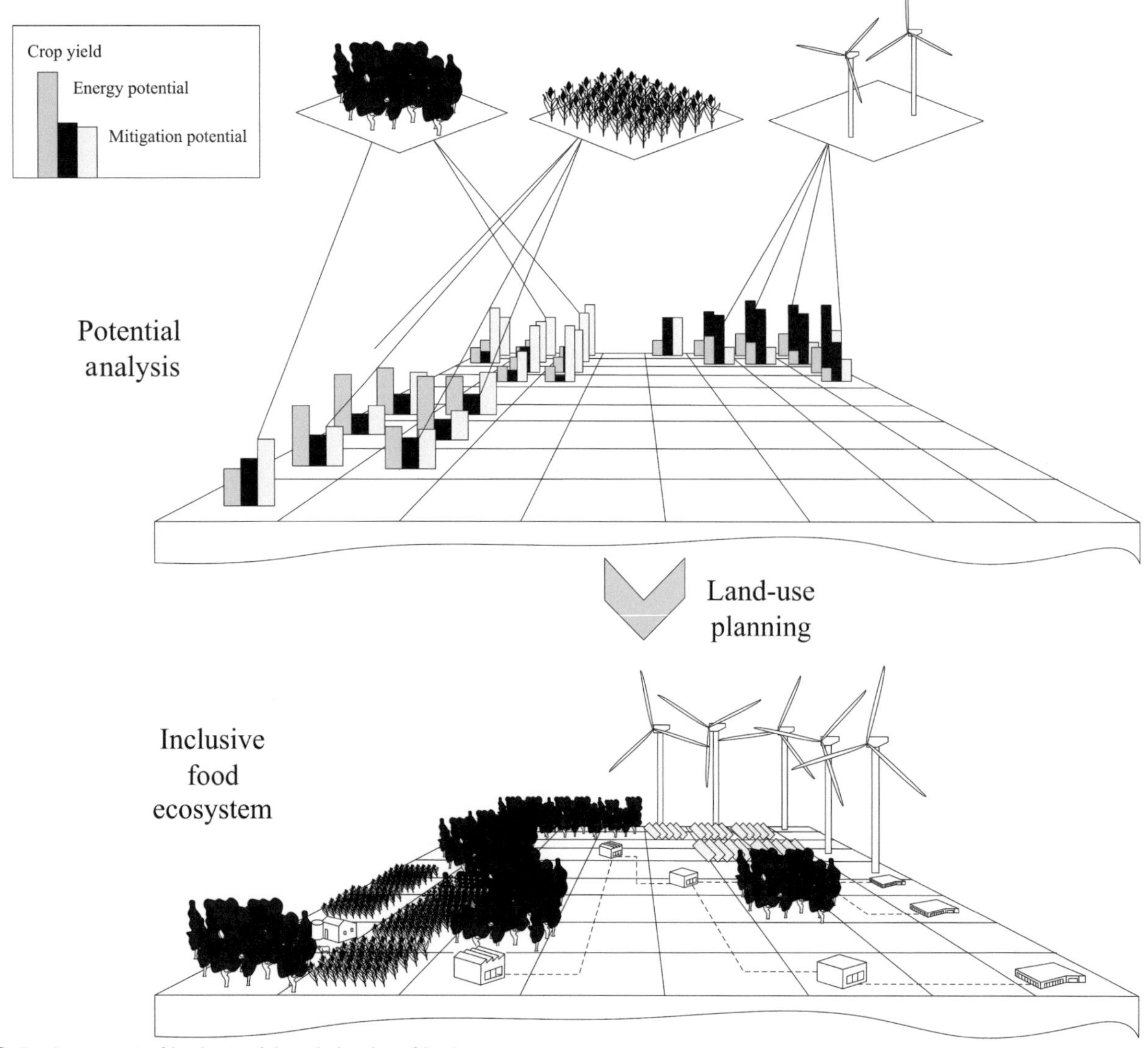

FIG. 5 Assessment of land potentials and planning of land use.

In equation system (1), x is the vector of decision variables that quantify the surface allocated to each spatial function; c^T is the transpose vector of linear costs (i.e., in our problem the carbon emissions associated with a spatial unit devoted to a use); A is the matrix of coefficients representing the weight of variables x; C is the matrix of coefficients representing the productivity of variables x; b is the vector of the system capacities (i.e., or potentials in our framework), and d is the demand vector.

The formulation (1) is well-known and not problem-oriented. The application of mathematical programming and optimization in agro-food production planning, particularly at the farm level, is indeed widely diffused in the literature (Glen, 1987; Ahumada and Villalobos, 2009). Mathematical programming in land-use planning is used to establish the combination of functions that optimize one or more functions subject to a series of constraints (Riveira and Maseda, 2006). In rural systems, the optimization of crop yields (e.g., in terms of profit and crop productivity as well) is typically constrained by the scarcity of resources and growing factors (e.g., land, water, energy). Linear programming (LP) and integer linear programming (ILP) models have commonly applied also to enhance the economic and environmental performance of rural areas. Some optimize economic and ecological performance independently but lack an investigation of the underlying relationship between such levers (Pacini et al., 2004).

By identifying the carbon footprint as a metric of environmental impact, this framework incorporates economic and environmental targets into the strategic design of an agro-food park. Two formulations of the problem are given: from a business-as-usual model to carbon-driven model. The following formulations are by no means comprehensive or punctual, but they highlight the key elements that the planning problem should take into account.

Let's assume that the production of product p can be described by considering two transformation functions (TFs). The first calculates the crop yield for a spatial unit (characterized by specific climate and soil conditions); the second quantifies its impact in terms of carbon emissions generated along with seeding, harvesting, fertilizing, irrigation, processing, packaging. Both functions depend by the allocation of land l to the function/use u, by the flow of resource r (e.g., energy, nutrients) exploited for the production, and by the properties (i.e., climate and soil) of land l. The problem sets, variables and parameters are defined in the following.

Sets and indices:

$l \in L$	Set of spatial units (e.g., land hectares) involved
$u \in U$	Set of land uses (e.g., spatial functions, crops)
$r \in R$	Set of resources (e.g., ingredients, supplies)
$p \in P$	Set of products
$t \in T$	Set of transportation or distribution modes

Fig. 6 illustrates the case where agriculture and food processing/transformation are considered as independent steps of FSC.

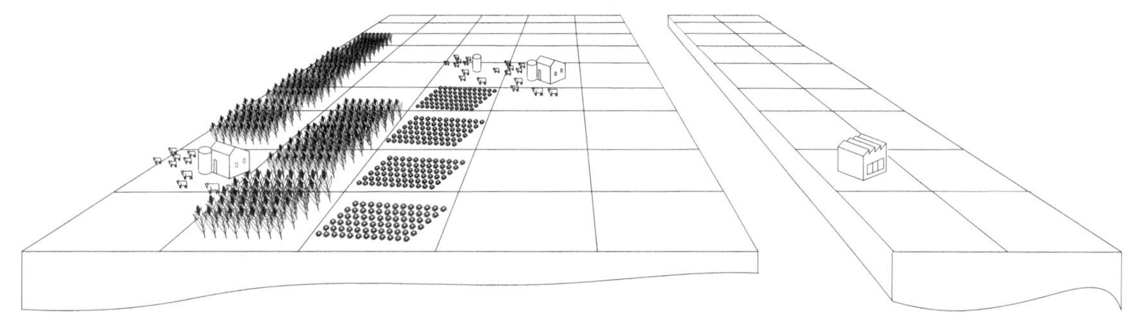

FIG. 6 *Business-as-usual* production system: agriculture and processing are separated steps.

In agreement with Fig. 6, the typical decision handled by growers and farmers can be formulated through the variable (2):

Decision variables (2):

y_{rlu}	Binary variables are 1 if land l allocated to use u to produce resource r; 0 otherwise.

The decision parameters result from the previous step, that is, primary climate soil data, secondary data as potentials, are the following:

Parameters:

$S_l(s_{l1}, s_{l2}, s_{l3}, \dots, s_{ln})$	Agro-climate conditions of land l
$X_{pr}(X_{p1}, X_{p2}, X_{p3}, \dots, X_{pr})$	Resources required for the production of a unit of p
$p_p^{'}(p_1, p_2, p_3, \dots, p_P)$	Sale prices of P set
$c_p^{'}(c_1, c_2, c_3, \dots, c_R)$	Resource costs
b_r	Capacity of resource r
$F_{pl}(X_{pr}, S_l)$	Production TF of product p on land l

An in-depth description of the TF used to quantify yield and impact of crops is not given in this chapter and is left to other papers (Accorsi et al., 2016a). Hence, assuming b_r the overall capacity of resource r within the planning area, the model is formulated in (3):

$$\max \sum_{p\in P}\sum_{l\in L}\sum_{u\in U}\left(F_{pl}\cdot p'_p - X_{pr}\cdot c'_p\right)\cdot y_{plu} \tag{3}$$

$$\text{s.t.} \sum_{u\in U}\sum_{p\in P} y_{plu} \le 1 \quad \forall l$$

$$\sum_{p\in P}\sum_{l\in L}\sum_{u\in U} X_{pr}\cdot y_{plu} \le b_r \quad \forall r$$

$$y_{plu} \in \{0,1\} \quad \forall p,l,u$$

In (3) the constraints are to limit the allocation of one use to each spatial unit, and to guarantee that the capacities of the resources (e.g., nutrients, water, energy, labor) are not violated. Under the intensive production scheme, a profit-oriented function motivates the adoption of high-revenue agricultural technology, the choice of variety with high return per hectare and fewer resources required. According to Fig. 6, this problem focuses on the agricultural phase and by no way involves handling and transport flows between rural areas and the production facility.

A new formulation of the same planning problem is then provided for a food production ecosystem that integrates agriculture and transformation phases, as illustrated in Fig. 7.

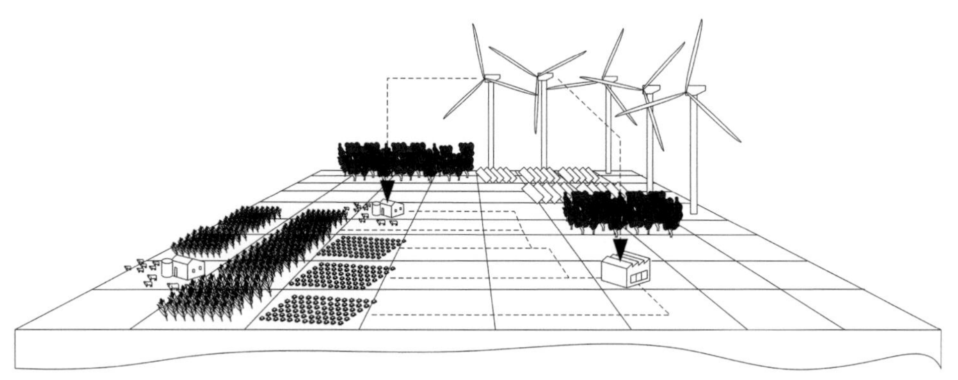

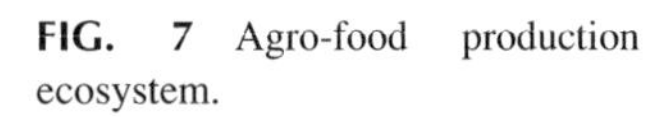

FIG. 7 Agro-food production ecosystem.

The inclusive production ecosystem is built involving the food processing facility as well as resource harvest and mitigation land-uses. It is worth noting that the distance between spatial units devoted to crops and the facility is not neglected. According to such a perspective, the transportation process between crops and the facility are accounted for in the function of the impacts to minimize. Moreover, spatial functions intended for resources harvest and mitigation are involved to provide ecosystem services. The new set of decision variables can be defined as follows:

Decision variables (4):

y_{rlu}	Binary variables that is 1 if land l allocated to use u to produce resource r (or product p); 0 otherwise.
$x_{l'lrt}$	Flow of resource r distributed from land l to l' by mode t

While the first variable is similar to that defined in (2), the second continuous variable quantifies the flow of resource that move from one spatial unit to another. The impact of resources supply is emphasized and each spatial unit can be considered as a source or destination of a set of resources. Furthermore, other two TFs quantify the environmental impacts associated with both resource generation and distribution. These are formulated as follows:

Parameters:

$E_{plu}(u, X_{pr}, S_l)$	Impact TF of product p produced on land l
$E_{l'lrt}(r, d_{l'l}, t)$	Impact TF of handling resource r from land l to land l' by mode t

The resulting planning problem is defined in (5).

$$\min \sum_{r\in R}\sum_{l\in L}\sum_{u\in U} E_{rlu}\cdot y_{rlu} + \sum_{r\in R}\sum_{l'\in L}\sum_{t\in T} E_{l'lrt}\cdot x_{l'lrt} \tag{5}$$

$$\text{s.t.} \sum_{l'\in L}\sum_{t\in T} x_{ll'rt} \le b_{lr}\cdot y_{rlu} \quad \forall r,l$$

$$\sum_{l'\in L}\sum_{t\in T} x_{l'lrt} \ge X_{pr}\cdot y_{rlu} \quad \forall r,l,p,u$$

$$\sum_{r\in R}\sum_{u\in U} y_{rlu} + \sum_{p\in P}\sum_{u\in U} y_{plu} \le 1 \quad \forall l$$

$$\sum_{u\in U}\sum_{l\in L}\sum_{p\in P}\left(F_{pl}\cdot p'_p - X_{pr}\cdot c'_p\right)\cdot y_{plu} - \sum_{r\in R}\sum_{t\in T}\sum_{l,l'\in L} c_{ll't}\cdot x_{l'lrt} \ge \phi_{obj} \quad \forall p,l,u$$

$$y_{plu}, y_{rlu} \in \{0, 1\} \quad \forall l,r,t$$

$$x_{ll'rr} \ge 0$$

The constraints of the model (5) impose that the flow of resources departing from a spatial unit is less than its capacity (5.c1); that the required resources to produce p are supplied to the facility (5.c2), at least in terms of energy and raw materials; that a spatial unit is assigned to no more than a user or p (5.c3), and that an expected profit, ϕ_{obj}, for the whole ecosystem is achieved (5.c4), while the overall carbon footprint the ecosystem minimized.

The design of sustainable production ecosystems is hence driven by the minimization of carbon footprint accounting for crops, processing, and resource distribution. To ensure the feasibility of the problem, the considered area has to be self-sufficient in resources generation (e.g., energy). The growers' profit works no longer as a leading driver but as a constraint. The higher the expected profit by the planner, the higher the risk of infeasibility will be. Two levers can be pursued in the case of infeasibility. First, he may relax the profit constraint, thereby reducing the resource requirements and their exploitation. When reducing the profit is not possible, the planner can involve a larger area and increase the boundaries of the ecosystem. This action leads to two consequences. First, a larger ecosystem provides more lands to allocate. Second, a larger area increases the influence of distribution activities in the planning problem. Therefore, the trade-off between the profit, land-use allocation, and resources distribution may be investigated within a larger ecosystem through the proposed model.

2.6 Modeling inclusive FSC

This step integrates the production ecosystem with the supply chain including freight storage and distribution operations (Flores and Villalobos, 2018).

The global trade intensifies the volumes of food shipped along supply chains. For this purpose, transportation and storage are widely considered as environmental stressors of the food sector (Weber and Matthews, 2008; Cholette and Venkat, 2009; Wakeland et al., 2012; Manzini et al., 2014a, 2016; Accorsi et al., 2017b; Banasik et al., 2018). The literature on transportation science provides several contributions adopting optimization to address environmental sustainability (Dekker et al., 2012; Mallidis et al., 2012; Manzini et al., 2014b; Accorsi et al., 2015b). De Keizer et al. (2017) formulate and solve mixed image linear programming (MILP) models to manage the distribution of perishable products with heterogeneous quality decay. Stellingwerf et al. (2018) explore the minimization of carbon emissions from refrigerated truck through vehicle routing models. Accorsi et al. (2017a) provide a decision-support model for the management of cold chains considering the weather conditions, while Gallo et al. (2017) formalize a model for the design of energy-effective long range cold chains. Other recent contributions (Soysal et al., 2018; Accorsi et al., 2018b) provide models to coordinate the delivery operations in a regional food retailer supply chain.

Focusing on the environmental impacts, transportation is the most influential process of the supply chains. GHG emissions from transportation are around 14% of total emissions at the EU level (Stern, 2006). The minimization of distances among logistic nodes and the optimization of flows have a crucial impact on providing environmentally friendly FSCs. Another polluting process along cold chains is storage. Indeed, optimization is also adopted to manage the storage of temperature-sensitive products such as food (Baruffaldi et al., 2018).

This step of the framework provides a strategy to reduce the environmental impacts of the whole FSC managed as an inclusive distributed ecosystem. The strategy intends that the production ecosystem and distribution network merge into a unified planning problem. Specifically, it integrates the land-use allocation problem into a location-allocation model, usually adopted to design logistics networks.

This new problem is named a land-network problem (LNP) and was first introduced by Accorsi et al. (2016a) and Accorsi (2019). The LNP appears as a use-location-allocation problem where production ecosystems (i.e., agro-food parks) are established instead of facilities, and the flows between these are optimized. Thus, it allows identifying the best configuration of production ecosystems and logistics systems that fulfills the food demand. The LNP aims at minimizing the total environmental impacts (e.g., carbon emissions) accounted to the FSC for both agriculture and distribution processes.

In the following a comparison between a generic location-allocation formulation and a version of the LNP is proposed.

Sets and indices:

$p \in P$	Set of processing/packaging plants
$r \in R$	Set of agro-food products
$d \in D$	Set of storage/consolidation facility
$c \in C$	Set of demand points (e.g., retailers, gross markets)

Parameters:

cf^n	Cost to establish a facility n
cv^t	Distribution costs per distance unit (e.g., km) by mode t
cv^p	Production costs per unit
cv^s	Storage costs per unit

The sets and parameters of the optimization model are defined in agreement with the well-known formulation of the location-allocation problem applied to the network represented in Fig. 8.

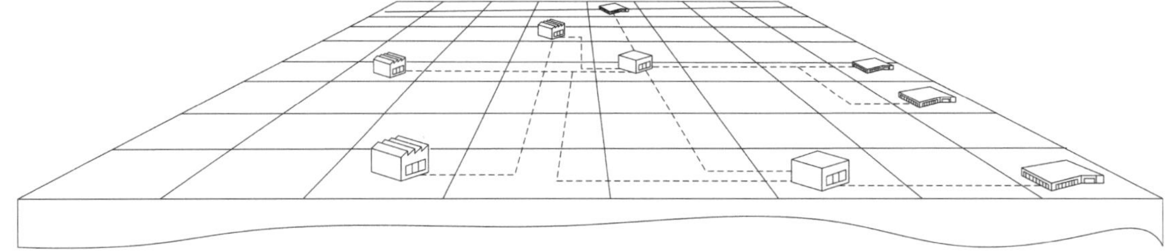

FIG. 8 Optimized food distribution network.

The set of decisions deals with the location of facilities devoted to production and storage, and the allocation of distribution flows throughout the network. These decisions can be formalized as follows:

Decision variables (6):

y_{pr}	Binary variable that is 1 if facility p is opened and devoted to product r; 0 otherwise.
y_d	Binary variable that is 1 if facility d is opened; 0 otherwise.
x_{ijrt}	Flow of product r distributed from node i to j by mode t

Then, the model results are as follows:

$$\max \sum_{p\in P}\sum_{d\in D}\sum_{t\in T}\sum_{r\in R} x_{pdrt}\cdot p'_r - \sum_{p\in P}\sum_{r\in R} y_{pr}\cdot cf^p - \sum_{d\in D} y_d\cdot cf^d - \sum_{p\in P}\sum_{d\in D}\sum_{t\in T}\sum_{r\in R}\left(x_{pdrt}\cdot cv^t\cdot d_{pd} + x_{pdrt}\cdot cv^p\right)$$
$$- \sum_{d,d'\in D}\sum_{r\in R}\sum_{t\in T}\left(x_{dd'rt}\cdot cv^t\cdot d_{dd'} + x_{dd'rt}\cdot cv^s\right) - \sum_{d\in D}\sum_{c\in C}\left(x_{dcrt}\cdot cv^t + x_{dcrt}\cdot cv^s\right) \tag{7}$$

$$\text{s.t.} \sum_{d\in D}\sum_{t\in T} x_{dcrt} = d_{cr} \quad \forall r,c$$

$$\sum_{d\in D}\sum_{t\in T} x_{pdrt} \le C_{pr}\cdot y_{pr} \quad \forall r,p$$

$$\sum_{p\in P}\sum_{t\in T}\sum_{r\in R} x_{pdrt} + \sum_{d'\in D}\sum_{t\in T}\sum_{r\in R} x_{d'drt} \le C_d\cdot y_d \quad \forall d$$

$$\sum_{p\in P}\sum_{t\in T} x_{pdrt} \ge \sum_{d'\in D}\sum_{t\in T} x_{dd'rt} + \sum_{c\in C}\sum_{t\in T} x_{dcrt} \quad \forall d,r$$

$$y_{pr} \in \{0,1\} \quad \forall p,r$$

$$x_{ijrt} \ge 0 \quad \forall i,j \in P\cup D\cup C, r, t$$

where the objective function maximizes the profit of the whole supply chain and the constraints ensure the demand fulfillment (7.c1), with respect to production (7.c2), and storage capacities (7.c3), and guarantee the flow balancing (7.c4).

As for the production ecosystem, a new formulation of the problem is possible. In the following, we integrate the production ecosystem nodes of the network. Assume the new sets, parameters, and decision variables are as follows:

Sets and indices:

$fp \in FP$	Set of processing parks/ecosystems.

Parameters:

ee^{fp}	Environmental impact generated by the agro-food park *fp* per unit of product.
ee^{d}	Environmental impact generated by the storage facility *d* per unit of product.
et^{t}	Environmental impact generated by mode *t* per distance unit and unit of product.

Decision variables:

y_{fpr}	Binary variable that is 1 if agro-park *fp* is chosen as source of product *r*; 0 otherwise.
x_{ijrt}	Flow of product r distributed from node *i* to *j* by mode *t*.

Then the model results are as follows:

$$\min \sum_{fp\in FP}\sum_{d\in D}\sum_{r\in R}\sum_{t\in T} x_{fpdrt}\cdot\left(ee^{fp}+et^{r}\right)+\sum_{d\in D}\sum_{c\in T}\sum_{r\in R}\sum_{t\in T} x_{dcrt}\cdot\left(ee^{dd}+et^{r}\right)+\sum_{t\in T}\sum_{r\in R}\sum_{d,d'\in D} x_{dd'rt}\cdot\left(ee^{dd}+et^{r}\right) \quad (8)$$

$$\text{s.t.} \sum_{d\in D}\sum_{t\in T} x_{dcrt}=d_{cr} \quad \forall r,c$$

$$\sum_{d\in D}\sum_{t\in T} x_{fpdrt}\leq C_{pr}\cdot y_{fpr} \quad \forall r,fp$$

$$\sum_{fp\in FP}\sum_{t\in T}\sum_{r\in R} x_{fpdrt}+\sum_{d'\in D}\sum_{t\in T}\sum_{r\in R} x_{d'drt}\leq C_{d}\cdot y_{d} \quad \forall d$$

$$\sum_{fp\in FP}\sum_{t\in T} x_{fpdrt}\geq \sum_{d'\in D}\sum_{t\in T} x_{dd'rt}+\sum_{c\in C}\sum_{t\in T} x_{dcrt} \quad \forall d,r$$

$$\sum_{fp\in FP}\sum_{d\in D}\sum_{t\in T}\sum_{r\in R} x_{fpdrt}\cdot p'_{r}-\sum_{fp\in FP}\sum_{r\in R} y_{pr}\cdot cf^{fp}-\sum_{d\in D} y_{d}\cdot cf^{d}-\sum_{fp\in FP}\sum_{d\in D}\sum_{t\in T}\sum_{r\in R}\left(x_{fpdrt}\cdot cv^{t}\cdot d_{fpd}+x_{fpdrt}\cdot cv^{p}\right)$$

$$-\sum_{d,d'\in D}\sum_{r\in R}\sum_{t\in T}\left(x_{dd'rt}\cdot cv^{t}\cdot d_{dd'}+x_{dd'rt}\cdot cv^{s}\right)-\sum_{d\in D}\sum_{c\in C}\left(x_{dcrt}\cdot cv^{t}+x_{dcrt}\cdot cv^{s}\right)\geq \psi_{obj}$$

$$y_{pr}\in\{0,1\} \quad \forall p,r$$

$$x_{ijrt}\geq 0 \quad \forall i,j\in P\cup D\cup C,r,t$$

According to model (8), the objective function minimizes the overall carbon emissions associated with the whole FSC, which entails both production ecosystems and the distribution network from crop to consumers. As for the production stage of model (5), the planning of the inclusive FSC (Fig. 9)imposes a profit-satisfaction constraint, that is, (8.c5), while minimizing the environmental externalities (8.1).

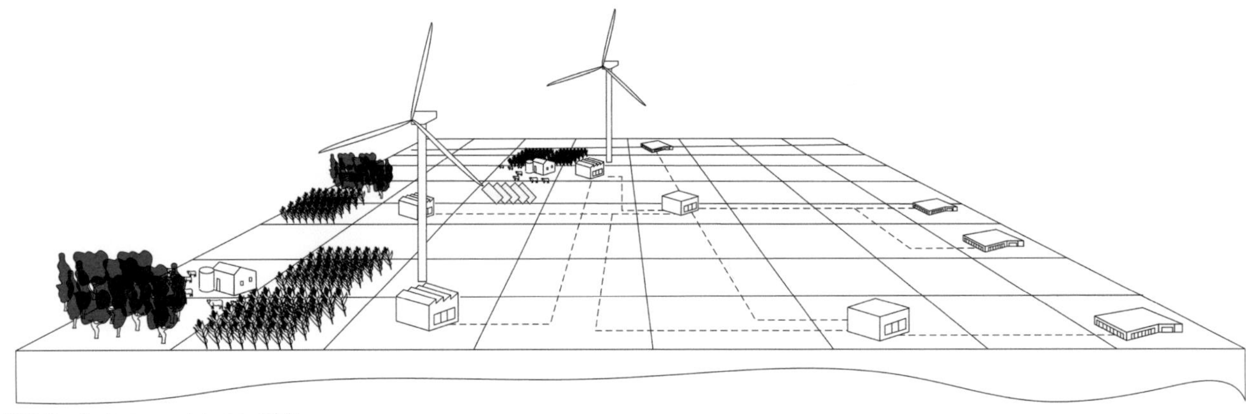

FIG. 9 Inclusive sustainable FSC.

As a consequence, the optimal solution will no longer be the most profitable from a purely economic perspective but will be the FSC that minimizes environmental externalities while satisfying the food demand. In view of this, the trade-off between organic food production ecosystems and intensive production systems, which mitigate their impacts through carbon plantings and renewable energy sources, can be identified even with regard to the transportation and logistics costs and impacts generated along the distribution network.

3 Discussion and applications

Global trade compels policymakers, planners, and food enterprises to carefully manage natural resources, and particularly to balance the impacts and the externalities generated along the supply chains from suppliers to consumers (Accorsi, 2019). Natural resources need to be preserved, and particularly those exploited by the food industry, i.e., land and soil, energy, and water, which give subsistence to humans, animals, and plants, and are the base of any ecosystem.

The proposed framework includes environmental externalities into decision-support models aiding the planning for sustainable FSCs. Thus, the framework provides a top-down methodology for planners and decision makers adaptable to many cases and different purposes.

First, given a generic geographical area, (1) it facilitates the analysis of the available resources, capacities, and potentials, over a quantitative perspective through the implementation of spatial data inventories. This allows increasing awareness of the characteristics, requirements, and potentials of the observed ecosystem.

Then, (2) it may support the design of FSCs from green field in developing and less developed countries, aiding the optimization of the available resources and the control of land-use change, responsible for most of the climate change effects. In view of this, the framework works as a diagnostic tool to manage the land-grabbing phenomenon, which is increasingly spreading in the last decades, especially in developing countries (Borras Jr. et al., 2011; Cotula et al., 2009). Land acquisitions increase the investment in agriculture in such areas, but they also affect social and environmental equilibrium due to intensive production patterns (Amler et al., 1999; World Bank, 2011) or nonfood crops such as biofuels. The proposed framework investigates this trade-off, suggesting a change of paradigm in the design of food systems driven by environmental sustainability goals rather than profit functions. It sustains the design of agro-food parks, to manage natural resources (e.g., forestry, soil, water) in responsible manner, and to mitigate global warming through carbon plantings and renewable energy fields.

Third, (3) the framework enables the analysis and quantification of the impacts generated by a food product throughout its life cycle from suppliers to consumers. The parameters and metrics collected and calculated draw a dashboard of economic and environmental key performance indicators (KPIs) (Accorsi et al., 2017c), which enable an assessment and description of an FSC over a quantitative dashboard and its comparison with others.

The literature demonstrates that food production and distribution are hotspots of global warming and provides tools such as LCAs to quantify such impacts (Andersson, 2000). The carbon footprint of food products still need a common metric to describe the sustainability of a product and its processes (Roy et al., 2009; Virtanen et al., 2011; Accorsi et al., 2015a). In comparison with an LCA, which is typically used to assess the environmental performance ex post, this framework is used to optimize such performance ex ante and to design a long-term sustainable food production and distribution ecosystem.

Although research has already debated the benefits provided by eco-industrial parks (Sheerr and McNeely, 2008; Roberts, 2004; Varga and Kuehr, 2007; Ometto et al., 2007), this framework applies the industrial ecology pattern to an FSC seen as a combination of production and distribution processes.

4 Conclusions and further development

The rising global food demand is pushing suppliers to find new strategies for food production. Among these, the adoption of intensive fossil-fueled agriculture and green chemistry has intensified. Globalization has bridged the barriers between production and consumption areas but has failed to ensure a balanced and responsible exploitation of natural resources, increasing the gap between developed and less developed countries. As examples, deforestation, land-grabbing, exploitation of water and soil, and air pollution are direct effects of the global food trade, and their mitigation represents challenges for all humanity.

The framework illustrated in this chapter aims at providing new formulations of FSC design models able to incorporate environmental externalities and to merge production and distribution processes within the same problem; it uses optimization to provide strategic food ecosystem planning.

Further research will be focused on finding real-world applications and case studies for the framework validation. The scope of further research is to shed light on the impact of integrating land-use and distribution network decisions onto economic and social sustainability of an FSC. This framework, which is by no means comprehensive, should elicit multidisciplinary research on the food sector and definitely provide support-decision tools for policymakers, planners, and managers on how to develop FSCs more sustainably.

References

Accorsi, R., 2019. Planning sustainable food supply chains to meet growing demands. In: Encyclopedia of Food Security and Sustainability. Reference Module in Food Sciences. vol. 3. Elsevier, pp. 45–53. https://doi.org/10.1016/B978-0-08-100596-5.22028-3.

Accorsi, R., Manzini, R., Mora, C., Cascini, A., Penazzi, S., Pini, C., Pilati, F., 2013. Life cycle modelling for sustainable food supply chain. In: Proceeding of the 22nd International Conference of Production Research (ICPR), Iguassu Falls, Brazil, 2013.

Accorsi, R., Manzini, R., Ferrari, E., 2014. A comparison of shipping containers from technical, economic and environmental perspectives. Transp. Res. Part D: Transp. Environ. 26, 52–59.

Accorsi, R., Versari, L., Manzini, R., 2015a. Glass vs. plastic: life cycle assessment of extra-virgin olive oil bottles across global supply chains. Sustainability 7 (3), 2818–2840.

Accorsi, R., Manzini, R., Pini, C., Penazzi, S., 2015b. On the design of closed-loop networks for product life cycle management: economic, environmental and geography considerations. J. Transp. Geogr. 48, 121–134.

Accorsi, R., Cholette, S., Manzini, R., Pini, C., Penazzi, S., 2016a. The land-network problem: ecosystem carbon balance in planning sustainable agro-food supply chains. J. Clean. Prod. 112 (1), 158–171.

Accorsi, R., Ferrari, E., Gamberi, M., Manzini, R., Regattieri, A., Montserrat Espiñeira, M., Santaclara, F.J., 2016b. A closed-loop traceability system to improve logistics decisions in food supply chains. A case study on dairy products. In: Advances in Food Traceability Techniques and Technologies: Improving Quality Throughout the Food Chain. Woodhead Publishing, Elsevier, pp. 337–351.

Accorsi, R., Gallo, A., Manzini, R., 2017a. A climate-driven decision-support model for the distribution of perishable products. J. Clean. Prod. 165, 917–929.

Accorsi, R., Manzini, R., Pini, C., 2017b. How logistics decisions affect the environmental sustainability of modern food supply chains. A case study from an Italian large scale-retailer. In: Bhat, R. (Ed.), Sustainability Challenges in the Agrofood Sector, first ed. 2017. John Wiley & Sons Ltd., pp. 179–200.

Accorsi, R., Bortolini, M., Baruffaldi, G., Pilati, F., Ferrari, E., 2017c. Internet-of-things paradigm in food supply chains control and management. Proc. Manuf. 11, 889–895.

Accorsi, R., Bortolini, M., Gamberi, M., Manzini, R., Pilati, F., 2017d. Multi-objective warehouse building design to optimize the cycle time, total cost, and carbon footprint. Int. J. Adv. Manuf. Technol. 92 (1–4), 839–854.

Accorsi, R., Baruffaldi, G., Manzini, R., 2017e. Design and manage deep lane storage system layout. An iterative decision-support model. Int. J. Adv. Manuf. Technol. 92 (1–4), 57–67.

Accorsi, R., Baruffaldi, G., Manzini, R., 2018a. Picking efficiency and stock safety: a bi-objective storage assignment policy for temperature-sensitive products. Comput. Ind. Eng. 115, 240–252.

Accorsi, R., Baruffaldi, G., Manzini, R., Tufano, A., 2018b. On the design of cooperative vendors' networks in retail food supply chains: a logistics-driven approach. Int. J. Log. Res. Appl. 21 (1), 35–52.

Accorsi, R., Cholette, S., Manzini, R., Tufano, A., 2018c. A hierarchical data architecture for sustainable food supply chain management and planning. J. Clean. Prod. 203, 1039–1054.

Ackerman, F., Stanton, E., 2013. Climate impacts on Agriculture: a challenge to complacency? In: Global Development and Environment Institute (GDAE) Working Paper No. 13-01. Tufts University.

Ahumada, O., Villalobos, R., 2009. Application of planning models in the agri-food supply chain: a review. Eur. J. Oper. Res. 195, 1–20.

Almer, C., Winkler, R., 2017. Analyzing the effectiveness of international environmental policies: the case of the Kyoto Protocol. J. Environ. Econ. Manag. 82, 125–151.

Amate, J., Gonzalez de Molina, M., 2013. 'Sustainable de-growth' in agriculture and food: an agro-ecological perspective on Spain's agri-food system (year 2000). J. Clean. Prod. 38, 27–35.

Amler, B., Betke, D., Eger, H., Ehrich, C., Kohler, A., Kutter, A., von Lossau, A., Müller, U., Seidemann, S., Steurer, R., Zimmermann, W., 1999. Land Use Planning: Methods, Strategies and Tools. Eschborn.

Andersson, K., 2000. LCA of food products and production systems. Int. J. Life Cycle Assess. 5 (4), 239–248.

Apaiah, R., Hendrix, E., Meerdink, G., Linnemann, A., 2005. Qualitative methodology for efficient food chain design. Trends Food Sci. Technol. 16, 204–214.

Ayyad, Z., Valli, E., Bendini, A., Accorsi, R., Manzini, R., Bortolini, M., Gamberi, M., Gallina Toschi, T., 2017. Simulating international shipments of vegetable oils: focus on quality changes. Ital. J. Food Sci. 29 (1), 38–49.

Banasik, A., Bloemhof-Ruwaard, J.M., Kanellopoulos, A., Claassen, G.D.H., van der Vorst, J.G.A.J., 2018. Multi-criteria decision making approaches for green supply chains: a review. Flex. Serv. Manuf. J. 30 (3), 366–396.

Baruffaldi, G., Accorsi, R., Manzini, R., 2019. Warehouse management system customization and information availability in 3pl companies: a decision-support tool. Ind. Manag. Data Syst. 119 (2), 251–273. https://doi.org/10.1108/IMDS-01-2018-0033.

Borras Jr., S., Hall, R., Scoones, I., White, B., Wolford, W., 2011. Towards a better understanding of global land grabbing: an editorial introduction. J. Peasant Stud. 38 (2), 209–216.

Bouman, B.A., Van Keulen, H., Van Laar, H., Rabbinge, R., 1996. The School of de Wit crop growth simulation models: a pedigree and historical overview. Agric. Syst. 52, 171–198.

Brussard, L., Caron, P., Campbell, B., Lipper, L., Mainka, S., Rabbinge, R., Babin, D., Pulleman, M., 2010. Reconciling biodiversity conservation and food security: scientific challenges for a new agriculture. Curr. Opin. Environ. Sustain. 2, 34–42.

Bryan, B., King, D., Ward, J.R., 2011. Modelling and mapping agricultural opportunity costs to guide landscape planning for natural resource management. Ecol. Indic. 11, 199–208.

Cascini, A., Mora, C., Gamberi, M., Bortolini, M., Accorsi, R., Manzini, R., 2013. Design for sustainability of agricultural machines. In: Proceeding of the 22nd International Conference of Production Research, Iguassu Falls, Brazil, 2013.

Cholette, S., Venkat, K., 2009. The energy and carbon intensity of wine distribution: a study of logistical options for delivering wine to consumers. J. Clean. Prod. 17 (16), 1401–1413.

Coehlo, C., Costa, S., 2010. Challenges for integrating seasonal climate forecasts in user applications. Curr. Opin. Environ. Sustain. 2, 317–325.

Conforti, P., Giampietro, M., 1997. Fossil energy use in agriculture: an international comparison. Agric. Ecosyst. Environ. 65, 231–243.

Cotula, L., Vermeulen, S., Leonard, R., Keeley, J., 2009. Land Grab or Development Opportunity? Agricultural Investment and International Land Deals in Africa. IIED/FAO/IFAD, London/Rome, ISBN: 978-1-84369-741-1.

De Keizer, M., Akkerman, R., Grunow, M., Bloemhof-Ruwaard, J.M., Haijema, R., van der Vorst, J.G.A.J., 2017. Logistics network design for perishable products with heterogeneous quality decay. Eur. J. Oper. Res. 262 (2), 535–549.

Dekker, R., Bloemhof, J., Mallidis, I., 2012. Operations research for green logistics—an overview of aspects, issues, contributions and challenges. Eur. J. Oper. Res. 219, 671–679.

Desjardins, R.L., Sivakumar, M.V.K., De Kimpe, C., 2007. The contribution of agriculture on the state of climate: workshop summary and recommendations. Agric. For. Meteorol. 142, 314–324.

Doralswamy, P., Moulin, S., Cook, P., Stern, A., 2003. Crop yield assessment from remote sensing. Photogramm. Eng. Remote Sens. 69, 665–674.

FAO, 1976. A Framework for Land Evalutation. FAO, Rome.

FAO, 1997. Agro-Ecological Zoning. FAO, Rome.

FAO, 2000. Land Resource Information System for Land Use Planning. FAO, Rome.

FAO, 2012a. Sustainability Assessment of Food and Agriculture Systems (SAFA). FAO, Rome.

FAO, 2012b. Towards the Future We Want: End Hunger and Make the Transition to Sustainable Agricultural and Food Systems. FAO, Rome.

FAO, 2013. The State of Food and Agriculture: Food Systems for Better Nutrition Argues. FAO, Rome, ISBN: 978-92-5-107672-9.

FAO, IFAD, and WFP, 2012. The State of Food Insecurity in the World 2012: Economic Growth is Necessary But Not Sufficient to Accelerate Reduction of Hunger and Malnutrition, Rome.

Fischer, G., Prieler, S., Van Velthuizen, H., Berndes, G., Faaij, A., Londo, M., De Wit, M., 2010. Biofuel production potentials in Europe: sustainable use of cultivated land and pastures, Part II: land use scenarios. Biomass Bioenergy 34, 173–187.

Flores, H., Villalobos, J.R., 2018. A modeling framework for the strategic design of local fresh-food systems. Agric. Syst. 161, 1–15.

Gallo, A., Accorsi, R., Baruffaldi, G., Manzini, R., 2017. Designing sustainable cold chains for long-range food distribution: energy-effective corridors on the Silk Road belt. Sustain. J. 9 (11), 2044. https://doi.org/10.3390/su9112044.

Glen, J., 1987. Mathematical models in farm planning: a survey. Oper. Res. 35 (5), 641–666.

Govindan, K., 2018. Sustainable consumption and production in the food supply chain: a conceptual framework. Int. J. Prod. Econ. 195, 419–431.

Jones, J.W., Hoogenboom, G., Porter, C.H., Boote, K.J., Batchelor, W.D., Hunt, L.A., Wilkens, P.W., Singh, U., Gijsman, A.J., Ritchie, J.T., 2003. The DSSAT cropping system model. Eur. J. Agron. 18, 235–265.

Jones, J.W., Antle, J.M., Basso, B., Boote, K.J., Conant, R.T., Foster, I., Godfray, H.C.J., Herrero, M., Howitt, R.E., Janssen, S., Keating, B.A., Munoz-Carpena, R., Porter, C.H., Rosenzweig, C., Wheeler, T.R., 2016. Brief history of agricultural systems modeling. Agric. Syst. 155, 240–254.

Koning, N., Ittersum, M., 2009. Will the world have enough to eat? Curr. Opin. Environ. Sustain. 1, 77–82.

Leduc, W., Van Kann, F., 2013. Spatial planning based on urban energy harvesting toward productive urban regions. J. Clean. Prod. 39, 180–190.

Lowe, T.J., Preckel, P.V., 2004. Decision technologies for agribusiness problems: a brief review of selected literature and a call for research. Manuf. Serv. Oper. Manag. 6 (3), 201–208.

Lowe, P., Philipson, J., Lee, R., 2008. Socio-technical innovation for sustainable food chains: role for social science. Trends Food Sci. Technol. 19, 226–233.

Lucas, M., Chhajed, D., 2004. Applications of location analysis in agriculture: a survey. J. Oper. Res. Soc. 55, 561–578.

Luning, P., Marcelis, W., 2009. A food quality management research methodology integrating technological and managerial theories. Trends Food Sci. Technol. 20, 35–44.

Mallidis, I., Dekker, R., Vlachos, D., 2012. The impact of greening on supply chain design and cost: a case for a developing region. J. Transp. Geogr. 22, 118–128.

Manzini, R., Accorsi, R., 2013. The new conceptual framework for food supply chain assessment. J. Food Eng. 115 (2), 251–263.

Manzini, R., Accorsi, R., Ziad, A., Bendini, A., Bortolini, M., Gamberi, M., Valli, E., Gallina Toschi, T., 2014a. Sustainability and quality in the food supply chain. A case study of shipment of edible oils. Br. Food J. 116 (12), 2069–2090.

Manzini, R., Accorsi, R., Bortolini, M., 2014b. Operational planning models for distribution networks. Int. J. Prod. Res. 52 (1), 89–116.

Manzini, R., Accorsi, R., Gamberi, M., Savino, M.M., 2016. Logistics and supply chain management. Issue and challenges for modern production systems. In: Handbook of Transportation. Taylor and Francis, Routledge, pp. 334–344.

Manzini, R., Accorsi, R., Piana, F., Regattieri, A., 2017. Accelerated life testing for packaging decisions in the edible oils distribution. Food Pack. Shelf Life 12, 114–127.

Massoud, M., Fayad, R., El-Fadel, M., Kamleh, R., 2010. Drivers, barriers and incentives to implementing environmental management systems in the food industry: a case of Lebanon. J. Clean. Prod. 18, 200–209.

Mathe-Gaspar, G., Fodor, N., Pokovai, K., Kovacs, G.J., 2005. Crop modelling as a tool to separate the influence of the soil and wheather on crop yields. Phys. Chem. Earth 30, 165–169.

McCown, R.L., 2002a. Changing systems for supporting farmers' decisions: problems, paradigms, and prospects. Agric. Syst. 74, 179–220.

McCown, R.L., 2002b. Locating agricultural decision support systems in the troubled past and socio-technical complexity of 'models for management'. Agric. Syst. 74, 11–25.

Meinke, H., Howden, M., Struik, P., Nelson, R., Rodriguez, D., Chapman, S., 2009. Adaption science for agriculture and natural resource management—urgency and theoretical basis. Curr. Opin. Environ. Sustain. 1, 69–76.

Multsch, S., Kraft, P., Frede, H.-G., Breuer, L., 2011. Development and application of generic plant growth modeling framework (PMF). In: 19th Proceedings of the International Congress on Modelling and Simulation, Perth, Australia, 2011.

Nilsson, H., 2004. What are the possible influences affecting the future environmental agricultural policy in the European Union? An investigation into the main factors. J. Clean. Prod. 12, 461–468.

Notarnicola, B., Hayashi, K., Curran, M.A., Huisingh, D., 2012. Progress in working towards a more sustainable agri-food industry. J. Clean. Prod. 28, 1–8.

O'Neill, B.C., Dalton, M., Fuchs, R., Jiang, L., Pachauri, S., Zigova, K., 2010. Global demographic trends and future carbon emissions. Proc. Natl. Acad. Sci. U. S. A. 107 (41), 17521–17526.

Ometto, A.R., Ramos, P., Lombardi, G., 2007. The benefits of a Brazilian agro-industrial symbiosis system and the strategies to make it happen. J. Clean. Prod. 15, 1253–1258.

Ostendorf, B., 2011. Overview: spatial information and indicators for sustainable management of natural resources. Ecol. Indic. 11, 97–102.

Ovando, P., Caparros, A., 2009. Land use and carbon mitigation in Europe: a survey of the potentials of different alternatives. Energy Policy 37, 992–1003.

Pacini, C., Wossink, A., Giesen, G., Huirne, R., 2004. Ecological-economic modelling to support multi-objective olicy making: a farming systems approach implemented for Tuscany. Agric. Ecosyst. Environ. 102, 349–364.

Partidario, P., Lambert, J., Evans, S., 2007. Building more sustainable solutions in production-consumption systems: the case of food for people with reduces access. J. Clean. Prod. 15, 513–524.

Penazzi, S., Accorsi, R., Ferrari, E., Manzini, R., Dunstall, S., 2017. Design and control of food job-shop processing systems: a simulation analysis in the catering industry. Int. J. Logist. Manag. 28 (3), 782–797.

Penning de Vries, F., Van Keulen, H., Rabbinge, R., 1995. Natural resources and limits of food production in 2040. In: Bouma, J., Kuyvenhoven, A., Luyten, J.C., Zandstra, H.G. (Eds.), Eco-Regional Approaches for Sustainable Land Use and Food Production. Kluwer Academic Publishers, pp. 65–87.

Raspor, P., 2008. Total food chain safety: how good practice can contribute? Trends Food Sci. Technol. 19, 405–412.

Riveira, I.S., Maseda, R.F., 2006. A review of rural land use planning models. Environ. Plan. B: Plan. Design 33, 165–183.

Roberts, B.H., 2004. The application of industrial ecology principles and planning guidelines for the development of eco-industrial parks: an Australian case study. J. Clean. Prod. 12, 997–1010.

Roy, P., Nei, D., Orikasa, T., Xu, Q., Okadome, H., Nakamura, N., Shiina, T., 2009. A review of life cycle assessment (LCA) on some food products. J. Food Eng. 90, 1–10.

Savino, M., Manzini, R., Della Selva, V., Accorsi, R., 2017. A new model for environmental and economic viability of renewable energy systems: the case of wind turbines. Appl. Energy 189, 739–752.

Schultink, G., Amaral, N., Mokma, D., 1987. User's guide to the cries agro-economic information system-yield model. In: Comprehensive Resource Inventory and Evaluation System (CRIES) Project. Michigan State University, East Lansing, Michigan.

Seguin, A., 2011. How could forest trees play an important role as feedstock for bioenergy production? Curr. Opin. Environ. Sustain. 3, 90–94.

Sheerr, S., McNeely, J., 2008. Biodiversity conservation and agricultural sustainability: towards a new paradigm of eco-agriculture landscapes. Philos. Trans. R. Soc. B 363, 477–494.

Smith, P., Martino, D., Cai, Z., Gwary, D., Janzen, H., Kumar, P., McCarl, B., Ogle, S., O'Mara, F., Rice, C., Scholes, B., Sirotenko, O., Howden, M., McAllister, T., Genxing, P., Romanenkov, V., Schneider, U., Towprayoon, S., 2007. Policy and technological constraints to implementation of greenhouse gas mitigation options in agriculture. Agriculture Ecosyst. Environ. 118, 6–28.

Soltani, A., Hoogenboom, G., 2007. Assessing crop management options with crop simulation models based on generated weather data. Field Crop Res. 103, 198–207.

Soysal, M., Bloemhof-Ruwaard, J.M., Haijema, R., van der Vorst, J.G.A.J., 2018. Modeling a green inventory routing problem for perishable products with horizontal collaboration. Comput. Oper. Res. 89, 168–182.

Spiertz, H., 2010. Food production, crops and sustainability: restoring confidence in science and technology. Curr. Opin. Environ. Sustain. 2, 439–443.

Steel, C., 2008. Hungry City: How Food Shapes Our Lives. Chatto & Windus, London.

Stellingwerf, H.M., Kanellopoulos, A., van der Vorst, J.G.A.J., Bloemhof, J.M., 2018. Reducing CO_2 emissions in temperature-controlled road transportation using the LDVRP model. Transp. Res. Part D: Transp. Environ. 58, 80–93.

Stephens, E.C., Jones, A.D., Parsons, D., 2017. Agricultural systems research and global food security in the 21st century: an overview and roadmap for future opportunities. Agricult. Syst. (in press).

Stern, N., 2006. The Stern Review: The Economics of Climate Change. HM Treasury, London.

Strengers, B.J., Van Minnen, J., Eickhout, B., 2008. The role of carbon plantations in mitigating climate change: potentials and costs. Climate Change 88, 343–366.

Tilman, D., Cassman, G., Matson, P., Naylor, R., Polasky, S., 2002. Agricultural sustainability and intensive production practices. Nature 418, 2002.

Tufano, A., Accorsi, R., Garbellini, F., Manzini, R., 2018. Plant design and control in food service industry. A multidisciplinary decision-support system. Comput. Ind. 103, 72–85.

Tukker, A., Emmert, S., Charter, M., Vezzoli, C., Sto, E., Andersen, M.M., Geerken, T., Tischner, U., Lahlou, S., 2008. Fostering change to sustainable consumption and production: an evidence based view. J. Clean. Prod. 16, 1218–1225.

Valli, E., Manzini, R., Accorsi, R., Bortolini, M., Gamberi, M., Bendini, A., Lercker, G., Gallina Toschi, T., 2013. Quality at destination: simulating shipment of three bottled edible oils from Italy to Taiwan. Rivista Italiana delle Sostanze Grasse 90 (3), 163–169.

Van Ittersum, M.K., Rabbinge, R., 1997. Concepts in production ecology for analysis and quantification of agricultural input-output combinations. Field Crop Res. 52, 197–208.

Van Ittersum, M.K., Leffelaar, P.A., Van Keulen, H., Kropff, M.J., Bastiaans, L., Goudriaan, J., 2003. On approaches and applications of the Wageningen crop models. Eur. J. Agron. 18, 201–234.

Varga, M., Kuehr, R., 2007. Integrative approaches towards zero emissions regional planning: synergies of concepts. J. Clean. Prod. 15, 1373–1381.

Virtanen, Y., Kurppa, S., Saarinen, M., Katajajuuri, J., Usva, K., Maenpaa, I., Makela, J., Gronroos, J., Nissinen, A., 2011. Carbon footprint of food-approaches from national input-output statistics and a LCA of a food portion. J. Clean. Prod. 19, 1849–1856.
Walters, B., Hansen, L., 2013. Farmed landscapes, tree and forest conservation in Saint Lucia (West Indies). Environ. Conserv. 40 (3), 211–221.
Wakeland, W., Cholette, S., Venkat, K., 2012. Food transportation issues and reducing carbon footprint. In: Boye, J., Arcand, Y. (Eds.), Green Technologies in Food Production and Processing.Food Engineering Series. Springer, Boston, MA.
Weber, C.L., Matthews, H.S., 2008. Food-miles and the relative climate impacts of food choices in the United States. Environ. Sci. Technol. 42 (10), 3508–3513.
West, P., Gibb, H., Monfreda, C., Wagner, J., Barford, C., Carpenter, S., Foley, J., 2010. Trading carbon for food: global comparison of carbon stocks vs. crop yields on agricultural land. Proc. Natl. Acad. Sci. U. S. A. 107 (46), 19645–19648.
Witlox, F., 2005. Expert systems in land-use planning: an overview. Expert Syst. Appl. 29, 437–445.
World Bank, 2011. In: Deininger, K., Byerlee, D., Lindsay, J., Norton, A., Selod, H., Stickler, M. (Eds.), Rising Global Interest in Farmland: Can it Yield Sustainable and Equitable?, ISBN: 978-0-8213-8592-0.
World Bank, 2013. Global Economics Prospects. Less Volatile, But Slower Growth. A World Bank Group Flagship Report, 2013, Vol. 7, ISBN: 978-1-4648-0036-8.
Wright, L., Kemp, S., Williams, I., 2011. 'Carbon footprinting': towards a universally accepted definition. Carbon Manag. 2 (1), 61–67.
Yewlett, C., 2001. OR in strategic land-use planning. J. Oper. Res. Soc. 52, 4–13.

Chapter 2

Emerging issues and challenges in agri-food supply chain

Rajeev Bhat and Ivi Jõudu

ERA Chair for Food (By-)Products Valorization Technologies, Estonian University of Life Sciences, Tartu, Estonia

Abstract

Globalization and free trade policies coupled with consumers' demand for safe and high-quality foods have created pressure on various stakeholders (key players) attached within the agri-food supply chain. Influence, contributions, and socioeconomic and environmental factors are major players in achieving a successful flow of supply chain. Globally, various techniques and conceptual models have been proposed to render the agri-food supply chain effective and profitable. However, there are still several gaps and emerging challenges in the supply chain hindering fruitful, sustainable food production. In this chapter, an attempt has been made to identify and highlight the present world scenario and challenges encountered along the agri-food supply chain and its future prospects.

1 Introduction

Consumer demand for high-quality and safer foods have led to proposals on innovative approaches and models to be adopted in the agri-food supply chain. In a global context, there are several sustainability challenges along the food supply chain. Some of the innovative solutions proposed (depending on the socioeconomic and environmental factors) are practically applicable only in certain regions of the world. Sustainability challenges in the "farm-to-fork" concept are well documented in the literature and available databases. However, several sustainability challenges in supply chain management still remain unresolved. These problems range from production and processing stages, all the way to the consumers. Managing the pressure exerted by consumers (regarding quality and safety) has led to potential negative and positive impacts. Even though food production has increased to meet the demands of ever-growing populations, rising prices and economic impacts/hardships on individual regions (low-income and high-income countries) remains as a recurring problems. On a global basis, with free-trade policies gaining a higher hand, the "trade revolution" has gained the upper hand over the "green revolution." Coupled with this are the changing weather patterns which have had a significant effect on the food supply chain. Nevertheless, the food-energy-water nexus is vital to effectively manage the agri-supply chain and environmental stewardship. Besides, in the food supply chain context, life cycle assessment is recognized as a reliable method to evaluate the environment impacts of various processes (such as farming, harvesting of raw materials, processing, packaging, etc.) (Thomassen et al., 2008). Not to forget, reduction in the carbon footprint should be one of the crucial factors to ensure and create a sustainable food supply chain system (Virtanen et al., 2011).

Of course, key logistics decisions can be of much help in meeting environmental sustainability challenges faced in the food supply chain. Certain regions of the world, though claiming to be economically stable, have never established a positive relationship between diversity and food self-sufficiency. Low- and middle-income groups of countries are having their own problems in the food supply chain that need to be resolved, such as feeding and meeting the demands of their own population as well as competing in the global food markets. Elementary sustainability issues valuable from a socioeconomic and environmental perspective have been discussed under the millennium and sustainable development goals of the United Nations (UN, 2014). Further, the food supply chain is recognized as a series of closely working interdependent firms/companies that manage the smooth movement of agri-food products (for value addition) to provide consumers with high-quality products at affordable prices (Folkerts and Koehorst, 1998).

In a recently published book (Bhat, 2017), a wide range of sustainability challenges faced in the agro-food sector (on a global scenario) has been discussed by leading experts. A wealth of additional literature has focused on various sustainability issues encountered in the agri-food supply chain (Akkerman et al., 2010; Beske et al., 2014; Brandenburg et al., 2014; Borodin et al., 2016; Kusumastuti et al., 2016). Moving forward, top priority and attention is being focused on

Sustainable Food Supply Chains. https://doi.org/10.1016/B978-0-12-813411-5.00002-8

sustainable food supply chains and sustainable supply chain management (Zhu et al., 2008; Ras and Vermeulen, 2009; Tseng and Chiu, 2013). The concept of sustainable supply chain management involves managing of materials, information, and capital flow between parties involved in the supply chain with a common intention of socioeconomic-environmental sustainable development (Seuring and Müller, 2008). Of late, Esfahbodi et al. (2016) has proposed a working model focusing on cost operations that encompass procurement and distribution, designing, and investment recovery options in a sustainable manner. The role of technological innovations and their impact on the agro-food sector (market, market structure, and supply chains) has been well documented by Reardon and Timmer (2007). Moreover, the concept of supply chain integration encompasses mutual understanding, tactical collaborations, and practical/operational strategic activities of companies along the chain. Supply chain integration has been envisaged to keep the entire network in a single loop together, thus helping in reduction of recurring supply chain challenges, such as deprived demand based management, forecasting, and maintaining of consumer and supplier understanding and relationships (Marzita Saidon et al., 2015). As a whole, the supply chain is independent and involves management and movement along various individual sectors like those of aquaculture, agriculture, livestock, food supplements, organic farming, sustainable farming, cold chain management, dairy industry, minimal processing, and many more.

Today's globalized food system is a highly complex mixture, with major roles being played by industrial personnel, NGOs, farming communities, and stakeholders. Nevertheless, the relationship along the food supply chain can be expressed as an understanding between individual stakeholders (players) to meet the interest of groups of consumers. Nonetheless, all the stakeholders must encourage sustainable development and should be aware of the supply chain flow (e.g., the origin of the raw material/products bought, marketed, etc.). It is also opined that if the stakeholders' approach in the supply chain is in line with sustainable behavior, the rapport of their firms will be much higher (Carter and Rogers, 2008). So maintenance of transparency is also vital, and both consumers and stakeholders should not be ignorant of sustainable production adopted along the supply chain (Wognum et al., 2011). Of course, apart from consumers, farming communities are also under great pressure to produce high-quality, safer foods at lower costs. Food diversification, food culture, and changing lifestyles have all led to several "tailor-made" solutions being proposed along the food supply chain. Linking producers and consumers is a major effort being undertaken on a global scale. At the international market levels, several glitches and risks have been identified along the food supply chain, such as food fraud, food recalls, foodborne illnesses (emergence of new resistant microbial pathogens), illegal production of "mimic" or "faked" foods and many others. Indeed, there are several schools of thought that claim that consumers, when buying their food items, might be the least interested in knowing what exactly happens along the food supply chain. These groups of consumers are more worried about price/cost rather than quality or health impacts. Further, there are demographic challenges in the low-income groups of countries (with household food insecurity problems), who tend to be rather unaware of the concept of the food quality supply chain. Conversely, to ensure food security, it is not the food production or consumption volumes (Barrett, 2010) but food diversity/diversification that has an impact. The rural population perceptions can be quite different from those of urban populations regarding food supply chain issues. Trust in the food supply chain varies among populations from region to region as well. In the supply chain context, the distance travelled by a agri-food commodity from farm (producers) up to table (consumers) is referred to as "food miles." In the majority of developed countries, consumer demand has risen for minimally processed foods that have undergone less processing along the food production chain. As a result, new agricultural policies and models being proposed for the food supply chain (globalization of food chain) have had reasonable success only in certain regions. So the challenges facing imports and exports in the international market chain are several (the need for certification, maintenance of international standards, quality assurance, etc.).

Some terms and concepts are being proposed by various researchers relevant to the supply chain. The term *supply chain responsibility* has been used by Spence and Bourlakis (2009) and *ethical supply chain* concepts have been referred to (by Carter and Easton, 2011; Walker and Phillips, 2009). Back in the 1980s, "the supply chain management" concept was introduced by Oliver and Webber (1982). Supply chain management is a direct representative of demand chain management wherein production, handling, processing manufacture, and distribution as well as marketing are tailored and designed to meet consumers' desires. There is a high level of uncertainty for food product quality and safety in comparison to the market-driven transaction costs like information, negotiation, or monitoring (Hobbs, 1996). Food trade and internationalization of supply chain management have led to competitive market approaches at the individual national levels. At international levels, for an imported food, consumers are much more interested to know the country (geographical origin), certifications of international standards, ethical issues followed, and overall whether sustainable food production has been adopted or not. In the following text, some vital parameters that influence a food supply chain are discussed.

2 Issues, advantages, and challenges

Several interesting sustainability challenges have arisen along the agri-food supply chains, which encompass various socio-economic and environmental aspects. Socioeconomic issues are mainly related to local populations or consumers, and this is related to household income, price variations of food commodities, gender inequality, health issues, and many others. Some of the other challenges like food shortages, pre-/postharvest losses, imbalance in demand versus supply gaps, small land holdings, price rise, consumer demand for safe and quality foods, nutritional security, access to market information for farmers, exposure to global opportunities for free trading or technology transfer, lack of updated information and intelligence, disorganized and inadequate market infrastructure and supply chains, climate change issues, unsustainable land-water-energy use practices, and many more. Globally, a wide range of heterogeneity in household food insecurity has been witnessed in various regions (Headey, 2013; Verpoorten et al., 2013), creating imbalance in the regional economy. With regard to environmental aspects, this covers those dealing with waste disposal (on farm or off farm), food industrial wastes, "on table" food wastes, etc. Li et al. (2014) have opined that in the low-income and high-income countries, the food industry contributes to elevated levels of wastes and greenhouse gas emissions. As a result, increased knowledge and concerns among consumers with available updated databases have pressured agri-food industry personnel to emphasize improving quality and safety along supply chains (Matopoulos et al., 2007). Partnerships between farming communities to work along with supply chain stakeholders (government organizations, NGOs, research scientists, academicians, industrialists, and others) are vital (Hamprecht et al., 2005). The rapid changes being witnessed in climate and weather conditions worldwide can have a significant impact on the agri-food supply chain. Imbalances in crop productivity, reductions in annual yields, depletions in water resources, impacts on soil quality, and variations in nutritional qualities are some of the expected climate-induced changes that can influence smooth functioning of the agri-food supply chain. These in turn can stress the overall marketing system with imbalances, in the supply versus demand systems in particular.

2.1 Food price volatility

Any change in the market food price can strongly influence a household's food security, especially in middle- and low-income groups of countries. The last decade has witnessed rapid changes in food prices in both agriculture and aquaculture sectors. Import and export businesses have suffered from detrimental long-term effects on normal lifestyles. Of course, it is a well-acclaimed fact that volatility in price can affect the continuous flow of the food supply chain and is a highly undesirable trait. According to Dawe and Timmer (2012), changes in price can lead to inefficiencies, low confidence in market/investments, and hampering of overall economic growth. Transaction cost theory and contract theory have been proposed and applied (Hobbs, 1997; Goodhue, 2000) to deliver various economic principles for designing the supply chain. In addition, contract theories have been proposed concerning providing contracts to agents (Bolton and Dewatripont, 2005), the pros and cons being well understood.

Price, cost reduction, and quality are main pointers to examine in the performance of a supply chain (Cai et al., 2009; Fynes et al., 2004). High prices have been linked with increased food insecurity and poverty, especially among the non-urban/rural consumers/populations (Headey and Martin, 2016). Moreover, changes in lifestyle, food habits, and preferences by consumers for vegetarian/plant-based diets have all led to a temporary imbalance being created in the supply chain and market demands. This in turn affects the normal price of a food commodity. Nevertheless, the movement of rural populations, especially those of farming communities to urban regions, can also contribute to price volatility. This holds true mainly in the Asian-Pacific, Southeast Asian, and North African regions. Converting of local agriculture farms and plantations to produce export-oriented crops is also of much concern, especially for small-scale farmers (Rullia et al., 2012). In addition, the emergence of new pests and vectors and the lack of appropriate logistics can also have an impact on price. Though several working models are being proposed, not much information is available on how these models contribute to the success of managing price volatility in the agri-food market supply chain. Besides, as indicated earlier, changes in the weather/climatic conditions can result in imbalances between supply and demand, thus pressurizing supply chain and marketing systems and subsequently giving rise to price alterations. The influence of climate variability can affect yield pathways as well as relative price changes. In Fig. 1, conceptual linkages between climate and price variability, adaptation options, and food security have been depicted (Source: Wossen et al., 2018). As per the author, climate-induced price variability can significantly affect household decisions. So also can price variability have an impact on food security wherein purchasing power in a household can be compromised.

2.2 Quality and safety issues

GAP (good agriculture practice), GHP (good handling practice), GMP (good manufacturing practice) along with HACCP (hazard analysis and critical control points) approaches and practices of international benchmark standards have been a success story in a majority of the countries worldwide. However, several issues still emerge from time to time related to quality and safety of various agri-food commodities (both for fresh and processed ones). Emerging issues of food quality and safety in the agri-food supply chain have been excellently documented by leading researchers in recent publications (Bhat et al., 2012; Bhat and Gómez Lopez, 2014).

Globalization of agri-food markets coupled with free-trade policies have tremendously enhanced inspection of food origin, their quality, safety, nutritional and health properties as well as ethics followed to achieve sustainable food production. Among all, labeling on the origin and processing of products can be considered as highly important from both consumer and stakeholder points of view. Implementation of strict policies, legislation, and trade agreements for safer agri-foods have their own detrimental effects. Cheap and low-quality food commodities have been marketed intercontinentally with many traceability issues. The presence of banned chemicals and contaminants, the presence of spoilage/pathogenic microorganisms, new insect vectors, etc. pose serious problems. The opportunistic behavior of middlemen or suppliers along the chain can be dangerous too. The management and functioning of the food supply chain emphasize food quality more (Rong et al., 2011). In the supply chain, the identified gaps are not entirely from the farming land, but contributions from other driving force institutions (like transport, storage, marketing, middle men, stake holders, suppliers, etc.) are equally high. Hence, building up of mutual trust and positive business partnerships is of paramount importance for the various stakeholders involved in the food supply chain. Mutual relationships, trust, extended partnerships, effective communications, commitment and consumer satisfaction are all more vital for maintaining high-quality standards and safety of foods along the supply chain than investing in modern-day technologies (Keller, 2002; Beth et al., 2003; Fynes et al., 2004; Ding et al., 2014; Swinnen, 2015). So, the bottom line is that food quality and safety are of paramount importance to individual countries, to overcome concerns relevant to increased cases of food recalls, economic losses, increased food fraud and faked foods, etc. along the supply chain. In addition, bioterrorism threats, increased reports of loss in consumer confidence, increased pressure for regulations and guidelines, and increased imports and trading of agri-food commodities from regions with less stringent food safety laws all pose serious concerns.

The issues of food frauds, blockchain technology, and traceability are discussed later.

2.3 Food wastage and loss

A major reason for concern in the agri-food supply chain is food wastage and loss. Alterations in the physical characteristics of a food (inedibility and reductions in weight) can be considered food loss while inability of consumption of edible foods can be considered waste. Nevertheless, food loss has been considered a subcategory of food waste. As per the United Nations Environment Programme, one of the main constraints for imbalance in the supply chain is the relation between production and consumption versus food waste and loss. Even though the Zero Hunger Challenge has been taken up seriously by the UN, if this imbalance occurs, then it can be an issue leading to food insecurity. The UN, by taking all of the stakeholders (involved in the food supply chain) into confidence, has developed a sustainable food system model to eliminate food wastage/loss and to ensure global food security (United Nations, 2012; Wickramasinghe, 2014). Food loss can occur at "on farm" or "off farm" levels. This might be due to inappropriate storage environments, logistic problems, pests and contaminants, mismatches in supply vs. demand, and much more. "On table" wastes can also be considered as food loss. Regarding food waste, this mainly happens due to negligence and mishandling and knowingly disposing of edible food stuffs (mainly by consumers in house or restaurants). In general, along the supply chain, food loss occurs during the initial stages, while food waste occurs later on (Parfitt et al., 2010). As per the FAO (2011), in developing and developed countries in the food supply chain, food loss and waste occur mainly at the consumer level, while in underdeveloped or low-income countries waste occurs during initial or middle stages. It is worthwhile to note that in developed countries, post-harvest loss in agri-food supplies are much less and this is owed to modernization of the supply chain. Over here, the loss can occur only at the consumer or retailer levels (Priefer et al., 2016; Teng and Trethewie, 2012). In the supply chain route, loss and wastage can occur irrespective of an agri-food commodity involved (fresh or processed). It is universally agreed that upgrading the conventional marketing system with modern technologies (waste disposal or reuse/value addition) can reduce wastage. Lack of market infrastructure and modernized facilities coupled with poor policy and regulatory frameworks and quality regulations can have a long-standing unwarranted influence on the supply chain. As per the Food and Agriculture Organization (FAO), nearly one-third of the world's food production is wasted (FAO, 2017). According to the United Nations Environment Programme (UNEP), nearly 300 million tons of food is wasted annually owing to irresponsible approaches either from producers, retailers, or consumers (UNEP, 2015). This food waste can lead to environmental issues/burdens and, if it remains unmanaged, then it can have an impact on achieving success in the smooth flow of the supply chain (leading to food insecurity) (FAO, 2014). Raak et al. (2017) have proposed a conceptual model of drivers for 'processing-related' food waste (Fig. 2).

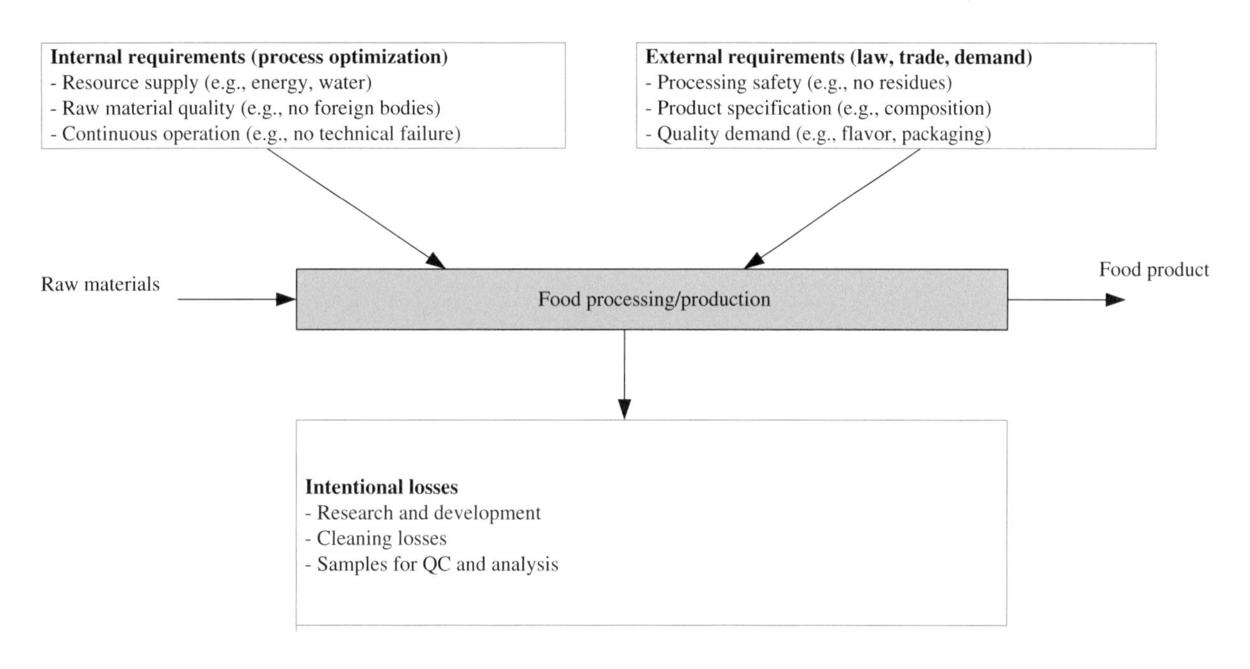

The model proposed is self-explanatory and it highlights processing-related food wastes and losses. Kibler et al. (2018) have recently proposed a conceptual model to characterize food waste within the food-energy-water nexus (see Fig. 3). Accordingly, the mechanism by which food waste is executed along the nexus is directly influenced by human behavior and the decision-making process (this nexus is discussed in detail by the author).

To prevent loss and wastage, many strategies have been put forth by researchers and institutions worldwide. In general, along the supply chain, it is opined that stakeholders need to work out and check for a balance in loss versus prevention costs, invest in setting up modernized marketing facilities, upgrade agri-food processing facilities, undertake measures to reduce carbon emissions, set up a working model for logistics such that the products reach consumers in minimal time (food miles), etc. It has been reported that vertical cooperation can contribute to reduction in food loss (deterioration) and carbon

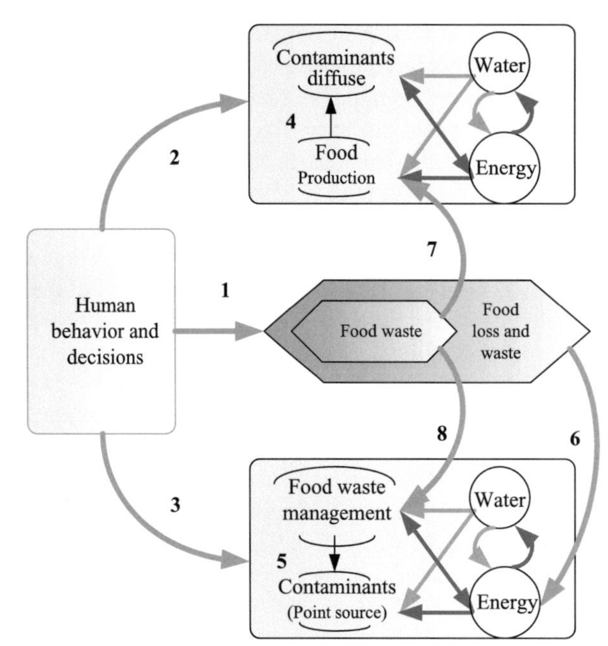

FIG. 3 Food loss and waste impact on the food-energy-water (FEW) nexus in food production and waste management phases. *(Reedited from Kibler, K. M., Debra, R., Christopher Hawkins, C., Motlagh A.M., Wright, J., 2018. Food waste and the food-energy-water nexus: a review of food waste management alternatives. Waste Manag. 74, 52–62; License Number: 4375641353554; Permission from Elsevier, dated: 24 June 2018.)*

emissions, wherein appropriate strategies of forward and backward integration can help a stakeholder (here retailer) in making investments in preservation technologies to reduce food loss or deterioration (Huang et al., 2018). So, developing models for optimization of the sales forecasting process can be a vital step that can have an impact on the expansion and growth of a representative supply chain (Dellino et al., 2018).

2.4 Food fraud

Food adulterations related to health issues are regularly reported from many regions of the world. Adulteration can be intentional or unintentional and not necessarily for economic gain. However, on the other hand, food fraud is defined as deliberate and intentional addition, substitution, tampering, mislabeling, and making false claims for a food product mainly aimed at economic gain. Food fraud being intentional, it is more about cheating consumers for economic gains in the market. In the majority of food fraud cases, serious health effects have been reported. Some recent examples of food fraud scandals include: marketing of fake eggs, plastic rice, intentional adulteration of herbs and spices, horsemeat scandals, melamine in milk, mislabeled olive oil, liquefying of fruit juice, reuse of drainage oil, mislabeling of the origin of fresh produce and processed foodstuffs, and more. Once the fraud is detected along the supply chain, the consequences are very damaging: loss of company rapport, loss of consumer confidence, product recalls, volatility/collapse in the local and international market, and in extreme cases imprisonment for serious criminal offenses. Understanding the individual agri-food supply chains at the primary producer levels can improve forecasting of fraud vulnerability. It is strongly suggested that evaluation of food fraud vulnerability along the supply chain can help in understanding key potential drivers (Spink et al., 2017). Of late, food fraud vulnerability and its key factors have been detailed by van Ruth et al. (2017). Accordingly, authors have identified food fraud vulnerability elements and have detailed various factors involved (see Fig. 4). The factors subgrouped were opportunity-related factors, motivation-related fraud factors, and control measures-related factors.

Moreover, assessments and data generation have been performed in milk, spices, seafood (fish), meat, and olive oil (Silvis et al., 2017). Recently, Moyer et al. (2017) opined that fraud vulnerability needs to be assessed both at the micro- and macrofactorial levels. According to Manning and Soon (2016), vulnerability can be minimized via implementation of appropriate control measures. Working on food fraud policy and the food chain, Manning (2016) has suggested that fraud in the food supply chain can be owed to the integrity of food items, various processes employed for production of the food items, personnel involved, as well as information (labeling) accompanying the food items. The author has also stated that food fraud mitigation can transfer from the standpoint of detecting fraud to preventing food fraud. As shown in Fig. 5, Manning (2016) has proposed three vital constituents, namely: food integrity (product, process, people, and data

Opportunities related factors

Technical opportunity
- Simplicity/complexity of adulteration
- Simplicity/complexity of counterfeiting
- Availability of detecting technology
- Knowledge of adulteration

In time and space
- Accessibility to materials in production / processing
- Transparency in supply chain network
- Historical evidence

Opportunities related factors

Economic drivers
- Supply and pricing of materials
- Special product attributes or value determining components of materials
- Economic health business
- Level of competition
- Financial strains imposed on suppliers

Culture and behavior
- Business strategy
- Ethical business culture
- Previous criminal offenses
- (Inter)national corruption level
- Victimization

Control and measures related factors

Technical measures
- Specificity and accuracy of fraud monitoring system
- Systematics and autonomy of verification fraud monitoring system
- Accuracy information system for mass balance control
- Extensiveness of tracking and tracing system
- Fraud contingency plan

Managerial measures
- Strictness ethical code conduct
- Application integrity screening
- Support whistle-blowing system
- Contractual requirements suppliers
- Social control and transparency across supply chain
- Established guidance for fraud prevention across supply chain
- Specificity national food policy
- Strictness enforcement for fraud prevention regulation/law

FIG. 4 Overview of the food fraud vulnerability main elements and detailed factors. *(Reedited from van Ruth, S.M., Huisman, W., Luning, P.A., 2017. Food fraud vulnerability and its key factors. Trends Food Sci. Technol. 67, 70–75; License Number: 4375650542424; Permission from Elsevier, dated: 24 June 2018.)*

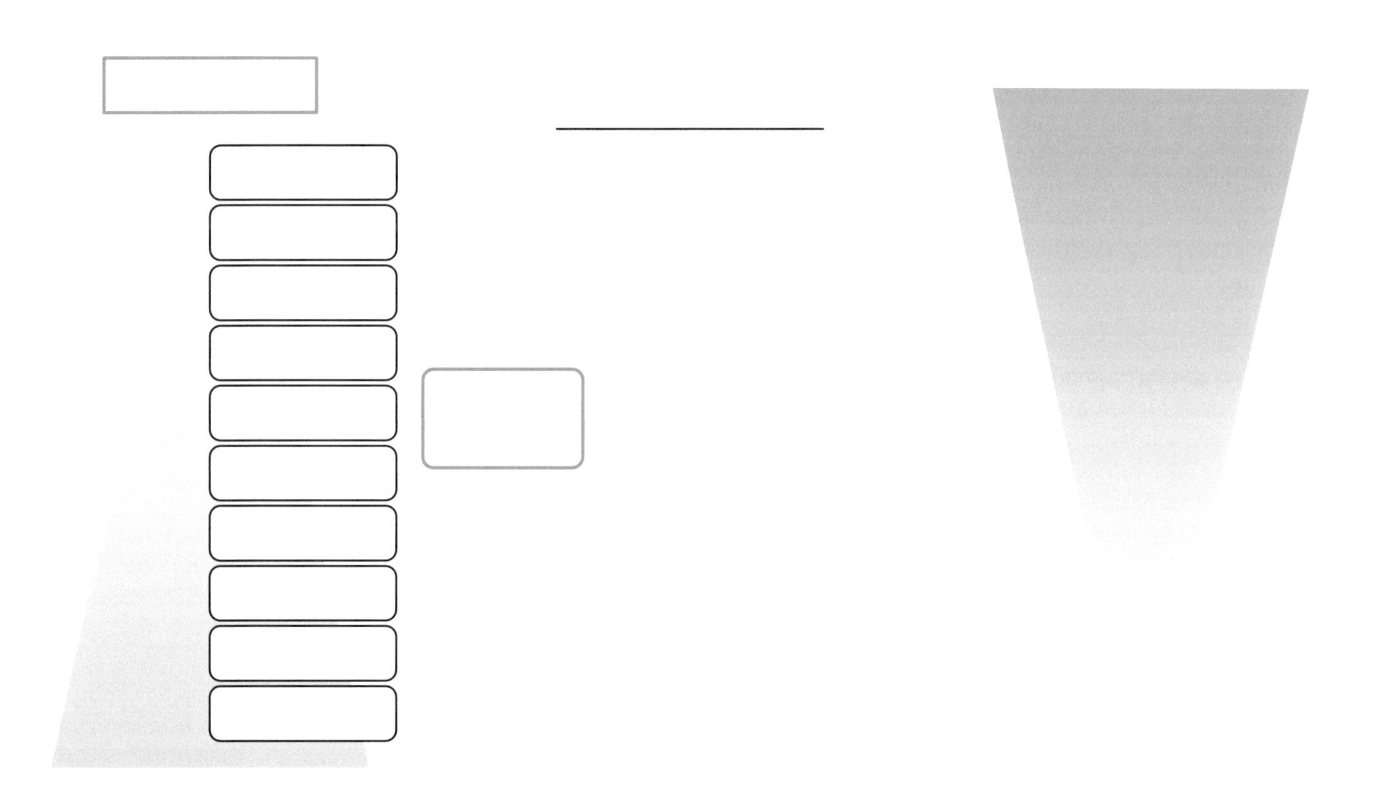

integrities), drivers of the supply chain linking food fraud, and mitigation methods (operating in organizational, supply chain, and global measures). Even though several invasive and noninvasive tools (including those of analytical and biotechnological methods) have been recommended for detecting food fraud, most of the time it has remained underexplored owing to cost effectiveness as well as the huge volumes of samples to be analyzed in a short time interval. On the other hand, gaining access to market intelligence can be a key factor in identifying food authentication and fraud.

Proposing innovative models to forecast risks involved in food fraud can be of much benefit. To protect global consumer interest, immediate efforts need to be made to develop universally acceptable models for each of the food commodities in high demand and most vulnerable for fraud. Maintenance of transparency and validations along the supply chain should be of paramount importance. Involvement of producers, suppliers, marketers, risk assessment researchers, food safety inspectors, and experts from industry and academia, and finally the consumers themselves, are important to prevent food fraud in the supply chain.

2.5 Food traceability and blockchain technology

Traceability literally means tracing the movement of an agri-food commodity along the supply chain. CODEX Alimentarius defines food traceability as the "ability to follow the movement of a food through specified stage(s) of production, processing and distribution" (Codex, 2006), while as per the ISO standard 22005:2007 food traceability is explained as "a technical tool to determine the history or location of a product or its relevant components" (ISO, 2016). The concept revolves around "one step ahead" and "one step behind" at any given point in the supply chain. Traceability in simple words is appropriately documented transparency of the supply chain (Opara and Mazaud, 2001) of an agri-food commodity to ensure success for sustainable agriculture and food systems.

As indicated by Petersen and Green (2005) tracing and tracking are two different concepts. Tracing involves a backward process wherein origin of a commodity can be determined via maintenance of records along the supply chain, whereas tracking involves the forward process wherein end users or the consumers are identified via location in the supply chain. According to Bechini et al. (2008), tracking follows a downstream path of an agri-food commodity along the supply chain, whereas tracing refers to upstream referring to records in the supply chain to identify origin and characteristic features of a commodity. According to Rábade and Alfaro (2006), both tracing and tracking require "managerial decisions" to be made along the supply chain to enhance efficiency, improve processing skills, manage risk, add value, and enhance relations between supplier and stakeholders. According to Rijswijk and Frewer (2008), product tracing helps to regulate and understand (physical and sensorial) conditions of produce at any given point of time. Traceability alone can help in assuring and overcoming sustainability challenges in the agri-food sector. Being transparent in the supply chain, a traceability system can be a valuable addition to maintain food quality and safety and overcome challenges of the management system (by providing direct communication linkages at micro levels with consumers and market).

Further detailing, traceability involves movement tracking from farm to table (raw materials originating from the farm, all of the geo-environmental conditions, harvesting, postharvest handling, processing, manufacturing, technological intercessions, any logistics/transportation involved, preservation, labeling and market sales) and is important to achieve success in the food security of a region. Tracking of agri-based foods at the production and processing stages as well as during the marketing and distribution stages along the entire supply chain can be of benefit to identity food fraud, maintain transparency, reduce food safety risks, minimize food recalls, and reduce wastes and economic losses. Agri-food supply chain traceability can involve a product (raw and finished) or process involved or origin (geographical source, genetic constituent, supplier, etc.). As an example (Lim et al., 2011), consumers are much more likely to opt for products from a particular geographical location of their preference compared to those which have their origin from an undesirable location. According to Larsen and Lees (2003), novel and well-organized traceability systems can manage human errors and create better awareness of food quality standards along the supply chain.

Aung and Chang (2014) have proposed various drivers or motivating factors of traceability in the food supply chain. A range of motivational factors is identified with reference to food safety, quality and visibility; traceability as a reliable tool can "answer simple queries raised, like "who (i.e., actor/product), what (actor/product's information), when (time), where (location) and why (cause/reasons)" (see Fig. 6).

Proceeding further, the same authors have proposed a conceptual framework for a food traceability system (see Fig. 7). As per this framework, all of the players (stakeholders or personnel involved) in the entire supply chain are required to have information on both internal and external traceability. Emphasis is laid on safety regulations and quality assurance systems that are enforced by all the players in order to achieve efficiency in a standardized way.

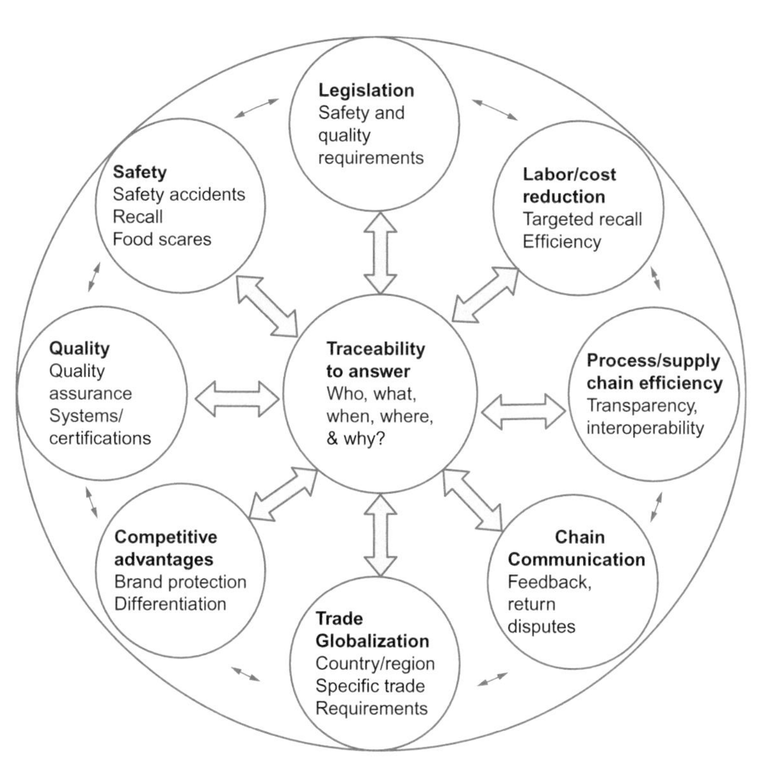

FIG. 6 Drivers for traceability of food supply chain. *(From Aung, M.M., Chang, Y.S., 2014. Traceability in a food supply chain: safety and quality perspectives. Food Control 39, 172–184; License Number: 4375630869300; Permission from Elsevier, dated: 24 Jun 2018.)*

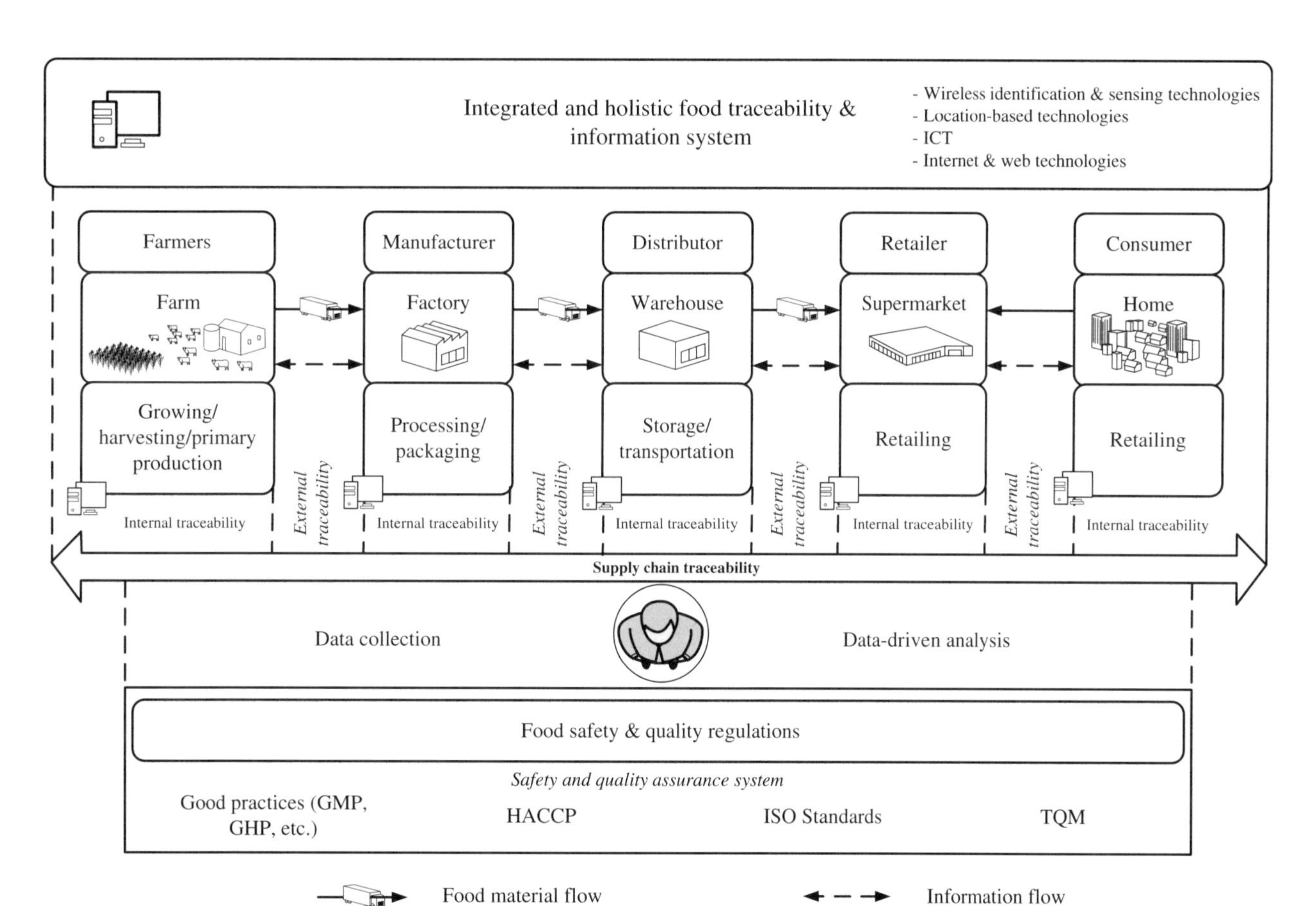

FIG. 7 Conceptual framework of food traceability system. *(Reelaborated from Aung, M.M., Chang, Y.S., 2014. Traceability in a food supply chain: safety and quality perspectives. Food Control 39, 172–184; License Number: 4375630869300; Permission from Elsevier, dated: 24 June 2018.)*

In addition, for successful operation and performance in the supply chain, modern technologies (wireless technology, sensor technology, location mapping technology, information and communication technology, internet/web technology) are referred to as facilitators and serve as a channel to all of the players (stakeholders involved), permitting them to have access to information systems of food traceability.

Traceability can be imperative to gain the confidence of consumers and stakeholders. Nevertheless, food traceability can also be considered as a preventative strategy in management of food quality and safety. On occurrence of health hazards or food scares in a society, maintaining a reliable food traceability system can enable timely recall of a particular batch of food products and minimize risks involved. Of late, many techniques have been routinely used to identify a particular product, like bar codes, QR codes, sensors, radio frequency tags, etc. Though consumers and public health personnel are demanding food traceability in the supply chain, at the end of the day it is the personnel involved in agri-food business (as the main economic drivers) who need to sustain this and are going to be benefited. Further, today the term *blockchain technology* is gaining much importance in minimizing food fraud.

2.5.1 Blockchain technology

Blockchain technology is a new innovation which is becoming prominent in supply chain management (mapping and scanning). This technique involves sharing a traceable and transparent document for record-keeping purposes. All of the vital information is collected at each of the transaction steps (traceability) in the food supply chain. This technology includes approval from all of the stakeholders involved in the supply chain. Blockchain technology is simpler and can be accessed (sharing of documents, intelligence, data mapping, recommendations, etc.) by all of the stakeholders in any given time frame. Blockchain technology enhances transparency by digitization of tracking and storage of information to reduce food fraud and enhance consumer confidence; it enhances food safety by reducing cross-contamination issues, product recalls, food waste and loss, and costs; and it improves overall efficiency of logistics, distribution, marketing, treatability, and consumer confidence. As this technology is still in its infancy stages, there are certain issues that needs to be sorted out, such as who will be the key person who retains or manages vital information, what exactly is the role of an individual stakeholder, how will cross-checking be handled for authenticity of the details documented or included, how ready is the market for this technology, and much more. Aschemann-Witzel et al. (2017) in work on a "planned behavior theory" concept (proposed originally by Ölander and Thøgersen, 2013) has provided an overview of how initiatives are localized in the up/downstream end of the supply chain, the main success key, and MAO factors involved (see Fig. 8). The MAO factors, originally proposed by Ölander and Thøgersen (2013), are motivation (M), personal ability (A), and externally identified opportunity (O).

2.5.2 Information and communication technologies in food supply chain

The revolution in information/digital technology has led to a newer phase in the agri-food technology field. Digital technology has gained prominence and has changed the way people communicate, interact, and exchange details with each other in society. These technological innovations are in the form of mobile phones, smartphones, smart watches, drones, notebooks, computers, broadband internet facilities, and much more. As of today, even the agri-food supply chain is highly influenced by the digital technology revolution. Information and communication technology (ICT) applications have been applied to monitor climate change and its impact on agriculture (Ospina and Heeks, 2011). ICT has benefited global food supply chains by providing vital data on innovative techniques that can be employed during pre- and postharvest operations (Coley et al., 2011). A wealth of literature is being published on ICT, artificial intelligence, geospatial science, etc. and their roles in the agri-food sector. Wang (2016) has described the importance and application of e-logistics in managing supply chains. ICT's efficiency has been best exploited in agriculture trade, extension programs, and for imposing good agricultural practices (Rao, 2007). Selecting appropriate planting seasons, disease and pest controls, irrigation management, livestock management, selecting the best seeds and plant varieties, and planning for storage facilities are just a few examples of the ICT role/advantages in the supply chain. Drone usage in agricultural fields is well known and popular. Sensors are also employed to get information and meteorological data from isolated or remote rural farming areas. According to Sylvester (2013), sensors can also help in conservation of highly valued agri-food products. Farming communities, especially in developing countries, have benefited by getting better prices for agri-food commodities in the international trade market

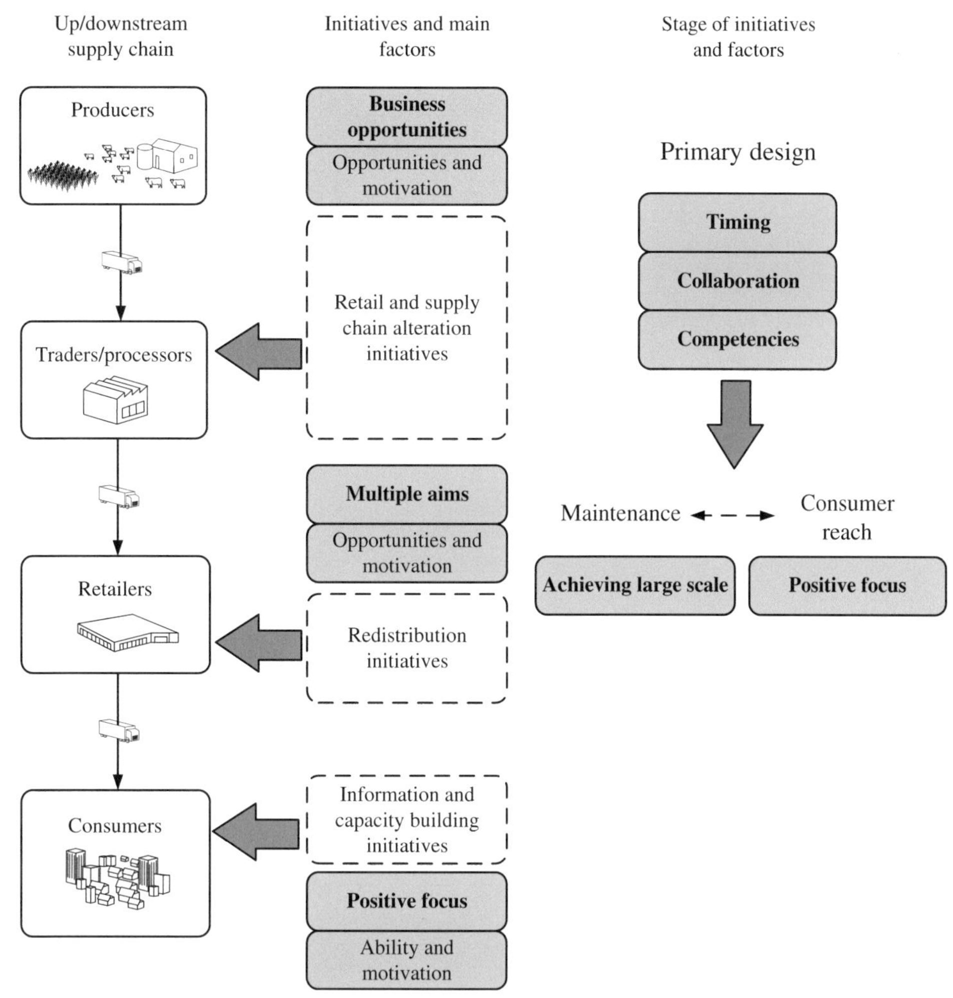

FIG. 8 An overview of how the three types of initiatives are localized within the up/downstream end of the supply chain; which primary key success and MAO factors are found for each of the three types specifically; and which key success factors were of particular importance in each of the stages of initiatives. *(Aschemann-Witzel, A., de Hooge, I.E., Rohm, H., Normann, A., Bossle, M.B., Grønhøj, A., Oostindjer, M., 2017. Key characteristics and success factors of supply chain initiatives tackling consumer-related food waste—a multiple case study. J. Clean. Prod. 155, 33–45; License Number 4375640844781; Permission from Elsevier, dated: 24 June 2018.)*

(Golan et al., 2004). Applications of ICT in rural farming areas have provided benefits due to raised awareness and technical skills in new policy frameworks (Grant, 2003a, b). Besides, the farming community is expected to gain better understanding of the demand for particular crop varieties, pest and disease forecasts, alternative market channels, decision-making processes, postharvest management practices, risk management of production and logistics, and other supply chain issues. ICT can also directly benefit and help in product identification, food fraud vulnerability, quality and safety measurements, etc. Büyüközkan and Göçer (2018) have recently proposed an integration framework for the development of a digital supply chain (DSC) (this is rather self-explanatory) with expected practical applications in the near future (see Fig. 9).

Some of the popular software programs developed to identify traceability are Enterprise Quality Management, Food Trak-2 and Qual-Trace. As with other technologies, ICT also has its own barriers, like shortages in technical experts and support staff, chances of miscommunication over long distances or in remote regions, lack of signal access (bandwidth), uncertainty in predicting agri-food supply chain trends such as demand versus supply, and much more (Huggins and Izushi, 2002; Smallbone et al., 2002; Deakins et al., 2003). However, in the coming days these hindrances are expected to be overcome.

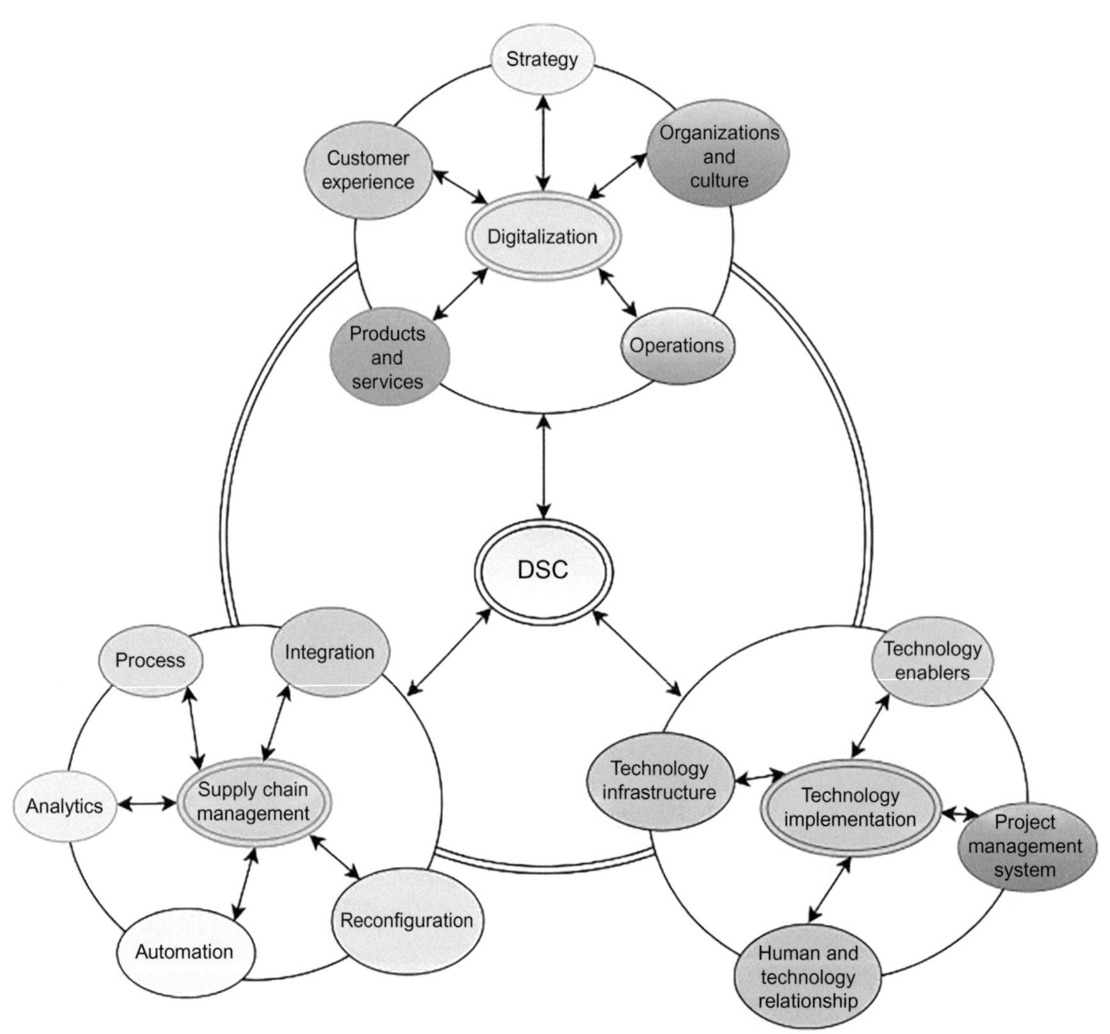

FIG. 9 Integration framework for the development of DSC. *(Büyüközkan, G., Göçer, F., 2018. Digital supply chain: literature review and a proposed framework for future research. Comput. Ind. 97, 157–177; License Number 4375660184129; Permission from Elsevier, dated: 24 June 2018.)*

3 Conclusions and future outlook

Agri-food supply chain success remains uncertain unless efforts are made by the involved stakeholders (from personnel involved in the farm up to the table) to seriously understand the underlying issues and problems. Some of the vital factors in supply chain management among the participants (stakeholders) are linkages, collaborations, belief, teamwork, and transparency. As long as unethical trading psychology persists, food fraud can be expected in the global market for the near future. However, emerging techniques like traceability and blockchain, food laws and legislation, and conceptual models are expected to contribute to a smoother flow of the agri-food supply chain, at least to some extent. In the following days, market-driven agriculture and food production needs to be given a higher priority by keeping in mind consumer demands. International support needs to be extended for low- and middle- income groups of countries for understanding and achieving success in their supply chain. Climate change, agricultural diversification, logistics issues, feeding the world's population, emerging pests and diseases, creating alternative rural markets, and overcoming terrorism-based threats are some of the challenges that are expected to influence the smooth flow of the agri-food supply chain.

Acknowledgment

This chapter theme is based on our ongoing project—VALORTECH, which has received funding from the European Union's Horizon 2020 research and innovation program under grant agreement No 810630.

References

Akkerman, R., Farahani, P., Grunow, M., 2010. Quality, safety and sustainability in food distribution: a review of quantitative operations management approaches and challenges. OR Spectr. 32 (4), 863–904.

Aschemann-Witzel, A., de Hooge, I.E., Rohm, H., Normann, A., Bossle, M.B., Grønhøj, A., Oostindjer, M., 2017. Key characteristics and success factors of supply chain initiatives tackling consumer-related food waste—a multiple case study. J. Clean. Prod. 155, 33–45.

Aung, M.M., Chang, Y.S., 2014. Traceability in a food supply chain: safety and quality perspectives. Food Control 39, 172–184.

Barrett, C.B., 2010. Measuring food insecurity. Science 327, 825–828.

Bechini, A., Cimino, M., Marcelloni, F., Tomasi, A., 2008. Patterns and technologies for enabling supply chain traceability through collaborative e-business. Inf. Softw. Technol. 50 (4), 342–359.

Beske, P., Land, A., Seuring, S., 2014. Sustainable supply chain management practices and dynamic capabilities in the food industry: a critical analysis of the literature. Int. J. Prod. Econ. 152, 131–143.

Beth, S., Burt, D.N., Copacino, W., Gopal, C., Lee, H.L., Lynch, R.P., Morris, S., 2003. Supply chain challenges—building relationships. Harv. Bus. Rev. 81, 64–73.

Bhat, R., 2017. In: Bhat, R. (Ed.), Sustainability Challenges in Agro-Food Sector. Wiley Blackwell Publishers, UK, ISBN 978-1-119-07276-8. 720 pp.

Bhat, R., Gómez Lopez, V.M., 2014. In: Bhat, R., Gómez Lopez, V.M. (Eds.), Practical Food Safety: Contemporary Issues and Future Directions. Wiley Blackwell Publishers, UK, ISBN 978-1-118-47460-0. 632 pp.

Bhat, R., Karim, A.A., Paliyath, G., 2012. In: Bhat, R., Karim, A.A., Paliyath, G. (Eds.), Progress in Food Preservation. Wiley Blackwell Publishers, UK. ISBN-13: 978-0-470-65585-6, 656 pp.

Bolton, P., Dewatripont, M., 2005. Contract Theory. MIT Press, Cambridge, MA, pp. 1–744.

Borodin, V., Bourtembourg, J., Hnaien, F., Labadie, N., 2016. Handling Uncertainty in Agricultural Supply Chain Management: A State of the Art. Eur. J. Oper. Res. 254 (2), 348–359.

Brandenburg, M., Govindan, K., Sarkis, J., Seuring, S., 2014. Quantitative models for sustainable supply chain management: developments and directions. Eur. J. Oper. Res. 233 (2), 299–312.

Büyüközkan, G., Göçer, F., 2018. Digital supply chain: literature review and a proposed framework for future research. Comput. Ind. 97, 157–177.

Cai, J., Liu, X., Xiao, Z., Liu, J., 2009. Improving supply chain performance management: a systematic approach to analyzing iterative KPI accomplishment. Decis. Support. Syst. 46, 512–521.

Carter, C.R., Easton, P.L., 2011. Sustainable supply chain management: evolution and future directions. Int. J. Phys. Distrib. Logist. Manag. 42, 4662.

Carter, C.R., Rogers, D.S., 2008. A framework of sustainable supply chain management: moving toward new theory. Int. J. Phys. Distrib. Logist. Manag. 38, 360–387.

CODEX Alimentarius, 2006. Principles Form Traceability/Product Tracing as a Tool Within a Food Inspection and Certification System. http://www.fao.org/fao-who-codexalimentarius/sh-proxy/ru/?lnk=1&url=https%253A%252F%252Fworkspace.fao.org%252Fsites%252Fcodex%252FStandards%252FCAC%2BGL%2B60-2006%252FCXG_060e.pdf. (accessed 26.06.18).

Coley, D.A., Howard, M., Winter, M., 2011. Food miles: time for a rethink? Br. Food J. 113 (7), 919–934.

Dawe, D.C., Timmer, P.C., 2012. Why stable food prices are a good thing: lessons from stabilizing prices in Asia. Glob. Food Secur 1, 127–133.

Deakins, D., Galloway, L., Mochrie, R., 2003. The Use and Effect of ICT on Scotland's Rural Business Community. Scottish Economists Network, Stirling, pp. 1–62.

Dellino, G., Laudadio, T., Mari, R., Mastronardi, N., Meloni, C., 2018. Microforecasting methods for fresh food supply chain management: a computational study. Math. Comput. Simul. 147, 100–120.

Ding, M.J., Matanda, M.J., Parton, K.A., Jie, F., 2014. Relationships between quality of information sharing and supply chain food quality in the Australian beef processing industry. Int. J. Logist. Manag. 25, 85–108.

Esfahbodi, A., Zhang, Y., Watson, G., 2016. Sustainable supply chain management in emerging economies: trade-offs between environmental and cost performance. Int. J. Prod. Econ. 181, 350–366.

FAO, 2011. Report on Global Food Losses and Food Waste. Study Conducted for the International Congress SAVE FOOD! at Interpack2011, Düsseldorf, Germany. http://www.fao.org/docrep/014/mb060e/mb060e02.pdf. (accessed 26.06.18).

FAO, 2017. Food Loss and Food Waste. http://www.fao.org/food-loss-and-food-waste/en/. (accessed 26.06.18).

FAO, 2014. The Food Wastage Footprint (FWF). http://www.fao.org/nr/sustainability/food-loss-and-waste and http://www.fao.org/3/a-i3991e.pdf (accessed 18.03.19).

Folkerts, H., Koehorst, H., 1998. Challenges in international food supply chains: vertical co-ordination in the European agribusiness and food industries. Br. Food J. 100 (8), 385–388.

Fynes, B., Búrca, S., Marshall, D., 2004. Environmental uncertainty, supply chain relationship quality and performance. J. Purch. Supply Manag. 10 (4–5), 179–190.

Golan, E., Krissoff, B., Kuchler, F., 2004. Food Traceability: One Ingredient in a Safe and Efficient Food Supply. https://www.ers.usda.gov/amber-waves/2004/april/food-traceability-one-ingredient-in-a-safe-and-efficient-food-supply/. (accesses 26.06.18).

Goodhue, R.E., 2000. Broiler production contracts as a multi-agent problem: common risk, incentives and heterogeneity. Am. J. Agric. Econ. 82 (3), 606–622.

Grant, J., 2003a. Growing rural female entrepreneurs: are they starved of ICT skills? In: Paper Presented at the ICSB World Conference, Belfast.

Grant, J., 2003b. Growing rural female entrepreneurs: are they starved of ICT skills? In: Paper Presented at the ICSB World Conference, Belfast.

Hamprecht, J., Corsten, D., Noll, M., Meier, E., 2005. Controlling the sustainability of food supply chains. Supply Chain Manag. Int. J. 10, 7–10.

Headey, D., 2013. The impact of the global food crisis on self-assessed food security. World Bank Econ. Rev. 27 (1), 1–27.
Headey, D., Martin, W.J., 2016. The impact of food prices on poverty and food security. Ann. Rev. Resour. Econ. 8, 329–351.
Hobbs, J.E., 1996. A transaction cost approach to supply chain management. Supply Chain Manag. 1 (2), 15–17.
Hobbs, J.E., 1997. Measuring the importance of transaction costs in cattle marketing. Am. J. Agric. Econ. 79 (4), 1083–1095.
Huang, H., He, Y., Li, D., 2018. Pricing and inventory decisions in the food supply chain with production disruption and controllable deterioration. J. Clean. Prod. 180, 280–296.
Huggins, R., Izushi, H., 2002. The digital divide and ICT learning in rural communities: examples of good practice service delivery. Local Econ. 17 (2), 111–122.
ISO 22005:2007, 2016. Traceability in the Feed and Food Chain—General Principles and Basic Requirements for System Design and Implementation. https://www.iso.org/standard/36297.html. (accessed 26.06.18).
Keller, S.B., 2002. Internal relationship marketing: a key to enhanced supply chain relations. Int. J. Phys. Distrib. Logist. Manag, 649–668.
Kibler, K.M., Debra, R., Christopher Hawkins, C., Motlagh, A.M., Wright, J., 2018. Food waste and the food-energy-water nexus: a review of food waste management alternatives. Waste Manag. 74, 52–62.
Kusumastuti, R.D., Donk, D.P.V., Teunter, R., 2016. Crop-related harvesting and processing planning: a review. Int. J. Prod. Econ. 174, 76–92.
Larsen, E., Lees, M., 2003. Traceability in Fish Processing. Food Authenticity and Traceability, pp. 507–517.
Li, D., Wang, X., Chan, H.K., Manzini, R., 2014. Sustainable food supply chain management. Int. J. Prod. Econ. 152, 1–8.
Lim, K.H., Maynard, L.J., Hu, W., Goddard, E., 2011. U.S. consumers' preference and willingness to pay for country of origin labeled beef steak and food safety enhancements. In: Paper Presented for Presentation at the Agricultural and Applied Economics Association's 2011 AAEA and NAREA Joint Annual Meeting. Pittsburgh, Pennsylvania, 24–26, July 2011.
Manning, L., 2016. Food fraud: policy and food chain. Curr. Opin. Food Sci. 10, 16–21.
Manning, L., Soon, J.M., 2016. Food safety, food fraud, and food defense: a fast evolving literature. J. Food Sci. 81, 823–834.
Marzita Saidon, I., Mat Radzi, R.M., Ab Ghani, N., 2015. Food supply chain integration: learning from the supply chain superpower. Int. J. Manag. Value Supply Chains 6 (4), 1–15.
Matopoulos, A., Doukidis, G.I., Vlachopoulou, M., Manthou, V., Manos, B., 2007. A conceptual framework for supply chain collaboration: empirical evidence from the agri-food industry. Int. J. Supply Chain Manag. 12, 177–186.
Moyer, D.C., DeVries, J.W., Spink, J., 2017. The economics of a food fraud incident e case studies and examples including melamine and wheat gluten. Food Control 71, 358–364.
Ölander, F., Thøgersen, J., 2013. Informing or nudging: which way to a more effective environmental policy? In: Scholderer, J., Brunsø, K. (Eds.), Marketing, Food and the Consumer. Festschrift in Honour of Klaus G. Grunert. Pearson Longman, London, pp. 141–155.
Oliver, R.K., Webber, M.D., 1982. Supply-chain management: logistics catches up with strategy. Outlook 5 (1), 42–47.
Opara, L.U., Mazaud, F., 2001. Food traceability from field to plate. Outlook Agric. 30 (4), 239–247.
Ospina, A.V., Heeks, R., 2011. ICTs and climate change adaptation: enabling innovative strategies. In: Strategy Brief 1. Climate Change, Innovations and ICTs Projects, pp. 1–9.
Parfitt, J., Barthel, M., Macnaughton, S., 2010. Food waste within food supply chains: quantification and potential change to 2050. Philos. Trans. R. Soc. B: Biol. Sci. 365 (1554), 3065–3081.
Petersen, A., Green, D., 2005. Seafood Traceability: A Practical Guide for the US Industry. http://seafood.oregonstate.edu/.pdf%20Links/Seafood%20Traceability%20-%20A%20Practical%20Guide.pdf. (accessed 26.06.18).
Raak, N., Symmank, C., Zahn, S., Aschemann-Witzel, J., Rohm, H., 2017. Processing-and product-related causes for food waste and implications for the food supply chain. Waste Manag. 61, 461–472.
Priefer, C., Jörissen, J., Bräutigam, K.-R., 2016. Food waste prevention in Europe—a cause-driven approach to identify the most relevant leverage points for action. Resour. Conserv. Recycl. 109, 155–165.
Rábade, L., Alfaro, J., 2006. Buyeresupplier relationship's influence on traceability implementation in the vegetable industry. J. Purch. Supply Manag. 12, 39–50.
Rao, N.H., 2007. A framework for implementing information and communication technologies in agricultural development in India. Technol. Forecast. Soc. Change 74, 491–518.
Ras, P.J., Vermeulen, J.V., 2009. Sustainable production and the performance of South African entrepreneurs in a global supply chain. The case of South African table grape producers. J. Sustain. Dev. 17, 325–340.
Reardon, T., Timmer, C.P., 2007. Transformation of markets for agricultural output in developing countries since 1950: how has thinking changed? Handb. Agric. Econ. 3, 2807–2855.
Rijswijk, W.V., Frewer, L.J., 2008. Consumer perceptions of food quality and safety and their relation to traceability. Br. Food J. 110 (10), 1034–1046.
Rong, A., Akkerman, R., Grunow, M., 2011. An optimization approach form managing fresh food quality throughout the supply chain. Int. J. Prod. Econ. 131, 421–429.
Rullia, M.C., Savoria, A., D 'Odorico, P., 2012. Global land and water grabbing. PNAS 110 (3), 892–897.
Seuring, S., Müller, M., 2008. From a literature review to a conceptual framework for sustainable supply chain management. J. Clean. Prod. 16 (15), 1699–1710.
Silvis, I.C.J., van Ruth, S.M., van der Fels-Klerx, H.J., Luning, P.A., 2017. Assessment of food fraud vulnerability in the spices chain: an explorative study. Food Control 81, 80–87.

Smallbone, D., North, D., Baldock, R., Ekanem, I., 2002. Encouraging and Supporting Enterprises in Rural Areas. London: Small Business Service/DTI. http://citeseerx.ist.psu.edu/viewdoc/download?doi=10.1.1.534.6380&rep=rep1&type=pdf. (accessed 26.06.18).

Spence, L., Bourlakis, M., 2009. The evolution from corporate social responsibility to supply chain responsibility: the case of Waitrose. Supply Chain Manag. 14 (4), 291–302.

Spink, J., Ortega, D.L., Chen, C., Wu, F., 2017. Food fraud prevention shifts the food risk focus to vulnerability. Trends Food Sci. Technol. 62, 215–220.

Swinnen, J. (Ed.), 2015. The Political Economy of the 2013 Reform of the CAP. Centre for European Policy Studies Publication, Brussels.

Sylvester, G., 2013. Information and Communication Technologies for Sustainable Agriculture Indicators From Asia and the Pacific. http://agris.fao.org/agris-search/search.do?recordID=XF2017001375. (accessed 26.06.18).

Teng, P., Trethewie, S., 2012. Tackling urban and rural food wastage in Southeast Asia: issues and interventions. In: Policy Brief No 17. Singapore: RSIS Centre for Non-Traditional Security (NTS) Studies. https://www.rsis.edu.sg/rsis-publication/nts/2377-tackling-urban-and-rural-food/#.WzHAr_l96M8. (accessed 26.06.18).

Thomassen, M.A., van Calker, K.J., Smits, M.C.J., et al., 2008. Life cycle assessment of conventional and organic milk production in the Netherlands. Agric. Syst. 96 (1-3), 95–107.

Tseng, M.L., Chiu, A.S., 2013. Evaluating firm's green supply chain management in linguistic preferences. J. Clean. Prod. 40, 22–31.

UNEP, 2015. Sustainable Consumption and Production and the SDGs. https://sustainabledevelopment.un.org/topics/sustainableconsumptionandproduction. (accessed 26.06.18).

United Nations, 2012. The Zero Hunger Challenge. http://www.un.org/en/zerohunger/challenge.shtml. (accessed 26.06.18).

United Nations, 2014. A/69/700 Follow-up to the outcome of the Millennium Summit, The road to dignity by 2030: ending poverty, transforming all lives and protecting the planet. In: Synthesis Report of the Secretary-General on the Post-2015 Sustainable Development Agenda.

van Ruth, S.M., Huisman, W., Luning, P.A., 2017. Food fraud vulnerability and its key factors. Trends Food Sci. Technol. 67, 70–75.

Verpoorten, M., Arora, A., Stoop, N., Swinnen, J., 2013. Self-reported food insecurity in Africa during the food price crisis. Food Policy 39, 51–63.

Virtanen, Y., Kurppa, S., Saarinen, M., et al., 2011. Carbon footprint of food: approaches from national input-output statistics and an LCA of a food portion. J. Clean. Prod. 19, 1849–1856.

Walker, H., Phillips, W., 2009. Sustainable procurement: emerging issues. Int. J. Procur. Manag. 2 (1), 41–61.

Wang, Y., 2016. E-logistics: Managing Your Digital Supply Chains for Competitive Advantage. Kogan Page, pp. 1–536.

Wickramasinghe, U., 2014. Realizing sustainable food security in the post-2015. In: Development Era: South Asia's Progress, Challenges and Opportunities. South and South-West Asia Development Papers 1402. http://www.unescap.org/sites/default/files/Realising%20sustainable%20food%20security%20in%20South%20Asia%20Upali%20Wickramasinghe%20_August%202014_FINAL.pdf. (accesses 26.06.18).

Wognum, P.M., Bremmers, J., Trienekens, J.H., et al., 2011. System for sustainability and transparency of food supply chains: current status and challenges. Adv. Eng. Inform. 25, 65–76.

Wossen, T., Berger, T., Haile, M.G., Troost, C., 2018. Impacts of climate variability and food price volatility on household income and food security of farm households in East and West Africa. Agric. Syst. 163, 7–15.

Zhu, Q., Sarkis, J., Lai, K., 2008. Confirmation of a measurement model for green supply chain management practices implementation. Int. J. Prod. Econ. 111, 261–273.

Chapter 3

The role of modeling and systems thinking in contemporary agriculture

Holger Meinke
Tasmanian Institute of Agriculture, University of Tasmania, Hobart, TAS, Australia

Abstract

The images and perceived roles of agriculture in our societies have changed over the last few decades. Today agriculture is regarded as an integral part of interconnected value chains that sit at the heart of our economies, providing invaluable services to society. In response, most governments around the world are now actively developing policies to support and grow their bio-economies. This increases the expectations that society and governments have in terms of agriculture's services and performance: agriculture is not only expected to generate food for our growing populations and income for farmers, it must be part of value chains that provide raw materials that can be incorporated or converted into feed, fiber, fuel, pharmaceuticals, and other industrial products. Farmers are expected to be responsible custodians of our landscapes and their farming practices must be economically, environmentally, and socially sustainable and aligned with the broader and changing values of our societies. Often these three objectives conflict and consequently societal expectations are not met. In a world that is increasingly data rich, practicing agriculture in a way that lives up to these expectations requires tools that can help to foresee the consequences of complex interactions. Hence, this chapter explores the role of modeling and systems thinking to manage this complexity by explicitly considering three attributes of complex, adaptive systems, whereby (i) order emerges rather than being predetermined; (ii) the system's future can only be assessed probabilistically rather than deterministically predicted; and (iii) the history of the system is largely irreversible. The chapter reflects on the contemporary use of models against these three systems characteristics and concludes that scientifically based and tested algorithms (i.e., models) are already a ubiquitous and indispensable management tool for modern farming. Countless apps are already in use for short-term, tactical decision making, while more complex modeling approaches are vital for strategic scenario planning and risk assessments for farmers, policymakers, and scientists.

1 Background

Agriculture plays a foundational role in ensuring the survival of humanity; we all need to eat, preferably three times a day. Increasingly, agriculture also contributes to the nonfood sectors of our economies (feed, fiber, fuel, pharmaceutical products, building materials, raw materials for industrial processes, etc.). This has led to a reframing of how we think about agriculture. The old, three-sector model thinking about our economies, which relegated agriculture to the lowest level of sophistication by viewing the primary sector as a hallmark of a primitive economy, is rapidly being replaced by much more appropriate, contemporary models. This more contemporary thinking views agriculture as a key driver of a broader bioeconomy for society (OECD, 2009). Although perspectives differ and debates about the details of such a new way of framing agriculture continue (e.g., see discussions about related concepts such as the "circular economy," the "green economy," and the "blue economy"; Bugge et al., 2016), governments around the world are now developing bioeconomy strategies. The following examples, shortlisted in Table 1, should suffice in demonstrating this ubiquitous trend, which recognizes the substantial added value that biological processes provide for our societies.

These new paradigms regard agriculture as an integral part of interlinked value chains that promote sustainable growth, food and energy security, and waste reduction. This perspective fundamentally changes the role and the status of agriculture in our societies. While we are still confronted with images of resource-poor farmers struggling to feed their families in the global South, or drought and poverty-stricken farmers walking off the land in developed economies such as Australia, the reality is that contemporary agriculture has become a knowledge-intensive endeavor requiring a broad range of skills and experience. Today's farmers are no longer simply growing crops or tending to their animals, they are managing highly complex businesses and enterprises. Politicians, policymakers, and civil societies are beginning to acknowledge this, which partly explains the resurgence of agricultural studies at our universities (Pratley, 2015).

Sustainable Food Supply Chains. https://doi.org/10.1016/B978-0-12-813411-5.00003-X

TABLE 1 Some regional examples of policymaking strategies on bioeconomy

Geographic area of interest	Paradigm source
European Union	https://ec.europa.eu/commission/news/new-bioeconomy-strategy-sustainable-europe-2018-oct-11-0_en
Germany	http://biooekonomierat.de/home-en.html
Sri Lanka and Malaysia	http://www.bioeconomycorporation.my/malaysia-sri-lanka-to-enhance-collaboration-in-bioeconomy/
USA	www.energy.gov/sites/prod/files/2016/12/f34/ beto_strategic_plan_december_2016_0.pdf
Africa	https://qz.com/africa/860598/africas-bioeconomy-will-be-as-transformative-as-the-mobile-phone-and-digital-technology/
China	https://gbs2018.com/fileadmin/gbs2018/Presentations/Workshops/Bioeconworldreg/Asia/Yin_Li__Bioeconomy_in_China.pdf

This reframed perspective, that is, a perspective that sees agriculture as an essential part of interconnected value chains, has also raised the expectations that societies rightfully have when it comes to agricultural practice and systems performance. Agriculture is not only expected to generate income for farmers and their communities, but it also must provide life-sustaining products for our growing, global populations in ways that are economically, environmentally, and socially sustainable and acceptable.

Rarely are these expectations met. Unsurprisingly, the three goals of agriculture often conflict, leading to suboptimal outcomes and contentious debates among stakeholders, including actors along the agriculture and food value chains. Seeking the best possible alignment and, ideally, triple-win solutions requires considerable vision, foresight, and political courage. It also requires a much clearer understanding of the roles and responsibilities that governments, industry, and consumers have in ensuring that these goals are achieved, or at least not irreversibly compromised.

This is where agricultural systems thinking, augmented by computer simulations of agricultural systems, comes in. Pioneered by C.T. de Wit in the late 1950s, simulation models have now become part of everyday decision making at farm and policy levels (De Wit, 1958; Jones et al., 2017a, b). Such systems thinking has been transformative and has contributed, for instance, to the global adoption of minimum tillage practices that are now regarded as fundamental components of sustainable agriculture. The plow, an iconic agricultural symbol for many centuries, is increasingly being replaced by the keyboard. This is but one example of the benefit that systems thinking, aided by simulation modeling, brings to modern agriculture.

This chapter is not a comprehensive review of modeling in agriculture. For such a review, I refer readers to a special issue of *Agricultural Systems*, Vol. 155 (2017) and some of the key papers within, such as Jones et al. (2017a, b) and Antle et al. (2017a, b). Instead, my objectives are to discuss current agricultural modeling capabilities and their roles in the context of

- decision making under uncertainty, including their foresighting capabilities;
- models as analytical scientific tools that help us in making sense of otherwise impenetrable system complexity; and
- models as heuristic tools for transdisciplinary research, discussion support tools, and for meaningful dialogues among diverse sets of stakeholders.

2 The power of agricultural systems thinking

Agricultural systems are complex, adaptive systems with three key characteristics:

i. Order emerges rather than being predetermined;
ii. The system's future can only be assessed probabilistically rather than deterministically predicted; and
iii. The history of the system is largely irreversible (Dooley, 1997).

As a system becomes more complex and open, cause-and-effect relationships become increasingly intangible (Nelson et al., 2007). In contrast to closed systems, open systems interact and exchange data and flows of information with their external environment. As outlined in the following sections, these characteristics have profound impacts on the way we think about and manage agricultural systems in the 21st century. Beyond that, they also influence how we think about our political systems and the policy environments that are created and governed by our political leaders. In the age of the Internet and instant information transfer, our political systems are extreme examples of such open systems, where memes (our basic building blocks of thought) are constantly created, exchanged, and mutated beyond any individual's control. These memes influence political actions and responses that are hard to foresee and impossible to predict.

2.1 Order emerges rather than being predetermined

Traditionally, agricultural research relied mainly on data collected from either lab- or field-based experimentation, that is, in vivo experimentation. While such traditional experimentation will always be an important part of agricultural research, it is, by design, backward looking. Treatments are imposed, crops are grown and monitored, data are collected and analyzed, and ultimately, once the experiments have concluded, the analyzed data provide insights that help us understand what has happened. This is a logical and useful approach based on the assumption that the immediate future will be governed by the same systems dynamics as the recent past. However, such empirical approaches based on direct experimentation on real systems quickly reach their limits in terms of complexity, time, and resources required.

During the 19th century and despite these impediments posed by agronomic experimentation, agricultural systems pioneers such as Justus von Liebig, Albrecht Thaer, and Carl von Wulffen managed to conceptualize some of the core principles underpinning agricultural systems dynamics. These early agricultural system scientists recognized the interdependence of variables in agricultural systems, which led to an understanding of, for instance, nutrient cycling (De Wit, 1992). Quantifying complex plant-soil-water-management interactions posed a major challenge for these early systems thinkers and often insights and plausible theories remained untested due to the complex nature of in vivo experimentation. While these pioneers of systems science could see the emerging order and were able to understand some of the component interactions leading to it, they had no way of foreseeing some of the more complex, emergent properties of these systems: while they might have been able to rationally explain some emergent patterns by interpreting the data they collected (i.e., looking back in time), they had only very limited capacity to foresee the future behavior of these systems.

It is only in the last few decades that a truly complementary approach to in vivo experimentation became feasible. Since the 1980s, technical innovations of computer hardware and software enabled natural and medical scientists to express their disciplinary domain knowledge through computer code. Once such computer programs had been tested against sufficient empirically derived data obtained from traditional experimental approaches, these algorithms could then be used for in silico experimentation, that is, simulations mimicking the real system in order to gain new insights into systems dynamics and behaviors (Meinke et al., 2009; see also Jones et al., 2017b for a comprehensive history of agricultural modeling). This is true not just for agriculture, but right across the life sciences and engineering; technological development has given rise to today's highly sophisticated simulation models that are an integral part of research methodologies regardless of discipline, be it medicine or astrophysics, in environmental or social sciences. Today all disciplines use modeling in some form in order to foresee the future behavior of their system of interest in response to management interventions. Such interventions can be medical treatment plans for patients or crop management plans for agriculture.

When a system becomes more complex and open, cause-and-effect relationships become increasingly intangible. Field-based agriculture (in contrast to glasshouse production or intensive, confined animal husbandry) is an excellent case study for a very open system: crops, soils, and animals are exposed to changing weather and climate conditions that profoundly affect the performance and behavior of the system. At the same time, market conditions vary daily, consumer preferences and expectations can change rapidly, and new policies are decreed, influencing the way agricultural systems can and should be managed.

Today farmers are much more than simply the producers of stable food. Their roles and responsibilities have considerably broadened; they are now expected to be the custodians of our landscapes as well as highly skilled managers who understand and comply with our legal systems, including workplace health and safety (WHS) and environmental policies and guidelines. In addition, they are expected to produce fresh, tasty, and healthy food at the lowest possible cost and minimal environmental footprint. This means that farmers need to have substantial operational knowledge of finances and marketing as well as the ability to negotiate acceptable terms of trade with their suppliers and retailers. A relentless drive for more efficiency within agricultural value chains means that, particularly in high labor-cost countries such as Australia, expensive labor is increasingly replaced by technology. Hence, farmers also must be extremely technology savvy. They need to know how to operate drones, variable rate irrigators, robotic dairies, self-driving tractors, and the latest software necessary to manage such automated systems. They need to communicate effectively using traditional communications channels as well as market their produce via social media and the Internet. They need to know how to set up and interrogate sensor networks or download the latest normalized difference vegetation index (NDVI) map to manage scarce water resources across all their enterprises. They need to do all this while having to cope with a flood of data and information that is often unstructured and untargeted. In fact, our most successful farmers are pioneers of "Industrie 4.0," the Fourth Industrial Revolution that is currently in progress (Hermann et al., 2015). Over the last four decades, agriculture has changed from a low-tech, low-skill activity into one of the most data-, information-, and knowledge-intensive sectors of our economies. This data revolution and the Internet of Things are currently transforming the practice of agriculture. Modeling and systems thinking can help farmers, policymakers, and scientists to foresee emerging trends and patterns and gain some advance understanding of the emerging order of this complex system. As one farmer quipped after a kitchen table discussion based on simulated data: "This tool enables me to gain hindsight in advance."

2.2 The system's future can only be assessed probabilistically rather than deterministically predicted

When farmers make decisions, the notion of "certainty" is always elusive. All decisions involve assumptions and many of these are either *a priori* unknowable or only knowable to a limited extent (e.g., future economic conditions, future climate conditions, details about the physical environment, and many more). In that regard, decision making on-farm is much like making other daily life choices: while many of the consequences of our choices are foreseeable, they are not predictable in a deterministic sense. There is always a large random element that can lead to unexpected or unintended outcomes. While we might be able to estimate the probabilities of an event or outcome occurring, we can never be absolutely certain about it. Models are ideal tools for such probabilistic risk assessments. This is what makes models useful: they are a simplified representation of real systems. This simplification allows for the investigation of systems properties and therefore enables scenario analyses based on the exploration of possible future outcomes, including the likelihood of these outcomes occurring (Meinke et al., 2001).

In other words, models can provide quantitative information about the operating environment and allow comparisons between alternative decisions. The uncertainty associated with the model output can be made explicit, but it can never be fully eliminated (Byerlee et al., 1982). It is therefore no coincidence that the application of models is dominated by cases that are characterized by high environmental variability, such as climate in the arid and semiarid regions of the world (Meinke and Stone, 2005). However, modeling is mainly used as a tool that assists in thinking about the future. This reduces the ambiguity surrounding alternative decision options. Occasionally, modeling can also assist in reducing uncertainty by probabilistically quantifying various levels of risk associated with decisions options (see Schrader et al., 1993 for a distinction between ambiguity, uncertainty, and risk).

Temporal and spatial scales need to be considered when choosing a model. These range from tactical decisions regarding the scheduling of planting or harvest operations to policy decisions regarding land use allocation (e.g., grazing systems versus cropping systems). Fig. 1 and Table 2 provide a few examples of these types of decisions.

Models need to be designed to suit the problem domain and the time frame under consideration (see, for instance, Fig. 1). Models that address daily tactical decisions such as crop protection measures or the management of cool chains rely on near real-time environmental input data, such as local temperature and humidity, to be useful; models that are used to address much longer time frames such as soil carbon balances or land-use planning require environmental data that mirror those timescales. Even a few decades ago, things were quiet on the farm; information took a long time before it reached the farm gate. In contrast, today's farmers and farm managers are drowning in a sea of data and information, finding it difficult to separate the salient from the trivial. This has given rise to a new type of model: apps that are accessible via mobile phones and that can interrogate real-time data via purposefully designed algorithms.

Such apps are designed to assist users in making sense of data for tactical decision making. As a consequence of the rapid uptake of this technology, modeling has become an everyday tool on the farm. For instance, farmers can choose from a multitude of apps to assist them with monitoring their yields on a square meter basis or developing their spray programs. Such apps are based on algorithms that interrogate real-time as well as historical data before making a recommendation to their users. Yet, the pervasive nature of such apps on the farm is not widely recognized by either the users of this technology (i.e., the farmers) or even the providers of the technology as "modeling" (we refer to them as "apps" rather than models). Most people associate the term modeling in agriculture with the type of traditional crop modeling conducted by scientists in the past using hardware (computers) and software (algorithms) that were generally not accessible or even of interest to farmers. Yet, from a technological and computational perspective, there is little if any difference between an app and a model. The differences are not so much in the algorithms themselves, but more in the data that these models access. Inputs used to drive "traditional" models are largely historical such as long-term historical weather data, measured soil nutrient content, or water status, as well as other, experimentally derived, input parameters. On occasion, this is combined with future-oriented data, such as climate forecasts, to test the likely outcome of alternative management interventions on economic or environmental performance indicators (e.g., crop yield, soil erosion, nutrient leaching). However, with a few notable exceptions (e.g., Hochman et al., 2009), traditional models do not use real-time data feeds as inputs.

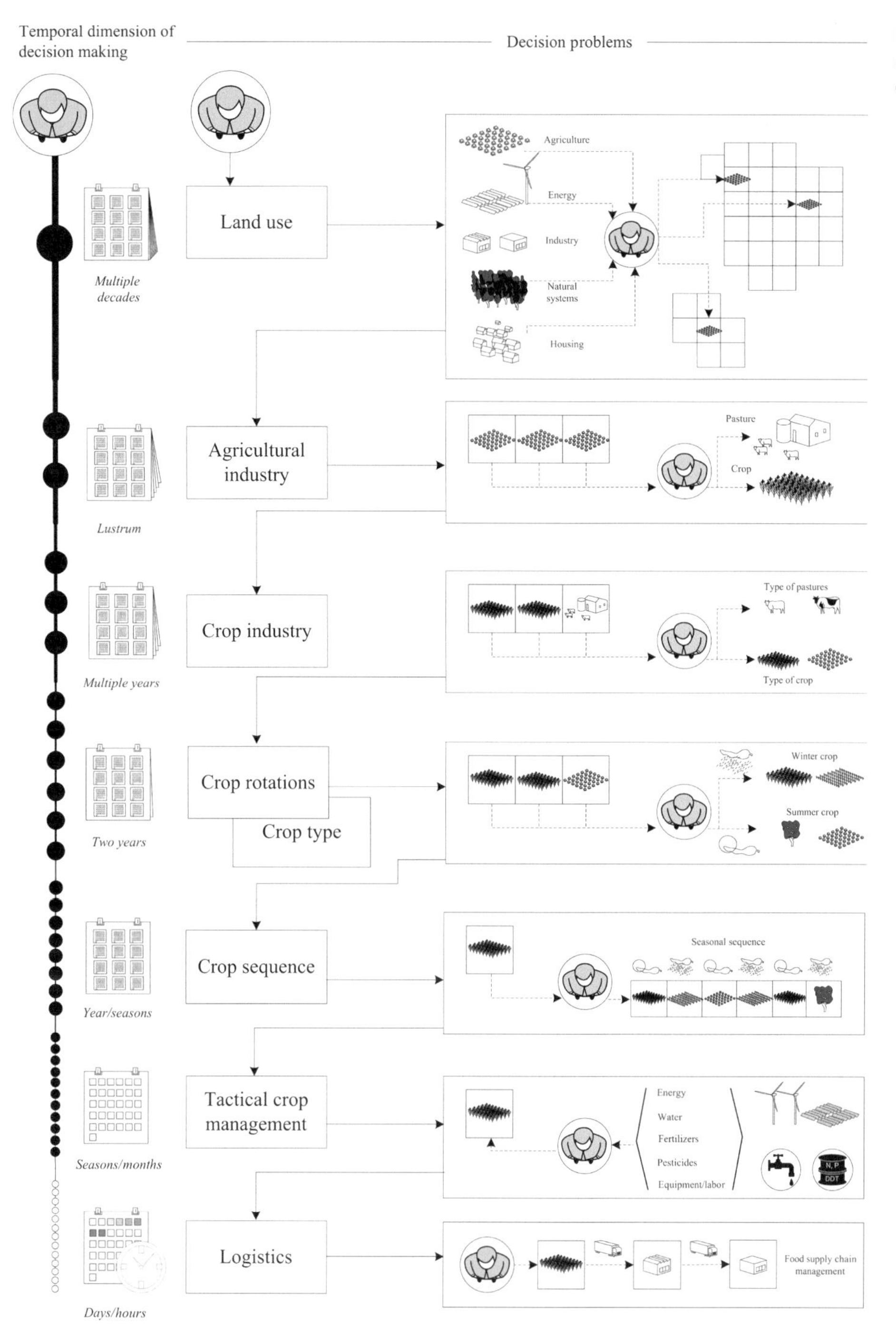

FIG. 1 Spatial and temporal dimensions of decision problems based on examples provided in Table 2.

Given the history of crop modeling, this is hardly surprising: particularly the large, systems-oriented modeling frameworks such as APSIM and DSSAT have their origin in the 1980s, long before the current data revolution became a reality (Keating et al., 2003; Jones et al., 2003) and before the concept of "Industrie 4.0" had emerged (Hermann et al., 2015). Responsibility for their development and maintenance rests with public institutions (universities or publicly funded research organizations) with little contribution from the private sector. Despite some rhetoric to the contrary, their primary purpose is either as a research tool to provide scenario analyses for management options, policy development, and industry engagement, or as a teaching tool. For on-farm decision-making, the models' strength is in providing long-term, strategic insight rather than short-term, tactical advice. For teaching and extension, Keating and McCown (2001) argued that the

TABLE 2 Agricultural decisions at a range of temporal and spatial scales

Example of decision types	Frequency (years)
Logistics (e.g., scheduling of planting/harvest operations)	Intraseasonal (<0.2)
Tactical crop management (e.g., fertilizer/pesticide use)	Intraseasonal (0.2–0.5)
Crop type (e.g., wheat or chickpeas) or herd management	Seasonal (0.5–1.0)
Crop sequence (e.g., long or short fallows) or stocking rates	Interannual (0.5–2.0)
Crop rotations (e.g., winter or summer crops)	Annual/biannual (1–2)
Crop industry (e.g., grain or cotton; native or improved pastures)	Decadal (~10)
Agricultural industry (e.g., crops or pastures)	Interdecadal (10–20)
Land use (e.g., agriculture, natural systems, housing, industrial)	Multidecadal (20+)

The table shows different time scales for decision making for issues relating to agricultural production.
Reproduced from Meinke, H., Stone, R., 2005. Seasonal and inter-annual climate forecasting: the new tool for increasing preparedness to climate variability and change in agricultural planning and operations. Climatic Change, 70, 221–253. doi:https://doi.org/10.1007/s10584-005-5948-6.

models' major value is heuristic, a value that remains underexploited in formal academic education as well as for the development of workplace or organizational skills.

The current data revolution is putting the spotlight on the much shorter tactical time frame of agricultural decision making (Table 2). The advent of mobile communication technologies has now provided us with algorithms that we already use routinely: every time we use an app, we are provided with information that is derived from algorithms using near real-time data. The use of such algorithms has become pervasive and now influences all aspects of our lives such as health, leisure, transport, and communication. Agriculture is no exception. Nearly all farmers, irrespective of wealth or location, have mobile phones, which they access routinely for communication, information, and data gathering. This technology seamlessly crosses the usual North/South divide. In fact, farmers in developing or emerging economies might see a more rapid uptake in mobile technologies than their richer counterparts in the global North (news.microsoft.com/en-in/features/ai-agriculture-icrisat-upl-india/). In fact, the 5G network is being rolled out in areas that were previously lacking in access to telecommunications, such as remote areas of Brazil or in the Arctic (https://www.oulu.fi/university/news/5g-in-remote-areas).

Big data is one of the catch phrases for voluminous amounts of data collected via multiple sources and often without a specific reason or use in mind. Environmental sensors are getting cheaper and increasingly common. The automation of farms is rapidly progressing and nearly every new piece of hardware purchased, be it the new tractor, a milking robot, or the latest Home Assistant, can relay and record data that emanates from the farm. Much of this data is collected but never used. Making sense of this flood of data and information that our always-on technologies now generate is a nontrivial task. Could traditional crop or cropping systems models be used more effectively to add value to such data? Could they detect trends that might contain important information for proactive management (e.g., changes in bacterial counts in dairies)? Could they uncover inefficiencies in resource use, such as a faulty pump in the irrigation network that, unbeknownst to the farmer or farm manager, drains the financial resources of the business? Can we use models to benchmark the current performance of the farm against its peers, thereby optimizing performance? The questions are endless, but each of these questions has at least a potential, model-based answer. Connecting the in-situ, continuous collection of data via sensors and mobile apps with the traditional models' capability to provide input into scenario analyses and planning seem like an obvious, but so far underused, potential that our agricultural software engineers have only recently begun to explore (Holzworth et al., 2018).

Models are tools: their real usefulness stems not only from their ability to predict the outcome of management interventions, but rather in their power to foresee the likelihood and severity of opportunities and consequences that might arise from a combination of climate, soil, plant, animal, and human interactions. The difference between prediction and foresight might appear subtle, yet it is profound: predicting an outcome transfers all power to the person making the prediction (usually a scientist), while foreseeing likelihoods and consequences empowers actors to choose and actively create the desirable future they envision while avoiding undesirable outcomes. This is increasingly recognized and evidenced by the emergence of *innovation platforms*, that is, fora where scientists and practitioners come together for constructive dialogues and joint capability development (Ekboir and Rajalahti, 2012). Innovation platforms are usually action-oriented communities of practice with a strong and common purpose. They respect formal as well as tacit knowledge and are based on the principles of codesign (Eastwood et al., 2012). By agreeing on the problem domain and the methods (including the models) to be employed, they empower all participants to act within their area of expertise. Used in such a way, the models serve as a means for "discussion support" rather than "decisions support" (Nelson et al., 2002).

The importance of such empowerment cannot be overestimated. When resources are scarce and in high demand, we need tools that facilitate dialogue and help navigate the messy space of contested goals, values, and outcomes (Moeller et al., 2014; Leith et al., 2017). This is where a foresight approach underpinned by good, quantitative modeling can help. Using computer models we can ask the "what if" questions that underpin scenario planning. We can investigate agricultural practices and resource uses we would never dare to explore in real life because they would be too risky and the potential consequences could be severe. Modeling such scenarios allows us to take these risks without having to manage real-life consequences. This provides us with benchmarks to feed into such innovation platforms that will engage their diverse participants (i.e., farmers, policymakers and scientists) in meaningful discussions. Everybody learns from this process; in many cases, partial solutions are identified and negotiated where previously none were in sight. This keeps the focus on the choices that need to be made continuously in order to realize a preferred future, rather than focusing on a single prediction that will inevitably be "wrong" when reality starts to diverge from the original assumptions that went into making the prediction in the first place.

2.3 The history of the system is largely irreversible

Modern, successful farm operators are not only excellent at growing crops or raising animals; they are also highly skilled at marketing and finance, they understand business risks and opportunities, and they are excellent managers of human resources. They understand that decisions they make today will fundamentally, and often irreversibly, affect the resource base and the future performance of their agricultural business or enterprise. They also understand the importance of maintaining a social license to operate.

For large infrastructure developments, social impact assessments (SIAs) are mandated industry standards that need to be conducted before planning permissions are granted by authorities (Boele, 2016). In agriculture, the use of SIAs is uncommon and social licenses are generally implicit. Actors along the value chain are often unaware of or are unwilling to acknowledge the huge risks associated with the loss of such a social license. Increasingly, changes in community attitudes challenge the status quo of existing agricultural practices. Recognizing these challenges and reacting quickly is important to maintain this social license to operate. Inaction or a lack of understanding of this challenge can lead to a rapid loss of a social license, often due to single, high-profile, unforeseen events that highlight some failures within the system. Causes that lead to a loss of social license are varied but can be broadly categorized by being either the result of a catastrophic, single systems failure (e.g., a major pollution event or negligence leading to death or injuries of humans or animals), endemic systems failure (continuously not meeting environmental or social expectations), or rapid changes in community attitudes or values (e.g., the treatment of animals). While the responsibility for maintaining these social licenses to operate rest with all the actors within the value chain, it is often farmers who mainly suffer from the adverse consequences following the loss of social license. As Boele (2016) points out: "... The exact boundaries of the social license can change overnight with community expectations being the key determinant. Ultimately, a social license is maintained by how responsive the project is to changing community concerns and challenges. It is a license that must be earned every day. Executives and project owners will know when they've lost it. Society has a powerful ability to tell companies when they are no longer wanted..."

The irreversible consequences of some management practices make agricultural systems modeling a very attractive and suitable tool for managing issues around social licenses (Moeller et al., 2014). They are not a panacea and they are certainly not beyond contention, but they are increasingly seen as the approach of choice when it comes to quantifying the future environmental performance of agricultural practices.

Models are inherently suitable for the development of alternative, sustainable management practices due to their ability to assess the long-term consequences of such practices. Examples include the use of models to estimate carbon emissions from agriculture, potential for carbon storage in soils, drought management strategies, or managing nutrient flows on farms for better environmental outcomes. An interesting case study is the use of the Overseer decision support software in New Zealand that is designed for farmers and growers to improve nutrient use on farms, delivering better environmental outcomes and better farm profitability (Ledgard and Waller, 2001; Cichota et al., 2010). The software is marketed as an essential farming tool, founded on decades of scientific investment (www.overseer.org.nz/). In some instances, Overseer is used as a regulatory instrument by local authorities to monitor and enforce compliance with emission standards. Although the use of such a tool for regulatory enforcement purposes is not without controversy, it is generally regarded as a very useful engagement and monitoring tool that helps bridge the gap between scientific knowledge, public opinion, policy intentions, and practical management options (FAR, 2013). Whether or not a model is appropriate for use as a regulatory instrument is governed by factors external to the model as well as the model itself, including the quality and availability of the necessary input data. Yet, controversies associated with such model use usually focus on model and data limitations rather than societal expectations and the values that underpin the rules and regulations in the first instance (http://www.scoop.co.nz/stories/SC1812/S00022/smc-bulletin-overseer-fit-for-regulation.htm). Michael Mann (https://www.yesmagazine.org/planet/the-scientization-of-politics) refers to this phenomenon as the "scientization of politics," whereby politicians and policymakers shy away from making contentious decisions (Sarewitz, 2004). Instead they "scientize" the problem by reframing a contentious issue in terms of scientific uncertainty, when the real issue is a lack of societal value consensus. Prime examples for such scientization are controversies associated with issues such as climate change and genetically modified food crops. These are no longer issues of data uncertainty or model quality; instead they go to the heart of our societies and the divergent values we hold.

2.4 Concluding remarks

Future model development and use will depend on their credibility and demonstrated ability of delivering business- or policy-relevant information. The design of these models, their governance, and their implementation strategies will depend on effective private-public partnerships for new data infrastructure (Antle et al., 2017a). Other chapters in this book provide further examples of how systems thinking and modeling has become an essential part of good management throughout the agriculture and food value chain. Akkerman (this publication), for instance, discusses modern operations research within the food-processing industries. These industries increasingly rely on food safety standards that are model-based (Tamplin et al., 2003). Predictive microbiology has become an essential risk management tool for the food industry, as demonstrated by the over 60,000 users of ComBase (https://www.combase.cc/index.php/en/). In fact, Accorsi et al. (this publication) refer to the agriculture and food value chain as an "inclusive food ecosystem" that inspires the collection of multidisciplinary, quantitative parameters that can be repurposed to investigate subcomponents of this ecosystem. Any such repurposing will inevitably involve some degree of modeling. Based on current trends, it is therefore highly likely that models will increasingly become part of the everyday toolkit used in agriculture, food processing, and distribution.

References

Antle, J.M., Basso, B., Conant, R.T., Godfray, H.C.J., Jones, J.W., Herrero, M., Howitt, R.E., Keating, B.A., Munoz-Carpena, R., Rosenzweig, C., Tittonell, P., Wheeler, T.R., 2017a. Towards a new generation of agricultural system data, models and knowledge products: design and improvement. Agric. Syst. 155, 255–268.

Antle, J.M., Jones, J.W., Rosenzweig, C.E., 2017b. Next generation agricultural system data, models and knowledge products: introduction. Agric. Syst. 155 (Suppl. C), 186–190. https://doi.org/10.1016/j.agsy.2016.09.003.

Boele, R., 2016. The power of social license. KPMG Insight 8, 27.

Bugge, M.M., Hansen, T., Klikou, A., 2016. What is the bioeconomy? A review of the literature. Sustainability 8, 691. https://doi.org/10.3390/su8070691.

Byerlee, D., Harrington, L., Winkelmann, D.L., 1982. Farming systems research—issues in research strategy and technology design. Am. J. Agric. Econ. 64 (5), 897–904. https://doi.org/10.2307/1240753.

Cichota, R., Brown, H., Snow, V.O., Wheeler, D.M., Hedderley, D., Zyskowski, R., Thomas, S., 2010. A nitrogen balance model for environmental accountability in cropping systems. New Zeal. J. Crop Horticult. Sci. 38, 189–207.

De Wit, C.T., 1958. Transpiration and crop yields. Volume 64 of Agricultural research report/Netherlands Volume 59 of Mededeling (Instituut voor Biologisch en Scheikundig Onderzoek va Landbouwgewasses) Verslagen van landbouwkundige onderzoekingen. Institute of Biological and Chemical Research on Field Crops and Herbage.

De Wit, C.T., 1992. Resource use efficiency in agriculture. Agric. Syst. 40, 125–151. https://doi.org/10.1016/0308-521x(92)90018-J.

Dooley, K.J., 1997. A complex adaptive systems model of organization change. Nonlin. Dyn. Psychol. Life Sci. 1 (1), 69–97. https://doi.org/10.1023/A:1022375910940.

Eastwood, C.R., Chapman, D.F., Paine, M.S., 2012. Networks of practice for co-construction of agricultural decision support systems: case studies of precision dairy farms in Australia. Agric. Syst. 108, 10–18.

Ekboir, J., Rajalahti, R., 2012. Agricultural Innovation Systems: An Investment Sourcebook. International Bank for Reconstruction and Development/International Development Association or the World Bank, Washington, DC.

FAR, 2013. A peer review of OVERSEER® in relation to modelling nutrient flows in arable crops. Report commissioned by the Foundation for Arable Research, 30 p.

Hermann, M., Pentek, T., Otto, B., 2015. Design Principles for Industrie 4.0 Scenarios: A Literature Review. https://doi.org/10.13140/RG.2.2.29269.22248.

Hochman, Z., van Rees, H., Carberry, P.S., Hunt, J.R., McCown, R.L., Gartmann, A., Holzworth, D., van Rees, S., Dalgliesh, N.P., Long, W., Peake, A.S., Poulton, P.L., McClelland, T., 2009. Re-inventing model-based decision support with Australian dryland farmers. 4. Yield Prophet (R) helps farmers monitor and manage crops in a variable climate. Crop Pasture Sci. 60, 1057–1070.

Holzworth, D., Huth, N.I., Fainges, J., Brown, H., Zurcher, E., Cichota, R., Verralla, S., Herrmann, N.I., Zheng, B., Snow, V., 2018. APSIM next generation: overcoming challenges in modernising a farming systems model. Environ. Model. Softw. 103, 43–51.

Jones, J.W., Hoogenboom, G., Porter, C.H., Boote, K.J., Batchelor, W.D., Hunt, A., Wilkens, W., Singh, U., Gijsman, A.J., Ritchie, T., 2003. The DSSAT cropping system model. Eur. J. Agron. 18, 235–265.

Jones, J.W., Antle, J.M., Basso, B., Boote, K.J., Conant, R.T., Foster, I., et al., 2017a. Brief history of agricultural systems modeling. Agric. Syst. 155, 240–254. https://doi.org/10.1016/j.agsy.2016.05.014.

Jones, J.W., Antle, J.M., Basso, B., Boote, K.J., Conant, R.T., Foster, I., et al., 2017b. Toward a new generation of agricultural system data, models, and knowledge products: state of agricultural systems science. Agric. Syst. 155 (Suppl. C), 269–288. https://doi.org/10.1016/j.agsy.2016.09.021.

Keating, B.A., McCown, R.L., 2001. Advances in farming systems analysis and intervention. Agric. Syst. 70, 555–579.

Keating, B.A., Carberry, P.S., Hammer, G.L., Probert, M.E., Robertson, M.J., Holzworth, D., Huth, N.I., Hargreaves, J.N.G., Meinke, H., Hochman, Z., McLean, G., Verburg, K., Snow, V., Dimes, J.P., Silburn, M., Wang, E., Brown, S., Bristow, K.L., Asseng, S., Chapman, S., McCown, R.L., Freebairn, D.M., Smith, C.J., 2003. An overview of APSIM, a model designed for farming systems simulation. Eur. J. Agron. 18, 267–288.

Ledgard S. F. and Waller J. E. 2001. Precision of estimates of nitrate leaching in OVERSEER®. Report to FertResearch. AgResearch Ruakura. 16 p.

Leith, P., O'Toole, K., Haward, M., Coffey, B., 2017. Enhancing Science Impact—Bridging Research, Policy and Practice for Sustainability. CSIRO Publishing. 206 p.

Meinke, H., Stone, R., 2005. Seasonal and inter-annual climate forecasting: the new tool for increasing preparedness to climate variability and change in agricultural planning and operations. Climatic Change 70, 221–253. https://doi.org/10.1007/s10584-005-5948-6.

Meinke, H., Baethgen, W.E., Carberry, P.S., Donatelli, M., Hammer, G.L., Selvaraju, R., Stöckle, C.O., 2001. Increasing profits and reducing risks in crop production using participatory systems simulation approaches. Ag Syst. 70, 493–513.

Meinke, H., Howden, S.M., Struik, P.C., Nelson, R., Rodriguez, D., Chapman, S.C., 2009. Adaptation science for agriculture and natural resource management—urgency and theoretical basis. Curr. Opin. Environ. Sustain. 1, 69–76. https://doi.org/10.1016/j.cosust.2009.07.007.

Moeller, C., Meinke, H., Manschadi, A.M., Pala, M., deVoil, P., Sauerborn, J., 2014. Assessing the sustainability of wheat-based cropping systems using APSIM: sustainability = 42? Sustain. Sci. 9, 1–16. https://doi.org/10.1007/s11625-013-0228-2.

Nelson, R.A., Holzworth, D.P., Hammer, G.L., Hayman, P.T., 2002. Infusing the use of seasonal climate forecasting into crop management practice in North East Australia using discussion support software. Agric. Syst. 74 (3), 393–414. pii:S0308-521x(02)00047-1.

Nelson, R., Kokic, P., Meinke, H., 2007. From rainfall to farm incomes-transforming advice for Australian drought policy. II. Forecasting farm incomes. Aust. J. Agric. Res. 58 (10), 1004–1012. https://doi.org/10.1071/Ar06195.

OECD, 2009. The Bioeconomy to 2030—Designing a Policy Agenda. 322 p https://www.oecd.org/futures/long-termtechnologicalsocietalchallenges/42837897.pdf.

Pratley, J., 2015. Agricultural education and damn stats II: graduate employment and salaries. Agric. Sci. 27, 51–55.

Sarewitz, D., 2004. How science makes environmental controversies worse. Environ. Sci. Policy 7, 385–403.

Schrader, S., Riggs, W.M., Smith, R.P., 1993. Choice over uncertainty and ambiguity in technical problem solving. J. Eng. Technol. Manag. 10, 13–99.

Tamplin, M., Baranyi, J., Paoli, G., 2003. Software programs to increase the utility of predictive microbiology information. In: McKellar, R.C., Lu, X. (Eds.), Modelling Microbial Responses in Foods. CRC, Boca Raton, FL.

Chapter 4

Food shelf-life models

Luciano Piergiovanni and Sara Limbo
Department of Food, Environmental and Nutritional Sciences—DeFENS, Università degli Studi di Milano, Milano, Italy

Abstract
A new approach to commercial life studies is clearly emerging in the most advanced scientific literature: combining the shelf-life assessment with environmental impact inventory data and the analysis of food losses to support adequate environmental strategies. To conduct this type of study, it is undoubtedly essential to have a thorough knowledge of how to face and manage a shelf life study, as well as of the different factors that influence the shelf life of any product. This chapter proposes general schemes and procedures useful for the planning and management of shelf-life studies, with particular attention paid to mathematical models that can help in their forecast and considering the most significant and most recent results in this field.

1 Food shelf-life meanings

Shelf life is a popular term in the food industry, where its basic meaning is generally well understood. It has long been a matter of great importance: production organization, transport and distribution, the choice of packaging characteristics, and the definition of the "best before" dates are all problems that can be dealt with correctly only by knowing what we define as the shelf life of products. Many different professionals have been interested in the kinetics of food-quality reduction and in experimental and predictive methodologies to estimate the shelf life: food technologists, obviously, but today also packaging, logistics, and sustainability experts. Actually, in dealing more closely with the problems of shelf life, several points of view emerge about the shelf life of foods and different definitions of the basic meaning of shelf life can be established.

Depending on the specific purposes and contexts, shelf life may be defined from diverse points of view (see Table 1). The most general definition of shelf life is "the period of time from the production and packaging of a product to the point at which the product first becomes unacceptable under defined environmental and distributing conditions" (Lee et al., 2008). Other and more detailed definitions have been proposed for the meaning of shelf life (Nicoli, 2012; Robertson, 2010). However, even in its generality, the simple and concise definition proposed emphasizes two fundamental points: the target is always the identification of a time (in days, weeks, or months) and there is no single shelf life for the same product. Depending on the defined boundary conditions, in fact, we can have different, even very different, shelf lives attributed to an identical food. Another point deserves, moreover, some attention. The reference to "shelf" clearly implies that the term is related to the commercial life of the product, thus to a packaged product, delivered through the common routes of distribution, not to a generic, "natural life."

What has been previously defined is often indicated as the primary shelf life, to distinguish it from the secondary shelf life, which is the period after package opening during which a food product maintains an acceptable quality level. The secondary shelf life and related issues have received very little attention in the literature, despite the important impact on consumers as well as on food wastage problems (Nicoli and Calligaris, 2018). It is worth noting that comparing primary and secondary shelf life can also be useful for a rough estimate of the protection provided by the packaging and the logistics used in specific circumstances.

In the product and packaging development processes, three related shelf-life concepts have been proposed (Marsh, 1986). Depending primarily on marketing and logistics considerations, the required shelf life has been defined as the minimum time that a food product must sustain in order to become viable in the marketplace. The required shelf life is usually the sum of the time for the product to travel till the distribution point, and the time for the product to be displayed on the store's shelves (Lee et al., 2008). If the primary shelf life is shorter than the required shelf life, better packaging solutions or better displaying conditions must be used to increase the product shelf life. Redesigning the distribution route might also be a possible solution. If, on the contrary, the primary shelf life is longer than the required shelf life, some savings

Sustainable Food Supply Chains. https://doi.org/10.1016/B978-0-12-813411-5.00004-1

solutions can be put in place: for instance, lightening the packaging or using lesser quality and less expensive packaging solutions. Looking at the possible available packaging solutions for a specific product, with different levels of protection against oxygen, moisture, or light, the ideal or maximum shelf life has been theorized. It is, evidently, the maximum shelf life obtainable with no limitations in the packaging choices, that is, with the best possible packaging. Finally, exclusively taking into account the specific needs of the point of sale, the display shelf life has been defined as the time for which products, in particular fresh produce, meats, and fish, can be stored under expected or specified conditions of store display. Nowadays, some packaging solutions have been designed expressly for this goal, particularly for refrigerated meat display cabinets.

TABLE 1 Common definitions of the various shelf-life concepts

Shelf-life type	Description
Primary shelf life	The period of time from the production and packaging of a product to the point at which the product first becomes unacceptable under defined environmental conditions
Secondary shelf life	The period of time after package opening during which a food product maintains an acceptable quality level
Required shelf life	The minimum period of storage time that a food product must guarantee to become profitable in the market
Maximum shelf life	The maximum period of time that a food product can achieve, using the best available packaging solution
Display shelf life	The period of time for which products (particularly fresh produce, meats, and fishes) can be stored under expected or specified conditions of store display

In all the proposed different definitions, two concepts have key and critical meanings: the concept of acceptability (or unacceptability) and the description of specific boundary conditions. It is certainly evident that the words acceptable and unacceptable may have different meaning to different people in different countries and that different boundary conditions can deeply affect the shelf-life quantification. In other words, as already noticed, there is always a certain amount of subjectivity and discretion in all the shelf-life assessments, however obtained.

2 Food shelf-life study motivations

Food shelf-life studies, and consequently shelf-life assessments, are carried out for different reasons and motivations, but certainly the most common one is the attainment of a commercial life extension. A longer time to manage distribution tasks and selling operations is normally assumed as a major advantage for economic sustainability and business effectiveness but, very likely, it also has great potential for reducing food losses and waste, and thus for social and environmental sustainability. The ways in which to obtain such an extension have to do with the food formulation and processing, its packaging, and the final product distribution. Every different strategy used to extend shelf life can affect the achievable time, as well as differently directing the manner of evaluation and prediction of that extension and hence the most appropriate model and methodology to use (see Fig. 1).

For instance, if an important change is made to the food formulation, such as different ingredients or additives to improve stability, it is essential to know the chemical and microbiological effects of these modifications and to take them into account in the forecasts. A change in processing conditions and/or equipment in order to improve stability generally affects the initial quality of the food product and, consequently, the time to reach an unacceptable level. The choice of different packaging materials and/or packaging processes to control the internal environment of the package (e.g., modified atmosphere, vacuum packaging, active packaging, etc.) can profoundly influence the kinetics of all the relevant phenomena related to the deterioration of the foods. In such cases, the shelf studies are always quite complex, as will be explained later; they must take into account many and different factors and, generally, lead to only approximate solutions. Finally, an intervention in the distribution or storage systems, to minimize the damages from the surrounding environment, is very difficult to model and it is usually evaluated by simulation tests.

Regrettably, in comparison to commercial life extension, much less attention is generally paid to the potential of shelf-life studies for different purposes that do not cover the extension of the validity period and the related economic benefits. Nevertheless, a well-designed experimental or theoretical study directed to the estimation of shelf life can be very useful for achieving different objectives.

In particular, an underestimated goal of shelf-life studies is addressed to a possible reduction of food loss and waste. According to FAO definitions (FAO, 2011), "food that gets spilled or spoilt before it reaches its final product or retail stage" is called food loss, while "food that is fit for human consumption, but is not consumed because it is left to spoil or discarded by retailers or consumers" is called food waste. Sometimes, the two negative phenomena are considered together as "food wastage." In order to conduct a shelf-life study addressed to a possible reduction of food wastage, it is essential to consider both the primary and secondary shelf life and to take into account reliable data about consumers' habits and about practices of commercial distribution. However, reliable data are lacking, and consequently the scientific literature on the subject is still limited (Conte et al., 2015; Williams and Wikström, 2011; Heller et al., 2018).

Possible packaging optimization by means of shelf-life studies has been a little more frequently investigated. Definitely, through the careful selection of packaging materials, package shape design, filling levels, modified atmosphere selection, and other important factors, it is possible to achieve better performance, fewer environmental impacts of the packaging, and greater economic savings by means of appropriate shelf-life studies.

A further reason for carrying out shelf-life studies is related to the objective definition of the legal durability of a packaged product. Although not strictly relevant to this text, it is certainly appropriate to mention the obligation for almost all packaged food products to report on the label, or on a designated part of their packaging, a date related to the so-called commercial durability. These obligations have long been regulated in European countries and in major countries worldwide, but the rules are very similar all over the world. They provide two possible and different indications. The first (generally more binding) is defined for products that can be rapidly perishable from a microbiological point of view and that may constitute a danger to human health; this takes the name of "use-by" or "expiration date" in the United States and many other places. The second is the date until which the foodstuff retains its specific properties under suitable storage conditions and is generally indicated with the expression "best before." It is quite clear that these indications will be correctly indicated only if they represent the result of an accurate study of shelf life and that, in the two cases, the survey approaches to be followed (microbiological or sensorial and merchandise only), will be different according to the factors that determine the conservation and on the basis of the quality indexes that must be evaluated.

Finally, it is worth mentioning that an extension of the durability of packaged food products is not always necessary and sometimes it is not even appropriate. In the case of foods and drinks that stand out for their freshness and naturalness, for example, long storage could possibly damage the product's image or reputation, discouraging the unwary consumer. Even in these cases, shelf-life studies are always very useful because, also when a lengthening of the commercial life is not essential, they allow measures to rationalize and optimize packaging, processes, and logistics.

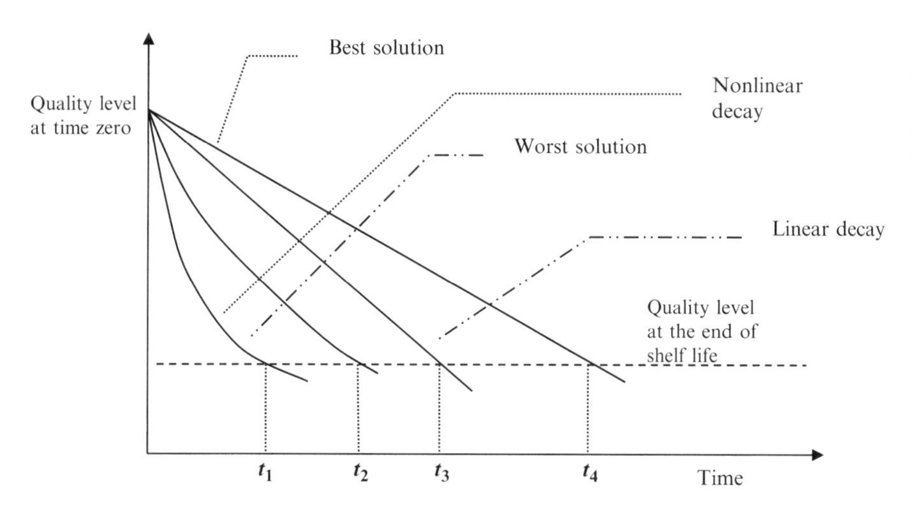

FIG. 1 Graphical representation of the shelf-life extension. *(Adapted from Lee, D. S., Yam, K.L., Piergiovanni, L., 2008. Food Packaging Science and Technology. CRC Press, Boca Raton, FL (Chapter 16).)*

3 Food shelf-life modeling

Using very simple words, modeling can be defined as an alternative to physical experimentation. With reference to shelf-life studies, modeling, which is basically the development and the use of a mathematical model, is a substitute for monitoring various quality indices of a food product during its simulated storage under the same conditions it will experience in real commercial life. It is always possible, evidently, to carry out practical experimentation in order to observe quality changes (normally quality decay) of the product over time and collect data that is very accurate and reliable, but it takes a lot of time and resources. The application of a mathematical model is much less expensive, faster in execution, and offers the opportunity to simulate. That is, it is possible to make different shelf-life estimates by modifying one or more parameters that are taken into account in the model, studying the effects of different components, and making predictions about the behaviors in different scenarios.

The complexity of the shelf-life problem, already underlined, is quite high and derives both from the large number of possible different situations (the same product can be packaged and stored in many different ways) and the fact that the durability of a packaged food is, in all circumstances, a function of many variables. These variables can be referred to as the composition/characteristics of the food (F), its packaging (P), and the environment (E) in which the product is stored. It is common and frequent, in fact, that the shelf-life problem is represented by equations like Eq. (1):

$$SL = f(Fi, Pi, Ei) \tag{1}$$

where

SL = the shelf life.
Fi = food variables.
Pi = packaging variables.
Ei = environmental variables.

The shelf life, therefore, appears as a function of different variables that belong to three distinct domains (see Fig. 2).

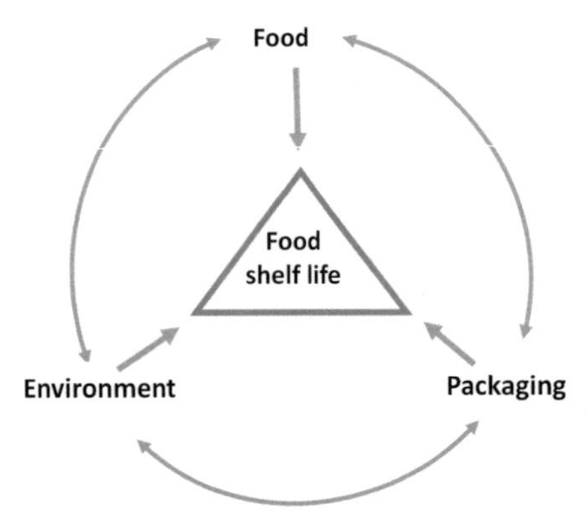

FIG. 2 The graphical representation of shelf-life study belonging to three domains. *(Adapted from Lee, D.S., Yam, K.L., Piergiovanni, L., 2008. Food Packaging Science and Technology. CRC Press, Boca Raton, FL (Chapter 16).)*

Expressions like Eq. (1) are effective in representing the complexity of the shelf-life issue but are of no practical use, as they do not allow any real quantification in this form. The solution of the problem must be sought in a more accurate and rigorous analysis of the factors that determine the shelf life and the modalities with which they intervene over time to modify the quality of the product.

3.1 Intrinsic and extrinsic variables

Food factors: The variable factors, related to the food characteristics that can play an important role in affecting the commercial durability are, for example, the microbial load, the enzymatic pattern, the values of pH and activity water (aw), the concentration of specific solutes, the presence of inhibitors and preservatives, promoters and catalysts. Each of these factors can occur at a certain level at the time of packaging and influences the shelf life of the product, changing over time. There is no doubt that these factors are the most relevant in modulating the preservation of packaged and not-packaged food. They are often defined as "intrinsic factors" that lead to different deterioration modes, shown in Table 2 where foods are classified as little, medium, and very perishable products. This is a very approximate classification from which, however, clearly emerges the fact that some foods have short preservation even at refrigerated temperatures and others are stable for long periods, even at room temperature. The former almost invariably correspond to fresh food with high water content, and the latter to processed foods or those with a low moisture content and low aw values.

Environmental factors: The effects of light, temperature, humidity, and concentration of environmental oxygen, the most important environmental variables external to the product, so described as extrinsic, can obviously change the typical shelf life, as indicated in Table 2. These changes might also be significant, but never to the point of transforming a shelf stable food into a very perishable one. Interestingly, these environmental factors can make the variables of food more or less critical. For example, the "microbial load" factor may be less important at lower temperatures, while other factors may

TABLE 2 Modes of deterioration, critical environmental factors for food, and approximate shelf lives of some products at different temperature

Food product	Deterioration modes (according to food factors and characteristics)	Critical environmental factors	Typical and approximate shelf life
Very perishable			
Milk and dairy products	Bacterial growth, oxidized flavor, hydrolytic rancidity	Oxygen, temperature	7–30 days at 0–7°C
Fresh bakery products	Staling, microbial growth, moisture loss causing hardening, oxidative rancidity	Oxygen, temperature, humidity	2–7 days at room temperature
Fresh red meat	Bacterial growth, loss of red color	Oxygen, temperature, light	3–4 days at 0–5°C
Fresh poultry	Bacterial growth, off-odor	Oxygen, temperature, light	2–7 days at 0–5°C
Fresh fish	Bacterial growth, off-odor	Temperature	3–14 days at ~ −2 to 0°C
Fresh fruits and vegetables	Respiration, compositional changes, nutrient loss, wilting, bruising, microbial growth	Temperature, relative humidity, light, oxygen, physical handling	7–30 days at 5–7°C
Medium perishables			
Fried snack food	Rancidity, loss of crispness, breakage	Oxygen, light, temperature, relative humidity, physical handling	3–12 weeks at room temperature
Cheese	Rancidity, browning, lactose crystallization, undesirable mold growth	Temperature, relative humidity	4–24 months at 5–7°C
Ice cream	Graininess caused by ice or lactose crystallization, texture	Fluctuating temperature (below freezing)	1–4 months at −25°C
Little perishable (shelf stable)			
Dehydrated foods	Browning, rancidity, loss of texture, loss of nutrients	Relative humidity, temperature, light, oxygen	1–24 months at room temperature
Nonfat dry milk	Flavor deterioration, loss of solubilization, caking, nutrient loss	Relative humidity, temperature	8–12 months at room temperature
Breakfast cereals	Rancidity, loss of crispness, nutrient loss	Relative humidity, temperature, rough handling	6–18 months at room temperature
Pasta	Texture changes, staling, vitamin and protein quality loss, breakage	Relative humidity, temperature, light, oxygen, rough handling	9–48 months at room temperature
Frozen concentrated juices	Loss of cloudiness, yeast growth, loss of vitamins, loss of color or flavor	Oxygen, temperature and thawing	18–30 months at −25°C
Frozen fruits and vegetables	Loss of nutrients, loss of texture, flavor, odor, color, and formation of package ice	Oxygen, temperature, temperature fluctuations	6–24 months at −25°C
Frozen meats, poultry and fish	Rancidity, protein denaturation, color change, desiccation (freezer burn), toughening	Oxygen, temperature, temperature fluctuations	2–14 months at −25°C
Frozen convenience foods	Rancidity in meat portions, weeping and curdling of flavor, loss of color, package ice	Oxygen, temperature, temperature fluctuations	6–12 months at −25°C
Canned fruits and vegetables	Loss of flavor, texture, color, nutrients	Temperature	12–36 months at room temperature
Coffee	Rancidity, loss of flavor and odor	Oxygen, temperature, light, relative humidity	9–36 months at room temperature
Tea	Loss of flavor, absorption of foreign odors	Oxygen, temperature, light, humidity	18 months at room temperature

Adapted from Lee, D.S., Yam, K.L., Piergiovanni, L., 2008. Food Packaging Science and Technology. CRC Press, Boca Raton, FL (Chapter 16).

become crucial; therefore, a change in environmental conditions can make a food factor determinant or negligible, depending on the circumstances.

It may also be important to underline that these environmental factors are often typical of confined environments, small in dimensions and changing over time. The cycle of temperatures in a fridge or the different lighting intensities during transport, for example, can be important as the factors of external environment. A very particular environment is certainly the one inside the packaging that changes due to the specific properties of the packaging.

Packaging variables: Variable factors that directly affect the packaging performance can be identified in the gas and vapor barrier offered by packaging materials, in its transparency to light, in its ability to withstand mechanical and thermal stresses, and in its inertia in contact with food. The variables of the packaging, in other words, have the effect of modulating those environmental variables already mentioned, creating a "microenvironment" different from the external one, the "macroenvironment"; they also are commonly indicated as extrinsic variables.

The three mentioned categories of factors, however, can rarely be considered independently because, in most cases, they interact by influencing each other; the microbial growth, for instance, may cause oxidative spoilage in food. The methods of deterioration of food and beverages and storage times are therefore different and very variable from case to case, making shelf-life modeling a critical and complex issue.

3.2 Quality indices and critical limits of acceptability

Whatever the factors to be considered for the shelf-life modeling, they will determine the variation of one or more specific attributes of product quality, reducing it to a critical value that will establish the end of the marketability of the food product. These quality attributes are perceptible and measurable parameters that describe food quality and allow us to quantify that minimum level of quality that we are willing to accept, according to the definition of shelf life. They are also called indices of failure (IOFs) and may have a very different nature and origin; they are often classified into chemical, physical, microbiological, and sensory indices (Robertson, 2010). Sometimes these indices decrease over the course of commercial life, such as the content of a vitamin or a typical aroma; other times they grow during the shelf life as the concentration of an undesirable migrating substance, the product of a decay reaction, or a microbial charge. Sometimes it is adequate to monitor only one factor and sometimes it is necessary to take into account an array of them, in a multifactorial analysis.

The choice of IOF to be considered in monitoring or modeling shelf life is a fundamental step of the shelf-life problem, and it is essential to give it proper attention. If the choice of the quality attribute can be relatively objective and simple (sometimes being conditioned by the analytical possibilities available), then establishing the critical level of this attribute becomes more complex and more subjective: that is, the level of the attribute that is not intended to be achieved. How and in what length of time this critical level of the chosen quality attribute is reached is the heart of the shelf-life problem, represented schematically in Fig. 1, which explains how the evolution of the IOF can be different (a linear or nonlinear decay with respect to time) and the effect of a shelf-life extension measure, which can be represented by a better packaging solution.

Even if a food product has many deterioration modes, it is not necessary to deal with all possible modes; it is better to focus on the one or two major deterioration modes that occur most rapidly, because the product shelf life is obviously limited by the fastest spoilage process. A shelf-life model, therefore, is basically a function that relates the IOF value (or a few IOFs) of the fastest deterioration modes to the storage time.

3.3 Mathematical models

A mathematical model is the description of a system using mathematical concepts and language. Mathematical models are used in almost all fields of knowledge: the natural sciences (the ones related to shelf-life modeling), the engineering disciplines, as well as the many branches of social sciences (such as economics, psychology, sociology, political science, etc.). Therefore, several classification criteria are used for them, according to their context, purpose, and structure. With reference to the shelf-life models in particular, it is worth remarking on only the following:

1. *Linear vs. nonlinear*: the mathematical model is defined as linear if all the operators exhibit linearity, in other words the mathematical equations used are first-degree equations. Otherwise it is considered to be nonlinear. In shelf-life studies both linear and nonlinear models are common and quite often the approach to nonlinear problems is, for simplicity, their linearization. This, for instance, happens in managing moisture isotherms in moisture-dependent shelf-life problems (Labuza, 1982).

2. *Static vs. dynamic*: a dynamic model accounts for time-dependent changes in the state of the system, while a static model is time-invariant and calculates the system in equilibrium. Obviously, shelf-life mathematical models are all dynamic.
3. *Discrete vs. continuous*: a model that treats variable entities is discrete, while a continuous model represents the figures in a continuous manner. Discrete variables are countable in a finite amount of time. For example, you can assess the appreciation of one product on the basis of a standard scale (the entity goes from excellent to very bad). Continuous variables change continuously their value. For example, you can measure the vitamin content day by day, hour by hour. Generally, shelf-life studies use continuous models, although when sensory tests are considered, a discrete approach is quite common.
4. *Deterministic vs. probabilistic*: the majority of shelf-life models are deterministic; this means that all the parameters are uniquely determined by variables and by their initial and final levels. Therefore, these models always perform in the same way and give the same result for a given set of initial conditions. In probabilistic (or stochastic) models, the variables are not described by unique values, but rather by probability distributions, hence randomness is taken into account. In shelf-life studies where survival analysis approaches are used for analyzing the expected time for one event (often a change in sensory evaluation), stochastic models are used (Robertson, 2010).

According to the deterioration mode and to the packaging solution adopted, one further classification of possible shelf-life models seems appropriate and useful.

When the food package is inert, opaque, and impermeable, the food is practically isolated from the surrounding environment and no significant mass transport phenomena can affect the shelf life. In these cases, the qualitative decay occurs as a result of the chemical and biochemical properties of the product and as a consequence of the process adopted. It means that the environmental factors (except the temperature) and the mass transport properties of packaging are not significantly influential. In these situations, it is appropriate to talk about product-dependent shelf-life problems. The role of packaging material in these cases is substantially passive and not relevant with respect to the shelf life of the product. Examples of such situations, which are found in many cases of packaging in metal or dark glass and rigid packages, are represented by wine in a hermetically sealed dark glass bottle, processed cheeses wrapped in aluminum foil, preserved foods canned in tinplate containers, vacuum-packed products in barrier packaging, and many others. The predictive mathematical models that apply in these cases are mainly based on the chemical kinetics of the alterations, the modeling of microbial growth, the effect of temperature, and the models that describe the variations of sensory attributes (Limbo et al., 2010; Manzocco and Lagazio, 2009). All these models can be classified, as is clarified in the following text, as direct models.

On the other hand, when the food quality preservation is strongly influenced by the performance of the packaging (i.e., depends on the packaging and environmental factors), it appears more appropriate to talk about packaging-dependent shelf-life problems. The predictive models that can be adopted to solve these problems are much more complex than those that can be adopted in the product-dependent problems, because they must combine the variables of the packaging with those of environment and the product. Furthermore, in these problems (which are very common since they are associated with the ubiquitous flexible packaging), the performance of the packaging affects the shelf life both in quantitative terms (length of durability), and in qualitative terms (the predominant degradation mode). For some foods, such as fat and dry products, for example, the barrier performance of the packaging material can change the critical quality attribute: in high moisture and low oxygen permeable packaging (e.g., made from a biopolymer material), the shelf life of the product will be conditioned by the moisture transmission and the IOF will be connected to the softening of the product. In a moisture-barrier but oxygen-permeable package (e.g., made from a thin polyolefinic material), conversely, the IOF would be related to rancidity, as the shelf-life results are conditioned by the permeability to oxygen.

To approach the models related to these last problems in particular, one further classification, of direct and mechanistic models, is opportune (Del Nobile and Conte, 2013).

The simplest ones are those called direct models. They generally do not have any particular physical or biological meaning, that is, they are not a true, objective description of the phenomenon involved, and they use an equation that relates the food quality (the IOF selected) to the storage time. These models can be linear and nonlinear, as the equations proposed are first-degree (rarely, Eq. 2) polynomial (Eq. 3), exponential (Eq. 4), or power law functions (Eq. 5):

$$IOF_{(t)} = a_0 + a_1 \cdot t \tag{2}$$

$$IOF_{(t)} = a_0 + a_1 \cdot t + a_2 \cdot t^2 + \cdots + a_n \cdot t^n \tag{3}$$

$$IOF_{(t)} = a_0 \cdot e^{(-a_1 \cdot t)} \tag{4}$$

$$IOF_{(t)} = a_0 \cdot t^{a_1} \tag{5}$$

In the previous equations, $IOF_{(t)}$ is the index of failure at time t, and a_i are fitting constants or equation parameters. These last are normally obtained by *fitting* experimental data, which come from limited experimental work. The time t corresponding to the shelf life is usually calculated by *interpolation* or small-scale *extrapolation*. These last terms (fitting, interpolation, extrapolation) deserve clear definitions as they are quite often used improperly.

The fitting process is the work done when, starting from experimental data (e.g., Y vs. X), a mathematical function (a curve) is derived that fits, at the best, the numbers obtained through the measures carried out. Many statistical packages permit a simple curve-fitting process that returns, on the input of experimental data, the required equation parameters (fitting constants) and a test that gives the appropriateness of those parameters, thus of the model chosen. The assessment as to whether or not a given mathematical model describes our system accurately is a crucial part of the modeling process, which is quite often underestimated or neglected. Interpolation is, somehow, the opposite work: that is, the method of obtaining a new data point within the range of the set of data points used for the fitting of the function (the model). Extrapolation refers to the use of a fitted curve beyond the range of the observed data.

The other types of models, representing another way to find a relationship between time and food quality, are represented by a *mechanistic* approach. Here it is assumed that the quantitative description of food spoilage processes and the mass transport properties of packages are known and balance equations are available that combine them. The shelf-life mechanistic model is, essentially, a set of differential equations related to three domains: the packaging performance, the food deterioration mechanism, and their balance. These models are quite complex to develop and to manage, and they are not accessible to everyone; it is worth noting, however, that they have been proposed and used in food science for almost 50 years (Labuza and Dugan, 1971). Several good readings that focus on these problems and on experimental applications of such models are available in the literature and make possible an in-depth analysis of these issues and their great variability (however, it seems inappropriate in this case) (Nicoli, 2012; Del Nobile and Conte, 2013).

4 Food shelf life and sustainability

The shelf-life assessment of foods is an expensive activity that has positive impacts on the economy and society only when the whole product-package-environment system is considered in its entirety and the interference or interactions among the domains are correctly predicted.

An improper definition of shelf life may lead to different scenarios: consumer complaints, product recalls, ineffective logistics impacts, and, most of all, expensive food losses and wastage that impact both economic and environmental sustainability. At the same time, expertise in shelf-life assessment of food offers an opportunity to achieve technological innovations that can contribute to the optimization of food shelf life itself and positively impact global sustainability, by acting on the formulation, process, and packaging of the product.

Recent studies emphasize that the precise knowledge of shelf-life requirements is very valuable for food companies and the different shelf-life values achievable by process technologies and packaging choices may be key drivers in the design and selection of various food products, reacting to market flexibility and responding to sustainability concepts. In this context, Stefansdottir and Grunow (2018) demonstrated that the development of a flexible process able to produce milk concentrates instead of milk powder may lead to the advantage of extensive energy savings in production but also to products with a "tailor-made" shelf life. Thus, using different concentration methods (evaporation, reverse osmosis, and evaporation plus reverse osmosis, which led to different dry matter contents) combined with different heat treatments after concentration (ultrahigh temperature (UHT), high-heat treatment (HHT), and pasteurization), the attainment of shelf-life values that vary from less than 10 days to 50 days are possible (see Fig. 3). In this way, a customization of both the product and the technology can be achieved, responding to market requests and variability, thus reducing the risk of failure and the food loss. For example, if a low shelf life is required, pasteurized products (N3) concentrated with reverse osmosis can be selected (P3-P4-P5, Fig. 3), but if customers have high shelf-life requirements, preheated products (N8-N9-N10) concentrated with combined reverse osmosis and evaporation to obtain the highest dry-matter content can be chosen (P16, Fig. 3).

Shelf-life optimization with a view to achieving sustainability goals can also involve the modification of the ingredients and/or the packaging solution of a food product. Sometimes the challenge is to prolong the shelf life of already stable products, like low-moisture baked products, coffee, etc., which are characterized by a certain stability due to the low availability of free water. However, market dynamics, globalization, and import-export policies impose technological efforts to reduce the major causes of deterioration that, for those products, are oxidation and water uptake. A study carried out by

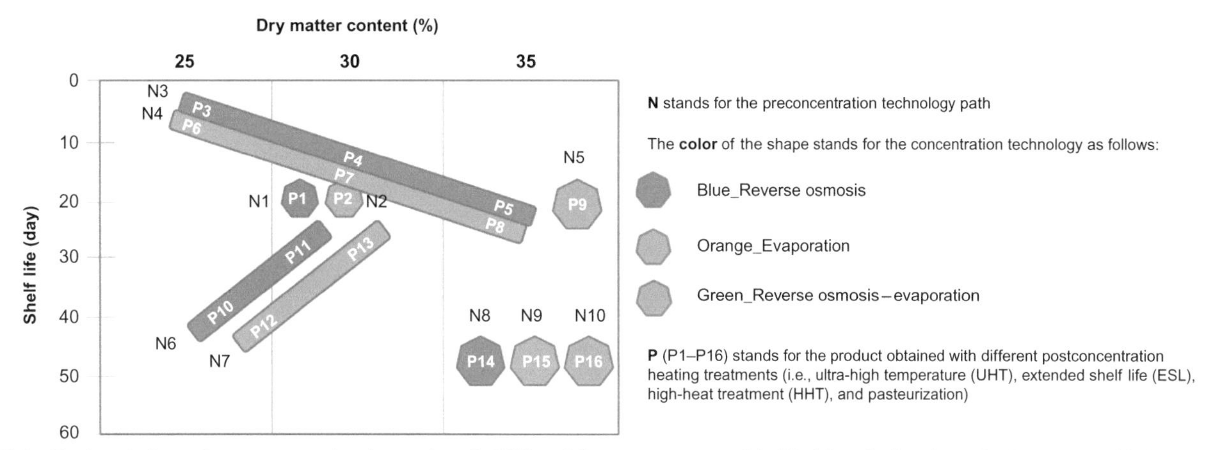

FIG. 3 Product designs of concentrates related to product shelf life and dry-matter content. *(Modified from Stefansdottir, B., Grunow, M., 2018. Selecting new product designs and processing technologies under uncertainty: two-stage stochastic model and application to a food supply chain. Int. J. Prod. Econ. 201, 89–101.)*

Alamprese et al. (2017) evaluated the role of rosemary extract in packaging under nitrogen in the shelf-life extension of whole-wheat breadsticks. The kinetic study of oxidation highlighted that the addition of a natural antioxidant yielded a 42% shelf-life extension, but the combination of the formulation and the packaging strategy allowed increasing the shelf life by up to 83% at 25°C. The environmental impacts of a change in formulation have also been investigated for the same whole-wheat breadsticks using the life cycle assessment (LCA) methodology (Bacenetti et al., 2018). This represents a good approach to finding a relationship between a shelf-life extension (or reduction) of a food and the economical or environmental sustainability. In the case referred to earlier, two scenarios were considered:

(a) the whole-wheat breadstick without rosmarinic acid and, consequently, without an extension of the shelf life;
(b) the whole-wheat breadstick with rosmarinic acid with an extended shelf life.

Table 3 contains the main questions that the authors tried to answer in this work (Bacenetti et al., 2018) and that can be taken into consideration when approaching an LCA study for the environmental impacts of foods with extended shelf life.

TABLE 3 Driving questions for the evaluation of environmental impacts as a function of food shelf-life extension due to a change in formulation

Food supply chain phase/step	Issue/question
Production process step	What is the environmental impact for product with and without the new ingredient?
Production process step	What are the main hotspots associated with the production system?
Shelf-life step	Can the shelf-life extension be an effective mitigation solution, even if it involves an increase of the environmental impact related to change in the production process?
Food loss	Can the reduction of product losses (due to shelf-life extension) overcome the impact increase (due to the process modification)?

Adapted from Bacenetti, J., Cavaliere, A., Falcone, G., Giovenzana, V., Banterle, A., Guidetti, R., 2018. Shelf-life extension as solution for environmental impact mitigation: a case study for bakery products. Sci. Total Environ. 627, 997–1007.

The main results showed that, in terms of climate change, the breadsticks with the antioxidant become less impactful than those without rosmarinic acid, and the food loss at the consumer level should be reduced from 15% to 8%, thanks to the shelf-life extension.

In most situations, the role of packaging in shelf-life extension is crucial. In addition, while it has been largely demonstrated that both packaging materials and technologies act positively on the quality of a food product, the contribution of packaging to the environmental impact is controversial and it is very often the focus of much criticism (Molina-Besch et al., 2018). There are few doubts about the protective function of packaging materials and packaging technologies against food spoilage. However, the environmental impacts of packaging at the end of the production and distribution food chain are very often important, because of the low recyclability of the materials and/or the lack of effective loop recovery systems. At the same time, packaging assumes an indirect but important role in environmental sustainability, especially when it is involved in the protection of food, avoiding food loss and waste along the supply chain up to the consumer's house. LCA has been largely used to determine the environmental impacts of food production and processing and/or packaging materials.

A recent study by Heller et al. (2018) considers that "a recalibration of packaging environmental assessment to include the indirect effects due to influences on food waste" is becoming more and more urgent, and this means being aware of the strict and unique relationship that exists between packaging and food shelf life. In this recent work, the authors illustrate the importance of considering food waste when different packaging solutions for a specific food type are proposed as viable alternatives. A methodology based on the estimation of the food-to-packaging (FTP) environmental impact relationship has been proposed as follows:

$$\mathrm{FTP}_E = \left[E\left(\frac{\text{agricultural (from gate) production}}{\text{kg food}}\right) + E\left(\frac{\text{food processing}}{\text{kg food}}\right)\right] / E\left[\frac{\text{packaging material}}{\text{kg food}}\right] \tag{6}$$

where E = environmental impact indicator of interest (e.g., GHG emissions, cumulative energy demand (CED), etc.) (Heller et al., 2018).

This practical index of Eq. (6) highlights that, for example, cereals, dairy, fish, seafood, and meats have larger FTP ratios than other food categories. This means that emissions or resource use of food production are much larger than those of the packaging. Therefore there may be room for changes in packaging configuration and, even when packaging impact increases, the food-waste reduction that can be obtained will lead to a net system decrease in environmental impact. On the other hand, at very low FTP ratios it is preferable to focus attention on reducing the packaging impacts, such as reducing thickness, changing materials, etc.

The Packaging Relative Environmental Impact (PREI) defined by Licciardello (2017) and associated with the contribution of packaging to the overall environmental load is consistent with these considerations. The very high impacts in primary production of meats and milk with dairy products influence the PREI values that remain very low (respectively, 1.2–6.5 and 2–14), indicating that packaging burdens represent only a small part of the overall burden. Also, from this point of view, the increase in packaging would be reasonable from an environmental point of view only if counterbalanced by a shelf-life optimization (extension or reduction) that offers a significant reduction of the food loss and waste along the chain up to the consumer level. However, there is always a strong need for more research into the relapse of a shelf-life strategy on food-loss reduction, taking into account the actual and more common scenario of food distribution and consumption (Molina-Besch et al., 2018).

In a study carried out by Adobati (2015), the environmental impacts of production and distribution of fresh strawberries packaged in different solutions, called "macro perforated PET tray" (base A scenario), "passive atmosphere modification solution in master bag" (B scenario), and "active atmosphere modification solution in master bag with devices" (C scenario), as two alternative scenarios of the base scenario, were estimated by means of an LCA approach. The shelf-life assessment was also carried out for all three solutions, leading to the situation shown in Fig. 4.

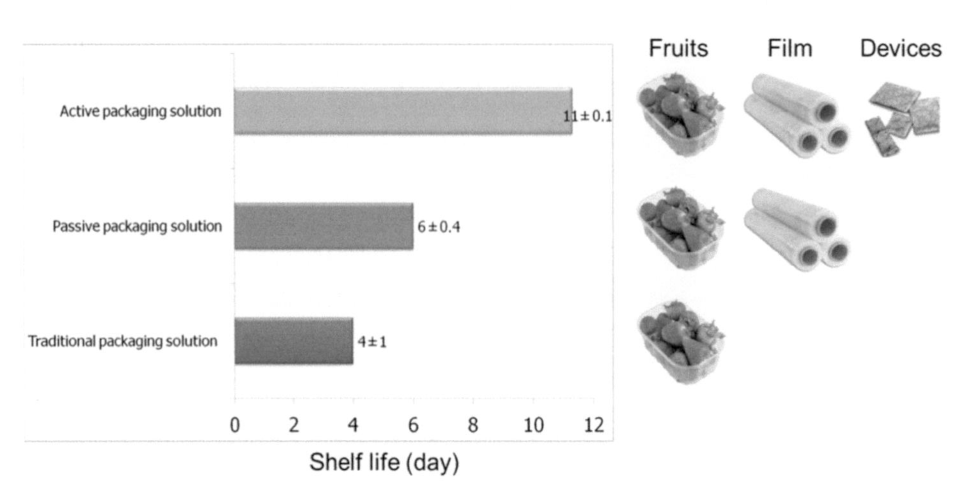

FIG. 4 Shelf-life scenarios for strawberries stored with different packaging materials and solutions.

As a result of this study, the environmental impact was assessed for the strawberries stored in different packaging solutions, identifying the best option in terms of lower environmental burn. The choice to use the days of shelf life as the functional unit meant that the environmental impacts could be shared along the lifetime. In this way, a "daily" impact was defined for each packaging solution adopted. Thus, the LCA methodology permitted defining the best solution for strawberry storage in terms of the lowest environmental impact: with the active packaging solution a reduction of the environmental load was reached, despite the use of more packaging materials.

A sustainable and efficient management of food products also involves the definition of shelf life in planning warehouse operations and avoiding food loss. The shelf-life modeling of fresh foods has been largely investigated considering the effects of temperature or humidity on some quality traits (Limbo et al., 2010; do Nascimento Nunes et al., 2014). More recently, the use of sensors for real-time monitoring of the emissions of volatiles from fruits has been investigated; numerical approaches like the Monte Carlo method have been implemented to simulate the emissions during the fruit storage steps and model the shelf life (La Scalia et al., 2019). In this case, the residual shelf life based on an exposure probability chart can be predicted and used to generate hypothetical scenarios in terms of the values attributed to the main quality parameter during all the segments of the supply chain, from production to final consumption. The new approach in shelf-life studies is to merge these data with inventory management systems, based, for example, on FIFO (first in, first out) or the FEFO (products first expired should be sold first) models, thus implementing a shelf life-based picking policy with economic and environmental data to obtain the corresponding impacts. In this way, an important decisional tool can be created to support suitable environmental strategies. In the study presented by La Scalia et al. (2019), it was demonstrated that the environmental impact in terms of carbon dioxide emissions decreased when moving from the FIFO to the FEFO policy in different pricing scenarios for strawberry management at the warehouse.

Among the different mitigation strategies adopted for the evaluation of environmental impacts, some studies highlighted that the optimization of packaging solutions can be an effective approach to decrease the environmental load of the food systems (Williams and Wikström, 2011). As stated before, only a few studies take into consideration the role of an accurately estimated food shelf-life extension or reduction due to a packaging solution on the impacts and method accounted for in the life cycle inventory. In addition, methodological aspects that take into consideration the time dimension and food losses, that is, the relationship between food loss probability and packaged food shelf life, are still lacking. An interesting approach has been recently proposed by Conte et al. (2015) that provided an eco-indicator able to quantify the indirect environmental effects related to the different choices in food packaging of a ripened sheep-milk cheese by using the LCA method.

However, future research needs to move towards the effect of shelf-life extension of the most critical food commodities on the supply chain and consumer habits. In fact, general knowledge of shelf-life extension should be improved as a response to sustainability in terms of environmental load and social/economic effects.

References

Adobati, A., 2015. Active Packaging in Master Bag Solutions and Shelf-Life Extension of Red Raspberries and Strawberries: A Reliable Strategy to Reduce Food Loss. (PhD Dissertation in Food Science). Technology and Biotechnology, XXVIII cycle, Università degli Studi di Milano.

Alamprese, C., Cappa, C., Ratti, S., Limbo, S., Signorelli, M., Fessas, D., Lucisano, M., 2017. Shelf-life extension of whole-wheat breadsticks: formulation and packaging strategies. Food Chem. 230, 532–539.

Bacenetti, J., Cavaliere, A., Falcone, G., Giovenzana, V., Banterle, A., Guidetti, R., 2018. Shelf-life extension as solution for environmental impact mitigation: a case study for bakery products. Sci. Total Environ. 627, 997–1007.

Conte, A., Cappelletti, G.M., Nicoletti, G.M., Russo, C., Del Nobile, M.A., 2015. Environmental implications of food loss probability in packaging design. Food Res. Int. 78, 11–17.

Del Nobile, M.A., Conte, A., 2013. Packaging for Food Preservation. vol. 572. Springer, Berlin, Germany.

do Nascimento Nunes, M.C., Nicometo, M., Emond, J.P., Melis, R.B., Uysal, I., 2014. Improvement in fresh fruit and vegetable logistics quality: berry logistics field studies. Philos. Trans. R. Soc. A: Math. Phys. Eng. Sci. 372 (2017), 1–19.

FAO, 2011. Global Food Losses and Food Waste—Extent, Causes and Prevention. FAO, Rome.

Heller, M.C., Selke, S.E.M., Koleian, G., 2018. Mapping the influence of food waste in food packaging environmental performance assessments. J. Ind. Ecol. https://doi.org/10.1111/jiec.12743.

La Scalia, G., Micale, R., Miglietta, P.P., Toma, P., 2019. Reducing waste and ecological impacts through a sustainable and efficient management of perishable food based on the Monte Carlo simulation. Ecol. Indic. 97, 363–371.

Labuza, T.P., 1982. Shelf-Life Dating of Foods. Food & Nutrition Press, Westport, CT pp. 1–87, 99–118.

Labuza, T.P., Dugan, L.R., 1971. Kinetics of lipid oxidation in foods. C R C Crit. Rev. Food Technol 2 (3), 355–405.

Lee, D.S., Yam, K.L., Piergiovanni, L., 2008. Food Packaging Science and Technology. CRC Press, Boca Raton, FL (Chapter 16).

Licciardello, F., 2017. Packaging, blessing in disguise. Review on its diverse contribution to food sustainability. Trends Food Sci. Technol. 65, 32–39.

Limbo, S., Torri, L., Sinelli, N., Franzetti, L., Casiraghi, E., 2010. Evaluation and predictive modeling of shelf-life of minced beef stored in high-oxygen modified atmosphere packaging at different temperatures. Meat Sci. 84 (1), 129–136. https://doi.org/10.1016/j.meatsci.2009.08.035.

Manzocco, L., Lagazio, C., 2009. Coffee brew shelf-life modelling by integration of acceptability and quality data. Food Qual. Prefer. 20, 24–29.
Marsh, K.S., 1986. Shelf-life. In: Bakker, M. (Ed.), The Wiley Encyclopedia of Packaging Technology. John Wiley, New York, pp. 578–582.
Molina-Besch, K., Wikström, F., Williams, H., 2018. The environmental impact of packaging in food supply chains—does life cycle assessment of food provide the full picture? Int. J. Life Cycle Assess. https://doi.org/10.1007/s11367-018-1500-6.
Nicoli, M.C., 2012. An introduction to food shelf-life: definitions, basic concepts, and regulatory aspects. In: Nicoli, M.C. (Ed.), Shelf-Life Assessment of Food, first ed. CRC Press, Boca Raton, FL, pp. 1–16.
Nicoli, M.C., Calligaris, S., 2018. Secondary shelf-life: an underestimated issue. Food Eng. Rev. 10 (2), 57–65.
Robertson, G.L., 2010. Food packaging and shelf-life. In: Robertson, G.L. (Ed.), Food Packaging and Shelf-Life—A Practical Guide, first ed. CRC Press, Boca Raton, FL, pp. 1–16.
Stefansdottir, B., Grunow, M., 2018. Selecting new product designs and processing technologies under uncertainty: two-stage stochastic model and application to a food supply chain. Int. J. Prod. Econ. 201, 89–101.
Williams, H., Wikström, F., 2011. Environmental impact of packaging and food losses in a life cycle perspective: a comparative analysis of five food items. J. Clean. Prod. 19 (1), 43–48.

Chapter 5

A support-design procedure for sustainable food product-packaging systems

Riccardo Accorsi
Department of Industrial Engineering, Alma Mater Studiorum—University of Bologna, Bologna, Italy

Abstract
This chapter discusses the drivers and characteristics of the product-package system in food supply chains and focuses on the role that the logistic performance of the package plays in the sustainability of the supply chain operations. The product-package system is first introduced through the definition of a hierarchical structure built upon the product. Then, a set of design parameters are introduced, each of them addressing a specific product-package function/performance according to a multidisciplinary approach. An in-depth discussion is provided about the logistic performance of food packaging, and some support-design quantitative models and KPIs are introduced to the scope.

Lastly, built upon the aforementioned entities and parameters, a hierarchical closed-loop support-design procedure for economic and environmental sustainable food product-packaging systems is formalized, illustrated, and discussed.

1 Background

The role of packaging is widely recognized as a substantial part of the food industry, since it represents not just the container for food during the handling, storage, and transportation processes but it also contributes significantly to the modifications of the product's state along the whole supply chain and to consumer perception of quality and taste (Manzini and Accorsi, 2013). Both in theory and in practical applications, the food product and its package represent an inseparable system from the processing and manufacturing phase through to storage, handling, and distribution, until the end-of-life phases (i.e., reuse, recycling, recovery, disposal) (Accorsi et al., 2014).

The main functions/scopes of food packaging can be summarized as follows:

- *Protect the product* from environmental stresses (e.g., temperature, vibrations, impacts, light, humidity);
- *Protect the user* (in the case of food with dangerous levels of pathogens or contaminated produce);
- *Ensure the shelf life* and the quality of the product along its nominal lifespan;
- *Facilitate the handling, storage, and transport* of the food products;
- *Facilitate the use phase*;
- *Generate as little material waste* as possible during the end-of-life phases.

Given the multidisciplinary nature of these scopes, the design of a food product-package system entails a wide set of parameters and data collected along the food supply chain (FSC), which should drive the study and design phase. Fig. 1 briefly illustrates some of the potential data and parameters to be collected at different FSC stages in order to perform coherent and data-driven food packaging design activities. For further details about data fields, please refer to Accorsi et al. (2018).

To address these functions/scopes and to properly design a product-package system in such a complex and multientities ecosystem, distinguishing the elements composing such a system is necessary. Indeed, packaging is organized through a hierarchy that identifies four levels: primary, secondary, tertiary, and shipping levels. According to this hierarchy, the design of a step can be the input of the following step, and different functions/scopes are addressed at different levels independently but through an integrated approach.

These four product-packaging layers are treated in detail in the following sections.

1.1 Primary package

The primary package wraps and covers the food item and represents the selling unit of the product. Its primary function is to protect the products from environmental stresses that may lead to a decay in quality before the product is consumed.

Sustainable Food Supply Chains. https://doi.org/10.1016/B978-0-12-813411-5.00005-3

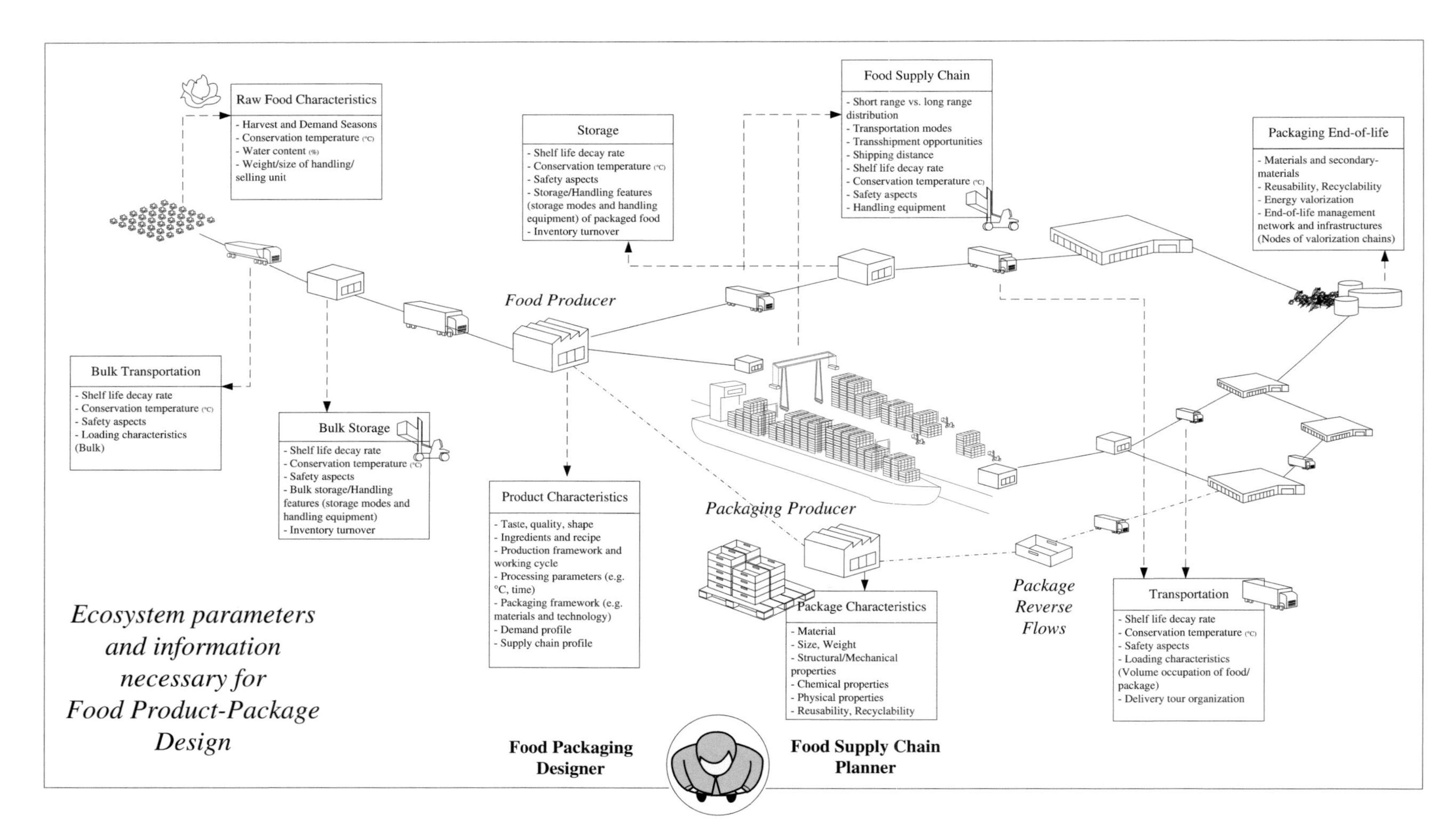

FIG. 1 Data and parameters collected on FSC systems to design sustainable food packaging.

The most common materials for the primary level package (in the following formalized as $m_{pkg}{}^{I}$) are glass or polyethylene (PET) for beverages and soft drinks; PET and Tetra Pak for dairy liquids (e.g., milk); polystyrene (PS) and PET as well as paper totes for fruit and vegetable items, or combined with polypropylene (PP) films for products of the fish and meat industries, and combined with aluminum foils for general dry foodstuffs and produce from the bakery industry.

Extended surveys of the most-used materials for food primary packages are provided by Piergiovanni and Limbo (2015) and Lee et al. (2008), who also illustrate their main characteristics and properties as well as their manufacturing and technological processes (Fig. 2).

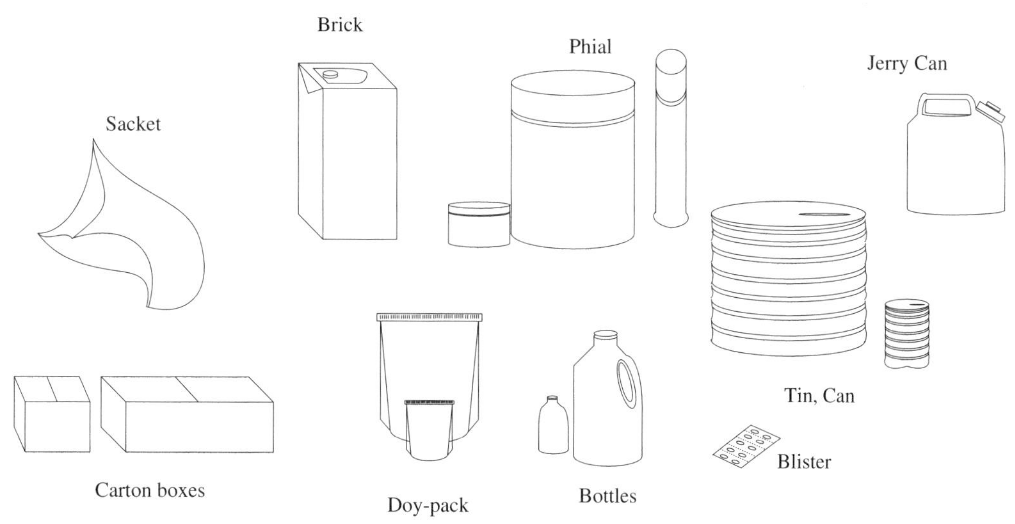

FIG. 2 Examples of food primary packages.

At this package layer, the main performance expected and to be optimized by the designer (in the following formalized as the performance $pf(p, pkg)$) based on the characteristics of product p and package pkg, $pf^{II}(p,pkg)$ involves the quality, materials, reliability, and sustainability of the package. The design of innovative primary packages is widely debated in the scientific literature, and particularly in the fields of food science and food packaging. The main research efforts are currently devoted to the design of edible primary packing able to prevent fast decay formation of the product and extend shelf life (i.e., edible coating) (Poverenova et al., 2014).

Other efforts, under the umbrella of smart packaging or intelligent food packaging, have been made to develop smart labels that track the state of conservation of the products, the experienced temperature profile, and other physical stresses occurring during handling/storage/transport activities; these detect if or when some stresses have affected the quality of the food or its freshness. Fig. 3 provides a representation of these innovative primary packaging solutions.

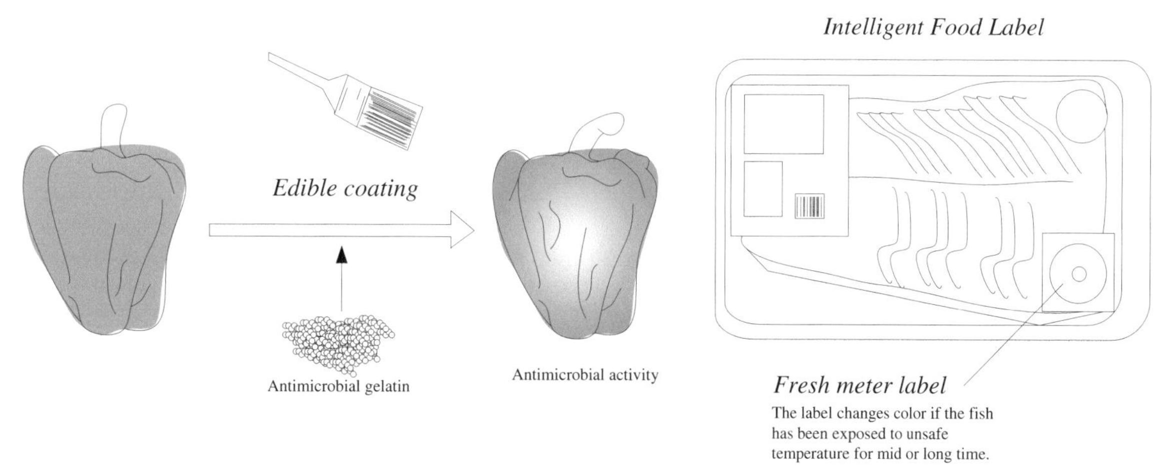

FIG. 3 Edible primary package (inspired by Poverenova et al., 2014) and intelligent food package labels.

Despite its primary and crucial role along the whole FSC, the primary package often becomes waste as soon as the product is consumed. As a consequence, despite its importance the design needs to take into account its environmental impact and address its environmental sustainability performance. This performance will be formulated later in the chapter as a numerical and quantitative function built upon the characteristics of the product p, the package pkg, their weight w, and the related materials, $CO_2(p, pkg, w, m_p, m_{pkg})$.

1.2 Secondary package

The secondary packaging contains single or multiple primary packages, and its role is to consolidate the units belonging to the same manufacturing lot, or more importantly to facilitate the handling operations at the manufacturing facility, the picking phase at the warehouse, and storage on the shelves at the selling point (i.e., market, shop, grocery). The most commonly used material for the secondary package (in the following formalized as m_{pkg}^{II}) is paper, usually paperboard, which can quickly assume the shape of cartons, boxes, and totes of different sizes according to the type of product and the logistic purposes (e.g., sizes consistent with the available space on the grocery stores' shelves) (Fig. 4).

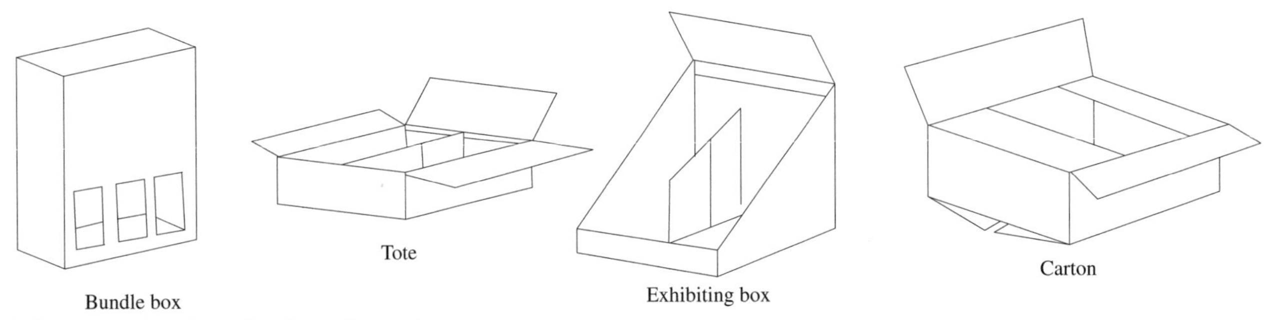

FIG. 4 Examples of paperboard secondary packages.

The volume (v), the size ($length \times width \times depth$) and shape ($s$), the weight ($w$), and more generally the ability of the secondary package to be handled/loaded (also known as logistic performance) all have a crucial importance in logistic operations, since they affect the level of utilization of unit loads, storage racks, and vehicle load, and thereby affect the optimization of handling and transport phases (Fig. 5).

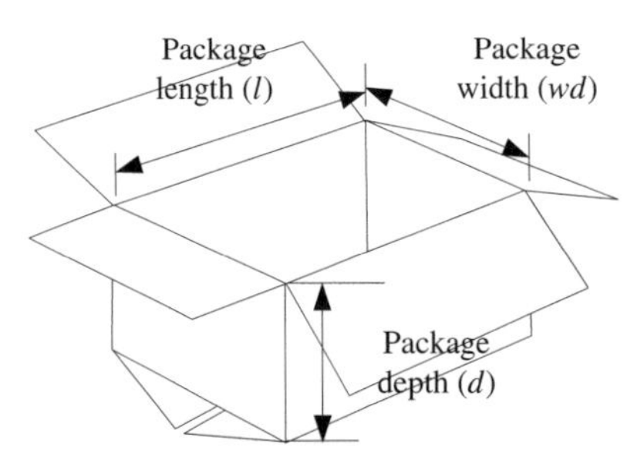

FIG. 5 Shapes and sizes of a paperboard box.

Other materials adopted for secondary packages are wood and PP, which are mostly used in the production of robust boxes for fresh fruit and vegetable products (see Fig. 6). In some supply chains, these boxes can replace the primary package, or can be used for the preproduction stages (see the left part of Fig. 1) when the food product is handled in bulk (i.e., harvest and consolidation phases).

Other examples of PP boxes are reusable plastic crates (as in the example of Fig. 7); these are characterized for being reused once the product is retrieved, instead of being disposed (see Fig. 1). Specific closeable banks allow minimizing of the occupied space when the crate is empty, thereby increasing the efficiency of the reverse collection flows (for further information about reusable crate logistics, refer to Chapter 20 of this book).

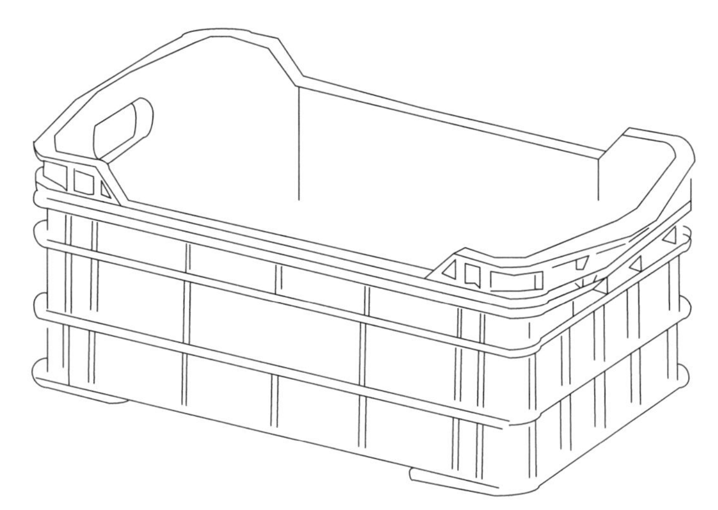

FIG. 6 Example of a rigid PP box/case for fruits and vegetables.

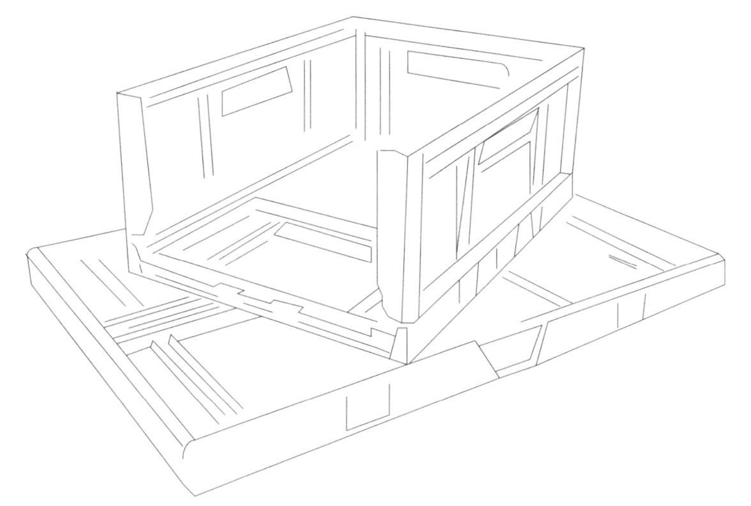

FIG. 7 Example of a flexible and closeable PP box/crate (open and closed) for food products.

1.3 Tertiary package

This packaging layer contains one or multiple secondary packages and is intended to facilitate the storage and transport of the product along the FSC. The tertiary package should be designed to be handled by forklifts, cranes, or other handling equipment and machinery. This is why it is commonly known as the *logistic* package. The materials used for tertiary packages (in the following formalized as $m_{pkg}{}^{III}$) are chosen for their robustness and ability to protect the load or facilitate handling activities. The most commonly used materials are wood and steel. Some examples of the most common tertiary packages used in the food industry are illustrated in Fig. 8.

As for the secondary layer of the packaging, the volume (v), the size ($length \times width \times depth$), and the weight ($w$) of the overall tertiary product–packaging system have a crucial importance in the logistic operations. Instead of being purely a part of the product, these packages are considered to be concrete logistic tools that join the products mostly during the storage and transportation phases.

In the view of its scope and function, the performance (again formalized as a function of the product and the package, $pf^{II}(p, pkg)$) to be optimized in the design of the tertiary packaging systems is the handling/loading capability (i.e., logistic performance) as well as the reliability in protecting food from mechanical stresses. Since it can be considered a service packaging, it is not delivered with food to the final consumers and typically returns to the FSC operators following a reverse packaging flow. In view of their reusability, the environmental sustainability performance of tertiary packages is implicitly considered at the design phase.

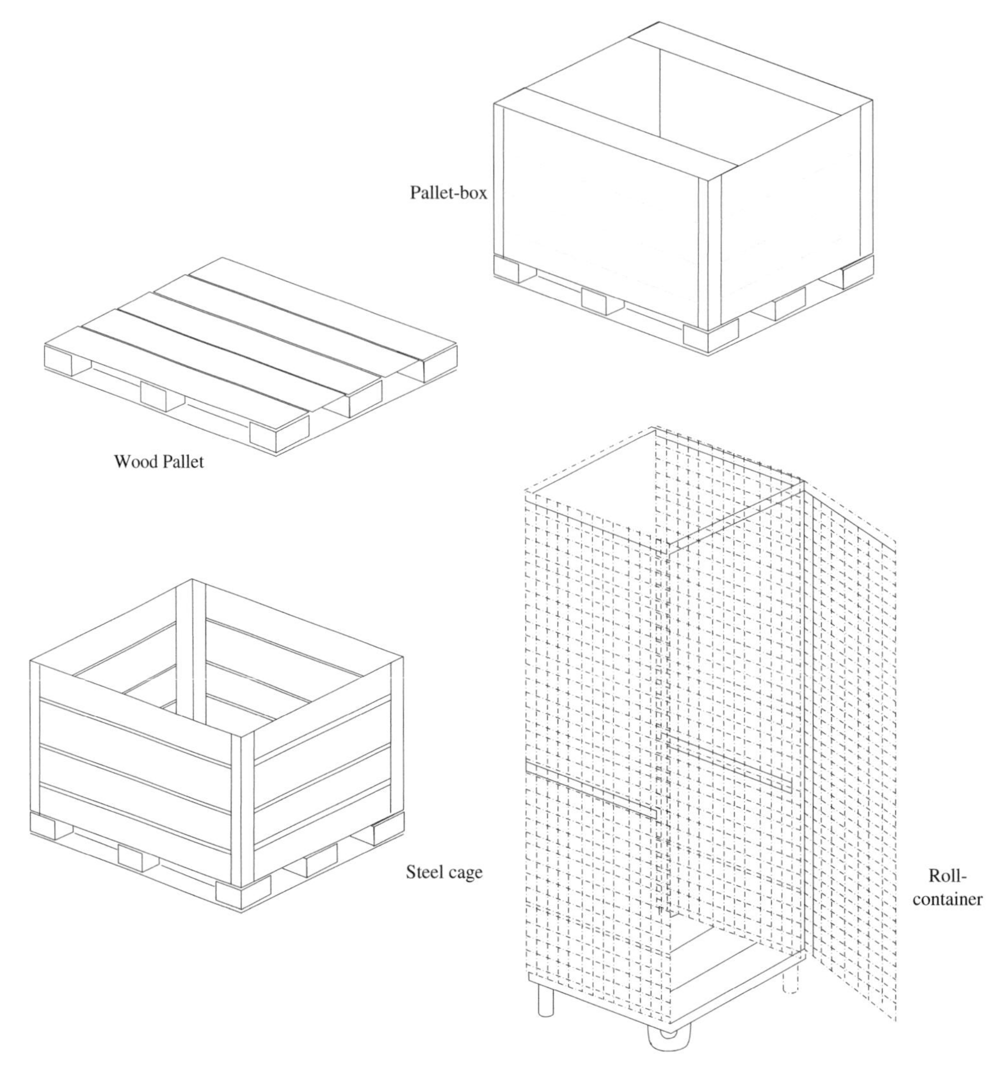

FIG. 8 Examples of tertiary packages used in FSC.

To this purpose, reverse logistic chains devoted to the so-called pooling of the tertiary packages exist (e.g., Chep, CPR operating in Italy). The pooling services include the supply of the packages (e.g., the pallet) to the manufacturers and vendors, their collection at the delivery selling nodes, their interchange and stock balancing among the nodes of the supply chain, and lastly their refurbishment and recovery in the end-of-life treatments. Fig. 9 provides a scheme of the most popular pallet circulation policies, i.e., exchange vs. pooling, that can also be used for other returnable tertiary packages.

1.3.1 Pallet

The most popular tertiary packaging is the pallet (Fig. 10). According to the rule UNI ISO 445, the pallet is defined as a horizontal plate able to be lifted and handled by hand-walker trucks, forklifts, or other generic handling machinery that enables sustaining, consolidating, collecting, storing, and moving goods and generic freight loads.

The pallet allows a flexible connection between the manufacturing phases and the storage and transport (i.e., logistics) operations and provides a standard that constrains and affects the design and optimization of the logistic processes. The pallet follows different standards (i.e., UNI 4121/88 and ISO 6780/78) that determine its sizes, as reported in Table 1.

Furthermore, the shape (s) and the weight (w) of the pallet can vary in accordance with the type of handling activities allowed. These alternatives are illustrated in Fig. 11.

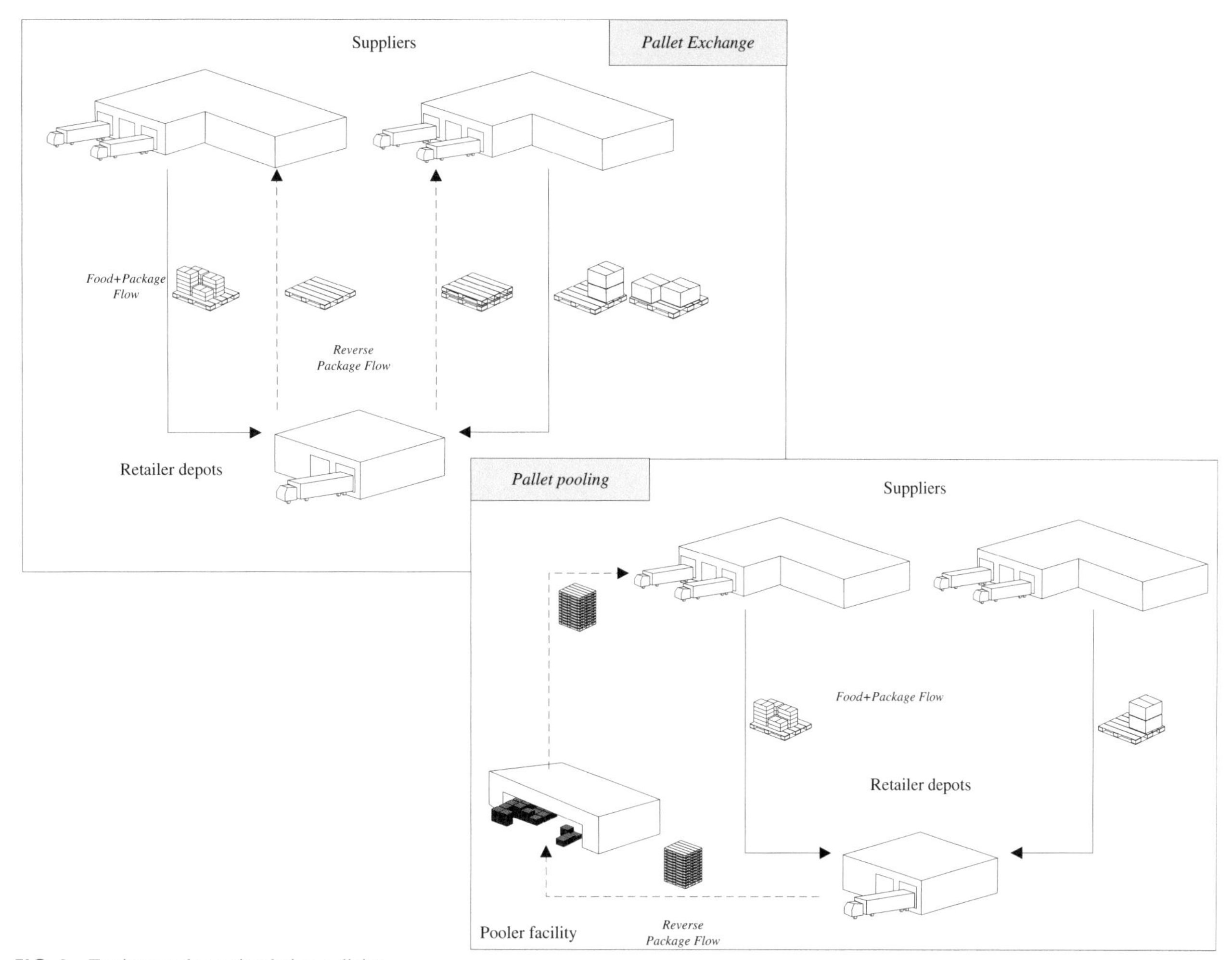

FIG. 9 Tertiary package circulation policies.

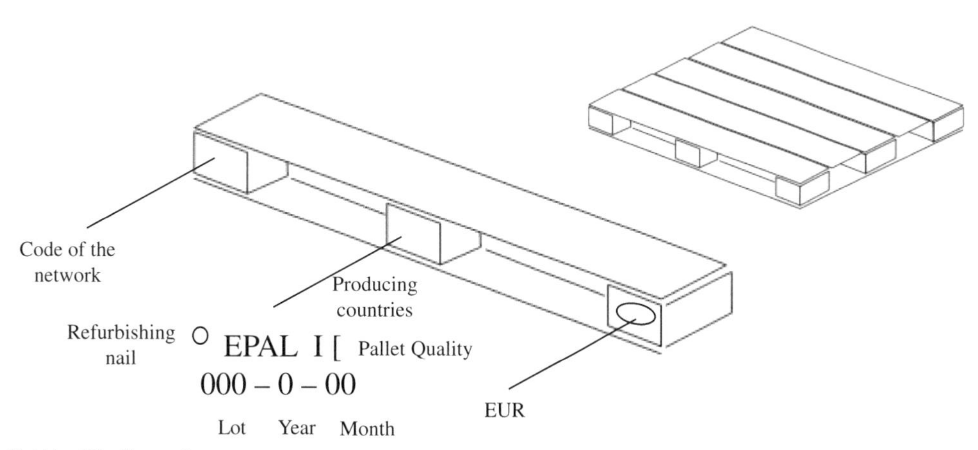

FIG. 10 The pallet identification codes.

TABLE 1 Pallet characteristics

Region	Size	Rules		EPAL characteristics		
		UNI 4121/88	ISO 6780/78	**Static load**	**4000**	**kg**
Europe	800 × 1000	X		Moving load	1500	kg
Europe (EPAL)—ISO1	800 × 1200	X	X	Pallet weight	20÷25	kg
U.K.—ISO2	1000 × 1200	X	X	Cost per pallet	8.5	€
Australia	1140 × 1240		X	Lifespan	7÷9	years
	1100 × 1100		X	Working cycles	50÷30	
USA	1216 × 1016		X			

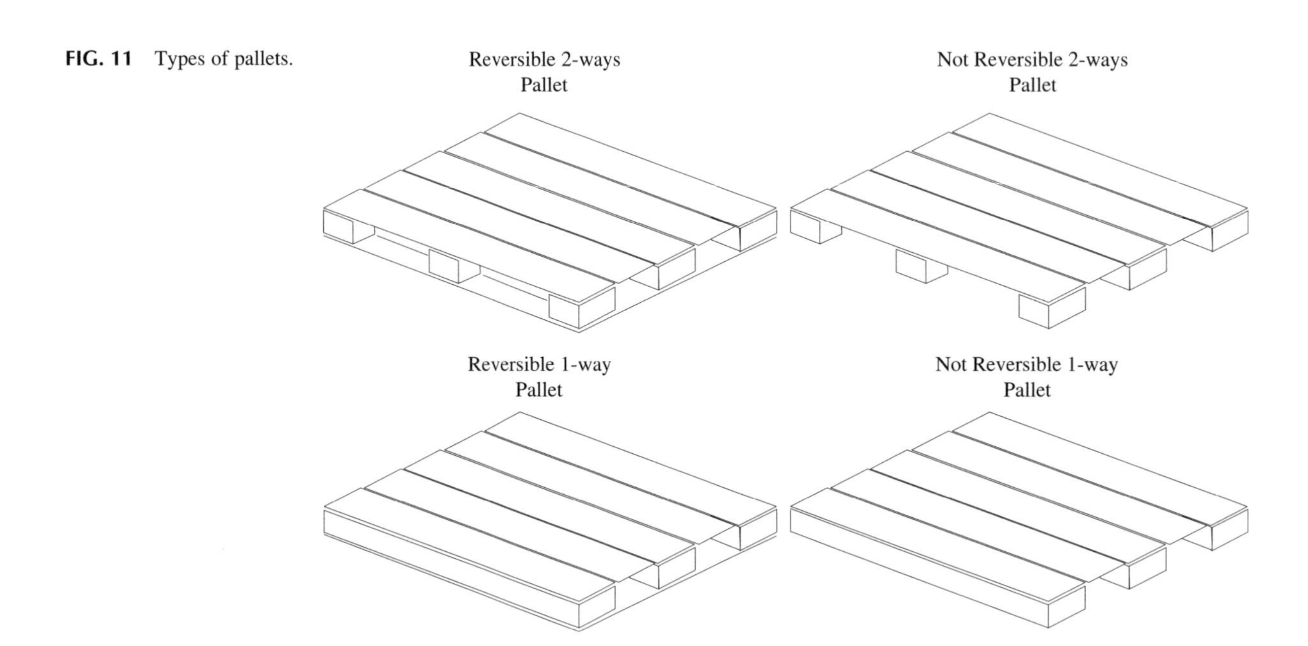

FIG. 11 Types of pallets.

1.4 Quaternary packaging (or shipping packaging)

The last package layer includes all the containment solutions that, filled by multiple loads (i.e., product-packaging tertiary systems) are shipped from a node (e.g., the manufacturing facility) to another (e.g., a warehouse or a demand point) along the supply chain. Like the tertiary package, this can be considered a sort of service package that supports, in this case, just the transport operations.

The shipping packaging requires the optimization of the loading/handling performance (formalized as $pf^{II}(p,pkg)$) that supports the minimization of the transportation costs as well as the protection of the shipped food from thefts, counterfeiting, contamination, and other environmental stresses (e.g., temperature, moisture, impacts and vibration, light exposure). The most common shipping containers are illustrated and described in Table 2, as proposed in Pareschi et al. (2002).

The illustrated four packaging layers combine to shape and configure the overall product-packaging system according to the packaging hierarchy, as in Fig. 12.

Specifically, the first three layers constitute the so-called unit load, which is the fundamental unit used for the design of the handling, storage, and transport equipment, systems, and infrastructures. The performance (i.e., $pf^{II}(p,pkg)$) of the unit load is determined by assessing how it matches with:

TABLE 2 Shipping packaging/containers

Dry TEU (twenty-foot equivalent unit) *container.* It is generally devoted to the shipment of dry food products without temperature conservation constraints. They can be *sealed* or *ventilated* and are usually made of steel (m_{pkg}^{IV}) This container has a: – size (*s*) of 8 × 8 × 20 feet – a volume (*v*) of 33 m^3 – a net load limit (*w*) of 18,300 kg – a pallet capacity of 11 ISO1 and 10 ISO2	
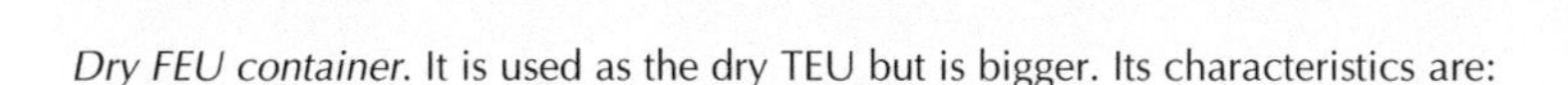 *Dry FEU container.* It is used as the dry TEU but is bigger. Its characteristics are: – size (*s*) of 8 × 8 × 40 feet – a volume (*v*) of 66 m^3	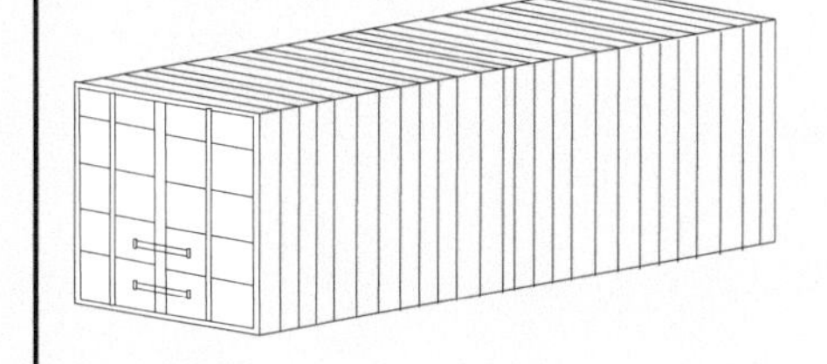
Bolster. It has the same size (*s*) as the dry containers but is used for voluminous commodities, typically transported in bulk (e.g., bales of hay)	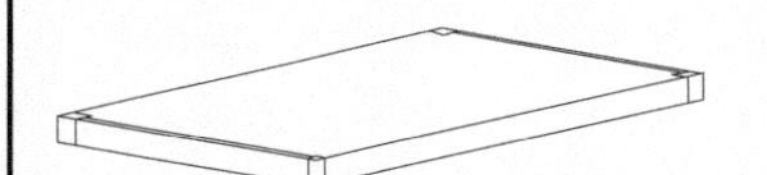
Open top container. These containers are used for bulk transportation. The top can be covered with a tarp to protect the contents from environmental stresses (i.e., rain and light exposure) It is generally used to transport *waste* and *scraps* collected during the reverse logistic operations of the end-of-life products	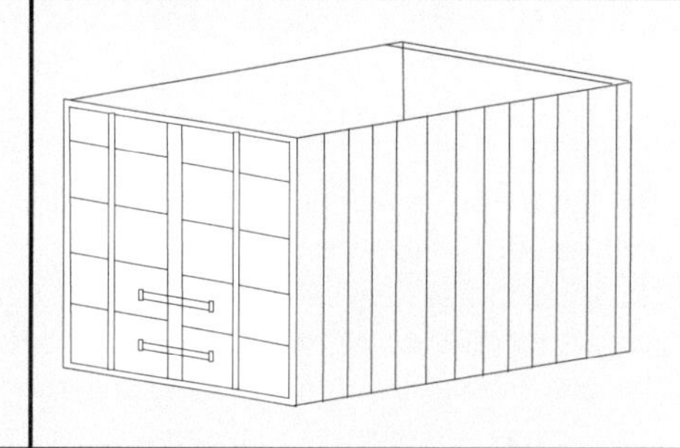
Flat container. These are used for food bulk distribution that exceeds the size of standard containers. Compared to the bolster, they are *stackable* and this property facilitates their management at the shipping docks (e.g., port yards, transhipping yards)	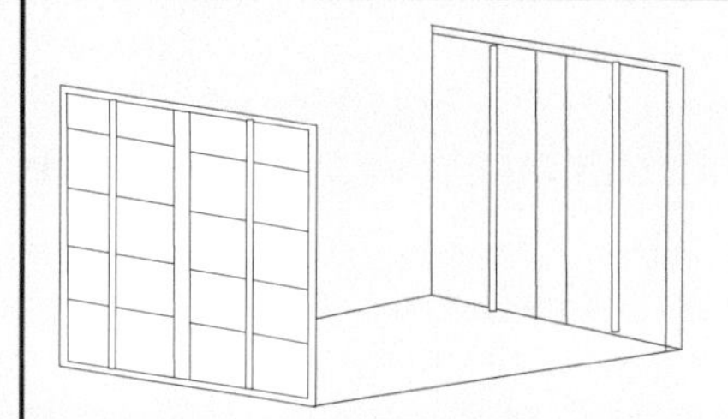
Refrigerated TEU/FEU container. These containers have the same characteristics in size and volume as the dry ones but are equipped with cooling systems that are able to conserve the fresh and frozen food at their proper conservation temperature. A detailed description of conservation temperatures for food items is given in Chapters 12 and 13 of this book These are also used for biomedical and pharmaceutical items	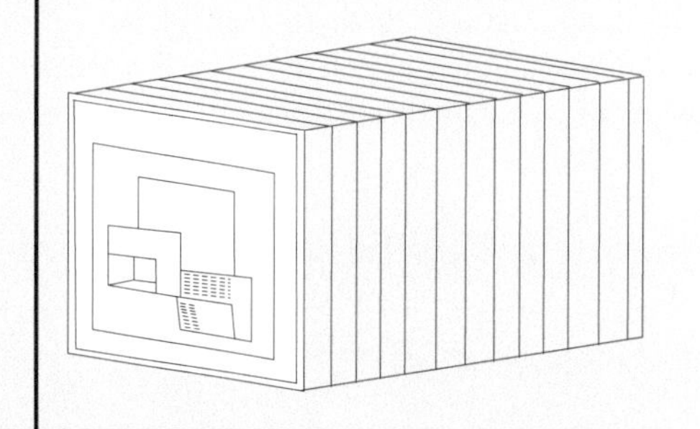

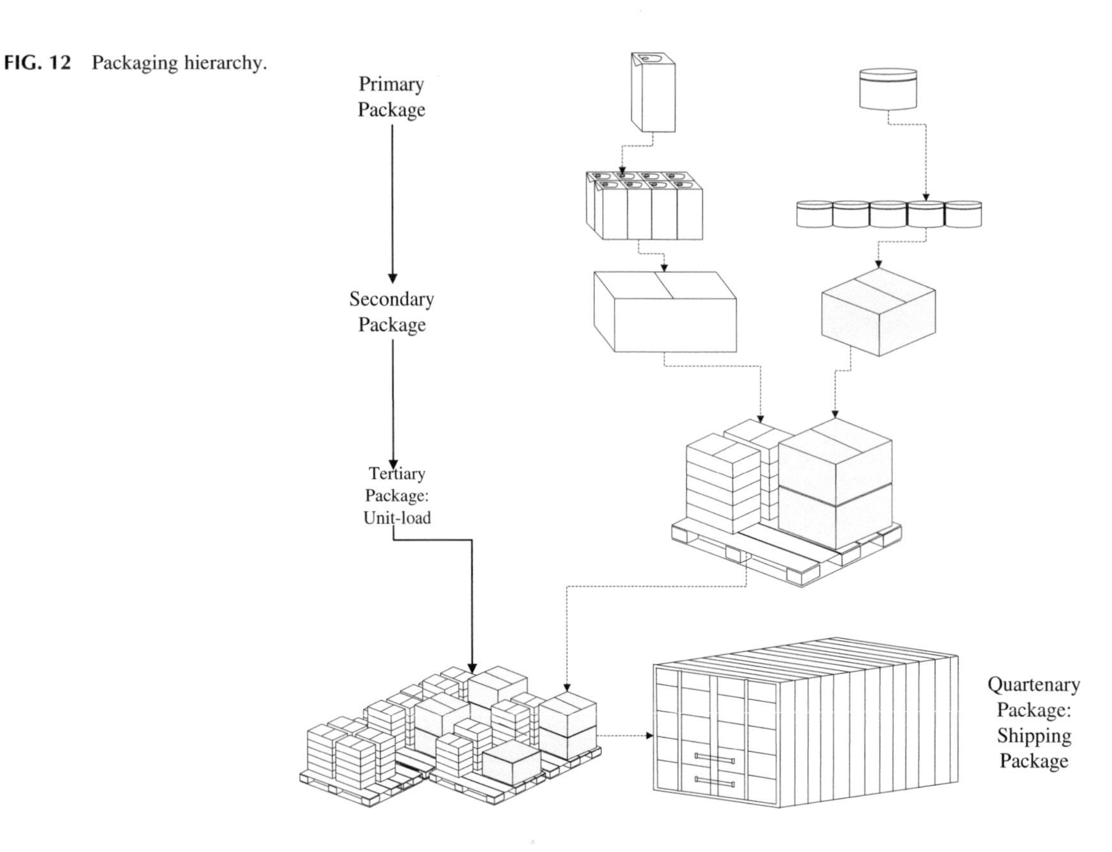

FIG. 12 Packaging hierarchy.

- the storage and warehousing infrastructures (e.g., height of the warehouse, maximum load tolerance of the racks, width and length of the storage locations);
- the transport systems and equipment (e.g., quaternary packaging and transport vehicles);
- the characteristics of product p and the secondary package (e.g., stackability, center of mass, safety, stability);
- the intrafacility handling systems and equipment (e.g., conveyor, forklifts, cranes, workers);
- the picking activities (i.e., manual retrieving by pickers).

2 Hierarchical support-design procedure

Given the importance that food packaging has in FSC planning, the study of a new sustainable food product-package system includes the whole set of activities carried out in the identification, definition, and study of the characteristics, the functionalities, the properties, and the requirements that the new product-package will have at each stage of the supply chain (see Fig. 1), starting from the potential demand profile and the objective consumer market. With so many aspects involved and decision drivers to consider, the study of a new product-package system could be aided by the use of multidisciplinary systemic methodologies that lead the designer through a set of subsequent tasks involved in the research, design, and engineering of food packaging systems (Azzi et al., 2012).

Here, an example of a closed-loop, top-down procedure is proposed and discussed in-depth. The tasks of the proposed procedure are classified into three macro-steps (as in Fig. 13):

i. Research (*i*)
ii. Development (*ii*)
iii. Engineering (*iii*).

The role of *Research* (*i*) is creating and inspiring innovation. New ideas for food products gathered from market analysis, consumer questionnaires, focus groups, suggestions from the company's research and development (R&D) department or personnel, or promoted by academic scientific research, are elaborated and assessed through a preliminary feasibility study.

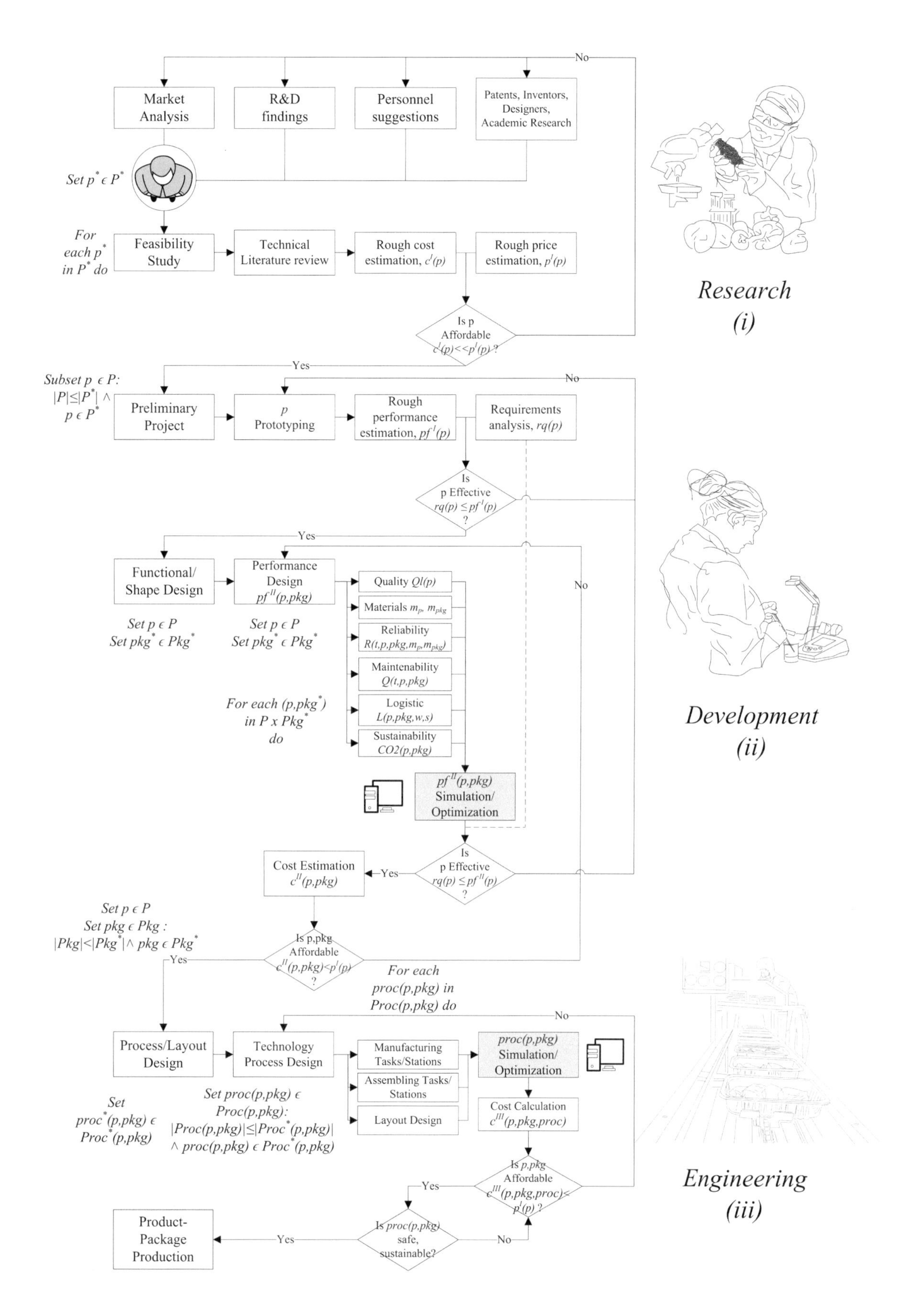

FIG. 13 Support-design closed-loop hierarchical procedure.

The feasibility study aims to identify a narrow range of concrete design choices (set $p \in P$) of a new product among wider opportunities and possibilities (set $p^* \in P^*$). The feasibility study consists of the accurate analysis of the state-of-art of a technical perspective (e.g., size, weight, features, market share, taste, flavor, shelf life, nutritional properties, harvest profile) of both the scientific databases (scientific journals, patent registers) and competitors' offerings.

This will result in identifying the main design boundaries and forming of a rough estimation (e.g., experience-driven, or comparison-driven) of the draft cost of product p, $c^I(p)$, to be compared with the expected selling price, $p^I(p)$, that the consumers would be willing to pay. The value of $p^I(p)$ can be estimated, for example, by the company's marketing department.

The feasibility study will be conducted for all the p^* opportunities to identify the minimum number of alternatives to be considered ($p \in P$: $|P| \leq |P^*| \wedge p \in P^*$) for the following steps of the closed-loop procedure.

The subset of the affordable design alternatives p (i.e., $p \in P^* : c^I(p) \ll p^I(p)$) that successfully make it through the feasibility study is handled through the *Development* (*ii*), and specifically, through the preliminary project.

The preliminary project is performed through the design of conceptual, virtual, and (sometimes) physical prototypes for each product-package p. The design and development of a product-package prototype will elicit the following aspects, which must be addressed and responded to by the designer:

- Functions and purpose (e.g., primary, secondary)
- Shape, size (s: l, wd, h), weight (w)
- Materials (m)
- Manufacturing/assembling tasks, also known as the working cycle.

By addressing these issues/decisions, designers first deal with the boundaries and limitations of the product-package design and exploit their feedback to roughly estimate the performance, $pf^I(p)$, that product p will provide.

The estimated level of $pf^I(p)$ is compared (qualitatively or quantitatively) with the expected requirements $rq(p)$ from the potential consumer of that product. These requirements are collected again via focus groups, questionnaires, or marketing analysis.

When the generic product $p \in P$ responds to the expected performance levels (i.e., $p \in P^* : rq(p) \leq pf^I(p)$), then the *Development* phase is improved by analyzing all the multidisciplinary aspects of the so-called product-packaging system. The product and its packaging are intended as two interdependent parts of the same object of design. Whereas the performances of the products are crucial in meeting consumer expectations, even the package has a central role for marketing purposes, for ensuring the conservation of the product's properties along the supply chain activities, and for the product handling during its manufacturing, storage, and transport.

In the functional/shape design, for each product $p \in P$ a set of packaging alternatives $pkg^* \in Pkg^*$ can be proved, tested, and compared. This step aims to assess all the feasible combinations of product-package (i.e., $P \times Pkg^*$) considering the following qualitative and quantitative panel of drivers and performance (illustrated in Fig. 14):

- Quality ($Ql(p)$)
- Materials ($m_p \in M^p$, $m_{pkg} \in M^{pkg}$)
- Reliability ($R(t, p, pkg, m_p, m_{pkg})$)
- Maintainability ($Q(t, p, pkg, m_p, m_{pkg})$)
- Sustainability ($CO_2(p, pkg, wt, m_p, m_{pkg})$)
- Handling/loading or logistics ($L(p, pkg, s, wt)$)

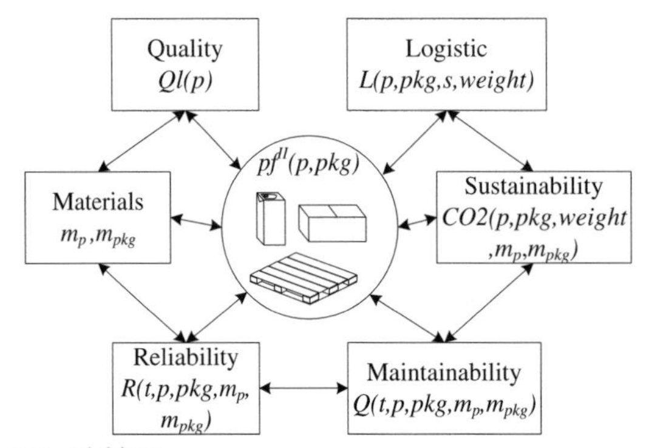

FIG. 14 Product-package design performance and drivers.

The set of decisions undertaken at this step deals with the quality, materials, reliability, maintainability, sustainability, and logistic performance of a generic product-package p, and allows the determination of a more detailed performance value, $pf^{II}(p,pkg)$, and the estimation of a more specific cost of the product, $c^{II}(p,pkg)$, for each pair of product and package solutions.

The quantification and assessment of both performance and cost values for each pair of p and pkg can be further facilitated and accelerated through simulation and optimization approaches. These can be implemented through software and computerized applications that use mathematical models to virtualize product-package systems and use algorithms and the computational capabilities of computers to simulate their behavior, study their response to external events, and optimize the performance. For a detailed survey of modeling and virtualization approaches and tools used in the development of food product-packaging solutions, refer to Chapter 6 of this book.

These approaches use an iterative method that generates and compares alternative design solutions and evaluates their quality, effectiveness, and affordability (i.e., $(p, pkg) \in P \times Pkg^*: rq(p) \leq pf^{II}(p,pkg) \wedge c^{II}(p,pkg) \leq p^I(p)$).

The adoption of mathematical modeling and virtualization is encouraged by the level of complexity of such comparative analysis, which usually increases with the number of combinations of $P \times Pkg^*$.

The last step of the procedure is the engineering phase, which is intended for the definition of the processing and manufacturing technologies and systems necessary for the production of the product-package system. This phase deals with the identification of the technological cycle and the facility layout required for the manufacturing of the product-packaging system made by p and pkg. The technological cycle is the set of processing, manufacturing, and assembling tasks necessary for the production from raw materials, while the facility layout comprises the type and number of working stations/resources, their location within the facility, and the handling systems and equipment for the management of the intrafacility material flow. Such decisions depend also on the expected demand profile that the new product-package will have. Three areas can be identified in the product-quantity curve (*P-Q* curve) of Fig. 15. This combines the company's production mix (the overall set of products realized by the company) with the demanded volumes of each product: the *I Area* is for small inventory mix produced in large volumes and is organized on rigid flow production lines; the *II Area* is for large inventory mix produced in small quantities through a food job-shop system (see Chapters 8 and 24 of this book); the *III Area* is intermediate and requires flexible and automated production and packing technologies, named flexible manufacturing systems (FMSs) built upon flexible manufacturing cells that operate on families of similar product-packages.

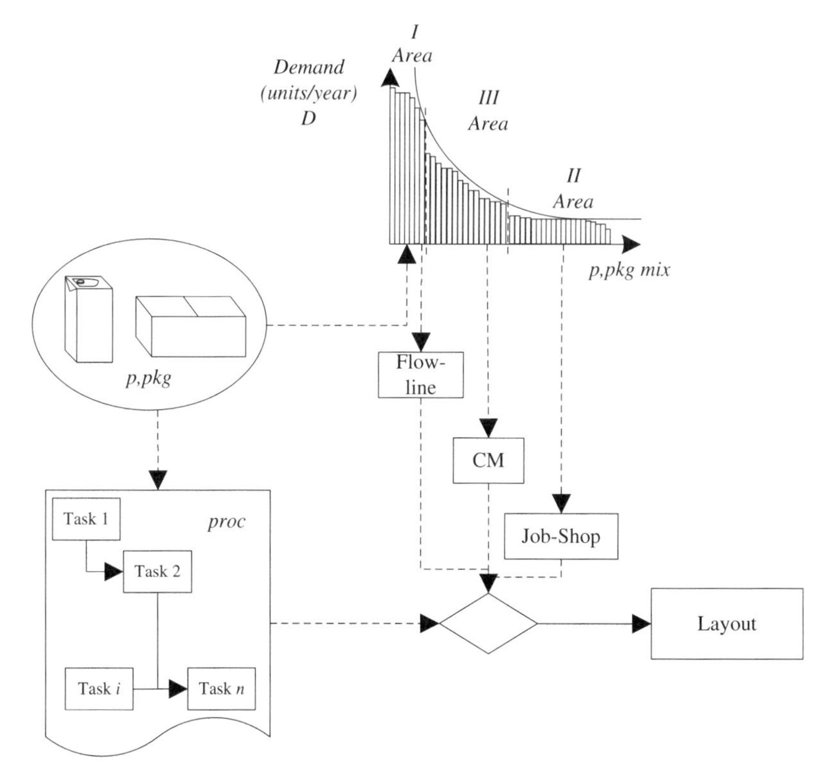

FIG. 15 Definition of manufacturing tasks and processes (*proc*) and the production system layout and configuration.

Once the production tasks are determined, a set of potential and alternative processes (i.e., $proc(p,pkg) \in Proc(p,pkg)$) can be proposed among a wider cluster of feasible solutions (i.e., $proc^*(p,pkg) \in Proc^*(p,pkg)$), and further evaluated in-depth. The combination of alternative manufacturing processes results in adopting different technologies, working stations, and facility layouts to address the requirements of the expected demand according to the well-known *P-Q* analysis.

Each manufacturing scenario $proc(p,pkg)$ is analyzed to calculate the final cost $c^{III}(p,pkg,proc)$, which could be further minimized again through optimization and simulation techniques via computerized approaches. The goal of this phase is to identify the best combination of product-package (p, pkg), materials (m_p, m_{pkg}), manufacturing, assembling and handling processes, and production layout ($proc$), which guarantees the satisfaction of efficiency, affordability (i.e., $(p, pkg, proc) \in P \times Pkg^* \times Proc$: $c^{III}(p,pkg,proc) \leq p^I(p)$), safety (i.e., product safety as well as for workers), and sustainability goals.

This step concludes the proposed hierarchical closed-loop procedure for the design of new product-packaging systems, which supports the identification of the optimal choice among an exponential set of design alternatives.

The educational purpose of this procedure is to provide a multidisciplinary tool to designers, engineers, practitioners, and marketing members that supports the design of new efficient, effective, affordable, safe, and sustainable food product-packaging. An in-depth discussion of the set of levers and design methodologies required at each step is far beyond the scope of this chapter.

With the hope of inspiring future debate on design methodologies and the improvement of the proposed framework, three general considerations on the nature of the proposed procedure follow:

- Hierarchical
- Multidisciplinary
- Closed-loop

The procedure is hierarchical to deal with the exponential complexity of the alternatives to be assessed and compared (i.e., $P \times Pkg^* \times Proc$). The top-down approach enables facing the decision levers by reducing, step by step, the number of alternatives to be evaluated. The procedure is multidisciplinary to incorporate different approaches, methodologies, tools, and goals from the different disciplines (e.g., mechanical engineers, marketing staff, environmental engineers, chemists and materials engineers, industrial engineers) involved in the study of the life cycle of the product-packaging system. Lastly, the procedure is closed-loop to feed back the decisions from the last step to the previous, when the obtained solutions are still not feasible, and to reiterate the decision process.

The following sections briefly discuss the levers of performance affecting the design of product-packaging systems, with particular focus on the handling/loading (i.e., logistics) properties.

2.1 Quality

The quality of the generic product p, $Ql(p)$, represents its goal market and reflects the expected standard of quality required by the consumers. It includes end results from a wide range of aspects and decisions, such as the quality of the raw materials and the product's components, the technology used for the manufacturing tasks, the precision and tolerance of the manufacturing stations, and so on. It also accounts for qualitative drivers, such as the company's brand or the market share of the competitors. The quality of product perceived by the final consumers is significantly affected also by the conditions experienced along storage and distribution phases (Valli et al., 2013; Ayyad et al., 2017).

2.2 Materials

The choice of the materials for the package, m_{pkg}, significantly affects the performance and the costs of a product-packaging system (p, pkg). The most common materials adopted for packaging, m_{pkg}, are paperboard, wood, and plastic polymers such as polyethylene (PET), polypropylene (PP), polyvinylchloride (PVC), metals, and glass.

2.2.1 Paper

The first paper material used in the packaging industry was corrugated cardboard that consists of a fluted corrugated sheet and one or two flat linerboards. It is used for boxes, cartons, and paper totes, and is classified by the structure (i.e., the number of faces) and the flute size and thickness, as reported in Table 3.

The main manufacturing tasks ($proc(pkg)$) for the production carton package are:

- Pasting
- Creasing/wedging/assembling
- Clipping

TABLE 3 Paperboard characteristics

Structure	Name	Flute Thickness (mm)	Flutes per linear m
	A Flute	4.8÷5	104÷125
	B Flute	3÷3.2	150÷184
Single wall	C Flute	4.0	120÷145
(Double faces)	E Flute	1.6	275÷310
	F Flute	0.8	>405
Double walls	AB Flute	6÷6.5	–
	BC Flute	7.5÷8	–

2.2.2 *Plastic*

Plastic packages are very popular for their lightness, robustness, reliability, and cost-efficiency, despite their massive environmental impacts, mainly due to the manufacturing phase (i.e., oil refinery and synthesis) and problematic end-of-life strategies. The main assembling processes ($proc(pkg)$) for plastic packages are:

- Pasting
- Heat sealing
- High-frequency assembling
- Ultrasound assembling
- Induction
- Spin-welding
- Wedging/mechanical assembling

2.2.3 *Wood*

Wood packages are still commonly used, despite the fact that many other materials currently have better performances from multiple aspects. They are mostly used as secondary packaging (i.e., boxes and cases) for unpacked food like vegetables and fruits, rather than for tertiary packaging (i.e., wood cases, pallets).

2.2.4 *Metal*

Among the metals, the most adopted for primary and secondary packaging are aluminum (Al); tinplate, which is a wrought iron or steel sheet that has passed through a process of coating and is further immersed into a stain liquid; and tinfoil. The important properties of aluminum include low density and therefore low weight, high strength, superior malleability, easy machining, excellent corrosion resistance, and good thermal and electrical conductivity. Aluminum is also chosen because it is completely recyclable.

2.2.5 *Glass*

Finally, *glass* is often used for liquids, particularly in the agro-food industry, for its robustness, safety, and sustainability properties (i.e. 100% recyclable).

2.3 Reliability

A product's lifespan is sensibly affected by the design phases, the selection of materials, the level of quality guaranteed by the manufacturing and assembling processes, and by the environmental stresses that occur during its life cycle. The combination of these aspects, and the resulting reliability, $R(t, p, pkg, m_p, m_{pkg})$, of a product-packaging system, usually influence the choices of consumers and their purchasing habits. The reliability of a product is achievable by:

- Reducing the complexity of the product-packaging system (e.g., the number of subcomponents);
- Adopting high-standard materials;
- Using back-up systems;

- Implementing a fair maintenance policy on the manufacturing and processing systems (i.e., a combination of corrective, predictive, inspective, and condition-based maintenance strategies);
- Adopting accelerated life test (ALT) analysis during the product-package design phase (Manzini et al., 2017).

The reliability of a product-packaging system, $R(t, p, pkg, m_p, m_{pkg})$, or simply $R(p, t)$ (i.e., $R_p(t)$) when the type of package pkg and the materials m_p and m_{pkg} are given for a generic product p, is the cumulative probability that a product produced at time 0 is still correctly working and performing at time t. The equations necessary to quantify the reliability of a generic product are reported from Manzini et al. (2009). These are based on the concept of failure, which is the event that breaks the nominal operating conditions (e.g., functionalities, performance) of a product. The failure event of the generic product p is, evidently, not deterministic, and we use the stochastic variable τ_p, called time-to-failure (TTF), to handle its nature.

Let $f_p(t)$ be the probability density function, for the generic product p, of the variable τ_p, defined as:

$$P\left(\tau_p \leq T\right) = \int_{-\infty}^{T} f_p(x)dx \tag{1}$$

where the normalization condition is as follows:

$$\int_{-\infty}^{\infty} f_p(x)dx = 1 \tag{2}$$

The function $f_p(t)$ is also called the unconditioned failure rate, since it measures, per each t, the velocity that the product p has in failing, if it starts working at time $t=0$. This definition leads to the following, given for the failure probability density, $f_p(t) \cdot dt$, the failure probability, $F_p(t)$, and the reliability function, $R_p(t)$, of the generic product p:

$$f_p(t) \cdot dt = P\left(t \leq \tau_p \leq t + dt\right) = \int_{t}^{t+dt} f(x)dx \tag{3}$$

$$F_p(t) = P\left(0 \leq \tau_p \leq t\right) = \int_{0}^{t} f(x)dx \tag{4}$$

$$R_p(t) = 1 - P\left(0 \leq \tau_p \leq t\right) = 1 - \int_{0}^{t} f(x)dx = \int_{t}^{\infty} f(x)dx \tag{5}$$

In the reliability theory, another fundamental metric is the hazard function $\lambda_p(t)$ defined for the generic product p and time t, as the velocity that the product p has in failing, conditioned to the fact that it is still working at t (i.e., $pf^I(p,t) = pf^I(p,0)$, which means that the performance of p is still as-good-as-new):

$$\lambda_p(t) \cdot \Delta t = P\left(t \leq \tau_p \leq t + \Delta t : pf^I(p,t) = pf^I(p,0)\right) \tag{6}$$

From its definition, it comes out that:

$$\lambda_p(t) = \frac{f_p(t)}{R_p(t)} \tag{7}$$

Furthermore, it can be demonstrated that:

$$R_p(t) = \mathrm{e}^{\left(-\int_0^t \lambda_p(t) \cdot dt\right)} \tag{8}$$

$$F_p(t) = 1 - \mathrm{e}^{\left(-\int_0^t \lambda_p(t) \cdot dt\right)} \tag{9}$$

2.4 Maintainability

The maintainability of a package pkg during its life cycle is the cumulative probability $Q(pkg, t)$ that a package broken at time 0 has been repaired and is correctly working at time t. This aspect directly affects the availability of a product p to be consumed and mostly to be used for a service (e.g., consider a working station of a manufacturing plant). If T_p is the total lifespan of a product p, the availability $A(p)$ is the ratio between the up-time UT_p (i.e., the time the product is ready for working) and its lifespan (where DT_p is the down-time):

$$A(p) = \frac{UT_p}{T_p} = \frac{UT_p}{UT_p + DT_p} \tag{10}$$

The higher the performance of a product-packaging system p, pkg in terms of reliability and maintainability, the higher will be the availability of the products and the satisfaction of the consumers.

2.5 Sustainability

The sustainability of a product-package system, $CO_2(p, pkg, wt, m_p, m_{pkg})$, is the measure of the impacts on the environment associated with a product along its lifespan. The chosen materials, m_p and m_{pkg}, have a clear effect on the sustainability of a product (e.g., plastic vs. glass bottles), but also the weight of a product-package system, wt, impacts the material flow as it is supplied, handled, stored, and transported along its life cycle (Accorsi et al., 2015; Manzini et al., 2014). There are several metrics and methodologies to measure and control the sustainability of a product (e.g., greenhouse gas emissions (GHGs)) and their proper adoption, case by case, is widely debated in the environmental engineering literature.

Among the others, life cycle assessment (LCA) is a renowned methodology to assess the environmental impacts associated with a product or process (i.e., a system) throughout its life cycle (i.e., "from cradle to grave"). The LCA methodology studies how the observed system affects the environment and natural resources, thereby supporting system improvements and strengthening more sustainable strategies. The LCA needs to comply with ISO standards (e.g., 14040:2006 and 14044:2006), which regulate the definition of the goal and scope of the analysis, the function unit (FU), the system boundaries, the collection of the overall input and output flows of the system, the data analysis, the quantification of the resulting environmental impacts, and the interpretation of the results. The LCA is conducted in accordance with a pattern or a set of rules (e.g., product category rules (PCR)), which provide the detailed specifications about what is included in the system boundaries, which FU to choose and which data to collect, in order to guarantee the comparability and replicability of the results.

The life cycle impact assessment (LCIA) is another step of this methodology, which deals with the definition of the categories of the methods and the impacts used to measure the sustainability of a product. These may include a panel of environmental metrics as follows (used by the method Environmental Design of industrial Products) (Hauschild et al., 2005):

- Global warming (i.e., gCO_2 eq./kg resulting from product p)
- Ozone depletion
- Photochemical ozone formation
- Acidification
- Eutrophication
- Nonrenewable energy consumption (i.e., kWh/kg consumed by product p)

The sustainability of a product-package system can also be assessed through the analysis of the end-of-life processes it will experience. Reuse, recycling, and recovery are certainly strategies that can enhance the sustainability of a product-package *p-pkg*, but these should be properly studied during the design process. Design approaches like design-for-disassembling (DFD) or design-for-recycling (DFR), if properly implemented at this design step, may result in more sustainable performances of the given product.

2.6 Logistic properties

The ability to handle and load a product-packaging system throughout the manufacturing, storage, and transport phases is crucial in the design of the sustainable logistic processes and the determination of the resulting costs. The handling/loading, i.e., logistic properties of a generic product-package system, are determined by the following decisions:

- Weight (wt)
- Volume (v)
- Shape and size (s)
- Property of being catchable
- Property of being damaged
- Physical and chemical properties
- Quantity per unit (q)

- Packaging hierarchy (discussed in Section 1)
- Packaging materials (m_{pkg})
- Package utilization

These factors are of fundamental importance in the design of the logistic processes and should be accurately taken into account during the illustrated hierarchical closed-loop design-support procedure.

2.6.1 Packaging hierarchy indicators

Among the logistic properties, some represent the constraints to be respected (i.e., weight wt, volume v, catchability, fragility), others represent the design factors (i.e., shape and size, quantity per unit q, materials), and others the so-called design performance (i.e., $pf^{II}(p,pkg)$).

The package utilization is indeed the most used quantitative metric for the performance of a product-packaging hierarchy. It is defined for each packaging layer as follows (the following notation is according to the definitions introduced in Section 1).

Sets

$p \in P$ products

$pkg^{II}{}_{c} \in Pkg^{II}P$ type ($_c$) of secondary packages (^{II})

$q^{I_p \to II_c}$ Number of units (q) of product p in secondary package $pkg^{II}{}_{c}$

$q_{pkg^{IV}}$ Number of units (q) (i.e., pallets) in quaternary package (e.g., TEU container)

$pl: 1, \ldots, q_{pkg^{IV}}$ Pallet pl in pkg^{IV} (e.g., TEU container)

Parameters

v_p: unit volume of product p

$v_{pkg^{I}{}_{p}}$: unit volume of primary package (^{I}) used for product p

$v_{pkg^{II}c}$: unit volume of secondary package (^{II}) of type c

Decision variables

$$y_{pc} = \begin{cases} 1, \text{ if product } p \text{ and related primary package } pkg^{I}{}_{p} \text{ are inserted into secondary package } c \\ 0, \text{Otherwise} \end{cases}$$

$$y_{cpl} = \begin{cases} 1, \text{ if secondary package } c \text{ is inserted into the unit } pl \text{ of the tertiary package } pkg^{III} \\ 0, \text{Otherwise} \end{cases}$$

$$y_{plpkg^{IV}} = \begin{cases} 1, \text{ if the unit } pl \text{ (i.e., pallet) is inserted into the quaternary package } pkg^{IV} \\ 0, \text{Otherwise} \end{cases}$$

Given the proposed notation a set of quantitative metrics/key performance indicators (KPIs) can be defined to describe the behavior of a product-package hierarchy system, as follows:

$$\eta_{p,pkg^{I}} = \frac{v_p}{v_{pkg^{I}{}_{p}}} \forall p \in P \tag{11}$$

$$\eta_{p,pkg^{II}{}_{c}} = \frac{q^{I_p \to II_c} \cdot v_{pkg^{I}{}_{p}}}{v_{pkg^{II}{}_{c}}} \forall p,c \in P \times Pkg^{II} : y_{pc} = 1 \tag{12}$$

$$\eta_{c,pkg^{III}{}_{pl}} = \frac{\sum_{c:y_{cpl}=1} q^{II_c \to III_{pl}} \cdot v_{pkg^{II}{}_{c}}}{v_{pkg^{III}{}_{pl}}} \forall pl : 1, \ldots, q_{pkg^{IV}} \tag{13}$$

$$\eta_{pkg^{IV}} = \frac{\sum_{l:y_{plpkg^{IV}}=1} v_{pkg^{III}{}_{pl}}}{v_{pkg^{IV}}} \tag{14}$$

The computation of the product-packaging system utilizations $\eta_{p,\ pkg^I}$, $\eta_{p,\ pkg^{II}c}$, $\eta_{c,\ pkg^{III}l}$, $\eta_{pkg^{IV}}$ allows assessment of the handling/loading performance in a quantitative, and thereby comparable, manner. This also contributes to better quantifying and refining the costs $c^{II}(p,pkg)$ and figuring out how they are affected by the logistics, handling, storage, and transport activities.

2.6.2 Packaging loading problems

Eqs. (10)–(13) are built upon the sets of indices, some design parameters (e.g., volumes v and unit quantities q), representing at the same time design levers and constraints, and three decision variables that represent the choices of the designer to combine different product and package alternatives into the same system. These decisions are simple when one or a few alternatives are allowed or merely taken into consideration. Nevertheless, when the number of alternatives increases, the problem complexity grows quickly until it is no longer approachable without support-decision tools.

Packing items (e.g., bricks) into larger objects (cartons) can be seen as similar to cutting objects (e.g., plates) to produce smaller items (ordered pieces). This topic has a long tradition in the scientific literature (Dowsland and Dowsland, 1992; Dyckhoff and Finke, 1992; Sweeney and Paternoster, 1992; Lim et al., 2000), which approaches to the so-called loading problem through the formulation of mathematical models and the development of solving algorithms to use in practice.

Solving the loading problem means to obtain the value, i.e., $\{0, 1\}$ of the decision variables y_{pc}, y_{cl}, $y_{lpkg^{IV}}$, that optimize the overall product-package system utilization.

Since these variables refer to decisions undertaken at different packaging layers, the associated mathematical models are usually formulated independently, and different independent problems result, as follows:

- Packager's items loading (PL) problem
- Manufacturer's pallet loading (MPL) problem
- Distributor's pallet loading (DPL) problem
- Single container loading (SCL) problem
- Multicontainer loading (MCL) problem

In the PL problem the decision to be made deals with the type of product p or primary package pkg^I to be packed in the secondary package pkg_c^{II} (i.e., y_{pc}), which pkg_c^{II} to adopt, and the position (p, q) of pkg^I in pkg_c^{II} with the purpose of maximizing the package utilization (i.e., $\eta_{p,\ pkg^{II}c}$). This problem is typical for packagers and is currently very common in the e-commerce supply chain (e.g., Amazon).

The MPL problem aims to optimize the utilization of the tertiary package pkg_{pl}^{III} when all secondary packages pkg_c^{II} are identical. This is the case of a manufacturer that produces goods packed in identical boxes of size (l, w, h), which are then arranged in horizontal layers on tertiary packages (e.g., pallets) of size (L, W, H). This problem decides which pkg_c^{II} is packed in the pkg_{pl}^{III} (i.e., y_{cpl}) and how to locate it. For a generic mathematical formulation for the two-dimensional MPL problem, please refer to Morabito and Morales (1998).

The DPL problem is similar to the MPL but occurs when we have multiple different secondary packages pkg_c^{II} not of the same size.

The SCL problem regulates how the tertiary packages pkg_{pl}^{III} are loaded into a quaternary shipping package (e.g., a truck, a container), while the MCL problem occurs when multiple pallets have to be loaded on multiple trucks with routing and delivery constraints.

The aforementioned loading problems, represented in Fig. 16, address the analysis and optimization of the product-packaging system according to the illustrated hierarchy. All food items have to be placed on pallets (i.e., pallet building) and all pallets have to be placed onto trucks (i.e., truck loading) or other shipping packaging. In the pallet-building phase the items are grouped in layers and the layers are stacked on the pallet base. A layer is an arrangement of items of the same product, composing a rectangle whose dimensions and number of items are known. A layer completely covers the pallet base in horizontal directions and other layers can be placed on top of it to make up the pallet. The layer composition of each product has been previously decided, so we have to stack layers to build pallets. Once the pallets are built, they have to be placed onto the trucks (or other quaternary shipping container) (Alonso et al., 2017; Junqueira et al., 2012).

Solving these problems in an optimal way produces a reduction in the handling, storage, and transportation costs for the company, thus increasing profits and also decreasing greenhouse gas emissions (i.e., $CO_2(p, pkg, wt, m_p, m_{pkg})$) and the other environmental impacts associated with food handling and transportation activities.

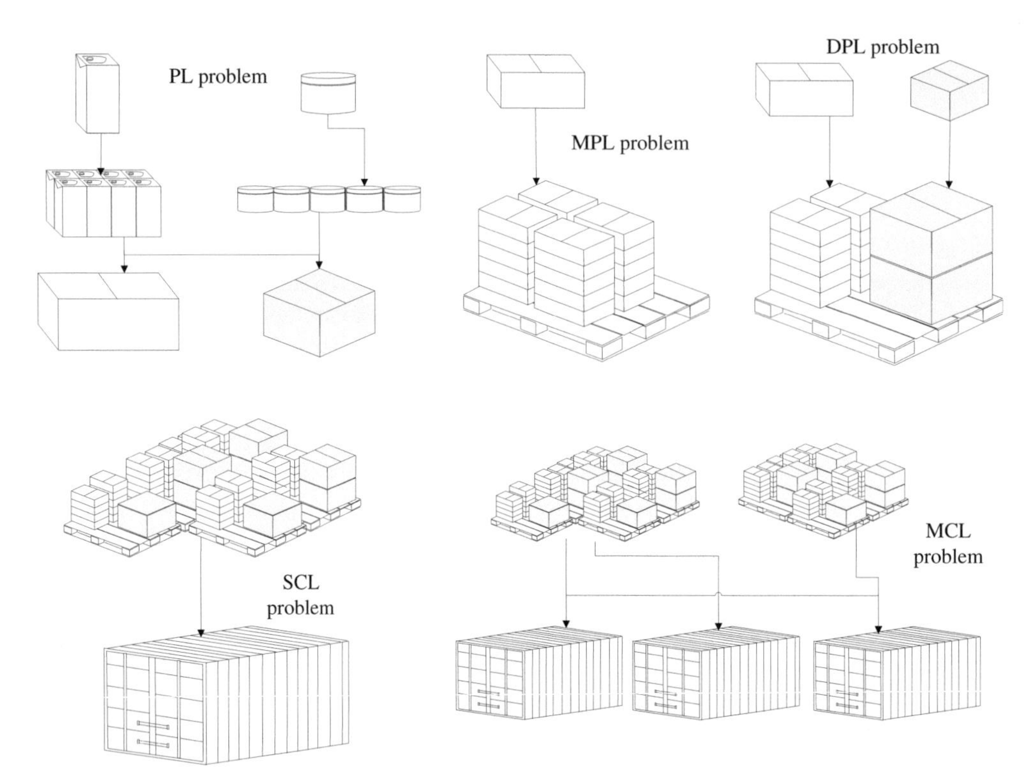

FIG. 16 Packaging loading problems.

3 Conclusion

The design of sustainable food packaging entails multiple and often conflicting decisional aspects. While the ability to protect the product and guarantee its shelf life is always critical, other drivers and measures of performance describing the behavior of a product-packaging system deserve attention with the purpose to guarantee the sustainability along the supply chain. These include the adopted materials, that can affect the waste management after its end-of-life, the reliability of package, and the logistic performance in term of volume utilization and ability to be handled. The latter influences the volume utilization during handling and transportation operations thereby affecting the environmental impact associated to food distribution. In order to tackle these aspects simultaneously, a hierarchical closed-loop support-design procedure has been introduced and described in this Chapter, which also illustrates in detail the decisional levers to involve in sustainable food package design.

References

Accorsi, R., Cascini, A., Cholette, S., Manzini, R., Mora, C., 2014. Economic and environmental assessment of reusable plastic containers: a food catering supply chain case study. Int. J. Prod. Econ. 152, 88–101.

Accorsi, R., Versari, L., Manzini, R., 2015. Glass vs. plastic: life cycle assessment of extra-virgin olive oil bottles across global supply chains. Sustainability 7 (3), 2818–2840.

Accorsi, R., Cholette, S., Manzini, R., Tufano, A., 2018. A hierarchical data architecture for sustainable food supply chain management and planning. J. Clean. Prod. 203, 1039–1054.

Alonso, M.T., Alvarez-Valdes, R., Iori, M., Parreño, F., Tamarit, J.M., 2017. Mathematical models for multicontainer loading problems. Omega 66, 106–117.

Ayyad, Z., Valli, E., Bendini, A., Accorsi, R., Manzini, R., Bortolini, M., Gamberi, M., Gallina Toschi, T., 2017. Simulating international shipments of vegetable oils: focus on quality changes. Ital. J. Food Sci. 29 (1), 38–49.

Azzi, A., Battini, D., Persona, A., Sgarbossa, F., 2012. Packaging design: general framework and research agenda. Packag. Technol. Sci. 25 (8), 435–456.

Dowsland, K.A., Dowsland, W.B., 1992. Packing problems. Eur. J. Oper. Res. 56, 2–14.

Dyckhoff, H., Finke, U., 1992. Cutting and Packing in Production and Distribution. Physica Verlag, Heidelberg, Germany.

Hauschild, M., Jeswiet, L., Alting, L., 2005. From life cycle assessment to sustainable production: status and perspectives. CIRP Ann. Manuf. Technol. 54 (2005), 1–21.

Junqueira, L., Morabito, R., Yamashita, D.S., 2012. Three-dimensional container loading models with cargo stability and load bearing constraints. Comput. Oper. Res. 39 (2012), 74–85.

Lee, D.S., Yam, K.L., Piergiovanni, L., 2008. Food Packaging Science and Technology. CRC Press.

Lim, A., Rodrigues, B., Wang, Y., 2000. A multi-faced buildup algorithm for three-dimensional packing problems. Omega 31 (2003), 471–481.

Manzini, R., Accorsi, R., 2013. The new conceptual framework for food supply chain assessment. J. Food Eng. 115 (2), 251–263.

Manzini, R., Regattieri, A., Pham, H., Ferrari, E., 2009. Maintenance for Industrial Systems. Series in Reliability Engineering, Springer.

Manzini, R., Accorsi, R., Ziad, A., Bendini, A., Bortolini, M., Gamberi, M., Valli, E., Gallina Toschi, T., 2014. Sustainability and quality in the food supply chain. A case study of shipment of edible oils. Br. Food J. 116 (12), 2069–2090.

Manzini, R., Accorsi, R., Piana, F., Regattieri, A., 2017. Accelerated life testing for packaging decisions in the edible oils distribution. Food Packag. Shelf Life 12, 114–127.

Morabito, R., Morales, S., 1998. A simple and effective recursive procedure for the manufacturer's pallet loading problem. J. Oper. Res. Soc. 49, 819–828.

Pareschi, A., Ferrari, E., Persona, A., Regattieri, A., 2002. Logistica Integrata e Flessibile. Progetto Leonardo, Esculapio, Bologna.

Piergiovanni, L., Limbo, S., 2015. Food Packaging Materials. Springer.

Poverenova, E., Zaitsev, Y., Arnon, H., Granit, R., Alkalai-Tuvia, S., Perzelan, Y., Weinberg, T., Fallik, E., 2014. Effects of a composite chitosan–gelatin edible coating on postharvest quality and storability of red bell peppers. Postharvest Biol. Technol. 96, 106–109.

Sweeney, P., Paternoster, S., 1992. Cutting and packing problems: a categorized, application-oriented research bibliography. J. Oper. Res. Soc. 43, 691–706.

Valli, E., Manzini, R., Accorsi, R., Bortolini, M., Gamberi, M., Bendini, A., Lercker, G., Gallina Toschi, T., 2013. Quality at destination: simulating shipment of three bottled edible oils from Italy to Taiwan. Riv. Ital. Sostanze Grasse 90 (3), 163–169.

Further reading

Hellström, D., Olsson, A., Nilsson, F., 2016. Sustainable development and packaging. In: Managing Packaging Design for Sustainable Development. Wiley. https://doi.org/10.1002/9781119151036.ch2.

Chapter 6

Designing advanced food packaging systems and technologies through modeling and virtualization

Fabrizio Sarghini*, Ferruh Erdogdu[†] and Riccardo Accorsi[‡]

*Department of Agriculture, University of Naples Federico II, Naples, Italy [†]Department of Food Engineering, Ankara University, Ankara, Turkey
[‡]Department of Industrial Engineering, Alma Mater Studiorum—University of Bologna, Bologna, Italy

Abstract

Finding an appropriate food package solution is a challenging task today, due to the necessity of integrating an increasing number of cross-linked and sometimes contradictory requirements related to the specific multiple purposes that a packaging system is supposed to provide. The complexity and variety of food products, stricter food quality requirements, rapidly changing consumer demand paradigms, increasing demands for sustainability, and the fast adaptations of multimodal food distribution networks to the modern way of living of our liquid society make food packaging development a difficult task in today's world. This chapter provides a taxonomy of existing research, design, and engineering approaches and technologies for food packaging and provides some examples of their application in practice. Such techniques provide wide support to designers and even food supply chain planners committed to proposing cost-effective, ergonomic logistics and environmentally friendly food packages.

1 Introduction

Modeling and virtualization (M&V) in any technological application have evolved to the level where they are able to support or even completely take over certain complex industrial processes, and packaging technology does not escape this rule.

Although M&V probably fall short of replacing completely the commonly used trial-and-error method, today they provide substantial support in research, design, and production of a food package (ten Klooster, 2008; Rundh, 2009; Svanes et al., 2010).

Finding an appropriate food package solution is a challenging task due to the necessity of integrating an increasing number of cross-linked and sometimes contradictory requirements, related to the specific multiple purposes that a packaging system is supposed to provide.

The complexity and variety of food products, the related stricter food quality requirements, the rapidly changing consumer demand paradigms, the increasing demands for sustainability, and the fast adaptation of multimodal food distribution networks to the modern way of living of our liquid society make development of a food package a quite difficult task. As the complexity of the system increases, traditional design and production methods become less and less suitable for providing a rapid response to the aforementioned issues and to the need to consider the interdependencies between the steps of research, design, and production (Mahalik and Nambiar, 2010).

2 Packaging research, preliminary design and production

The requirements and performance of modern packaging systems impose a balance between functional aspects, like the need for food protection (e.g., specific barrier properties), and regulations on packaging waste oriented towards a circular economy approach (European Union, 2004, European Packaging and Packaging Waste Directive 2004/12/EC). Indeed, there is a growing demand for innovative packaging systems designed to be lighter, thinner, and with reduced environmental impacts (Sirkett et al., 2007).

Sustainable Food Supply Chains. https://doi.org/10.1016/B978-0-12-813411-5.00006-5

Fast development of innovative systems requires appropriate methods and reliable tools, for example, introducing virtual frameworks resembling physical systems in silico or using modern manufacturing technologies (Chwastyk and Kolosowski, 2012).

The term "computer-aided" in food manufacturing systems was introduced by Saguy (1983). The computerization of mathematical models in food process engineering, however, started with the study of Teixeira et al. (1969) in which a finite difference numerical approach was used to predict the temperature and related changes in food products. Nowadays, advanced and integrated computerized platforms incorporate these models and provide several functional targets for the food package designer:

- Facilitate, optimize, and validate preliminary packaging design via computer-aided design (CAD) and virtual reality (VR), i.e., digitally defining, creating, and analyzing (the acceptance of) the system's geometry (shape, dimensions) and appearance (artwork, labels, text, colors).
- Predict and analyze the performance and behavior of new food packaging or packages using computer simulation and optimize production process efficiency and determine optimal production process parameters via computer-aided engineering (CAE).
- Support the multiple target research, design, and production approach and enhance the structural and functional performance of food packaging.
- Possibly improve package design, testing, and production processes, eventually resulting in fully digitized processes under the framework of computer-integrated manufacturing (CIM) where data file formats and transfer protocols are standardized so that easy file and data exchange between different computer systems is possible (Cubonova and Kumicakova, 2005).

Fig. 1 presents a description of the different computer systems applied in the research, design, and production stage, their role in the realization of the different packaging functions, and their interrelationships.

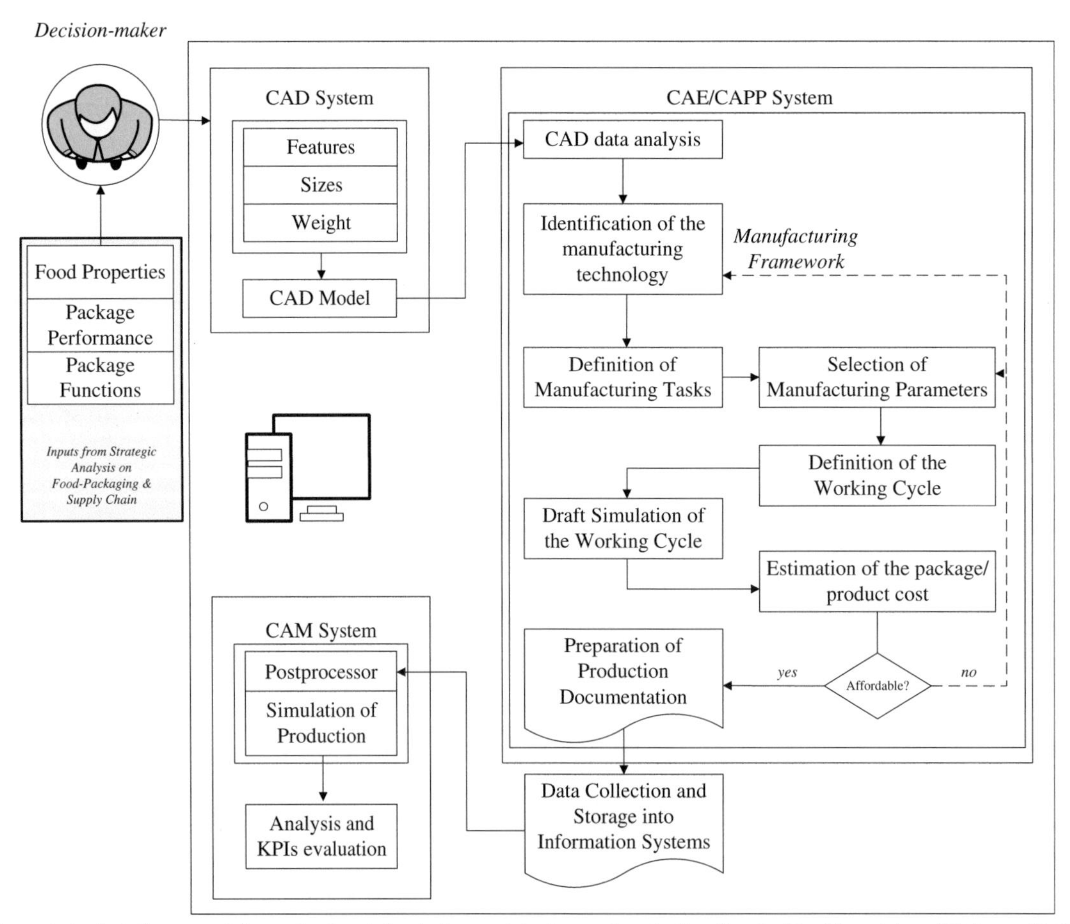

FIG. 1 Computerized platforms used for packaging study, design, prototyping, and production (inspired by Chwastyk and Kolosowski, 2012).

Integration of technologies is a key factor in making multitarget approaches easier to implement. For example, currently an increasing number of computer systems and architectures provide integration of CAD, CAE, and/or computer-aided manufacturing (CAM) functionalities to fulfill certain requirements. In such integrated systems, CAD models are used in CAE systems, for example, to calculate mechanical strength, while CAE results are used in CAM to automatically configure and control its related production processes.

2.1 Modeling and technological tools for packaging systems

After food production, the packaging is essential for containment, protection, and convenience of design and communication; therefore, optimization of the package design in a specific way for each food product is required. The four essential functions of food packaging can be defined as:

- Protection,
- Communication,
- Convenience, and
- Containment.

However, society has been demanding further innovative and creative approaches to guarantee safety and traceability as well (Vanferrost et al., 2014). Given the wide range of options and solutions, mathematical modeling approaches for the design of effective process-packaging play a significant role.

In the cycle of research, design, and production of packaging, CAD and CAE, through the implementation of mathematical modeling, have led to investigating the effects of changing mechanical/structural/physical parameters and optimizing two- and three-dimensional structural features, thereby addressing the mechanical complexity of the package. In addition to the given complexity of the packaging, changes in the food product during storage as a function of the packaging material (due to the creation of a microenvironment within the package and the relationship with the changes in the atmospheric conditions) might be further predicted. Package environmental considerations include the physical conditions that the package might encounter, i.e., the ambient environment including the presence of oxygen, moisture, odors, and microorganisms, and the human environment that is concerned with the human aspects (Rutgers, 2012). The first two cases directly become potential subjects for use of mathematical modeling approaches and virtual tools. Within such a framework, many similarities exist between food processing and packaging manufacturing. Mahalik and Nambiar (2010) underline this similarity as well as the challenges for planning food supply chain decisions in agreement with the trends of food packaging and food processing technologies.

The use of CAD and CAE with a mathematical model in the background digitally creates a virtual environment in which the changes in the product and package might be determined/predicted for different conditions (i.e., multiscenario, what-if analysis).

Vanferrost et al. (2017) presented a detailed review of the digitization of food packaging focusing on research/design based on CAE for manufacturing and quality control. As summarized by Paine and Paine (1992), the eventual ideas behind design of packaging using virtualization/CAE might be to reduce the unit cost in the production line, promote product sales, increase turnover/profit, improve shelf life (increase control of safety of quality), extend the market, provide customers a better method of using the product, and improve handling in transport. Within this concept, designing packaging material with optimum size/geometry, production ease, and convenience demonstrate the challenges in design considerations. Hence, modeling and technological tools become a requirement. A relatively high number of modeling tools are available for the aforementioned activities. Some of them are interdependent, as the output of one tool can be used as input for the following one, in agreement with the food supply chain stream illustrated in Fig. 2. The following sections explore the main virtualization and modeling techniques and approaches used in food packaging design (see Fig. 3) and describe the characteristics that constrain and elicit or limit their use.

2.1.1 Virtual reality

VR describes a three-dimensional, computer-generated environment that can be explored and interacted with by a person. The decision maker here becomes part of the virtual world or is immersed within this environment and, while there, is able to manipulate objects or perform a series of actions (Fauster, 2007; McMahan et al., 2012).

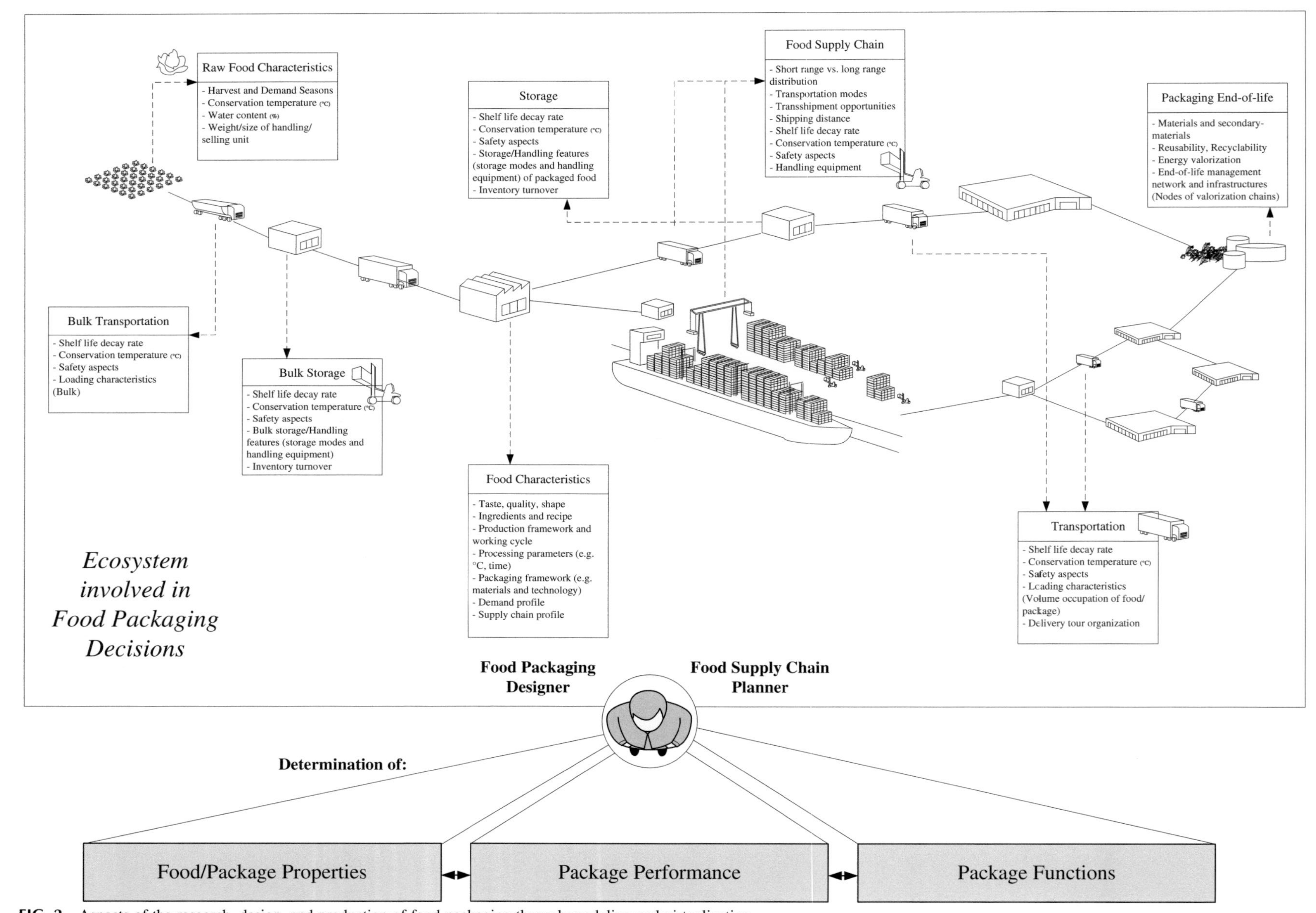

FIG. 2 Aspects of the research, design, and production of food packaging through modeling and virtualization.

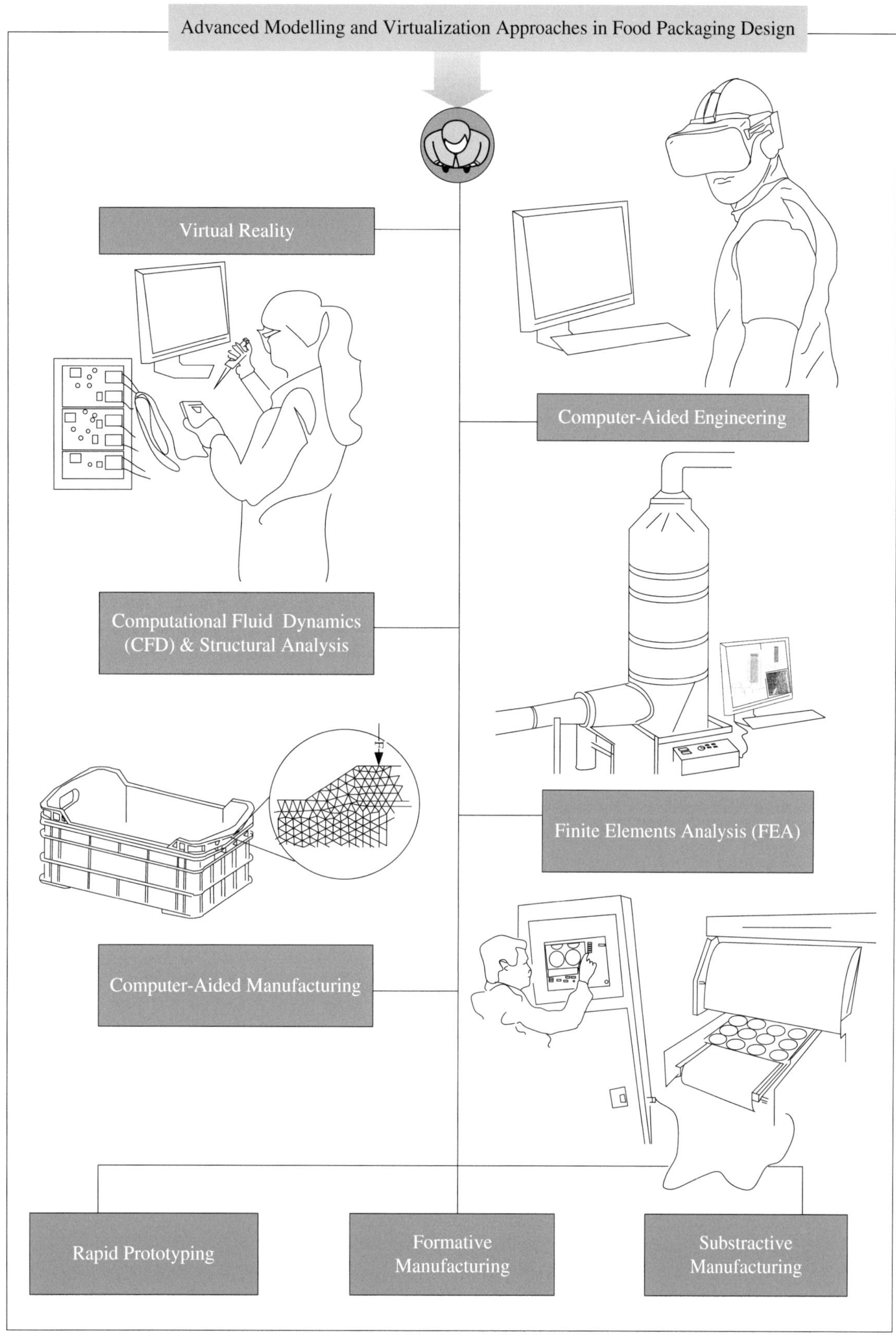

FIG. 3 Taxonomy of main modeling and virtualization approaches used in food packaging design.

Virtualization has been a popular term in information technology and computer science, while in food process engineering, a wide range of activities, from simple equations of curve fitting to complex physics-based models, have been included under this umbrella (Erdogdu et al., 2017a, b). Hence, VR in food packaging design might be considered an advanced tool for CAD and CAE with the support of comprehensive mathematical models. With the development of food process modeling, VR in food process engineering can be implemented on physics-based mathematical models in which the required transport equations are solved by applying initial/boundary conditions with physical properties set on the base of the virtualized process.

The use of comprehensive mathematical models allows changes in the food product to be explored with respect to the variation of the package microenvironment, so that predicting the various impact factors in terms of food quality and safety is feasible at an earlier design phase. As a consequence, optimization of the product-package performance and parameters becomes possible.

The mathematical models implemented through VR, more than with other techniques, must be interdisciplinary, as they should be built using a combination of bioprocess engineering (e.g., physiology of food in some cases, such as modified atmosphere packaging), physical and chemical properties of the packaging materials, and optimized design data for the packaging system (Belay et al., 2016). Various modeling approaches, combining qualitative and quantitative methods and focusing on the design of a food system as a whole, are illustrated by Lara et al. (2018), who focus on alternative food packaging scenarios.

For modeling purposes, depending upon the complexity of the problem, exact solutions for the given differential equation set (with simplified assumptions on the initial and boundary conditions) and numerical methods provide efficient and powerful tools to simulate the process and predict the changes in the required variables. For example, the diffusion of a substance from packaging material to the food might adopt the Fickian diffusion model to provide the exact solutions within and for a simplified geometry.

Among the numerical solutions, while finite difference methods have been widely applied, especially for regular geometries with isotropic properties, finite element and finite volume methods are more flexible in addressing such issues. All these numerical methods can be described under the weighted residuals umbrella of the Petrov-Galerkin methods, and their mathematical implementation commonly rules out most of the design commercial software. Computational fluid dynamics (CFD) has become the general term for the commercial and open-source codes, whether or not the applied problem includes a fluid problem, but all these might be assumed to fall under the definition of virtualization with CAE.

With the aid of advanced imaging techniques and modern display, sound, and position tracking technologies, VR is currently able to offer designers the experience of being totally immersed in a virtual environment, isolated from the real world, where they can move around and interact with different objects (Bowman and McMahan, 2007).

VR applications in the packaging design process introduce the possibility of obtaining direct feedback about the design without the need for a physical prototype. Effects related to shapes, dimensions, and color can be analyzed by designers, engineers, or even consumers by a direct interaction in order to optimize appearance and functionalities. Notwithstanding the modeling aspects, parametric CAD design is a prerequisite to enable a better and more straightforward adoption of VR systems in the food packaging industry (Lorenz et al., 2016; Whyte et al., 2000; Marks et al., 2014).

For the design of food packages, VR has been successfully applied by scholars and academic researchers to investigate the influence of a package's features (e.g., shape, color) on users' preferences or perceptions (Hsiao and Chen, 2006; Katicic et al., 2011, 2013; Ayanogolu et al., 2013). Moreover, VR allows investigation of consumer behavior by capturing physiology and brain response (Noldus Information Technology, 2012), using technologies that are able to track, monitor, and log human behavior and responses, such as electroencephalography (EEG) or eye tracking (ET).

Applications of such approaches have already been introduced in the industrial world. Nonetheless, applications of VR in the research and design process of a food package can be considered still at the preliminary stages, waiting to be integrated with other tools and equipment. Some examples include haptic devices like sensing gloves, allowing a more realistic immersion into the virtual environment by more realistic tactile and force feedbacks (Falcão and Soares, 2013, 2014; Smurfit, 2013).

2.1.2 Computer-aided engineering

This class of tools includes computer software used to analyze the performance of real objects or systems represented through a CAD geometry/file. CAE tools are being used, for example, to analyze the structural robustness and performance of components and assemblies. The term includes simulation, validation, and optimization of products and manufacturing tools.

CAE systems are mostly used for examining a parametric design's performance and behavior at a macrolevel, determining the production process efficiency and optimal production process parameters through in-silico analysis, for different conditions. For food packages, two main applications of CAE can be considered:

1. Process analysis,
2. Phenomena analysis.

2.1.2.1 Process analysis

In the packaging sector, the virtualization of process analysis allows investigation in silico of the effects of different operative parameters in packaging production. This approach is gaining momentum in the plastic and metal food package industry, where a growing number of commercial software applications are available. These include ANSYS Polyflow, Moldex 3D, Moldflow, PTI Virtual Prototyping for plastic, ADINA, Inventium, LSTC LS-DYNA, HyperForm, and AutoForm for metal-based processes, just to mention the most important current tools for *a-priori* production process analysis. The information obtained from these tools can be used by manufacturers to change or optimize/upgrade existing production processes or to develop new ones.

2.1.2.2 Phenomena analysis

Phenomena analysis through CAE simulations introduces the possibility of making predictions about a design's structural and functional performance, which are carried out by physical phenomena such as fluid flow, heat or mass transfer, and mechanical deformation. Today's extensive and cheap computational power allows rapid responses to the changing market requirements (Sirkett et al., 2007).

For example, integrated approaches in secondary packaging allow the study of air flows by using computational fluid dynamics (CFD) and at the same time the study of the mechanical deformation of a package due to an external force by using finite element analysis (FEA). The most important CAE problems can thus be solved by using either CFD (modeling fluid flows) or FEA (modeling stress, deformation, or heat diffusion), or a combination of both.

Although CAE systems are commercially available with friendly graphical user interfaces, they should be used appropriately, and some engineering skills and technical background are necessary. The reliability of a numerically assisted simulation is strictly linked to verification and validation, requiring educated and skilled personnel to ensure that a certain model is a good representation of the real world and to avoid results being blindly accepted. The acceptance of CAE as a valuable alternative to physical testing (Itech-soft, PALLET-Express, 2014) is strongly related to the possibility of assessing results critically. If the verification and validation stage is correctly performed, then CAE offers the possibility of dedicating more effort to the definition stage (early stage) of the design process, rather than using a trial-and-error approach.

A balance between an ultra-detailed representation of the real world and computational/solving of the model is often required. A golden rule is to model the problem with the minimum complexity required with respect to the proposed target, but no less. This is because the more detailed the model, the more time and computational power needed to obtain accurate results through simulation.

In using CAE tools, two major sources of errors are reported: one is linked to material properties characterization, and the second one is strictly linked to the capacity of technical knowledge of the personnel involved to properly use the numerical tools. This is because the problem of results reliability is linked not only to the software tool adopted but also to some extent to modeling choices like grid resolution, or even to mathematical properties of discretization schemes.

Regarding the first one, each model requires input parameters (e.g., packaging material properties) that are either estimated or experimentally measured with a certain degree of uncertainty, sometimes characterized by a distribution of values due to natural variation. Thus, the uncertainty of the input parameters propagates through the model, adding uncertainty to the simulation results.

In order to consider computer simulations as a reliable full alternative to physical testing, it is thus of utmost importance that the models first be thoroughly validated with experimental data for a wide range of input parameters. An innovative solution is provided by the multiscale approach, where global physical parameters are obtained by modeling a small incremental subset of the packaging system at a lower scale.

2.1.3 Computational fluid dynamics and structural analysis

A powerful tool in packaging research and development (if/when properly used) is provided by CFD. In this tool, transport equations including heat transfer and species transport are directly solved using a numerical discretization of partial differential equations governing the process. Fluid (gas and liquid) flows are governed by partial differential equations

(PDEs), which represent conservation laws for the mass, momentum, and energy. CFD is the art of replacing such PDE systems by a set of algebraic equations that can be solved using digital computers. CFD provides a qualitative, and sometimes even quantitative, prediction of fluid flows by means of

- *Mathematical modeling* (partial differential equations),
- *Numerical methods* (discretization and solution techniques),
- *Software tools* (solvers, pre- and postprocessing utilities).

CFD enables scientists and engineers to perform "numerical experiments" (i.e., computer simulations) in a "virtual flow laboratory." CFD does not completely replace the experimental approach, but it gives an insight into flow patterns that are difficult, expensive, or impossible to study using traditional (experimental) techniques. Application of CFD has been reported for analyzing the performance of a food package design, especially for thermal processes in which packaging is involved (e.g., chilling a package's contents via external airflow, internal package sterilization).

PDEs describing the problem can be discretized either by a control volume approach or by a finite element approach. The latter is described in the following section as more suited for elliptic problems, often related to solid mechanics, although application of the control volume approach can be found as well. The reported list of applications of CFD is massive, and a limited number of them are described in the following text.

2.1.3.1 Cooling processes

Cooling is a common thermal process to preserve fresh products and extend their shelf life: for example, horticultural products are generally cooled by forced-air ventilation through ventilated packages to achieve efficient and uniform cooling of the internal food product. Several factors affect cooling efficiency and uniformity, like the food characteristics, the package design, the packaging material, and the airflow pattern within the package (Xia and Sun, 2002; Norton and Sun, 2006).

Effects of design parameters had been investigated by several authors. Zou et al. (2006a, b), Ferrua and Singh (2009), Tutar et al. (2009), Ngcobo (2013), and Delele et al. (2013) applied a CFD approach to study the effect of the package design and the ventilation hole configuration (shape, area, number, and position) on the internal airflow pattern and temperature distribution inside the package during cooling.

Another topic of application is related to digital retail logistics, as today traditional retailers of food products partially shift or extend their business to digital sales by means of e-commerce. Compared to traditional retail channels, digital retail logistics is extremely dynamic, more diversified, and thus less predictable (Teixeira, 2005). Pathways to reach a customer and to deliver food products in digital retail channels are sometimes uncertain, making it much more difficult to assure that the cold chain is preserved. In such a challenge, CFD can also be applied, for example, to analyze the insulating performance of a food package when subjected to the temperature variations occurring during digital retail logistics.

2.1.3.2 Heating processes

Another common application of CFD in packaging is linked to the heating process. For example, the so-called *in-container sterilization* is currently a widely used food preservation technique that can be applied to all types of food products. With this processing technique, food packages such as cans, jars, or pouches are first filled with food (liquid or solid), then sealed, and finally heated in a retort using saturated steam or pressurized hot water, resulting in heat penetrating into the food from the package wall inward.

Using this approach, both the food product and inner package wall together are sterilized at the same time (Teixeira, 2005; Anandharamakrishnan, 2013). Given the complexity of the experimental set-up, to date little is known about the temperature and velocity profiles inside a package during in-container sterilization and the relationship between these profiles and the design of the packaging. This results in poorly designed packages and mostly overdimensioned process parameters (time and temperature), resulting in a decrease in the sensorial and nutritional characteristics of the food product. In the last decade, CFD has also been applied in several investigations to obtain insight into the relationships between the package design and the heating process, for example, for sterilization of food packages (Kannan and Sandaka, 2008; Augusto et al., 2010; Augusto and Cristianini, 2012). Some applications of CFD to this purpose were also reported by Sarghini and Erdogdu (2016) and Erdogdu et al. (2017a, b).

2.1.4 Finite element analysis for structural analysis

Originally developed in the field of mechanical engineering, FEA is a powerful numerical technique that has been transformed into an indispensable modern tool for modeling and simulation in the food packaging industries. FEA is the simulation of a physical phenomenon using the numerical technique called the finite element method (FEM), and it is used to reduce the number of physical prototypes and experiments and optimize components in the design phase to develop products better and faster.

The comprehensive understanding and quantification of a physical phenomena related to the packaging industry, such as structural behavior, thermal transport, or vibration analysis, require the use of mathematics. Most of these processes are described by a set of PDEs. Numerical techniques for a computer to solve these PDEs have been developed over the last few decades, and one of the prominent ones, especially for structural analysis, is FEA. These PDEs are complicated equations that need to be solved in order to compute relevant quantities of a structure (like stresses, strains, thermal load, etc.) in order to estimate a certain behavior of the investigated component under a given load.

FEA only gives an approximate solution to the problem. Simplified, FEA is a numerical method used for the prediction of the behavior of a part or assembly under given conditions. The results of a simulation based on FEA are usually depicted via a color scale that shows, for example, the pressure distribution over the object.

To create a simulation, a mesh is first created, consisting of up to millions of small elements that together form the shape of the structure, and calculations are made for every single element. Combining the individual results of each mesh point gives us the final result of the structure. The approximations mentioned are usually polynomial and, in fact, are interpolations over the elements.

The advent of cheap computational power and easy-to-use FEA software packages, coupled with the relatively reduced cost of these software packages, has noticeably increased FEA use for simulating and studying the structural design of packages. This numerical technique, nowadays validated by experimental results, helps package designers improve the mechanical integrity and strength of packages in order to protect the produce and package against damage.

FEA is actually a general method to discretize PDEs (also used in CFD) widely used by researchers, in particular to study migration and diffusion problems through multilayer polymeric food packaging and to investigate and predict the structural performance of food packages. With the strong push toward environmentally friendly packaging within the context of resource conservation, this type of analysis has become a very important tool for designing lighter food packages requiring less, and different, material, without hampering the mechanical barrier of the packaging.

FEA is often used as an alternative to expensive and time-consuming experimental tests (Delele et al., 2010) and it has been used in simulating almost any type of design concept and in determining the behavior of different packaging configurations in variable environments (Rao, 2010; Srirekha and Bashetty, 2010). As a consequence, FEA allows evaluation of the possibilities and behavior of a new design over a wide range of operational parameters once a model has been developed and validated, without the need to produce a real prototype, reducing time and resources (Chandrupatla et al., 2002; Delele et al., 2010; Rao, 2010; Reh et al., 2006; Srirekha and Bashetty, 2010). The overall effect of mathematical modeling is to enhance performance, improve reliability, provide more confidence in scale-up, and improve consistency of the product (Xia and Sun, 2002; Bakker et al., 2001)

2.1.4.1 Structural performance

Geometry and material composition strongly affect the mechanical properties of food packaging. FEA has been used to study the effect of the geometry of cardboard (Biancolini et al., 2005; Han and Park, 2007; Pathare et al., 2012), plastic (Abunawas, 2010; Suvanjumrat and Puttapitukporn, 2011), and metal (Reid et al., 2001; Yoon et al., 2010) packages in terms of load capacity, package compressibility, and impact resistance. The resistance of wooden pallets has been subjected to FEA as well (Waseem et al., 2012), to analyze their behavior under different types of loads. FEA has also been used to obtain a more detailed knowledge of the interaction between cardboard and machine tooling, used to transform flattened cardboard into a boxed package (Sirkett et al., 2007).

A critical application of FEA can be identified in paper and paperboard designs, due to their peculiar response to loads, moisture, and temperature, making them quite complex engineering materials (Alava and Niskanen, 2006; Bandyopadhyay et al., 2000; Fadiji et al., 2017, 2018a, b; Haslach Jr., 2000; Hubbe, 2013). As a matter of fact, paper and paperboard are two commonly used materials in almost every industry, including the food industry (Xia et al., 2002), and they can be utilized for storage purposes in two principal ways, either as a wrapping material or converted into a container. The paper used for container manufacturing is usually a thick, hard paper (paperboard), with an average grammage above 200 g/m^2 (Hagman, 2013; Huang et al., 2014).

2.1.4.1.1 Structural/mechanical performance of paperboard In fresh fruit industries, where packaging is playing a continuously growing role, paper and paperboard are increasingly considered reliable and important (Ahmed and Alam, 2012; Kibirkstis et al., 2007; Koutsimanis et al., 2012; Mahalik and Nambiar, 2010; Pré, 1992; Raheem, 2013; Smith et al., 1990). Analysis of the structural integrity of the various components to evaluate the strength and stiffness properties is required in designing paperboard packages, and understanding the structural responses of paperboard is a crucial step in the design of an entire paperboard package. Over the years, FEA has been used widely to better understand the material properties of paper (Biancolini, 2005; Talbi et al., 2009).

From a structural point of view, paper is an anisotropic, nonhomogeneous network of cellulose fibers surrounded by different types of fillers (Hagman, 2013). The anisotropic and nonhomogeneous properties of paper are consequences of the manufacturing process, where a fiber suspension is deployed onto a moving web, which is then pressed between rolling cylinders and dried (Hagman, 2013). The synthetic result of this production process is highlighted by different mechanical properties in three principal directions (Hagman, 2013; Harrysson and Ristinmaa, 2008; Jimenez-Caballero et al., 2009; Pathare and Opara, 2014; Stenberg, 2003; Xia, 2002; Xia et al., 2002). This directional dependence of paper and paperboard are the longitudinal direction that the paper moves through the machine (LD), transversal direction perpendicular to the machine direction (TD), and the normal out of plane direction (ZD) (see Fig. 4).

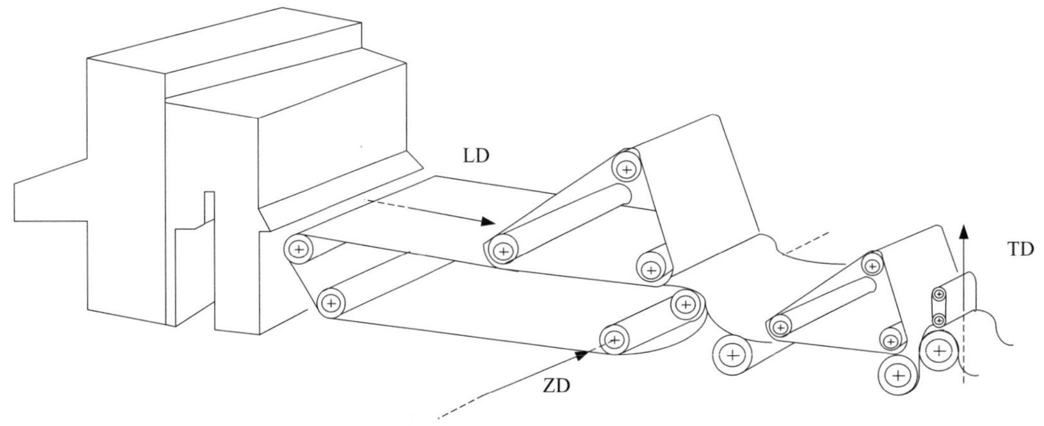

FIG. 4 Paperboard processing line.

Constitutive models of mechanical properties of paper and paperboard have been studied by several authors (Beex and Peerlings, 2009; Borgqvist et al., 2016; Domaneschi et al., 2017; Giampieri et al., 2011; Hallbaack et al., 2006; Huang and Nygards, 2010; Linvill and Ostlund, 2016; Ostlund and Nygards, 2009; Pradier et al., 2016; Ramasubramanian and Wang, 2007; Stenberg, 2003; Xia, 2002). In more detail, the required properties of paper and paperboard are typically high compression strength, high bending stiffness, foldability, crease-ability, and, last but not least, aesthetic properties such as exterior appearance and feel.

In paperboard conversion, some operations such as gluing, folding, and cutting are straightforward (Beex and Peerlings, 2009), while a possible cracking of high grammage paperboard, sometimes due to the quality of the fold at the creases (Beex and Peerlings, 2009), can arise during the conversion process. Creasing is an important mechanical behavior of paperboard die cutting (Nagasawa et al., 2003). Creasing (Fig. 5) is often necessary to facilitate folding of paperboard, and the folding quality of paperboard is defined by the crease.

FIG. 5 Paperboard creasing process.

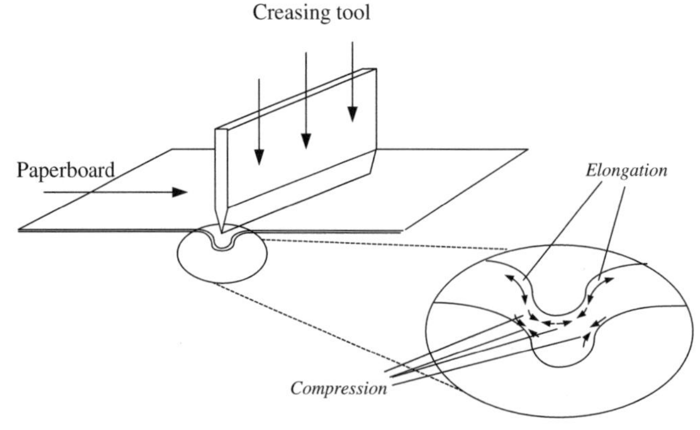

The bending stiffness of the paperboard around the folding line is lowered when creased due to a shear-induced delamination into the paperboard structure, which consequently weakens the paper fibers (Giampieri et al., 2011). In the case of cracked folds, the strength of the paperboard and the paperboard package's attractiveness to customers may be affected (Giampieri et al., 2011).

A detailed delamination and material mode was proposed by Xia et al. (2002) in the framework of a computational hybrid model for paperboard creasing, where an in-plane elastoplastic model and an elastic behavior in the thickness direction were used to simulate the in-plane behavior of paperboard. In their work, the damaging creasing and folding behavior in the out-of-plane direction of the paperboard was investigated by adopting an interface model. Beex and Peerlings (2009) also developed a numerical model to predict the behavior of paperboard in a virtual environment for variable creasing properties.

The development of numerical techniques to model and understand the damage mechanisms of paperboard properties suggests the integration of experimental and numerical studies to provide insight into the understanding of paperboard properties such as creasing and folding, eventually improving the end-use of paperboard (Huang and Nygards, 2010). The creasing and folding behavior of three commercially produced paperboards was studied by Huang et al. (2014) both numerically and experimentally, with a good agreement among predicted and experimental force displacement curves for all paperboards.

Li et al. (2016) used the FEA approach to investigate the pure crack opening mode and sliding mode of multi-ply paperboards. In their work, the authors adopted four techniques to characterize the delamination property of the paperboard: z-directional tensile test (ZDT), double-notch shear test (DNS), double-cantilever beam test (DCB), and end-notched flexure (ENF). Results showed that fracture properties are strongly affected by the fiber orientation bias.

Numerical techniques were also used to study structural properties of corrugated paperboard. Corrugated paperboard is obtained by assembling laminating paperboard layers (two or more) using a conversion process. The external layers are known as the liners and the corrugated core is called fluting (Fadiji et al., 2018a, b; Harrysson and Ristinmaa, 2008). This packaging presents numerous advantages, such as recyclability, low cost-to-weight ratio, and high stiffness per unit weight. For the aforementioned reasons, corrugated paperboard is an attractive material commonly used in manufacturing corrugated paperboard boxes, used for transportation of various products (Fadiji et al., 2017, 2018a, b; Harrysson and Ristinmaa, 2008).

As illustrated in Fig. 2, food packages are exposed to different hazards and mechanical loadings during their lifetime in both transportation and storage, affecting the structural performance of the packaging components and the whole package (Fadiji et al., 2016, 2017, 2018a, b; Harrysson and Ristinmaa, 2008; Opara and Fadiji, 2018; Pathare and Opara, 2014). Mechanical properties such as elastic behavior, ultimate failure, collapse, creasing, stability, buckling, and transverse shear of corrugated paperboard have been extensively studied using FEA (Aboura et al., 2004; Biancolini et al., 2005; Fadiji et al., 2017; Gilchrist et al., 1999; Haj-Ali et al., 2009; Jimenez-Caballero et al., 2009; Peterson, 1983; Pommier and Poustis, 1990; Talbi et al., 2009; Thakkar et al., 2009; Zhang et al., 2014).

An initial study of the mechanical performance of corrugated paperboard was published by Peterson (1983), adopting the use of FEA to identify the stress developed under three-point bending in the machine direction on corrugated paperboard. In this case, a symmetric model with linear elastic behavior was considered, and the fluting medium was assumed to have a sinusoidal shape. In the same way, a linear model was proposed by Pommier and Poustis (1990) to investigate the bending stiffness of a single-wall corrugated paperboard. In the latter, perfect bonding between the liners and the fluting was simulated by merging nodes at contact points and bending stresses on the board were obtained and compared with experimental results, resulting in a model flow to determine the bending flexibility matrix of an orthotropic material.

Nordstrand and Carlsson (1997) used FEA to determine the transverse shear of corrugated paperboard using a three-point bending method, while Patel et al. (1997) used FEA to evaluate the local buckling and the failure on the facings of cylindrical corrugated paperboard. The mechanical performance of corrugated paperboard was also investigated by Gilchrist et al. (1999). In their work they included the geometries of anticlastic bending tests, four-point bending, and edge compression tests. In general, the model showed a good agreement in the load-displacement responses for both experimental tests and numerical analysis.

The structural performance of paper material is strongly dependent on environmental conditions such as temperature, paper moisture content, and humidity, which may vary significantly throughout the food supply chain processes, illustrated in Fig. 2. Such environmental conditions could affect the strength of the paperboard (Haslach Jr., 2000; Pathare and Opara, 2014; Vishtal and Retulainen, 2012).

Permeability of moisture through the layers (linerboards and flutings) of corrugated paperboard was investigated by Rahman et al. (2006) using a finite elements model. The model was able to predict the response of the board to creep and hygroexpansion to determine the overall structural behavior of the board, and it was experimentally validated.

Navaranjan and Johnson (2006) introduced a finite element model to predict the nonlinear creep behavior of corrugated paperboard, with the capability of predicting the creep performance of corrugated paperboard for a given set of properties of the constituents and the geometry of the board, when the board was uniaxially loaded under compression at a constant relative humidity.

Prediction of the steady-state moisture transport through corrugated paperboard was investigated by Bronlund et al. (2013) using a finite element model, showing reasonable agreement with experimentally measured moisture fluxes.

Fadiji et al. (2017) used NASTRAN for the FEA, reporting a significant difference in the edge compression resistance of a C-fluted corrugated paperboard at standard conditions (23°C and 50% RH) and refrigerated conditions (0°C and 90% RH); model results were in good agreement with experimental data (less than 10% difference).

An extensive description of FEA applications in the packaging industries can be found in Fadiji et al. (2018a, b).

2.1.4.2 Diffusion process

FEA has been applied to the diffusion process, as gas permeability strongly affects the shelf life of packaged products, in terms of conservation of temperature variation and initial microbial contamination. If material properties are correctly identified, FEA allows a rapid development of optimal packaging through a combination of materials and thickness that results in the desired permeability of a certain package design (Rennie and Tavoularis, 2009).

The use of FEA enables prediction of O_2, CO_2, N_2, and H_2O concentrations within microperforated, polymeric modified atmosphere packaging during cold-storage (Xanthopoulos et al., 2012) or to study the diffusion process of a gas through plastic bottles (Profaizer, 2007).

2.1.4.3 Migration

Another issue in packaging design is related to contamination of the food product by chemical components (additives, residues, inks, etc.) originating from food packaging, as almost all packaged food products are in (direct or indirect) contact with food packaging during production, distribution, storage, and final use. Since the beginning of the 21st century, specific FEA systems have been developed (ATKS-SML, MIGRATEST©, SFPP3) and applied for in-silico migration analysis of chemicals in multilayer polymeric packaging of food products (Roduit et al., 2005; Nguyen et al., 2013). European regulation is still nonspecific concerning migration of certain components (coatings, printing, inks, and adhesives) (European Union, 2011 CR 10/2011), but FEA provides a substantial help, especially in the preliminary design phase of packaging.

2.1.4.4 Electronic and photonic design automation

In the design of intelligent food packaging, electronic and photonic design automation is gaining momentum as an innovative technological improvement. Furthermore, the rapid development of specialized low-cost sensors and automatic identification technologies results in growing applications for electronic design automation (EDA) and photonic design automation (PDA) systems.

With the aim to provide food packages with a certain degree of intelligence, EDA and PDA applications facilitate the design and in-silico analysis of the electronic and photonic integrated circuits that can be found in electrical and optical devices for food packaging. Although there is wide use of EDA systems in the packaging industry (Brayton and Cong, 2009), only recently have PDA systems been made available (Korthorst and Stoffer, 2012).

2.1.5 Computer-aided manufacturing and cyber-physical systems

CAM is used as a broad term embracing production processes carried out with the aid of computer systems (British Nutrition Foundation and Design Technology Association, 2002).

Today's generation of CAM systems can be considered a first-generation cyber-physical system (CPS), because of the tight coupling currently present between CAD models (i.e., for in silico design), the production process parameters (e.g., machine settings, sequence of operations and working cycle, raw material flow, process and environmental conditions, etc.), and the manufactured product (Xu and He, 2004). A broad classification of CAM systems currently used in the food package industry requires a distinction between CAM systems used for subtractive manufacturing, formative manufacturing, and additive manufacturing.

2.1.5.1 Subtractive manufacturing

Subtractive manufacturing (SM) techniques are used extensively, for example, in rapid prototyping (RP) and small, medium, or high volume production of relatively straightforward geometries. They can be described as a production process with a top-down logic that relies on the manual or automatic subtraction of unwanted material from a large solid block of raw material or sheet through the use of cutting tools like power CO_2 lasers, water jets, or plasma torches.

Integration in the production system is obtained by coupling a CAD/CAM system and computer numerically controlled (CNC) machine tools for the production of a wide variety of CAD models (Heynick and Stotz, 2006). Small- and large-scale production of packages can be realized by means of advanced CNC machines that are specifically developed to handle large sheets of packaging material (Packsize, BCS AutoBOX, Panotec, Versaflex Coveris, Autobag). These machines can cut, fold, and seal the packaging material "on demand" to the desired shape and dimensions. Shapes can either be chosen from a huge library of standard package types (e.g., standard FEFCO12 box types), or be specifically tuned to the size and shape of a particular product, fitting requirements for digital retail purposes, or offering a suitable adaptation to e-commerce orders (INCPEN, 2012) to reduce void fill and optimize transport costs.

Another important application of automated SM in the injection molding process of food packaging is the automated RP and/or software modeling of new molds to test the functionality and efficiency of a certain mold design. In injection molding, fused plastic packaging material is forced into a mold cavity under pressure and then cooled down and removed from the mold cavity.

Moreover, they can be used to evaluate new package designs to study the shortcomings or consumer acceptance of a certain design in a preliminary stage of the design process.

2.1.5.2 Additive manufacturing

The food packaging industry started to use additive manufacturing (AM) in the design and production process of new food packages only recently, in particular AM-based RP and digital printing of graphic designs and electronic systems. AM includes all processes in which designs are performed according to a bottom-up logic by assembling very thin layers of different materials in a successive way by using a computer system (Berger, 2013). Additive manufacturing is already widely used to rapidly create prototypes and small batches of geometrically complex products, and to digitally print graphic designs and electronic circuits on a broad variety of substrates with little material loss and without the use of specific tools, such as molds or printing plates (Santos et al., 2006; Harris, 2012).

These applications can provide advanced tools for package designers, engineers, and manufacturers to improve the package design and production process and to rapidly bring new and advanced package designs to market (Beaman et al., 2004).

General descriptions of the most common additive manufacturing processes for each working principle, along with lists of major applications, are provided in recently published papers (Harris, 2012; Beaman et al., 2004; Berger, 2013; Wong and Hernandez, 2012).

2.1.5.3 Rapid prototyping

In RP, a CAD model is translated into source code (e.g., in STL language) that automatically programs an AM machine to manufacture, either layerwise (i.e., fused deposition approach) or by laser sintering of polymeric liquids, a prototype of the design (Topçu et al., 2011; Volpato et al., 2013). AM-based RP of food packages can be realized in two different ways, either via the creation of an AM mold prototype for short-run package production, or directly via a paper- or polymer-based prototype of the package.

While commercial polymer-based 3D printers (e.g. MakerBot, 3DSystems, Stratasys) have already been available on the market for several years, the industry of paper-based commercial 3D printers is still at the early stages (see, e.g., Mcor Technologies). The latter approach provides a less expensive and more sustainable alternative to polymer-based 3D printers (Mcor Technologies, 2013)

Subtractive manufacturing and additive manufacturing represent two complementary and in some cases overlapping technologies, and while the first one is considered the future, in some areas SM-based RP currently provides a competitive solution to AM-based RP due to greater dimensional accuracy and material flexibility, together with smoother surface finishes (Destefani, 2005).

2.1.5.4 Formative manufacturing

Formative manufacturing (FM) describes all packaging production processes in which materials are shaped, deformed, or printed by applying external mechanical forces, restricting forming tools, and machines like molds, dies, printing plates, printing presses, and/or air, heat, or steam (Afify and Elghaffar, 2007), generally resulting in less material loss than in SM. CAM systems linked to a network of sensors and actuators for FM are mainly applied in real-time monitoring and control of production and printing processes (Szentgyörgyvölgyi, 2005), while in SM and AM they are mostly used before the start of a production process for automated machine programming. FM is the most widely used food package production process on an industrial scale, and it is used for the production and printing of a wide variety of trays, cans, jars, bottles, mono- or multilayer foils in paper, plastic, glass, or metal materials.

Moreover, CAM systems for SM and AM are also often indirectly adopted in FM for the production of process tools such as molds or printing plates. In such cases, CAM systems are typically controlled by one or more programmable logic controllers (PLCs), offering the possibility to change in real time the behavior of the actuators based on measured sensor data like temperature, pressure, dimensions, or images in order to maintain a required set-point of the final product quality (Xu and Shi, 1999; Bliesener et al., 2002). Sometimes they are used in combination with artificial intelligence algorithms or statistical process control (SPC) tools in which data-driven methods can compare historical and current sensor data and drive the optimization of the process parameters accordingly.

3 Applications of theoretical and experimental modeling techniques

3.1 Mathematical modeling of modified atmosphere packaging for fresh products

Respiration of fresh foods continues after harvest, and along their production and distribution chains they continue to consume O_2 and produce CO_2. Factors associated with product characteristics, packaging material, and the environmental conditions experienced can affect the respiration rate (RR), which can be slowed down or accelerated (Al-Ati and Hotchkiss, 2003).

In more detail, the dynamics of atmosphere generation inside modified atmosphere packaging (MAP) depends on product physiological characteristics (respiration/transpiration), environmental conditions (temperature and relative humidity (RH)), and properties of the packaging material (thickness, perforation, permeability to water vapor and gases).

Several studies have reported modeling approaches for describing and predicting these interactions using mathematical equations for successful MAP design (Mahajan et al., 2008; Gomes et al., 2010). Most of the model applications aim at extending the shelf life and maintaining the quality of packaged produce (Oms-Oliu et al., 2008). Others assess the effects of the design parameters on physicochemical and microbial attributes of produce (Caleb et al., 2013), while some studies focus on models used to test different package configurations to achieve optimum postharvest storage conditions (Barrios et al., 2014). These studies have used experimental data on the product characteristics, environmental factors, and packaging material properties to predict both physiological responses of produce and properties of packaging material in order to identify the optimum packaging solutions. This indicates that a successful MAP design requires the integration of these essential parameters, as illustrated in Fig. 2.

Fresh and minimally processed produce continue physiological and metabolic processes after harvest; hence, they are susceptible to quality deterioration and reduced shelf life. MAP is a well-proven postharvest technology for preserving the natural quality of fresh and minimally processed produce, and extending the storage life under optimum MAP design and storage conditions. Successful MAP design is achieved by the mathematical integration of dynamic produce physiological characteristics, properties of the packaging material, coupled with optimum equilibrium atmospheric conditions for the given product.

The dynamics of atmosphere generation inside MAP depend on product physiological characteristics (respiration/transpiration), environmental conditions (temperature and relative humidity (RH)), and properties of the packaging materials (thickness, perforation, permeability to water vapor and gases). Various studies have reported modeling approaches for describing and predicting these interactions using mathematical equations for successful MAP design (Mahajan et al., 2008; Gomes et al., 2010). Most of the model applications aimed to extend shelf life and maintain quality of packaged produce (Oms-Oliu et al., 2008).

An extensive review of this topic is provided by Belay et al. (2016).

3.2 Quality control in packaging production and fault detection

Quality standards in a food package in terms of size, shape, sealing, graphics, and content are typically checked and assessed after the filling and sealing processes just before the distribution activities (i.e., storage, transport). Offline testing procedures are replaced today by nondestructive and online image analysis systems (IASs), gradually replacing conventional and time-consuming systems and allowing fast detection of irregularities and failures.

Nowadays, in the production phase of food packages, IASs are generally integrated into the previously discussed CAM systems to enable online product control and process management. Package parameters like surface, shape, color, and dimensions as well as the dynamic behavior of mechanical parts can be analyzed using nondestructive technologies (Leadbeater, 2009; Chiffre et al., 2014).

A typical IAS is composed of hardware to capture the images (today through ordinary cameras). Advanced image-capturing systems include three-dimensional laser scanners, interferometers, ultrasonic sensors, electromagnetic field detectors, high-speed cameras, hyperspectral cameras, and, depending on the application, even medical imaging scanners (e.g., X-ray imaging, computed tomography (CT), positron emission tomography (PET), and magnetic resonance imaging (MRI)). These tools are completed by software that processes the signals, generates the images, and automates image analysis (e.g., through AI and machine learning).

Both the hardware and software implementation depend on the purpose of the analysis: shape and size determination, crack or leak detection, contaminants identification, object recognition, or optical character recognition. Each analysis has its specific system requirements like camera acquisition speed, resolution, and field of view, that can seldom be provided by off-the-shelf image analysis systems. At this writing, X-rays, electromagnetic fields, and ultrasound technologies are applied to detect foreign objects or materials like insects, metals, glass, or plastics in food packages. Existing destructive food package methods for leak detection can be replaced by infrared imaging (Al-Habaibeh et al., 2004), ultrasonic imaging (Pascall et al., 2002), and also vibration analysis. They are recognized as beneficial and faster alternatives (Patankar, 2005).

Regarding food package size and handling properties, accurate shape and dimension values can be obtained after production by using three-dimensional laser scanners and lateral chromatic imaging. These are also used for reverse engineering purposes by first scanning a physical object and then converting the image into a digital representation in CAD, while high-speed cameras can be used to control the quality of the graphic design and the correct printing of the text on the food package label.

3.3 Emerging applications of modeling tools

As previously described, modeling techniques are already widely used in preliminary design and industrial production of food packages and, in this section, the focus will be shifted onto computer-based technologies that are currently still under development within the context of food packaging.

3.3.1 Image analysis (IA) systems and biomimetics

Analysis of existing literature suggests that image analysis has been widely used with the purpose of quality control of fault detection, but the potential has been underestimated in the prelogistic phase of a food package's life cycle.

New applications of IA have been emerging recently, and among them we can find the development of biomimetic food packaging. This practice can be defined as the development of new food packaging or packages with properties similar to the structural and functional properties of entities created by nature (Carmona-Ribeiro et al., 2011; Tian et al., 2013; Morganti, 2014).

Other new applications deal with the analysis of the inner structure of food packaging based on high-resolution techniques like the scanning electron microscope (SEM) and computer tomography (CT). Using SEM, a resolution of less than 1 nm can be reached, providing a view into the packaging microstructure. CT applications, on the other hand, can identify critical points or detect structural weaknesses in a packaging material, like folds or seals, that are experiencing mechanical stress, and they can also relate microstructure such as porosity, composition, and homogeneity to some macroscopic properties.

Even though such applications have been proven useful for other industries and applications like construction and textiles (Eadie and Ghosh, 2011; Bhushan, 2012), the food industry has not yet invested enough effort to explore the potential of these applications in food packaging development.

3.3.2 Data-driven and knowledge-based internet of everything design and production

With respect to the technologies introduced in the previous sections, all supported by computer systems and aimed at significantly reducing packaging design, testing, and production times, the rapid and continuous advances in information systems and technology have made this framework much more dynamic. The integration of information and knowledge coming from all stages of the food package's life cycle (i.e., prelogistic, logistic, and end-of-life phases), as illustrated in Fig. 2, can contribute to setting up a full-fledged holistic research, design, and production approach for food package conceptualization and design in which decisions are based on data and consumer preferences and are linked to the perceived product quality.

A paradigm shift has already started, from Internet of Things (IoT) to Internet of Everything (IoE). To be more precise, when in addition to physical objects and entities also data, processes, and people are mutually connected, we are within the IoE paradigm. The IoE provides for different types of connections:

- Machine to Machine (M2M);
- People to Machine (P2M);
- Machine to People (M2P).

Specifically, the IoE combines a series of sensors, incorporated into the various entities of a supply chain ecosystem and capable of recording certain parameters. These objects/entities, in our case food packages, connecting to the smartphone or directly to the network wirelessly, send the tracked values or the parameters they represent.

This task is also known as intelligence on the cloud or mobile applications on smartphones, involving collection, interpretation, and manipulation of the data tracked, and dialogue with the various objects/entities on the basis of the results processed. Such continuous and real-time dialogue is capable of adapting the way all the objects are connected to the ecosystem (i.e., these are both source and users of the information) according to a peer-to-peer paradigm.

The emergence of the IoE and the exponential growth of real-time quantitative and qualitative data have introduced an opportunity to develop:

- new cyber-physical systems (CPSs), i.e., intelligent computer systems that establish a connection between the digital and the physical world, and
- Product lifecycle management (PLM), defined as a business approach encompassing all managerial and collaborative skills, practices, and technological tools for the design and manufacturing of new sustainable products. Such systems are able to automate, control, or improve business processes and to meet consumer, transporter, and/or retailer needs and requirements by extensively analyzing and manipulating the wide spectrum of data and knowledge (e.g., ideas, feedback, or records from sensors) behind the food supply chain ecosystem (Manyika et al., 2011; Ane et al., 2012; Verdouw et al., 2016; Accorsi et al., 2018).

Fig. 2 shows a theoretical scheme of the possible data, information, and knowledge flows between the identified phases and stages in the digitization of a food package's life cycle; this is further detailed in Accorsi et al. (2018). This scheme can be considered a blueprint of the envisioned cyber-physical system to automate and control the prelogistic (manufacturing), logistic, and postlogistic operations related to food packages.

As an example of data-driven and knowledge-based design of food packages, the first example of a decision support system (designed during the French research project TailorPack) has been developed by Destercke et al. (2011). This system determines optimal permeabilities of packaging and identifies the best packaging material for a given fruit or vegetable.

New challenges in the food packaging design/production industry include better linking postlogistic data to the prelogistic design phase. For example, the relationship between food waste possibly caused by a poorly designed package and the environmental impact of the packaging itself needs to be carefully investigated. This can be evaluated by performing an ex-ante life cycle assessment (LCA), i.e., an analysis of the environmental performance and impact of a food package during its life cycle, which is usually carried out during the design step in the prelogistic phase. The principles and the framework of LCA are described in ISO standard 14040:2006, and several software applications can be used to perform an LCA of a food package (e.g., Instant LCA Packaging, PackageSmart, SimaPro, OpenLCA, Ecochain, GABI).

The main focus of all these applications is to evaluate the environmental impacts associated with packaging materials and the food package production process, as well as to recover resources at the end of the life cycle and to reduce waste. As a matter of fact, it is reasonable to speculate that there exists a causal relationship between food packages and food waste, possibly implying that food waste can be partially linked to poor design or low consumer appreciation of food packages. Disregarding this reality in an LCA implies that packaging with a lower ecological impact may have an overall impact (including food waste) higher than packaging with a higher environmental impact but with reduced food waste (Wikström et al., 2014).

Joining in silico data with the tremendous amount of information from the postlogistic phase (e.g., the quantity of packages damaged because of poor design and/or transportation) and the buyer criticism about a specific packaging offers the possibility of further enhancing the precision of such effects on the LCA.

Data analysis could also highlight the link between other phenomena that have a causal relationship with food packaging, increasing the environmental impact of the latter. For example, it would be possible to ensure that potential environmental impacts at all life cycle stages are considered by designers, manufacturers, or other decision makers to include all food package's life cycle stages that could potentially result in a high environmental impact (Flanigan et al., 2013). This would eventually enable designers and manufacturers to consider package designs not only in terms of functionality and cost-effectiveness but also in terms of environmental performance.

References

Aboura, Z., Talbi, N., Allaoui, S., Benzeggagh, M.L., 2004. Elastic behavior of corrugated cardboard: experiments and modeling. Compos. Struct. 63, 53–62.

Abunawas, M., 2010. A suitable mathematical model of PET for FEA drop-test. Int. J. Comput. Sci. Network Secur. 10, 215–219.

Accorsi, R., Cholette, S., Manzini, R., Tufano, A., 2018. A hierarchical data architecture for sustainable food supply chain management and planning. J. Clean. Prod. 203, 1039–1054.

Afify, H.M., Elghaffar, Z.A., 2007. Advanced digital manufacturing techniques (CAM) in architecture. In: International ASCAAD Conference on Em'body'ing Virtual Architecture, Alexandria, pp. 67–80.

Ahmed, J., Alam, T., 2012. An overview of food packaging: material selection and the future of packaging. In: Handbook of Food Process Design. Wiley-Blackwell, pp. 1237–1283.

Alava, M., Niskanen, K., 2006. The physics of paper. Rep. Prog. Phys. 69 (3), 669–723.

Al-Ati, T., Hotchkiss, J.H., 2003. The role of packaging film permselectivity in modified atmosphere packaging. J. Agric. Food Chem. 51 (14), 4133–4138.

Al-Habaibeh, A., Shi, F., Brown, N., Kerr, D., Jackson, M., Parkin, R., 2004. A novel approach for quality control system using sensor fusion of infrared and visual image processing for laser sealing of food containers. Meas. Sci. Technol. 14, 1995–2000.

Anandharamakrishnan, C., 2013. Applications of computational fluid dynamics in the thermal processing of canned foods. In: Anandharamakrishnan, C. (Ed.), Computational Fluid Dynamics Applications in Food Processing. Springer Briefs in Food, Health, and Nutrition, pp. 27–36.

Ane, B.K., Roller, D., Abraham, A., 2012. Multiagent based product data communication system for computer supported collaborative design. In: Rückemann, C.-P. (Ed.), Integrated Information and Computing Systems for Natural, Spatial, and Social Sciences. Information Science Reference (IGI Global), p. 532.

Augusto, P.E., Cristianini, M., 2012. Computational fluid dynamics evaluation of liquid food thermal process in a brick shaped package. Cienc. Tecnol. Aliment, 134–141. https://doi.org/10.1590/s0101-20612012005000014.

Augusto, P.E., Pinheiro, T.F., Cristianini, M., 2010. Using computational fluid-dynamics (CFD) for the evaluation of beer pasteurization: effect of orientation of cans. Cienc. Tecnol. Aliment. 30, 980–986.

Ayanogolu, H., Rebelo, F., Duarte, E., Noriega, P., Teixeira, L., 2013. Using virtual reality to examine hazard perceptionin package design. In: Second International Conference, DUXU 2013, Held as Part of HCI International 2013, Proceedings Part III, LNCS 8014, pp. 30–39.

Bakker, A., Haidari, A.H., Oshinowo, L.M., 2001. Realize greater benefits from CFD. Chem. Eng. Prog. 97 (3), 45–53.

Bandyopadhyay, A., Radhakrishnan, H., Ramarao, B.V., Chatterjee, S.G., 2000. Moisture sorption response of paper subjected to ramp humidity changes: modelling and experiments. Ind. Eng. Chem. Res. 39 (1), 219–226.

Barrios, S., Lema, P., Lareo, C., 2014. Modelling respiration rate of strawberry (cv. San Andreas) for modified atmosphere packaging design. J. Food Prop. https://doi.org/10.1080/10942912.2013.784328.

Beaman, J.J., Atwood, C., Bergman, T.L., Bourell, D., Hollister, S., Rosen, D., 2004. Additive/Substractive Manufacturing Research and Development in Europe. World Technology Evaluation Center, Inc., Baltimore.

Beex, L.A.A., Peerlings, R.H.J., 2009. An experimental and computational study of laminated paperboard creasing and folding. Int. J. Solids Struct. 46 (24), 4192–4207.

Belay, Z.A., Caleb, O.J., Opara, U.L., 2016. Modeling approaches for designing and evaluating the performance of modified atmosphere packaging (MAP) systems for fresh produce: a review. Food Packag. Shelf Life 10, 1–15.

Berger, R., 2013. Additive Manufacturing: A Game Changer for the Manufacturing Industry? Roland Berger Strategy Consultants, Retrieved from https://www.rolandberger.com/publications/publication_pdf/roland_berger_additive_manufacturing_1.pdf.

Bhushan, B., 2012. Biomimetics—Bioinspired Hierarchical-Structured Surfaces for Green Science and Technology. Springer, Berlin, Heidelberg. https://doi.org/10.1007/978-3-642-25408-6.

Biancolini, M.E., 2005. Evaluation of equivalent stiffness properties of corrugated board. Compos. Struct. 69 (3), 322–328.

Biancolini, M.E., Brutti, C., Mottola, E., Porziani, S., 2005. Numerical evaluation of buckling and post-buckling behaviour of corrugated board containers. In: Associazine Italiana Per l'analisi Delle Sollecitazioni—XXXIV Convengno Nazionale, Milano.

Bliesener, R., Ebel, F., Löffler, C., Plagemann, B., Regber, H., Terzi, E.V., Winter, A., 2002. Programmable Logic Controllers: Basic Level. Festo Didactic.

Borgqvist, E., Wallin, M., Tryding, J., Ristinmaa, M., Tudisco, E., 2016. Localized deformation in compression and folding of paperboard. Packag. Technol. Sci. 29, 397–414.

Bowman, D.A., McMahan, R.P., 2007. Virtual reality: how much immersion is enough? IEEE Comput. Soc., 36–43.

Brayton, R., Cong, J., 2009. Electronic design automation: past, present, and future. In: NSF Workshop Report, Arlington, Virginia.

British Nutrition Foundation, Design Technology Association, 2002 Implementing CAD and CAM, BNF/DATA. Retrieved from http://www.foodafactoflife.org.uk/attachments/161e4fb4-e95d-4bc5b464ecbf.pdf.

Bronlund, J.E., Redding, G.P., Robertson, T.R., 2013. Modelling steady-state moisture transport through corrugated fibreboard packaging. Packag. Technol. Sci. 27 (3), 193–201.

Caleb, O.J., Mahajan, P.V., Al-Said, F.A.-J., Opara, U.L., 2013. Modified atmosphere packaging technology of fresh and fresh-cut produce and the microbial consequences—a review. Food Bioprocess Technol 6 (2), 303–329.

Carmona-Ribeiro, A.M., Barbassa, L., Melo, L.D., 2011. Antimicrobial biomimetics. In: Cavrak, M. (Ed.), Biomimetic Based Applications. InTech, pp. 228–284.

Chandrupatla, T.R., Belegundu, A.D., Ramesh, T., Ray, C., 2002. Introduction to finite Elements in Engineering. vol. 2 Prentice Hall, Upper Saddle River, NJ.

Chiffre, L.D., Carmignato, S., Kruth, J.-P., Schmitt, R., Weckenmann, A., 2014. Industrial applications of computed tomography. CIRP Ann. Manuf. Technol. 2, 655–677.

Chwastyk, P., Kolosowski, M., 2012. Integration CAD/CAPP/CAM systems in design process of innovative products. In: Annals & Proceedings of DAAAM International 2012, DAAAM International, Vienna, Austria, EU, pp. 397–400.

Cubonova, N., Kumicakova, D., 2005. Tools and applications of computer aided manufacturing and robot virtual model designing. Sci. Bull. Ser. C 19.

Delele, M.A., Verboven, P., Ho, Q.T., Nicolaı, B.M., 2010. Advances in mathematical modelling of postharvest refrigeration processes. Stewart Postharvest Rev. 6 (2), 1–8.

Delele, M., Ngcobo, M., Getahun, S., Chen, L., Mellmannd, J., Opara, U.L., 2013. Studying airflow and heat transfer characteristics of a horticultural-produce packaging system using a 3-D CFD model. Part II: effect ofpackage design. Postharvest Biol. Technol. 86, 546–555.

Destefani, J., 2005. Additive or subtractive? Which rapid prototyping process is right for your job. Manuf. Eng., 2–5.

Destercke, S., Buche, P., Guillaume, C., Guillard, V., 2011. A decision support system to optimize fresh food packaging. In: 6th International CIGR Technical Symposium—Section VI: Towards a Sustainable Food Chain. Food Process, Bioprocessing and Food Quality Management, Nantes, France.

Domaneschi, M., Perego, U., Borgqvist, E., Borsari, R., 2017. An industry-oriented strategy for the finite element simulation of paperboard creasing and folding. Packag. Technol. Sci. 30 (6), 269–294.

Eadie, L., Ghosh, T.K., 2011. Biomimicry in textiles: past, present and potential. An overview. J. R. Soc. Interface 8, 761–775. https://doi.org/10.1098/rsif.2010.0487.

Erdogdu, F., Tutar, M., Sarghini, F., Skipnes, D., 2017a. Effects of viscosity and agitation rate on temperature and flow field in cans during reciprocal agitation. J. Food Eng. 213, 76–88.

Erdogdu, F., Sarghini, S., Marra, F., 2017b. Mathematical modeling for virtualization in food processing. Food Eng. Rev. 9, 295–313.

European Union, 2004, European Packaging and Packaging Waste Directive 2004/12/EC of 11 February 2004 Amending 94/62/EC11.

European Union, 2011, Commission Regulation (EU) No 10/2011.

Fadiji, T., Coetzee, C., Chen, L., Chukwu, O., Opara, U.L., 2016. Susceptibility of apples to bruising inside ventilated corrugated paperboard packages during simulated transport damage. Postharvest Biol. Technol. 118, 111–119.

Fadiji, T., Berry, T., Coetzee, C.J., Opara, L., 2017. Investigating the mechanical properties of paperboard packaging material for handling fresh produce under different environmental conditions: experimental analysis and finite element modelling. J. Appl. Packag. Res. 9 (2), 20–34.

Fadiji, T., Coetzee, C.J., Berry, T.M., Ambaw, A., Opara, U.L., 2018a. The efficacy of finite element analysis (FEA) as a design tool for food packaging: a review. Biosyst. Eng. 1537-5110. 174, 20–40. https://doi.org/10.1016/j.biosystemseng.2018.06.015.

Fadiji, T., Berry, T.M., Coetzee, C.J., Opara, U.L., 2018b. Mechanical design and performance testing of corrugated paperboard packaging for the post-harvest handling of horticultural produce. Biosyst. Eng. 171, 220–244.

Falcão, C.S., Soares, M., 2013. Application of virtual reality technologies in consumer product usability. In: Marcus, A. (Ed.), Design, User Experience, and Usability. Web, Mobile, and Product Design. vol. 8015, pp. 342–351.

Falcão, C.S., Soares, M., 2014. Applications of haptic devices & virtual reality in consumer products usability evaluation. In: Marcelo Soares, F.R. (Ed.), Advances in Ergonomics in Design, Usability & Special Populations Part I. AHFE Conference Publisher, pp. 377–383.

Fauster, L., 2007. Spectroscopic Techniques in Computer Graphics, TUWien, Retrieved 12 2014, from http://www.cg.tuwien.ac.at/courses/Seminar/WS2006/stereo_techniques.pdf.

Ferrua, M., Singh, R., 2009. Modeling the forced-air cooling process of fresh strawberry packages, part I: numerical model. Int. J. Refrig. 32, 335–348.

Flanigan, L., Frischknecht, R., Montalbo, T., 2013. An Analysis of Life Cycle Assessment in Packaging for Food and Beverage Applications, United Nations Environment Programme and SETAC, Retrieved from http://www.lifecycleinitiative.org/wp-content/uploads/2013/11/food_packaging_11.11.13_web.pdf.

Giampieri, A., Perego, U., Borsari, R., 2011. A constitutive model for the mechanical response of the folding of creased paperboard. Int. J. Solids Struct. 48 (16), 2275–2287.

Gilchrist, A.C., Suhling, J.C., Urbanik, T.J., 1999. Nonlinear finite element modeling of corrugated board. Mech. Cellulosic Mater. 231/85, 101–106.

Gomes, M.H., Beaudry, R.M., Almeida, D.P.F., Malcata, F.X., 2010. Modelling respiration of packaged fresh-cut "Rocha" pear as affected by oxygen concentration and temperature. J. Food Eng. 96, 74–79.

Hagman, A., 2013, Investigations of In-Plane Properties of Paperboard. Licentiate Dissertation, KTH Royal Institute of Technology.

Haj-Ali, R., Choi, J., Wei, B.-S., Popil, R., Schaepe, M., 2009. Refined nonlinear finite element models for corrugated fiberboards. Data Compos. Struct. 87 (4), 321–333.

Hallbaack, N., Girlanda, O., Tryding, J., 2006. Finite element analysis of ink-tack delamination of paperboard. Int. J. Solids Struct. 43 (5), 899–912.

Han, J., Park, J.M., 2007. Finite element analysis of vent/hand hole designs for corrugated fibreboard boxes. Packag. Technol. Sci. 20, 39–47.

Harris, I.D., 2012. Additive manufacturing: a transformational advanced manufacturing technology. Adv. Mater. Process. 170, 25–29. Retrieved from, http://www.asminternational.org/documents/10192/1900715/amp17005p25.pdf/489dfd7b-a9e6-4a92-8417-f5bd114f8eff/AMP17005P25.

Harrysson, A., Ristinmaa, M., 2008. Large strain elasto-plastic model of paper and corrugated board. Int. J. Solids Struct. 45 (11), 3334–3352.

Haslach Jr., H.W., 2000. The moisture and rate-dependent mechanical properties of paper: a review. Mech. Time-Depend. Mater. 4 (3), 169–210.

Heynick, M., Stotz, I., 2006, 3D CAD, CAM and Rapid Prototyping, LAPA Digital Technology Seminar. Retrieved from http://enac-oc.epfl.ch/files/content/sites/enacco/files/3D%20CAD%20CAM%20and%20Rapid%20PrototypingV1.1.pdf.

Hsiao, K.-A., Chen, L.-L., 2006. Fundamental dimensions of affective responses to product shapes. Int. J. Ind. Ergon. 36, 553–564.

Huang, H., Nygards, M., 2010. A simplified material model for finite element analysis of paperboard creasing. Nordic Pulp Paper Res. J. 25 (4), 505–512.

Huang, H., Hagman, A., Nygards, M., 2014. Quasi static analysis of creasing and folding for three paperboards. Mech. Mater. 69 (1), 11–34.

Hubbe, M.A., 2013. Prospects for maintaining strength of paper and paperboard products while using less forest resources: a review. Bioresources 9 (1), 1634–1763.

INCPEN, 2012, Packaging and the Internet: A Guide to Packaging Goods for Multi-Channel Delivery Systems, INCPEN, Retrieved from http://www.incpen.org/docs/packagingandtheinternet.pdf.

Itech-soft, PALLET-Express. 2014, Computation of the Mechanic Performances, from Itech: http://www.itech-soft.com/.

Jimenez-Caballero, M.A., Conde, I., Garcia, B., Liarte, E., 2009. Design of different types of corrugated board packages using finite element tools. In: SIMULIA Customer Conference.

Kannan, A., Sandaka, P.G., 2008. Heat transfer analysis of canned food sterilization in a still retort. J. Food Eng. 88, 213–228. https://doi.org/10.1016/j.jfoodeng.2008.02.007.

Katicic, J., Bachvarov, A., Walter, S., Hrabal, D., 2011. Emotional customer feedback on virtual products in immersive environment. In: XX MNTK Automation of Discrete Production 2011–June 2011, Sozopol, pp. 665–675.

Katicic, J., Häfner, P., Ovtcharova, J., 2013. Methodology for immersive emotional assessment of virtual product design by customers. In: Joint Virtual Reality Conference of EGVE, pp. 77–82.

Kibirkstis, E., Lebedys, A., Kabelkaite, A., Havenko, S., 2007. Experimental study of paperboard package resistance to compression. Mechanics 63 (1), 27–33.

Korthorst, T., Stoffer, R., 2012, Integrated Photonics Design Flow Automation, PhoeniX Software. Retrieved from http://www.phoenixbv.com/appnote/Photonic_Design_Automation.pdf.

Koutsimanis, G., Getter, K., Behe, B., Harte, J., Almenar, E., 2012. Influences of packaging attributes on consumer purchase decisions for fresh produce. Appetite 59 (2), 270–280.

Lara, C.A.A., Athes, V., Buche, P., Valle, G.D., Farines, V., Fonseca, F., Guillard, V., Kansou, K., Kristiawan, M., Monclus, V., Mouret, J.-R., Ndiaye, A., Neveu, P., Passot, S., Penicaud, C., Sablayrolles, J.-M., Salmon, J.-M., Thomopoulos, R., Trelea, C., 2018. The virtual food system: innovative models and experiential feedback in technologies for winemaking, the cereals chain, food packaging and eco-designed started production. Innov. Food Sci. Emerg. Technol. 46, 54–64.

Leadbeater, T.W., 2009. The Development of Positron Imaging Systems for Applications in Industrial Process Tomography (PhD Thesis), The University of Birmingham, Birmingham, Retrieved from http://etheses.bham.ac.uk/521/1/Leadbeater09PhD_A1b.pdf.

Li, Y., Stapleton, S.E., Reese, S., Simon, J.-W., 2016. Anisotropic elastic-plastic deformation of paper: in-plane model. Int. J. Solids Struct. 100–101, 286–296.

Linvill, E., Ostlund, S., 2016. Parametric study of hydroforming of paper materials using the explicit finite element method with a moisture-dependent and temperature-dependent constitutive model. Packag. Technol. Sci. 29 (3), 145–160.

Lorenz, M., Spranger, M., Riedel, T., Pürzel, F., Wittstock, V., Klimant, P., 2016. CAD to VR—a methodology for the automated conversion of kinematic CAD models to virtual reality. Research and Innovation in manufacturing: key Enabling Technologies for the Factories of the Future, Procedia CIRP 41, 358–363.

Mahajan, P.V., Oliveira, F.A.R., Montanez, J.C., Iqbal, T., 2008. Packaging design for fresh produce: an engineering approach. New Food 1, 35–36.

Mahalik, N.P., Nambiar, A.N., 2010. Trends in food packaging and manufacturing systems and technology. Trends Food Sci. Technol. 21, 117–128.

Manyika, J., Chui, M., Brown, B., Bughin, J., Dobbs, R., Roxburgh, C., Byers, A.H., 2011. Big Data: The Next Frontier for Innovation, Competition, and Productivity. McKinsey & Company.

Marks, S., Estevez, J.E., Connor, A.M., 2014. Towards the holodeck: fully immersive virtual reality visualisation of scientific and engineering data. In: Proceedings of the 29th International Conference on Image and Vision Computing New Zealand. https://doi.org/10.1145/2683405.2683424.

McMahan, R., Bowman, D., Zielinski, D., Brady, R., 2012. Evaluating display fidelity and interaction fidelity in a virtual reality game. IEEE Trans. Vis. Comput. Graph. 18, 626–633.

Mcor Technologies, 2013, How Paper-Based 3D Printing Works: The Technology and Advantages, Mcor Technologies Ltd., Retrieved from http://mcortechnologies.com/how-paper-based-3d-printing-works-the-technology-and-advantages/.

Morganti, P., 2014. Biomimetic materials mimicking nature at the base of EU projects. J. Sci. Res. Rep. 3, 532–544.

Nagasawa, S., Fukuzawa, Y., Yamaguchi, T., Tsukatani, S., Katayama, I., 2003. Effect of crease depth and crease deviation on folding deformation characteristics of coated paperboard. J. Mater. Process. Technol. 140 (1), 157–162.

Navaranjan, N., Johnson, B., 2006. Modelling and experimental study of creep behaviour of corrugated paperboard. In: 60th Appita Annual Conference and Exhibition, Melbourne, Australia, pp. 3–5. Proceedings (p. 43). Appita Inc.

Ngcobo, M.E., 2013. Resistance to Airflow and Moisture Loss of Table Grapes Inside Multi-Scale Packaging (PhD Dissertation), Stellenbosch University.

Nguyen, P.-M., Goujon, A., Sauvegrain, P., Vitrac, O., 2013. A computer-aided methodology to design safe food packaging and related systems. Am. Inst. Chem. Eng. J. 53, 1183–1212. https://doi.org/10.1002/aic.14056.

Noldus Information Technology, 2012. Retrieved June 2014, from http://www.noldus.com/news/focom-new-technology-measurement-product-perception-and-choice-behavior.

Nordstrand, T., Carlsson, L.A., 1997. Evaluation of transverse shear stiffness of structural core sandwich plates. Compos. Struct. 38, 145–153.

Norton, T., Sun, D.-W., 2006. Computational fluid dynamics (CFD)—an effective and efficient design and analysis tool for the food industry: a review. Trends Food Sci. Technol. 17, 600–620.

Oms-Oliü, G., Soliva-Fortuny, R., Martín-Belloso, O., 2008. Modelling changes of headscape gas concentration to describe the respiration of fresh-cut melon under low or super atmospheric oxygen atmospheres. J. Food Eng. 85, 401–409.

Opara, U.L., Fadiji, T., 2018. Compression damage susceptibility of apple fruit packed inside ventilated corrugated paperboard package. Sci. Horticult. 227, 154–161.

Ostlund, S., Nygards, M., 2009. Through-thickness mechanical testing and computational modelling of paper and board for efficient materials design. In: Hannu Paulapuro Symposium, Esbo, Finlan, pp. 69–82.

Paine, F.A., Paine, H.Y., 1992. A Handbook of Food Packaging. Chapman and Hall. Chapter 14.

Pascall, M.A., Richtsmeier, J., Riemer, J., Farahbakhsh, B., 2002. Non-destructive packaging seal strength analysis and leak detection using ultrasonic imaging. Packag. Technol. Sci. 15, 275–285.

Patel, P., Nordstrand, T., Carlsson, L.A., 1997. Local buckling and collapse of corrugated board under biaxial stress. Compos. Struct. 39 (1), 93–110.

Patankar, K.A., 2005, Evaluation of Seal Integrity of Flexible Food Polytrays by Destructive and Non-Destructive Techniques, Master's Thesis, University of Tennessee, Knoxville, Retrieved 2016, from http://trace.tennessee.edu/utk_gradthes/2308/.

Pathare, P.B., Opara, U.L., 2014. Structural design of corrugated boxes for horticultural produce: a review. Biosyst. Eng. 125, 128–140.

Pathare, P.B., Opara, U.L., Vigneault, C., Delele, M.A., Al-Said, F.A.-J., 2012. Design of packaging vents for cooling fresh horticultural produce. Food Bioprocess Technol. 5, 2031–2045. https://doi.org/10.1007/s11947-012-0883-9.

Peterson, J., 1983. Bibliography on lumber and wood panel diaphragms. J. Struct. Eng. 109 (12), 2838–2852.

Pommier, J.-C., Poustis, J., 1990. Bending stiffness of corrugated board prediction using the finite elements method. Am. Soc. Mech. Eng. Appl. Mech. Div. 112, 67–70.

Pradier, C., Cavoret, J., Dureisseix, D., Jean-Mistral, C., Ville, F., 2016. An experimental study and model determination of the mechanical stiffness of paper folds. J. Mech. Des. 138(4).

Pré, G., 1992. Trends in food processing and packaging technologies. Packag. Technol. Sci. 5 (5), 265–269.

Profaizer, M., 2007. Shelf life of pet bottles estimated via a finite elements method simulation of carbon dioxide and oxygen permeability. Ital. Food Beverage Technol., 1–6.

Raheem, D., 2013. Application of plastics and paper as food packaging materials—an overview. Emir. J. Food Agric. 25 (3), 177–188.

Rahman, A.A., Urbanik, T.J., Mahamid, M., 2006. FE analysis of creep and hygroexpansion response of a corrugated fibreboard to a moisture flow: a transient nonlinear analysis. Wood Fibre Sci. 38 (2), 268–277.

Ramasubramanian, M.K., Wang, Y., 2007. A computational micromechanics constitutive model for the unloading behaviour of paper. Int. J. Solids Struct. 44 (22), 7615–7632.

Rao, S.S., 2010. The Finite Element Method in Engineering. Elsevier.

Reh, S., Beley, J.D., Mukherjee, S., Khor, E.H., 2006. Probabilistic finite element analysis using ANSYS. Struct. Saf. 28 (1), 17–43.

Reid, J.D., Bielenberg, R.W., Coon, B.A., 2001. Indenting: buckling and piercing of aluminum beverage cans. Finite Elem. Anal. Des. 37, 131–144.

Rennie, T., Tavoularis, S., 2009. Perforation-mediated modified atmosphere packaging. Part II. Implementation and numerical solution of a mathematical model. Postharvest Biol. Technol. 51, 10–20.

Roduit, B., Borgeat, C.H., Cavin, S., Fragnière, C., Dudler, V., 2005. Application of Finite Element Analysis (FEA) for the simulation of release of additives from multilayer polymeric packaging structures. Food Addit. Contam. 22, 945–955.

Rundh, B., 2009. Packaging design: creating competitive advantage with product packaging. Br. Food J. 111, 988–1002. https://doi.org/10.1108/00070700910992880.

Rutgers, L.Y., 2012, Emerging food packaging technologies: an overview. In: Emerging Food Packaging Technologies, Ed. By Yam, L.K. and Lee, D.S., Woodhead Publishing Limited, Cambridge, UK, Chapter 1.

Saguy, I., 1983. Computer-Aided Techniques in Food Technology. M. Dekker, New York.

Santos, E.C., Shiomi, M., Osakada, K., Laoui, T., 2006. Rapid manufacturing of metal components by laser forming. Int. J. Mach. Tools Manuf. 46, 1459–1468.

Sarghini, F., Erdogdu, F., 2016. A computational study on heat transfer characteristics of particulate canned foods during end-over-end rotational agitation: effect of rotation rate and viscosity. Food Bioprod. Process., 100. https://doi.org/10.1016/j.fbp.2016.07.009.

Sirkett, D., Hicks, B., Mullineux, G., Medland, T., 2007. Simulation based design in the packaging and processing industry. In: International Conference on Engineering Design, Paris, France.

Smith, J.P., Ramaswamy, H.S., Simpson, B.K., 1990. Developments in food packaging technology. Part II. Storage aspects. Trends Food Sci. Technol. 1, 111–118.

Smurfit, T., 2013. Smurfit Kappa, retrieved from http://www.smurfitkappa.com/vhome/ie/newsroom/pressreleases/pages/smurfit-kappa-steps-into-the-shoppers-mind.aspx.

Srirekha, A., Bashetty, K., 2010. Infinite to finite: an overview of finite element analysis. Indian J. Dent. Res. 21 (3), 425–432.

Stenberg, N., 2003. A model for the through-thickness elasticeplastic behaviour of paper. Int. J. Solids Struct. 40 (26), 7483–7498.

Suvanjumrat, C., Puttapitukporn, T., 2011. Determination of drop-impact resistance of plastic bottles using computer aided engineering. Kasetsart J. (Nat. Sci.) 45, 932–942.

Svanes, E., Vold, M., Møller, H., Kvalvåg Pettersen, M., Larsen, H., Jørgen Hanssen, O., 2010. Sustainable packaging design: a holistic methodology. Packag. Technol. Sci. 23, 161–175. https://doi.org/10.1002/pts.887.

Szentgyörgyvölgyi, R., 2005. Effect of the digital technology to the print production processes. Acta Polytech. Hung. 5, 125–134.

Talbi, N., Batti, A., Ayad, R., Guo, Y. Q, 2009, An analytical homogenization model for finite element modelling of corrugated cardboard. Compos. Struct., 88(2), pp. 280-289.

Teixeira, A., 2005. Conventional thermal processing. In: Encyclopedia of Life Support Systems, Food Engineering Theme. vol. 3. EOLSS Publishers Co. Ltd., Oxford, UK, pp. 415–424.

Teixeira, A.A., Dixon, J.R., Zahradnik, J.W., Zinsmeister, G.E., 1969. Computer aided optimization of nutrient retention in the thermal processing of conduction-heated foods. Food Technol. 23 (6), 137–142.

ten Klooster, R., 2008. Packaging and Dilemmas in Packaging Development. CurTec Int.

Tian, F., Decker, E., Goddard, J., 2013. Controlling lipid oxidation via a biomimetic iron chelating active packaging material. J. Agric. Food Chem. 61, 12397–12404.

Thakkar, B.K., Gooren, L.G.J., Peerlings, R., Geers, M., 2009. Experimental and numerical investigation of creasing in corrugated paperboard. Philos. Mag. 88 (28–29), 3299–3310. https://doi.org/10.1080/14786430802342576.

Topçu, O., Taşcıogolu, Y., Ünver, H.Ö., 2011. A method for slicing CAD models. In: 6th International Advanced Technologies Symposium (IATS'11), pp. 141–145.

Tutar, M., Erdogdu, F., Toka, B., 2009. Computational modeling of airflow patterns and heat transfer prediction through stacked layers' products in a vented box during cooling. Int. J. Refrig. 32 (2), 295–306.

Vanferrost, M., Ragaert, P., Devlieghere, F., De Meulenaer, B., 2014. Intelligent food packaging: the next generation. Trends Food Sci. Technol. 39, 47–62.

Vanferrost, M., Ragaert, P., Verwaeren, J., De Meulenaer, B., De Baets, B., Devlieghere, F., 2017. Comput. Ind. 87, 1–14.

Verdouw, C., Wolfert, J., Beulens, A., Rialland, A., 2016. Virtualization of food supply chains with the internet of things. J. Food Eng. 176, 128–136. https://doi.org/10.1016/j.jfoodeng.2015.11.009.

Vishtal, A., Retulainen, E., 2012. Deep-drawing of paper and paperboard: the role of material properties. Bioresources 7 (3), 4424–4450.

Volpato, N., Franzoni, A., Luvizon, D.C., Schramm, J.M., 2013. Identifying the directions of a set of 2D contours for additive manufacturing process planning. Int. J. Adv. Manuf. Technol. 68, 33–43.

Waseem, A., Nawaz, A., Munir, N., Islam, B., Noor, S., 2012. Comparative analysis of different materials for pallet design using ANSYS. Int. J. Mech. Mechatron. Eng. (IJMME-IJENS) 13, 26–32.

Whyte, J., Bouchlaghem, N., Thorpe, A., McCaffer, R., 2000. From CAD to virtual reality: modelling approaches: data exchange and interactive 3D building design tools. Autom. Constr. 10, 43–55.

Wikström, F., Williams, H., Verghese, K., Clune, S., 2014. The influence of packaging attributes on consumer behaviour in food-packaging life cycle assessment studies—a neglected topic. J. Clean. Prod. 73, 100–108.

Wong, K.V., Hernandez, A., 2012. A review of additive manufacturing. ISRN Mech. Eng. https://doi.org/10.5402/2012/208760.

Xanthopoulos, G., Koronaki, E., Boudouvis, A., 2012. Mass transport analysis in perforation-mediated modified atmosphere packaging of strawberries. J. Food Eng. 111, 326–335.

Xia, Q., 2002. Mechanics of Inelastic Deformation and Delamination in Paperboard. Doctoral Dissertation. Massachusetts Institute of Technology.

Xia, B., Sun, D.-W., 2002. Applications of computational fluid dynamics (CFD) in the food industry: a review. Comput. Electron. Agric. 34, 5–24.

Xia, Q.S., Boyce, M.C., Parks, D.M., 2002. A constitutive model for the anisotropic elasticeplastic deformation of paper and paperboard. Int. J. Solids Struct. 39 (15), 4053–4071.

Xu, X., He, Q., 2004. Striving for a totalintegration of CAD, CAPP: CAM and CNC. Robot. Comput. Integr. Manuf. 20, 101–109.

Xu, X.W., Shi, Q., 1999. Plastics forming processes and the applications of CAD/CAM technology. IPENZ Trans. 26, 33–42.

Yoon, J.W., Cardoso, R.P., Dick, R.E., 2010. Puncture fracture in an aluminum beverage can. Int. J. Impact Eng. 37, 150–160.

Zhang, Z., Qiu, T., Song, R., Sun, Y., 2014. Nonlinear finite element analysis of the fluted corrugated sheet in the corrugated cardboard. Adv. Mater. Sci. Eng. 2014, 654012. https://doi.org/10.1155/2014/654012.

Zou, Q., Opara, L.U., McKibbin, R., 2006a. A CFD modeling system for airflow and heat transfer in ventilated packaging for fresh foods: I. Initial analysis and development of mathematical models. J. Food Eng. 77, 1037–1047.

Zou, Q., Opara, L.U., McKibbin, R., 2006b. A CFD modeling system for airflow and heat transfer in ventilated packaging for fresh foods: II. Computational solution, software development, and model testing. J. Food Eng. 77, 1048–1058.

Further reading

Ampuja, J., Reddy, J., Sterling, P.J., 2009, Packaging Value Chain Optimization: A New Frontier for Supply Chain Excellence, CombineNet Supply Chain Optimizers.

Bilalis, N., 2000, Computer Aided Design—CAD, Technical University of Crete, Retrieved from http://www.adi.pt/docs/innoregio_cad-en.pdf.

Ceppa, C., Marino, G.P., 2012. Food-pack waste systemic management. Alternative ways to reuse materials and to develop new business, products and local markets. In: The 7th International Conference on Waste Management and Technology, pp. 616–623.

Corallo, A., Latino, M.E., Lazoi, M., Lettera, S., 2013. Defining product lifecycle management: a journey across features, definitions, and concepts. ISRN Ind. Eng. https://doi.org/10.1155/2013/170812.

European Commission, 2011. Retail Forum for Sustainability, Issue Paper 8: Packaging Optimisation, European Commission, Retrieved from http://ec.europa.eu/environment/industry/retail/pdf/issue_paper_packaging.pdf.

International Electrotechnical Commission, 2015, Factory of the Future, International Electrotechnical Commission, Retrieved from http://www.iec.ch/whitepaper/pdf/iecWP-futurefactory-LR-en.pd.

Mandal, S., 2013. Brief introduction of virtual reality & its challenges. Int. J. Sci. Eng. Res., 304–309.

PAC NEXT 2014. Ecommerce Packaging Optimization Guidelines, PAC Packaging Consortium, Retrieved from http://www.pac.ca/assets/ecommerce-document-online-version.pdf.

Park, S.-I., Lee, D.S., Han, J., 2013. Eco-design for food packaging innovations. In: Han, J.H. (Ed.), Innovations in Food Packaging. Academic Press Elsevier, p. 624.4.

TechFront, 2012. Subtractive Rapid Prototyping Advances. TechFront, pp. 31–37. Retrieved from, www.sme.org/manufacturingengineering.

Wismans, J.G., 2011. Computed Tomography-based Modeling of Structured Polymers (PhD Dissertation), Technische Universiteit Eindhoven, Eindhoven, https://doi.org/10.6100/IR716799.

Chapter 7

Sustainable food processing: A production planning and scheduling perspective

Renzo Akkerman

Operations Research and Logistics Group, Wageningen University, Wageningen, The Netherlands

Abstract

In food-processing industries, agricultural raw materials are processed into consumer products or food ingredients. Within this context, efficiency and sustainability are mainly impacted by losses of valuable food products that already caused significant monetary and environmental impacts, as well as by the inefficient use of utilities such as energy, water, and cleaning agents. In turn, these factors are mostly driven by product changeovers and cleaning of production and storage equipment. Efforts to improve the efficiency and sustainability of food processing, therefore, emphasize managerial and technological solutions to decrease the number and impact of changeovers. This chapter first distinguishes technological and managerial perspectives on product losses and utility consumption. Elaborating on the managerial perspective, we subsequently provide an overview of cyclic production planning and scheduling approaches that can be used to improve efficiency and sustainability. The intuitive nature of cyclic planning frameworks also provides a lean perspective on planning that facilitates implementation processes.

1 Introduction

The food-processing industry is a key part of food supply chains, covering the transformation of agricultural raw materials in food products. Like most of the actors in food supply chains, the food-processing industry has significant focus on concerns related to quality, safety, and sustainability (Akkerman et al., 2010). In this chapter, the focus is on sustainability and its key underlying concern: the efficient use of resources. In terms of improving resource efficiency and improving sustainability, industries are focusing on both efficient use of raw materials and efficient use of utilities such as water, energy, and cleaning agents.

Raw material efficiency is mainly determined by processing yields and product losses, which can be distinguished by the amount of raw material lost in the actual processing activities, and the amount of raw material lost around the processing activities. In this context, yields are mostly determined by the technological characteristics of the processes that are used. However, product losses are determined by a combination of technological and managerial characteristics as they are often the result of product changeovers and the cleaning of production and storage equipment.

The efficient use of utilities has a similar background. Their use in production and changeovers is partly a technological characteristic of the processes that are used, but the extent of their use (and potential reuse) is also a function of production sequences and batch sizes, which are direct results of production planning and scheduling decisions.

Solutions for more sustainable food processing therefore also focus on either technological or managerial aspects (or both):

- Technological solutions aim to reduce the environmental impact of product losses or utility consumption by changing individual production technologies or changeover operations. This is, for example, achieved by reducing the amount of off-spec products produced at the start of a production batch, by changing to a different (more environmentally friendly) cleaning agent in clean-in-place (CIP) operations, or reducing the amount of water used in rinsing equipment.
- Managerial solutions aim to reduce product losses or utility consumption without changing the underlying technology, but by changing the number and type of changeovers and cleanings that are required. For example, this is achieved here by a reduction of changeover activities through intelligent lot sizing, or the identification of production sequences that avoid more intensive cleaning operations.

Even though changing technologies or processing characteristics is highly relevant, it is beyond the scope of this work. This chapter focuses on the reduction of product losses and utility consumption from a managerial perspective. In this context, production planning and scheduling are important drivers of efficiency, both in terms of traditional measures of operational

Sustainable Food Supply Chains. https://doi.org/10.1016/B978-0-12-813411-5.00007-7

efficiency (e.g., time- or cost-related measures) as well as the resource efficiency underlying sustainable performance. In this chapter, the aim is therefore to (i) outline the key decisions in production planning and scheduling, (ii) identify their impact on the sustainability of food processing, and (iii) discuss how several cyclic planning and scheduling procedures can be used to improve the sustainability of food processing.

2 Production planning and scheduling

Production planning and scheduling are important parts of supply chain management, mostly considering tactical and operational decisions at the interface between sales and operations. They deal with the determination of production and inventory quantities for the upcoming time periods (e.g., days or weeks), as well as the determination of the actual task assignments, timing, and sequencing of production quantities within those time periods (e.g., Garcia and You, 2015; Puigjaner et al., 2012). These two interrelated decision problems are often referred to as production planning and detailed scheduling, providing the decision-making hierarchy illustrated in Fig. 1. To further illustrate these two decision levels, we briefly outline the main characteristics of the decision problem on each level. We also provide simple archetype optimization models that are meant to illustrate the core of the decision problems, and capture the main trade-offs that have to be dealt with on the specific decision level.

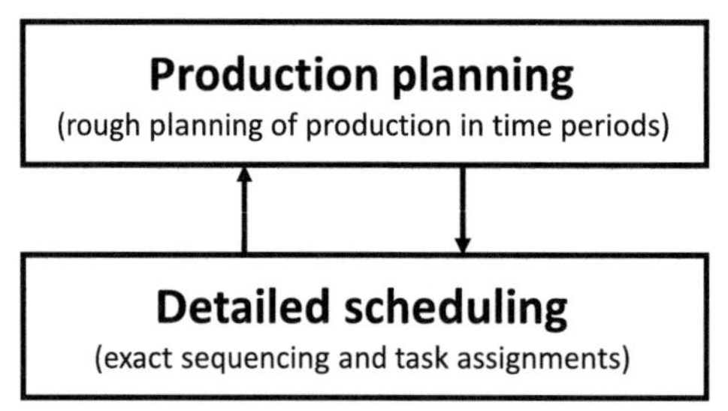

FIG. 1 Hierarchical view on production planning and detailed scheduling.

2.1 Production planning

The top level, production planning, deals with the aggregate planning of production and inventory quantities and is often the core of the sales and operations planning function in organizations. Mathematically, it minimizes production costs (including setup costs) and inventory costs. Using the notation presented in Table 1, the archetype decision problem can be formulated as follows.

$$\min \quad \sum_{i \in \mathcal{I}} \sum_{t \in \mathcal{T}} h_i I_{it} + c_{it} X_{it} + s_{it} Y_{it} \tag{1}$$

$$\text{s.t.} \quad \sum_{i \in \mathcal{I}} \alpha_i X_{it} \leq k_t \qquad \forall t \in \mathcal{T} \tag{2}$$

$$I_{it} = I_{i,t-1} + X_{it} - d_{it} \qquad \forall i \in \mathcal{I}, t \in \mathcal{T} \tag{3}$$

$$X_{it} \leq M \cdot Y_{it} \qquad \forall i \in \mathcal{I}, t \in \mathcal{T} \tag{4}$$

$$X_{it}, I_{it} \geq 0, Y_{it} \in \{0,1\} \qquad \forall i \in \mathcal{I}, t \in \mathcal{T} \tag{5}$$

This model minimizes the sum of holding costs, production costs, and setup costs in Eq. (1). The included constraints make sure that production capacity is not exceeded (Eq. 2), that inventory is tracked over the time horizon (Eq. 3), that setup activities are initiated if there is production of the related product (Eq. 4), and that the variable domains are defined properly (Eq. 5).

Since these decision problems typically deal with the distribution of a fixed production quantity over time, production costs are often not considered and the key trade-off is between holding costs and setup costs, similar to traditional economic order quantity models (see, e.g., Andriolo et al., 2014).

From a sustainable food production perspective, the decisions in the previous model can each be related to their environmental impacts.

In food production, setup activities mostly include significant cleaning activities, using energy, water, and cleaning agents (Stefansdottir et al., 2017) to assure product quality and safety. Furthermore, setup activities are often linked to product waste, if, for example, the first product after starting production does not meet specifications, or remaining products

TABLE 1 Notation for the archetype production planning model (indices and sets, parameters, and decision variables)

Sets and indices	
$i \in \mathcal{I}$	Set of products
$t \in \mathcal{T}$	Set of time periods
Parameters	
d_{it}	(Expected) demand for product i in period t
k_t	(Expected) production capacity in period t
α_i	Capacity required to produce product i
h_i	Holding costs per period for product i
c_{it}	Production costs for product i in period t
s_{it}	Setup costs for product i in period t
M	A relatively large number
Decision variables	
X_{it}	Production quantity for product i in period t
Y_{it}	Binary variable representing the production of product i in period t
I_{it}	Inventory of product i at the end of period t

from previous batches have to be removed from production equipment (e.g., Akkerman and van Donk, 2008). Minimizing the number of setups therefore often becomes increasingly important, as it directly reduces resource consumption and product waste. It should, however, be noted that the actual resource consumption resulting from setups can only be determined in the detailed scheduling level, as sequence dependency often has a significant impact. This also means that a good estimation of the aggregate resource consumption resulting from setups is required on the production planning level, even though actual sequencing is not known yet.

Holding inventory does, however, also lead to increased environmental impact, especially when perishable products have to be stored in temperature-controlled environments. More specifically, this concerns the environmental impact of warehousing (e.g., energy consumption for cold or frozen storage) and obsolescence (i.e., potentially wasted perishable products).

Together, the environmental impacts related to setups and inventory may or may not change the trade-off between setups and inventory significantly. Battini et al. (2014) for instance show that for more expensive products, the trade-off changes: the economically preferred small production lots that reduce the amount of capital bound in inventories (reflected in the holding cost parameter) conflict with an environmental preference for a lower number of setups.

Finally, it should be noted that neglecting the production costs in these production planning problems might not be possible anymore if environmental objectives are to be considered. For instance, the resource consumption related to production activities might depend on the mix of production orders assigned to a time period (e.g., Duflou et al., 2012). The consideration of the environmental impacts related to this resource consumption increases the relevance of considering this in the decision problem, and the dynamic behavior resulting from product mix assignments might also have to be modeled in detail (which would likely require consideration on the detailed scheduling level).

2.2 Detailed scheduling

The production planning discussed previously only provides the rough outline for the process of detailed scheduling. Here, the production orders have to be sequenced and assigned to production equipment. This detailed scheduling step usually covers one or more of the time periods dealt with on the production planning level. Fig. 2 illustrates this relationship.

Where the production planning is often based on cost considerations, the detailed scheduling is often more concerned with time considerations. There are many ways to model the required timing and sequencing constraints. The formulation we use here as an illustration uses the concept of immediate precedence, where a sequencing variable ($Z_{kk'}$) connects the production orders that immediately follow each other. For more details on possible ways to formulate timing and

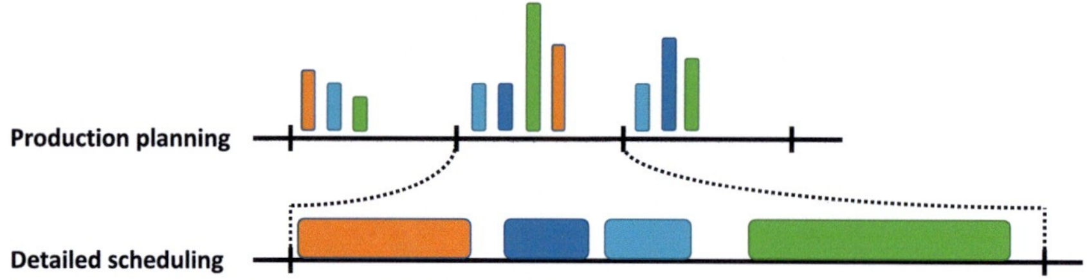

FIG. 2 Key differences between production planning and detailed scheduling.

TABLE 2 Notation for the archetype detailed scheduling model (indices and sets, parameters, and decision variables)

Sets and indices	
$k, k' \in \mathcal{K}$	Set of production orders
$l \in \mathcal{L}$	Set of machines
Parameters	
p_{kl}	Processing time for production order k on machine l
$\tau_{kk'l}$	Setup time between production order k and production order k' on machine l
M	A relatively large number
Decision variables	
W_{kl}^0	Binary variable representing the assignment of production order k as the first activity on machine l
W_{kl}	Binary variable representing the assignment of production order k to machine l
$Z_{kk'}$	Binary variable representing that production order k' is sequenced immediately after production order k
S_k	Starting time of production order k
C_k	Completion time of production order k
$C_{\max}$	Makespan: completion time of the last production order

sequencing constraints, we refer to Mendez et al. (2006). Using the notation introduced in Table 2, the following model formulation can be made.

$$\min \quad C_{\max} \tag{6}$$

$$\text{s.t.} \quad \sum_{k \in \mathcal{K}} W_{kl}^0 \leq 1 \qquad \forall l \in \mathcal{L} \tag{7}$$

$$\sum_{l \in \mathcal{L}} W_{kl}^0 + \sum_{l \in \mathcal{L}} W_{kl} = 1 \qquad \forall k \in \mathcal{K} \tag{8}$$

$$\sum_{l \in \mathcal{L}} W_{kl}^0 + \sum_{k' \in \mathcal{K}} Z_{kk'} = 1 \qquad \forall k \in \mathcal{K} \tag{9}$$

$$\sum_{k' \in \mathcal{K}} Z_{kk'} \leq 1 \qquad \forall k \in \mathcal{K} \tag{10}$$

$$W_{kl}^0 + W_{kl} \leq W_{k'l} - Z_{kk'} + 1 \qquad \forall k, k' \in \mathcal{K} : k \neq k', \forall l \in \mathcal{L} \tag{11}$$

$$C_k = S_k + \sum_{l \in \mathcal{L}} p_{kl}(W_{kl}^0 + W_{kl}) \qquad \forall k \in \mathcal{K}, l \in \mathcal{L} \tag{12}$$

$$S_{k'} \geq C_k + \sum_{l \in \mathcal{L}} \tau_{kk'l} W_{k'l} - M(1 - Z_{kk'}) \qquad \forall k, k' \in \mathcal{K} \tag{13}$$

$$S_k, C_k, C_{\max} \geq 0, Z_{kk'}, W_{kl}^0, W_{kl} \in \{0, 1\} \qquad \forall k, k' \in \mathcal{K}, \forall l \in \mathcal{L} \tag{14}$$

This formulation minimizes the makespan, that is, the time it takes to complete all production orders, in Eq. (6). Constraints (7) assure that at most one production order is assigned as the first order on a machine. Constraints (8) subsequently make sure that each production order gets assigned to one of the machines, and constraints (9) assure that the production orders all follow another order in a production sequence (or are assigned the first slot on a machine). Also, each production order can at most have one other order immediately following it, which constraints (10) ensure. To link the assignment decisions to the sequencing decisions, constraints (11) make sure that production orders that are immediately following each other are also assigned to the same machine. Constraints (12), (13) ensure that the start times and completion times of the production orders are determined, taking into account the processing times and setup times, respectively. Finally, constraints (14) set the variable domains for the decision variables.

As in the previous discussion on production planning, the decisions in the previously mentioned scheduling model can also be related to their environmental impacts. On this level, the key decisions are the allocation and sequencing of production orders. The sequence-dependent setups resulting from these decisions play the most important role here, as they determine the actual resource consumption (i.e., energy, water, cleaning agents) for the setup activities. The amount of resources consumed resulting from, for example, cleaning operations can be reduced significantly if food products are scheduled from light to dark colors or from mild to strong flavors. Also, taking into account product family considerations in the assignment of products to machines can make a big difference.

In addition to the direct result of sequence-dependent setup, the production sequence can also have other implications. First, a more balanced workload over time or over machines might lead to more stable resource requirements in terms of energy and water. For instance, avoiding energy usage peaks might be both economically and environmentally interesting for a company, not even considering the potential environmental benefits upstream in the energy production chain (see also Gahm et al., 2016). In some cases, the scheduling model could also be extended with a variable speed for production, impacting the energy consumption (Mansouri et al., 2016).

The positioning of production and setup activities in schedules also impacts possible reuse of utilities. When machines or processes have different resource usage characteristics, heat exchange between processes might lead to significant savings (e.g., Miah et al., 2014). Also, water used in cleaning operations might be reused if the water quality is sufficient (or could be made sufficient) and the timing of the water supply and water demand would allow for reuse (e.g., Pulluru and Akkerman, 2018; Pulluru et al., 2017).

3 Cyclic scheduling

The food-processing industry can be distinguished by several product and process characteristics. The most important characteristics identified by Akkerman and van Donk (2008) are the presence of (sequence-dependent) setup times, a divergent product structure, perishability of the products, shared resources, and a variable demand for end products.

Related to planning and scheduling, these environments often see the use of cyclic schedules. Such schedules have several advantages in these environments:

- They inherently focus on efficient production sequences.
- They can easily be built around the product mix structures resulting from a divergent product mix.
- They provide the opportunity for regular production of the perishable products.
- They facilitate the sharing of production resources by different products.
- They can often easily be adapted for demand patterns that fluctuate over time.

Several of these industry-specific characteristics do, however, favor large efficient production batches, leading to decreased flexibility and large inventories or long lead times. This is one of the main reasons that production flexibility is challenging in process industries such as food processing.

Flexibility is, however, a concept that can be interpreted in various ways (Upton, 1995), and increased flexibility can therefore also be achieved in different ways. To improve the flexibility and create a more "lean" production environment, cyclic schedules have traditionally been proposed (Hall, 1988). Also for the process industries, concepts like just-in-time (JIT) and lean often boil down to the development of a good cyclic schedule (Pool et al., 2011): the production regularity and coordination simplicity that result from cyclic schedules are the main drivers behind process improvement, inventory reductions, customer service improvements, as well as an increased efficiency in production planning and scheduling activities. As such, cyclic schedules contribute to various key performance indicators at the same time, as elaborated upon in Table 3 (see also Dora et al., 2013, for an empirical perspective on these relationships).

TABLE 3 Key performance indicators for operational performance and the way cyclic scheduling aims to improve them

Key performance indicator	Main link to cyclic scheduling
Stock/inventory reduction	Regular production of each product, leading to reduced cycle stock
Productivity improvement	Repetition of similar setup activities efficiently use labor resources
Cycle time reduction	Focusing on setup reduction in cycles eventually allows for cycle length reduction
Quality improvement	Regularity in production and changeovers allows for improved process control
On-time delivery improvement	Predictable schedules increase coordination with customers

Source: Selection of indicators based on Dora, M., Kumar, M., Van Goubergen, D., Molnar, A., Gellynck, X., 2013. Operational performance and critical success factors of lean manufacturing in European food processing SMEs. Trends Food Sci. Technol. 31 (2), 156–164.

From an operations management perspective, there are several methodologies that support the development of cyclic schedules. The practitioner literature on lean in the process industries often refers to cyclic schedules as fixed sequence variable volume (FSVV) (e.g., Floyd, 2010). The repetitive structure in the resulting schedule is often referred to as a "product wheel" or "rhythm wheel" (King, 2009; King and King, 2013; Packowski, 2013; Packowski and Francas, 2013; Wilson and Ali, 2014), resembling a regularly repeating sequence of products being produced.

The quantitative operations management literature also provides methods to support the development of cyclic schedules. For instance, approaches to the well-known economic lot scheduling problem (ELSP) normally lead to schedules of a cyclic nature (Maxwell, 1964; Pinedo, 2009), and the block planning approach specifically aims to support the planning activities in regularly repeating sequences (Günther, 2014).

The remainder of this chapter will focus on some of the main methodologies that facilitate the creation of cyclic schedules. Here, we distinguish two types of approaches: conceptual modeling approaches and quantitative modeling approaches. Both of these types of approaches build on the same principles, but have been focusing on different audiences. Whereas the conceptual modeling approaches are mostly presented in books targeted at practitioners and consultants, the quantitative modeling approaches are mostly presented in scientific journal papers. Following the hierarchical perspective on planning and scheduling, Fig. 3 includes a hierarchical view of the modeling approaches, where the conceptual modeling approaches provide the upper level and the quantitative modeling approaches the lower level. The interpretation of this relationship is the identification of the main characteristics of a cyclic schedule on the upper level, complemented with the quantitative support on the lower level. The conceptual modeling approaches typically take the form of guidelines or principles, where the quantitative modeling approaches aim to capture the mathematical relationships behind cyclic schedules so as to determine their optimal use.

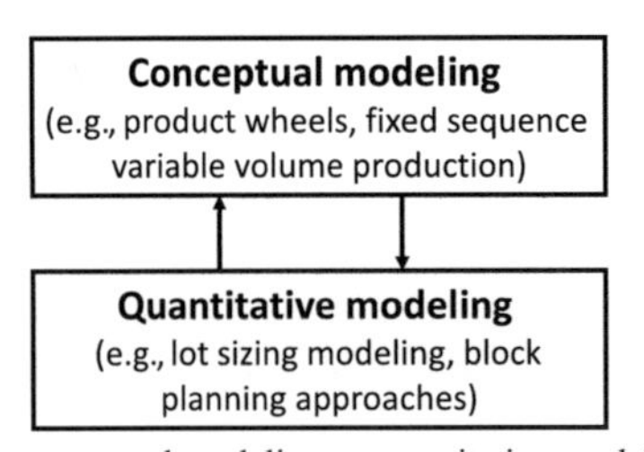

FIG. 3 Decision support for cyclic scheduling: from conceptual modeling to quantitative modeling.

3.1 Fixed sequence variable volume

The concept of FSVV production relates to fixed production cycles that sequence products in such a way that high-cost changeovers are minimized and low-cost changeovers are maximized (Floyd, 2010). As described in detail by Floyd (2010), FSVV is composed of four components:

1. a *fixed sequence* that aims at minimizing changeover costs;
2. an *inventory policy* with production quantities that are able to cover demand during the production cycle;
3. a *variable volume scheduling* approach, that enables a more flexible use of the production cycle; and
4. a *continuous improvement* focus, aiming to improve the (limited number of) changeover operations that are used in the fixed sequence.

Obviously, these key components might be affected by product and process characteristics that, for instance, require minimum run lengths. In such cases, products can also move in and out of the production cycles. Also, following lean and JIT principles, the continuous improvement focus is also expected to contain a drive for continuously shorter cycles, where possible.

3.2 Product wheels

The concept of product wheels (or rhythm wheels) is a way to represent fixed sequences and an intuitive way to design schedules based on FSVV logic. It uses the analogy of the wheel for the structure of a cyclic schedule: it cycles around until it arrives back at the production of the first product (i.e., after the cycle time). As such the wheel consists of production batches and changeovers between these batches, as illustrated in Fig. 4. Please note that these production batches do not necessarily have to be of the same size; the size of the batches could also be represented graphically.

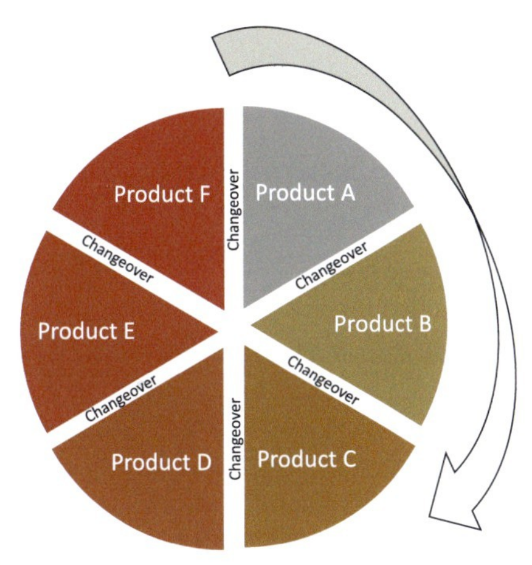

FIG. 4 Illustration of the product wheel concept.

The construction of product wheels can be supported by stepwise procedures, such as the 10-step procedure presented by King (2009) or the 17-step procedure presented by King and King (2013). These procedures focus on the determination of wheel characteristics such as sequence, cycle time, and inventory levels for the make-to-order products. Also, it should be considered whether products should be made in every cycle or should skip one or more cycles (for low-volume products). In addition, part of the wheel may intentionally be left open to allow for the later addition of make-to-order products.

The construction of the wheel is based on many traditional production and inventory control principles. For instance, even though lean thinking would prefer a cycle time that is as short as possible, the cycle time still has to consider the classic trade-off between inventory costs and setup costs (in addition to possible technical constraints as well as shelf life requirements). The inclusion of products in cycles is typically based on demand levels: high-demand products are included in every cycle, medium-demand products skip one or more cycles, and low-demand products can only be included in cycles after certain inventory triggers (or these could be included as make-to-order products that are only included after an order arrives).

Finally, additional steps in the mentioned procedures focus on the organizational embedding of the cyclic schedule, for example, related to visualization, implementation, periodic revision of the product wheel, as well as contingency planning for situations in which the wheel structure has to be abandoned.

For more details on the principles behind product wheels, we refer to King (2009), Packowski (2013), Packowski and Francas (2013), and Wilson and Ali (2014). For a hands-on stepwise procedure, we refer to King and King (2013).

3.3 Planning: Economic lot scheduling models

One of the quantitative modeling approaches that can support the construction of cyclic schedules is the ELSP. There is an abundance of literature on modeling approaches that address this problem, illustrated by early reviews (e.g., Elmaghraby, 1978), inclusion in textbooks (e.g., Pinedo, 2009), and more recent reviews (e.g., Santander-Mercado and Jubiz-Diaz, 2016).

The core of ELSP is the concurrent determination of lot sizes, a production sequence, and cycle length with the objective to minimize the sum of setup costs and holding costs. The classic version of this problem uses averages for demand rates d_i and production rates p_i to analyze the trade-off between the mentioned cost factors. The determination of the optimal cycle time T_i for a single product $i \in \mathcal{I}$ is based on the demand rate and production rate, as well as setup cost c_i, setup time u_i, and holding cost h_i. Based on the resulting inventory levels (as illustrated in Fig. 5), the total setup and holding costs (for the case that all products use the same cycle time T) can easily be determined to be as follows.

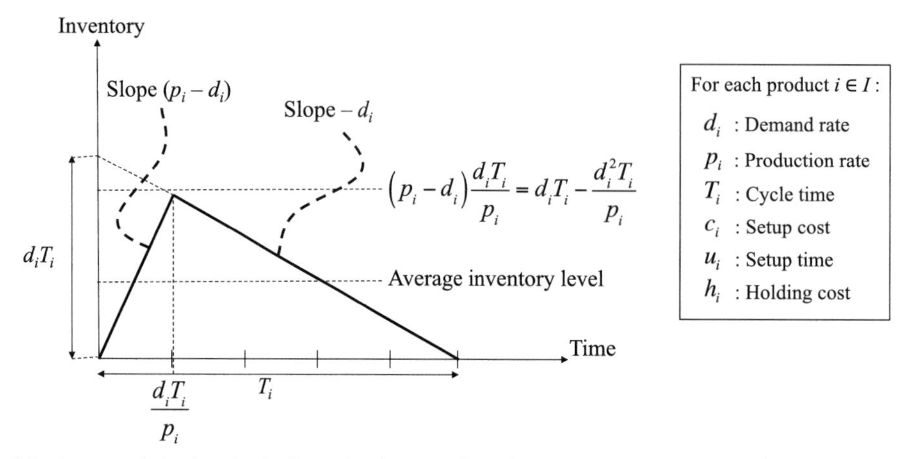

FIG. 5 Illustration of the inventory behavior of a single product in an ELSP setting, including relevant time points and inventory levels.

$$\sum_{i \in \mathcal{I}} \left(\frac{1}{2} h_i \left(d_i T - \frac{d_i^2 T}{p_i} \right) + \frac{c_i}{T} \right) \tag{15}$$

From this equation, the optimal cycle time T and related lot sizes ($d_i T$) can be derived. Extensions of this logic would, for example, lead to consideration of setup time as well as sequence-dependent setup costs (see Pinedo, 2009), as well as more specific aspects such as product shelf life (Soman et al., 2004).

As incarnations of the ELSP, especially when different considerations are added, can be computationally challenging to solve, simplified solution procedures are often used. The advantage of these simplifications is that they often have a natural fit with the simplicity and regularity that cyclic scheduling approaches advocate. Examples of this are the basic period approach (in which cycles can only be integer multiples of a certain basic period, such as a week), and the power-of-two restriction (in which only integers that are powers of two are applied in the multiplication of the basic period). These simplifications, for instance, lead to a cyclic schedule where some products are produced every week, while other products could be produced every 2 or 4 weeks.

Using average production rates and demand rates, the solution of the ELSP would be the production lot sizes and cycle time related to this average. This result would then be used to construct a schedule that can be used as a rough cyclic plan that is executed repetitively for some time (until conditions change to the extent that a revision is warranted). As such, it is an obvious choice of quantitative model to support the various steps in the determination of product wheels.

The inclusion of environmental impacts in ELSP is similar to the discussion in Section 2.1, which could lead to significant differences in lot sizes, as recently shown in relation to energy considerations by Beck et al. (2018).

3.4 Scheduling: Block planning models

Another quantitative modeling approach that can be used to support the use of cyclic schedules is block planning (e.g., Günther, 2014; Günther et al., 2006; Lutke Entrup et al., 2005). This methodology is more focused on the scheduling level, and more appropriate to support the operational decision-making happening while using a cyclic schedule determined by, for example, ELSP methodology. Block planning models typically focus on the minimization of makespan or setup time, and support decision making of the production lot sizes as well as the exact timing of the production lots. This last aspect emphasizes the increase in level of detail compared to the approaches presented in the previous sections. It should also be noted that it has similarities with the scheduling approach described in Section 2.2.

The basic premise of block planning is the definition of blocks that are assigned to time periods (e.g., weeks). Each of these blocks represents a production cycle, without predetermining the exact production lot sizes within the block. The blocks do, however, contain a predetermined production sequence, which is determined up front and used within each block.

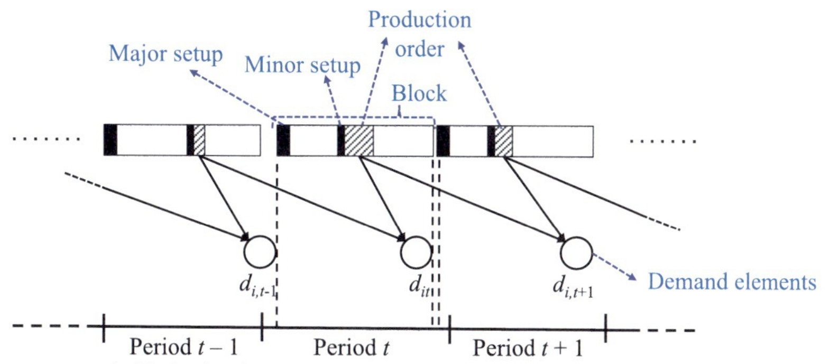

FIG. 6 Illustration of the key concepts in block planning models (e.g., Günther, 2014; Günther et al., 2006; Lutke Entrup et al., 2005).

Based on the data available at the time of scheduling, upcoming blocks get assigned actual production quantities to fulfill demand elements (which could be production orders triggered by actual customer orders or inventory replenishment orders). Often, a distinction is made between major setups in the beginning of a block and minor setups between production orders inside a block. Fig. 6 illustrates these key concepts graphically.

For every period, there are decisions to make on production quantities (also determining the setup activities), exact production timing, and inventory levels (as production could also be related to future demand elements). As such, the approach combines a planning logic based on discrete time periods with a continuous time scheduling logic to determine the exact timing of production batches. In the literature on block planning (e.g., Günther et al., 2006), one of the main arguments is that this planning logic is a natural fit with planning practice, especially when using advanced planning systems.

Facilitating the consideration of environmental objectives in block planning is possible in different ways, similar to the discussion in Section 2.2. First, environmental aspects can be included in the determination of the fixed production sequence used. Sequencing preferences could result from the resource efficiency of the changeovers involved, but could also result from differences in resource consumption during production activities (e.g., more gradual changes in utility consumption might be preferable). Second, as setups are normally explicitly modeled with binary variables in block planning models, the environmental impact of setup activities can easily be modeled and included in an environmental objective function or in constraints representing environmental considerations.

4 Conclusion and discussion

This chapter presented a production planning and scheduling perspective on sustainable food processing. After identifying the key decision problems and their link to sustainable performance of food production systems, it discussed several conceptual and quantitative modeling approaches that can be used to support the design and operation of cyclic planning and scheduling procedures.

The many positive effects that cyclic planning and scheduling have on operational performance translate well to the sustainability drivers in food processing. In addition, the intuitive planning frameworks resulting from the lean perspective on planning in process industries facilitate implementation processes.

This chapter only briefly discussed the main quantitative modeling approaches for cyclic scheduling, as more detailed implementations would have to contain many industry-specific aspects that would go beyond the scope of this chapter. However, it does provide a framework with the core decisions that form the basis of a production management perspective on sustainable food processing, and as such the basis for improvement projects. When also combined with a technological perspective on resource-efficient production and changeovers, it provides a comprehensive roadmap to more sustainable food processing.

References

Akkerman, R., van Donk, D.P., 2008. Development and application of a decision support tool for reduction of product losses in the food-processing industry. J. Clean. Prod. 16 (3), 335–342.

Akkerman, R., Farahani, P., Grunow, M., 2010. Quality, safety and sustainability in food distribution: a review of quantitative operations management approaches and challenges. OR Spectr. 32 (4), 863–904.

Andriolo, A., Battini, D., Grubbström, R.W., Persona, A., Sgarbossa, F., 2014. A century of evolution from Harris's basic lot size model: survey and research agenda. Int. J. Prod. Econ. 155, 16–38.

Battini, D., Persona, A., Sgarbossa, F., 2014. A sustainable EOQ model: theoretical formulation and applications. Int. J. Prod. Econ. 149, 145–153.

Beck, F.G., Biel, K., Glock, C.H., 2019. Integration of energy aspects into the economic lot scheduling problem. Int. J. Prod. Econ. 209, 399–410. Available from https://www.sciencedirect.com/science/article/pii/S0925527318301014.

Dora, M., Kumar, M., Van Goubergen, D., Molnar, A., Gellynck, X., 2013. Operational performance and critical success factors of lean manufacturing in European food processing SMEs. Trends Food Sci. Technol. 31 (2), 156–164.

Duflou, J.R., Sutherland, J.W., Dornfeld, D., Herrmann, C., Jeswiet, J., Kara, S., Hauschild, M., Kellens, K., 2012. Towards energy and resource efficient manufacturing: a processes and systems approach. CIRP Ann. Manuf. Technol. 61 (2), 587–609.

Elmaghraby, S.E., 1978. The economic lot scheduling problem (ELSP): review and extensions. Manag. Sci. 24 (6), 587–598.

Floyd, R., 2010. Liquid Lean: Developing Lean Culture in the Process Industries. CRC Press, Boca Raton, FL.

Gahm, C., Denz, F., Dirr, M., Tuma, A., 2016. Energy-efficient scheduling in manufacturing companies: a review and research framework. Eur. J. Oper. Res. 248 (3), 744–757.

Garcia, D.J., You, F., 2015. Supply chain design and optimization: challenges and opportunities. Comput. Chem. Eng. 81, 153–170.

Günther, H.O., 2014. The block planning approach for continuous time-based dynamic lot sizing and scheduling. Bus. Res. 7 (1), 51–76.

Günther, H.O., Grunow, M., Neuhaus, U., 2006. Realizing block planning concepts in make-and-pack production using MILP modelling and SAP APO(c). Int. J. Prod. Res. 44 (18–19), 3711–3726.

Hall, R.W., 1988. Cyclic scheduling for improvement. Int. J. Prod. Res. 26 (3), 457–472.

King, P.L., 2009. Lean for the Process Industries: Dealing With Complexity. CRC Press, Boca Raton, FL.

King, P.L., King, J.S., 2013. The Product Wheel Handbook: Creating Balanced Flow in High-Mix Process Operations. CRC Press, Boca Raton, FL.

Lutke Entrup, M., Günther, H.O., van Beek, P., Grunow, M., Seiler, T., 2005. Mixed-integer linear programming approaches to shelf-life-integrated planning and scheduling in yoghurt production. Int. J. Prod. Res. 43 (23), 5071–5100.

Mansouri, S.A., Aktas, E., Besikci, U., 2016. Green scheduling of a two-machine flowshop: trade-off between makespan and energy consumption. Eur. J. Oper. Res. 248 (3), 772–788.

Maxwell, W.L., 1964. The scheduling of economic lot sizes. Nav. Res. Log. Quart. 11 (2), 89–124.

Mendez, C.A., Cerdá, J., Grossmann, I.E., Harjunkoski, I., Fahl, M., 2006. State-of-the-art review of optimization methods for short-term scheduling of batch processes. Comput. Chem. Eng. 30 (6–7), 913–946.

Miah, J.H., Griffiths, A., McNeill, R., Poonaji, I., Martin, R., Yang, A., Morse, S., 2014. Heat integration in processes with diverse production lines: a comprehensive framework and an application in food industry. Appl. Energy 132, 452–464.

Packowski, J., 2013. LEAN Supply Chain Planning: The New Supply Chain Management Paradigm for Process Industries to Master Today's VUCA World. CRC Press, Boca Raton, FL.

Packowski, J., Francas, D., 2013. LEAN SCM: a paradigm shift in supply chain management. J. Bus. Chem. 10 (3), 131–137.

Pinedo, M.L., 2009. Economic lot scheduling. In: Planning and Scheduling in Manufacturing and Services, Springer, New York, NY, pp. 143–171.

Pool, A., Wijngaard, J., van der Zee, D.J., 2011. Lean planning in the semi-process industry, a case study. Int. J. Prod. Econ. 131 (1), 194–203.

Puigjaner, L., Laínez, J.M., Reklaitis, G.V., 2012. Process systems engineering. 8. Plant operation, integration, planning, scheduling, and supply chain. In: Ullmann's Encyclopedia of Industrial ChemistryWiley, Weinheim, Germany.

Pulluru, S.J., Akkerman, R., 2018. Water-integrated scheduling of batch process plants: modelling approach and application in technology selection. Eur. J. Oper. Res. 269 (1), 227–243.

Pulluru, S.J., Akkerman, R., Hottenrott, A., 2017. Integrated production planning and water management in the food industry: a cheese production case study. Comput. Aided Chem. Eng. 40, 2677–2682.

Santander-Mercado, A., Jubiz-Diaz, M., 2016. The economic lot scheduling problem: a survey. Int. J. Prod. Res. 54 (16), 4973–4992.

Soman, C.A., Van Donk, D.P., Gaalman, G.J.C., 2004. A basic period approach to the economic lot scheduling problem with shelf life considerations. Int. J. Prod. Res. 42 (8), 1677–1689.

Stefansdottir, B., Grunow, M., Akkerman, R., 2017. Classifying and modeling setups and cleanings in lot sizing and scheduling. Eur. J. Oper. Res. 261 (3), 849–865.

Upton, D.M., 1995. Flexibility as process mobility: the management of plant capabilities for quick response manufacturing. J. Oper. Manag. 12 (3–4), 205–224.

Wilson, S., Ali, N., 2014. Product wheels to achieve mix flexibility in process industries. J. Manuf. Technol. Manag. 25 (3), 371–392.

Chapter 8

Design-support methodologies for job-shop production system in the food industry

Alessandro Tufano, Riccardo Accorsi, Giulia Baruffaldi and Riccardo Manzini

Department of Industrial Engineering, Alma Mater Studiorum—University of Bologna, Bologna, Italy

Abstract

To meet the increasing demand for functional food and changing consumer habits, the food processing industry is shifting from flow manufacturing to batch manufacturing. As a consequence, job-shop (JS) systems are progressively developing to replace traditional food-processing systems. Such systems are characterized by high complexity in their design and management due to the high number of entities involved (i.e., products, components and parts, resources or workshops, operators, handling tools), the technological and operational constraints (i.e., working cycle, shop throughput, set-up tasks), and the layout issues (i.e., flow lines, congestions, bottlenecks). This chapter explores the impact of logistics and handling tasks in the design of food JS processing facilities. The aims of the JS designer are the minimization of the infrastructural costs, the optimization of the products and labor flows, and the enhancement of the safety of food products, affecting concurrently the layouts, operations, and related performances.

The chapter illustrates a set of methodologies and quantitative indicators that aid the design of a JS system involving logistic efficiency, infrastructure cost minimization, and food safety targets. We assess the layout of the manufacturing system through a multidisciplinary dashboard of key performance indicators (KPIs) (1) and a design methodology addressing the resource dimensioning problem (2). A numerical example gathered from an Italian catering company showcases the application of the proposed tools and elicits debate on the best practice for facility design in the food service (catering) industry.

1 Introduction

As a part of the food industry, food service is involved in the processing and delivery of ready-to-eat meals to consumers in canteens, schools, hospitals, and office buildings. These services are typically outsourced to food catering companies. This industry has grown significantly in the last decade, achieving an annual turnover of approximately €24 billion and a market coverage of approximately 33% of European companies. This sector employs 600,000 people in Europe and delivers approximately 6 billion meals each year (Europe Food Service, 2017). The catering sector satisfies a broad demand and offers services to childcare, schools, hospitals, businesses, and nursing homes (Garayoa et al., 2011). Due to the steady aging of the population and changes in lifestyles and consumption habits, the need for catered meals is increasing rapidly.

The catering sector is one of the most complex in the food industry (Farahani et al., 2012). Caterer processes include delivering fresh food, mostly hot, to many customers and preserving product quality while minimizing production and logistic costs. Caterers have to cope with a high demand variability (1), sequence-dependent set-up times of the working centers (2), and a rapid decay of product quality (3). Furthermore, higher consumer expectations of quality as well as religious and ideological reasons increase demand for different diets and specific meals, raising the number of items/products and associated processing cycles.

In order to meet such a wide production mix, the catering industry usually adopts job-shop (JS) production systems (Curt et al., 2007) instead of flow processing systems. JS systems are more complex than typical flow-line systems, as they are able to process hundreds of different recipes each day with small lot sizes. Cooking, frying, chilling, baking, boiling, and packing are typical processing tasks performed in these systems. Two main decision problems affect the design of a JS production system:

1. The definition of the number of machines (i.e., the number of resources);
2. The location of resources in the plant layout.

Sustainable Food Supply Chains. https://doi.org/10.1016/B978-0-12-813411-5.00008-9

The determination of the number of machines (1) for each production department is a crucial activity, since a small number of machines may lead to production bottlenecks and delays. On the other side, a high number of machines requires a huge investment and larger spaces and leads to higher production costs.

The system layout (2) deeply affects the performance and the throughput of a food JS system. The layout planning requires the determination of the location of the processing equipment and resources, including machines, storage areas, and manual working centers.

In a JS food facility, the type and sequence of the processing tasks vary in accordance with the product's recipe. As different machines are located in different departments, intense flows of operators, materials, and handling tools throughout the plant departments are necessary to complete a recipe.

This chapter presents an original methodology to deal with the aforementioned two issues and to aid decision makers in the design of food JS processing systems. An original dashboard of key performance indicators (KPIs) is presented and used to evaluate alternative design scenarios and to assess the appropriateness of the number of machines and the plant layout compared with the expected workload of the facility. Finally, a case study from a leading Italian food catering company is illustrated. The potential of the methodology and the KPI dashboard is tested in the redesign of an extant food service facility. A multiscenario analysis is performed to analyze the outputs and to compare the system behavior in the case of an increase in product quantities.

The remainder of this chapter is organized as follows: Section 2 introduces a brief literature review on the design of a food service facility; Section 3 illustrates the decision support methodology and the KPI dashboard; Section 4 presents a case study and a multiscenario analysis of the application in the selected case study; Section 5 discusses the results and outlines further developments; and Section 6 concludes the chapter.

2 Background

This section presents a short literature review of the design of food processing facilities. Each of the following sections overviews the existing literature according to the hierarchical design framework adapted from (Heragu, 2008) and represented in Fig. 1. In Fig. 1, the blocks from 1 to 9 represent the steps needed to implement a feasible and sustainable design project of a processing facility.

The analysis of the market demand (*Block 1*) is a preliminary activity to pinpoint the required amount of resources and the workload. Like most of the service industries, the food service sector experiences a highly volatile and unpredictable demand that varies with the turnover of the client portfolio (Connell et al., 1984; Minner and Transchel, 2017). In order to meet customer expectations and to avoid food losses, advanced computerized techniques are used to predict the market demand (Ji and Tan, 2017) and to determine an expected production plan (Urbani et al., 2010).

To reduce distribution costs, a suitable location for the production facility is to be identified (*Block 2*). In the case of a food service facility, it is necessary to consider the shelf-life threshold of products required by clients together with the distance (de Keizer et al., 2017). For example, a school or a hospital food service requires the production facility to be located near the point of demand, while services for companies usually do not have this requirement. For this reason, advanced techniques allow the embedding of perishability issues in the decision problem (Accorsi et al., 2017a; Amorim and Almada-Lobo, 2014) and concurrently defining the location, the inventory level, and the distribution network of a node in a food supply chain (Hiassat et al., 2017).

The choice of the processing technology (*Block 3*) is deeply connected to the final shelf life of the product. Food products can be delivered ready-to-eat (i.e., with few hours of shelf life) or they can be blasted and chilled, being warmed just before consumption (i.e., guaranteeing a few days of shelf life) (Henderson et al., 1997; Williams, 1996). In addition to such criticality, marketing strategies typically enlarge the product portfolio in order to meet new food trends and consumer expectations, but this significantly affects the complexity of the production system and its efficiency (e.g., local, low-fat, functional food) (Betoret et al., 2011; Van Kampen and Van Donk, 2013).

In the production facility the processing resources/machines often are required to be powered by auxiliary mechanical or thermal plants (*Block 4*). Indeed, a crucial issue for the food service facilities is the energy consumption, whose cost may contribute up to 20% of the finished product cost (Mudie et al., 2014). Kitchen ventilation and heat production are the activities requiring a significant amount of energy (CIBSE, 2009; Clark, 2009). The choice of the energy source to produce heat (e.g., gas or electricity) should be carefully considered together with the design of production processes, to reduce energy inefficiencies and costs (Fusi et al., 2015; Tufano et al., 2018a).

The determination of the number of resources and of the space to allocate to them (*Block 5*) is a key design step that contributes to avoiding production bottlenecks and delivery delays. In particular, a suitable number of resources is necessary to concurrently complete the production and the shipping operations, since people are used to having lunch or dinner

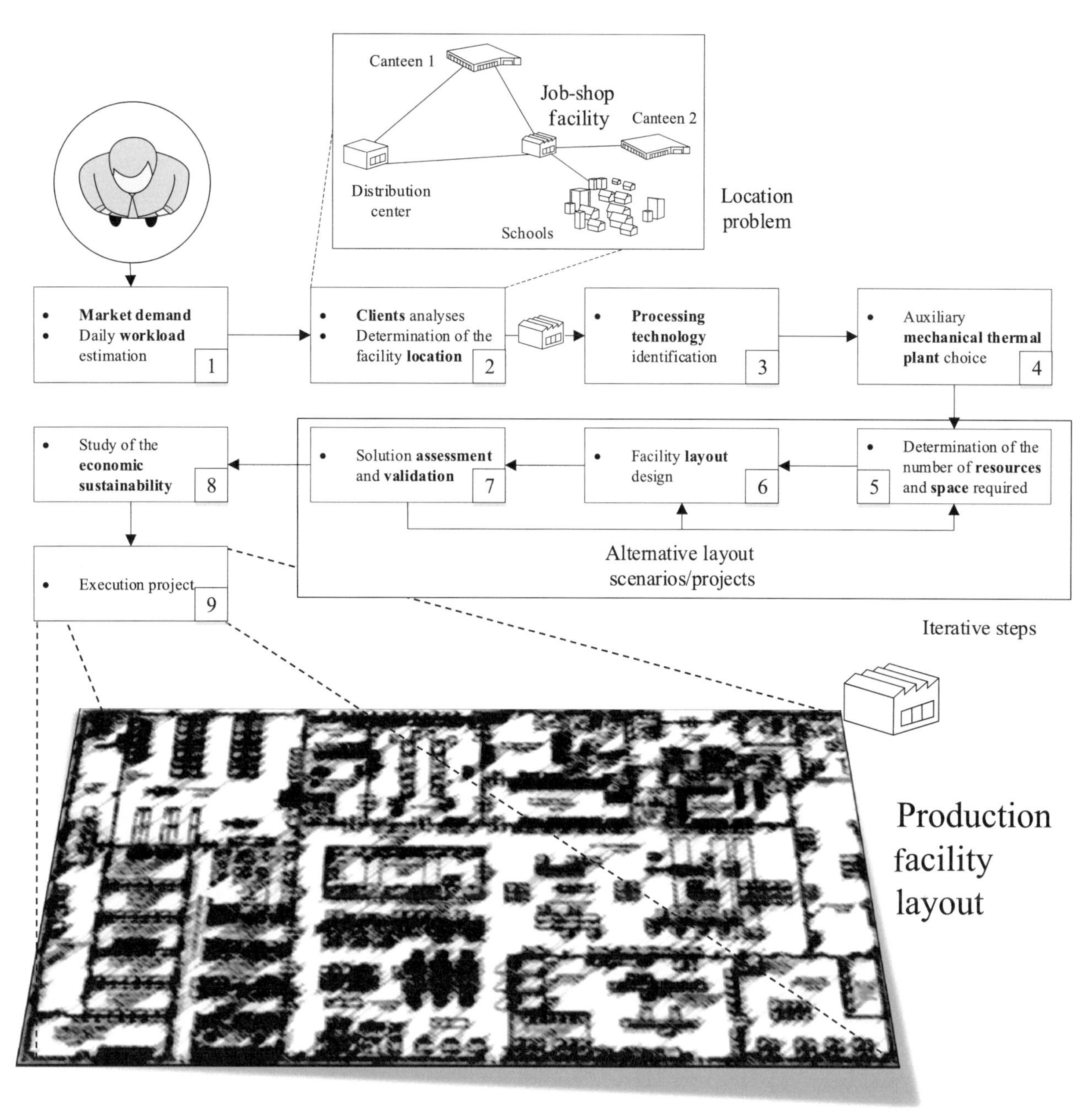

FIG. 1 Hierarchical framework for facility design. *(Based on Heragu, S.S., 2008. Facilities Design, third ed.)*

at the same time (Devapriya et al., 2017; Farahani et al., 2009). On the other hand, since product price should be maintained low to gain market share, high investments in equipment are discouraged. To this purpose, workload and resource-usage simulation techniques are used to virtualize the production processes and to identify performance benchmarks (Penazzi et al., 2017).

An adequate location of the processing resources (*Block 6*) over the plant layout can reduce the traveling and handling cost from 10% to 30% (Amit et al., 2012). Many studies aim to find the optimal facility layout using either heuristic methods or simulation techniques (Drira et al., 2007; Tufano et al., 2018b).

A set of KPIs (*Block 7*) is necessary to evaluate the effectiveness of the introduced design steps (i.e., blocks) and to compare several alternatives and design configurations resulting from these levers (*Block 8*). The following sections of the chapter particularly explore the design steps belonging to *Blocks 5, 6,* and *7*. A comprehensive dashboard of KPIs is proposed to evaluate the performances of a food service facility. A methodology to identify a suitable number of resources and to evaluate their locations on the plant layout is introduced and explored with a case study.

3 Methodology

The methodology proposed in this section is intended to support the decision maker throughout the iterative (i.e., multi-scenarios) steps and decisions of Fig. 1. Specifically, it includes a method for the determination of the number of resources and the assessment of the effectiveness of the proposed layout solution. Three policies devoted to the design of the number of processing resources (1) are illustrated with the purpose of reducing the bottlenecks and delays throughout the production system. Furthermore, the methodology provides a multidisciplinary dashboard of KPIs (1) suitable to assess the overall performance of a production facility intended for food service.

This section concludes with some insights into the data architecture needed to propel the method and dashboard with real-world datasets. Fig. 2 illustrates the proposed methodological framework and shows how it builds upon existing frameworks (Heragu, 2008). The output from the application of this methodology is the number of processing resources to allocate to control points (CPs) in the plant. CPs are defined as significant physical locations and areas where the products (raw materials, semifinished products, or final products) are handled or processed.

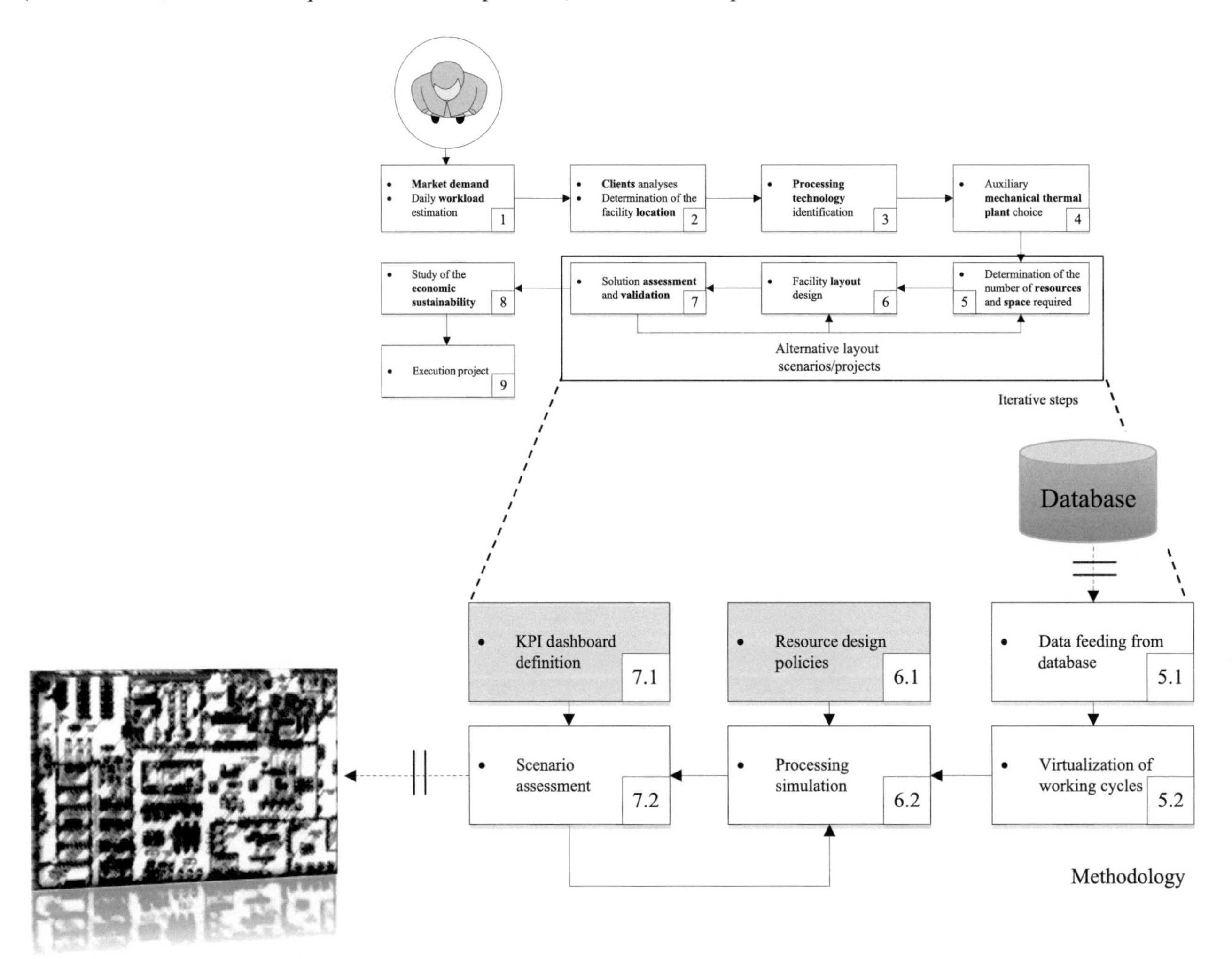

FIG. 2 Methodological framework.

3.1 Design procedure and performance indicators

The methodology for the virtualization of working cycles and the assessment of the effectiveness of the layout solution are illustrated in this section. Given the complexity of such a production system, several drivers should lead its design. As a consequence, a multidisciplinary dashboard of quantitative indicators and performance metrics has to be defined to this purpose. The KPIs are classified into four families:

1. *Cost indicators*: connected to the return on investment in infrastructure and in the production resources;
2. *Time indicators*: regarding labor time (i.e., the workload) allocated to each resource;
3. *Energy indicators*: connected to the power/thermal load of each resource;
4. *Environmental indicators*: related to the environmental impact associated with the energy consumption by each resource.

The set of KIPs is summarized in the scheme of Fig. 3, which also links such indicators to targets of the food service manager.

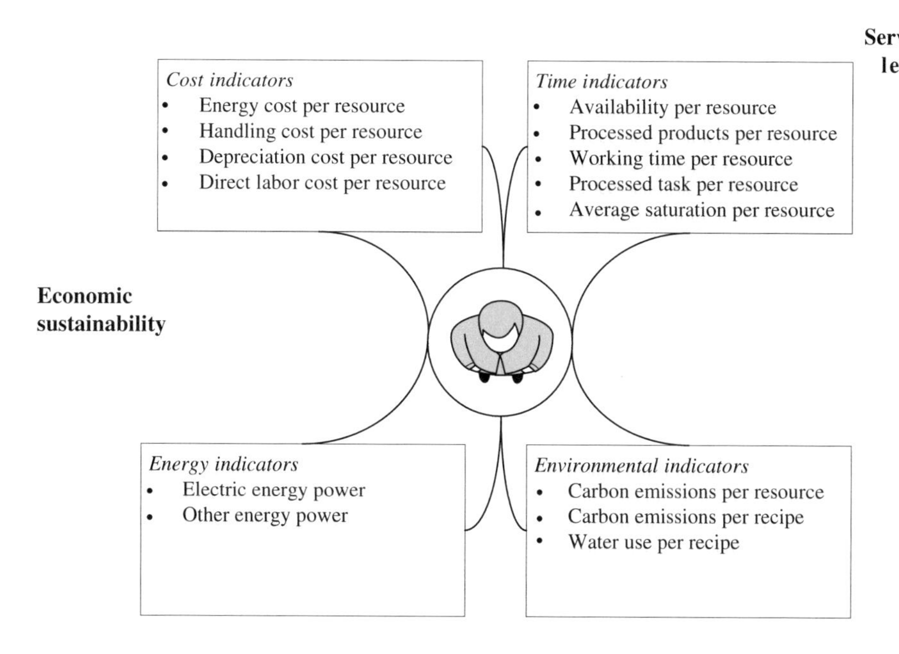

FIG. 3 KPI dashboard to lead the determination of the number of resources.

TABLE 1 KPI notation and formulae

Notation	KPI	Equations
M: set of machines. $j \in M$, $j = 1, \ldots, k$	Energy cost per machine j:	$\sum_{i \in T_j} \frac{t_i}{60} \cdot \left(c_{en}^{el} \cdot k_i^{el} + c_{en}^{gas} \cdot k_i^{gas}\right)$
T_j: set of task i processed by machine j. $i \in T_j : j \in M$	Handling cost per machine j:	$\sum_{i \in T_j} \frac{t_i^h}{60} \cdot c_{dl}$
H: time horizon considered. $t \in H$, $t = 1, \ldots, m$	Depreciation cost per machine j:	$\sum_{i \in T_j} \frac{t_i}{60} \cdot c_{dp(j)}$
c_{en}^{el}: cost of 1 kWh of electric power (€)	Direct labor cost per machine j:	$\sum_{i \in T_j} \frac{t_i}{60} \cdot c_{dl}$
c_{en}^{gas}: cost of 1 kWh of thermal power (€)	Number of tasks per machine j:	$\sum_{i \in T_j} n_{i,t}$
c_{dl}: direct labor cost per hour (€)	Working time per machine j:	$\sum_{i \in T_j} n_{i,t} \cdot t_i$
$c_{dp(j)}$: cost (depreciation) per hour of machine j (€)	Saturation per machine j:	$\frac{1}{\lvert i \in T_j \rvert} \sum_{i \in T_j} p_{i,j}$
t_i: working time of a task (min)	Electric energy power per machine j:	$\sum_{i \in T_j} n_{i,t} \cdot t_i \cdot k_i^{el}$
d_i: traveled distance to perform task i	Gas energy power per machine j:	$\sum_{i \in T_j} n_{i,t} \cdot t_i \cdot k_i^{gas}$
t_i^h: handling time of a task (min)	CO_2 generated per machine j:	$\sum_{i \in T_j} n_{i,t} \cdot t_i \cdot \left(q^{el} \cdot k_i^{el} + q^{gas} \cdot k_i^{gas}\right)$
c_{en}: cost of 1 kWh of energy		
$n_{i,t}$: number of task type i processed in the time horizon t		
k_i^{el}: electric power to perform task i (kWh)		
k_i^{gas}: thermal power to perform task i (kWh)		
q^{el}: CO_2 per kWh of electricity (kg)		
q^{gas}: CO_2 per kWh of gas (kg)		
$p_{i,j}$: percentage saturation to perform task i on machine j		

The four families of KPIs give the designer four different perspectives from which to evaluate the layout scenario according to the resulting number of processing resources established. Table 1 presents the mathematical formulation used for the calculation of each of these.

Each recipe is created by a working cycle defined as a list of tasks performed by different processing resources. According to the set of equations in Table 1, the metrics for each resource (i.e., machine) are obtained as the sum of the contributions resulting from all the processing tasks performed for different recipes on that resource.

Given a generic product, the sequence of tasks is fixed and given from the working cycle, and preemption of tasks is not allowed. For each task, the cost, time, and energy impacts/contributions are calculated in agreement with the characteristics/capacity of the resources required by that task. At each machine, a task is tracked by two time-stamps (i.e., start and end time) to detect the time spent by the resource to perform a task.

The recipes are scheduled during the day according to an "at-the-latest" policy, driven by the delivery due time imposed by the contract with the customer. This approach is exemplified in Fig. 4, which shows the whole production cycle of two recipes and how the utilization of a processing resource (e.g., oven) is quantified.

This approach considers all the products and related working cycles independently. The production batches are single product/recipe and are managed by a single operator. Whenever a resource is allocated to a task of one recipe, no other tasks can be executed by the machine until the processing of that batch is completed.

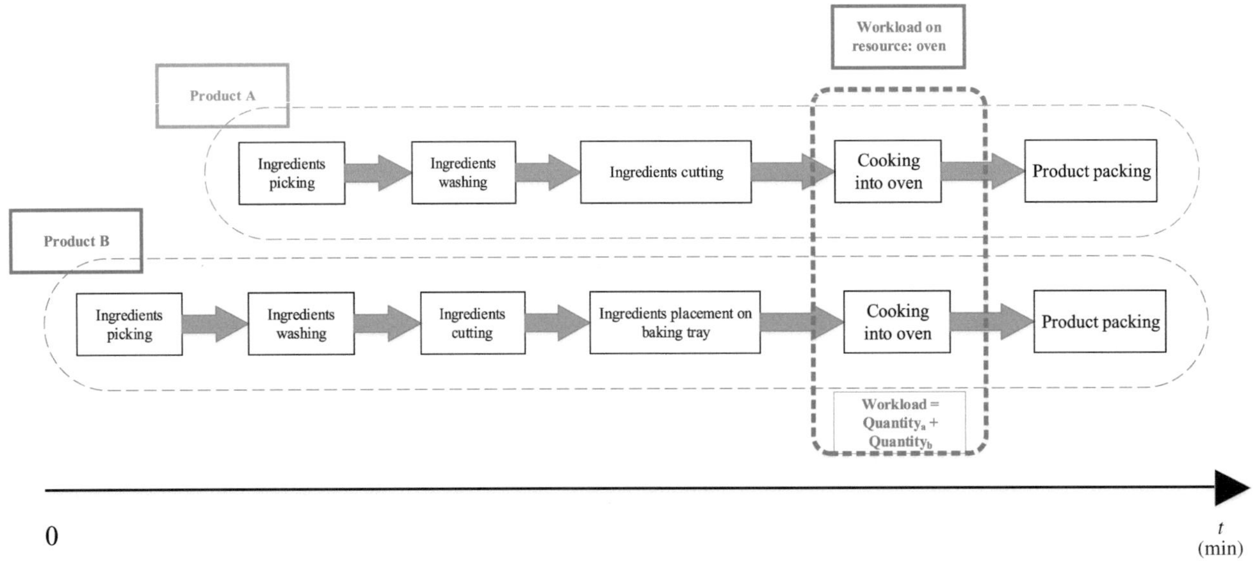

FIG. 4 Procedure to calculate KPIs.

This approach allows the study and quantification of the cost, time, and energy impacts associated with each single task, processing resource, and recipe. In addition, this procedure allows analyzing the material flows exchanged between processing resources over the production bucket and quantifying the impact of logistics and handling activities on the production process. For example, the time spent to travel between the machines can be used as a metric for an assessment of the effectiveness of the facility layout.

Different scenarios can be generated and virtualized (i.e., simulated) according to different production mixes, production schedules, and resource layouts. Then, calculating these KPIs/impacts per each different scenario can aid decision makers in identifying the most efficient one.

3.2 Resources design policies

Other scenarios can be built upon the number of resources to establish per each type. The same panel of KPIs can then be used to assess the performance of each scenario and determine what number of machines is most adequate. Three alternative policies are defined to decide on the number of resources to adopt:

1. *Machine-based*: "exploratory" policy; set all the resources equal to one.
2. *Layout-based*: "current" policy; set all the resources equal to the current number of resources.
3. *Recipe-based*: "workload-driven" policy; set the number of resources different for each recipe. For each recipe, the number of resources is that necessary to process a whole production batch without generating a queue or a delay.

The latter policy can be formulated as (1) assuming the following sets and parameters:

$$M: \text{set of machines}. j \in M, j = 1, \ldots, k$$

$$R: \text{set of recipes}. r \in R, h = 1, \ldots, l$$

$$q_{jr}: \text{maximum quantity of recipe } r \text{ processable in a single cycle of } j$$

$$q_r: \text{maximum quantity of recipe } r \text{ required in the production plan}$$

$$N_{rb(jr)} = \left\lceil \frac{q_{jr}}{q_r} \right\rceil \quad (1)$$

Fig. 5 illustrates and compares the characteristics of the three resource design policies.

The machine-based policy (1) is a lower bound of the number of resources of type j, while $Max_r(N_{rb(j,r)})$ gives an upper bound of the number of resources per type j. The layout-based policy (2) indicates the current number of resources established (or set by the designer). The difference between the number of resources obtained by the layout-based and the recipe-based policies indicates the level of risk of experiencing bottlenecks and queues throughout the production.

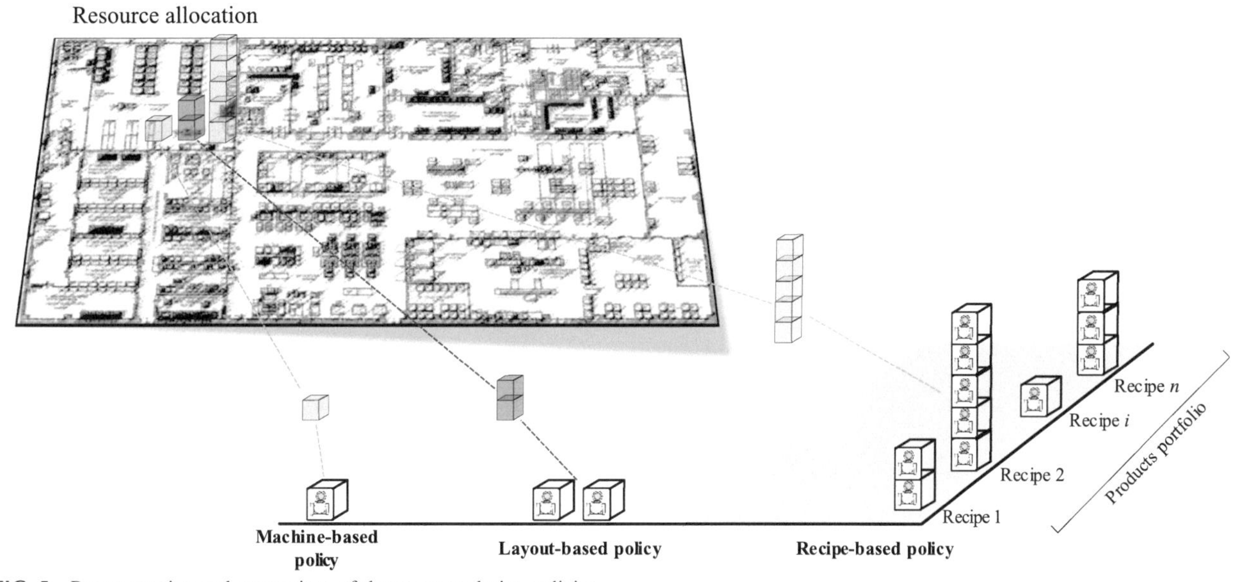

FIG. 5 Representation and comparison of the resource design policies.

3.3 Data architecture

In order to implement the methodology through the quantification of the KPIs dashboard and the implementation of the three design policies, a data architecture able to store and link the information on working cycles, tasks, and machines is proposed. Table 2 shortlists the set of tables belonging to an operations-based database that can be quickly developed on a commercial DBMS such as the Microsoft SQL Server database.

TABLE 2 Data structure

Input tables	Content description	Attributes
Menu	Order list with the information about product code, production due date, quantity, and customer.	• Dateperiod, • Ordercode, • Productcode, • Quantity, • Duedate, • Client
Ingredient	Bill of material of each product, containing raw materials, semifinished, and finished products.	• ProgrRaw, • CodProdFamily, • CodRawMat1, CodRawMat2, • DescRawMat1, DescRawMat2, • RawType
Machine	List of machines and their properties (e.g., capacity, number of machines, electric and thermal power, initial cost, remaining cost).	• MachName, • Manufacturer, Type • MachCategory, • Cost, • PowerLoad
CP	List of control points for each department.	• CodCP, • CPDecription, • CoordX, CoordY, CoordZ, • Department
Handling	List of handling equipment (e.g., baking tray, casserole, trolley for handling the baking trays) and their characteristics.	• VehicleCode, • Description, • Weight, Load, Speed
Cycles	List of the working cycles (i.e., the sequence of tasks) necessary for producing a final food product.	• ProgrTask, DescTask, • Product, • CodRaw1, CodRaw2, • Qty, • WorkingTime

The required records deal with the product and packaging characteristics, the type of machines, the product recipes, the equipment (e.g., baking trays, trolleys, totes), the layout zones and departments, the bill of materials (BOM) of the ingredients, and the recipes' working cycles.

4 An industrial catering application

A case study inspired by an Italian food service company is illustrated and discussed to test the proposed methodology. This is a leading company operating in collective catering for 70 years. It has about 1300 points of demand (e.g., restaurants and canteens), producing more than 66 million meals per year. To enhance the level of service to consumers and the quality standards, as well as to improve production efficiency, the company selects about two processing facilities per year out of a network of 50 facilities for a layout redesign. In 2016, one of these improvement projects regarded the relayout planning of a facility located in Tuscany. This facility is composed of 119 processing machines, 11 departments, and 4 refrigerated chambers. The production system produces meals for schools, companies, and hospitals not equipped with on-site kitchens.

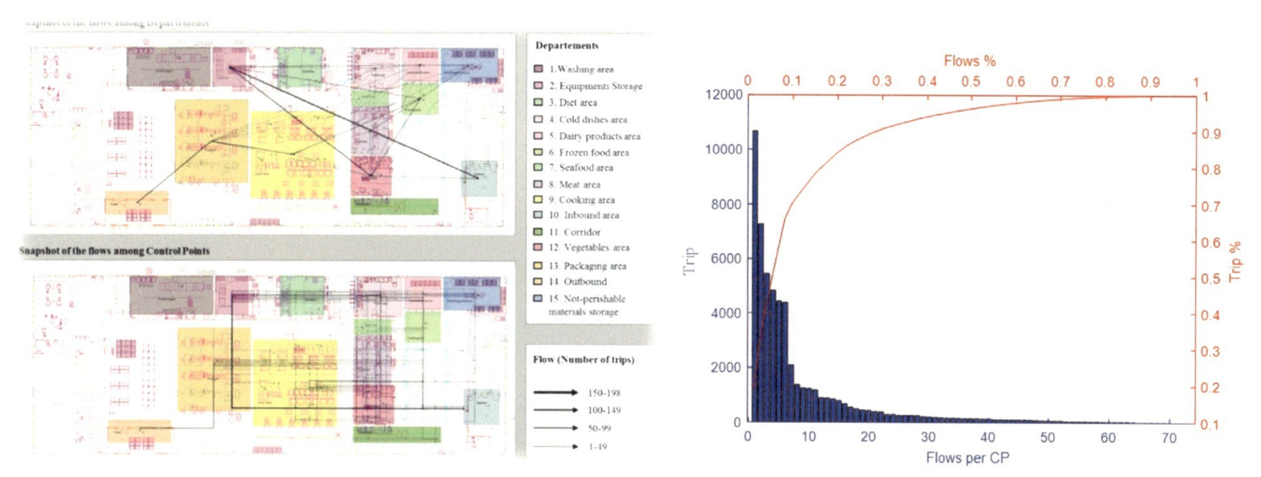

FIG. 6 Snapshots of the flows view and Pareto analysis of the number of trips.

Table 3 illustrates the number of input entities and the output of the process virtualization and simulation used to design the multiscenario analysis.

TABLE 3 Input and output of the simulation

Input entities	Value	Output values	Value
Control points	60	Traveling time (min)	935
Products	42	Processing time (min)	6356
Production quantity (5 days)	53,420	Traveled distance (km)	37,726
Working cycles	785		
Handling tasks	292		
Working tasks	493		

Fig. 6 shows the layout of the production system and highlights the material flows, where the line thickness is proportional to the number of trips between departments or control points in the selected horizon of time. For example, most of the trips are (largest line) between the refrigerated cells, typically considered as the beginning of any recipe, and the tool washing department. This conclusion would encourage establishing a washing machine close to the inbound area, in order to reduce labor-intensive handling activities for tray replenishment.

Fig. 6 also demonstrates how each couple of CPs accumulate trips necessary to complete the production, and quantifies, for example, that 20% of the connections accumulates about 85% of the trips.

This first finding indicates how to use the proposed methodology to track the material flow over the plant layout and support its redesign accordingly. Furthermore, the results obtained are influenced by the production batches, the produced volumes to fulfill the demand, and the capacities and number of the production resources involved.

Based on these assumptions, a multiscenario analysis is conducted in order to study the system responsiveness to stressing conditions generated by growth of the number of meals demanded.

Fig. 7 presents two different analyses of the impact of production volumes on costs and material handling of the production processes.

In Fig. 7A, a scatter-plot shows the costs per recipe associated with handling, i.e., logistics, labor and energy consumption (i.e., see the indicators proposed in Table 1) with respect to the quantity (i.e., number of meals) of a production batch. By observing Fig. 7, a clear linear relationship between such dots is hard to find. A nonlinear behavior of the system is, indeed, confirmed by the multiscenario sensitivity analysis shown in Fig. 7B.

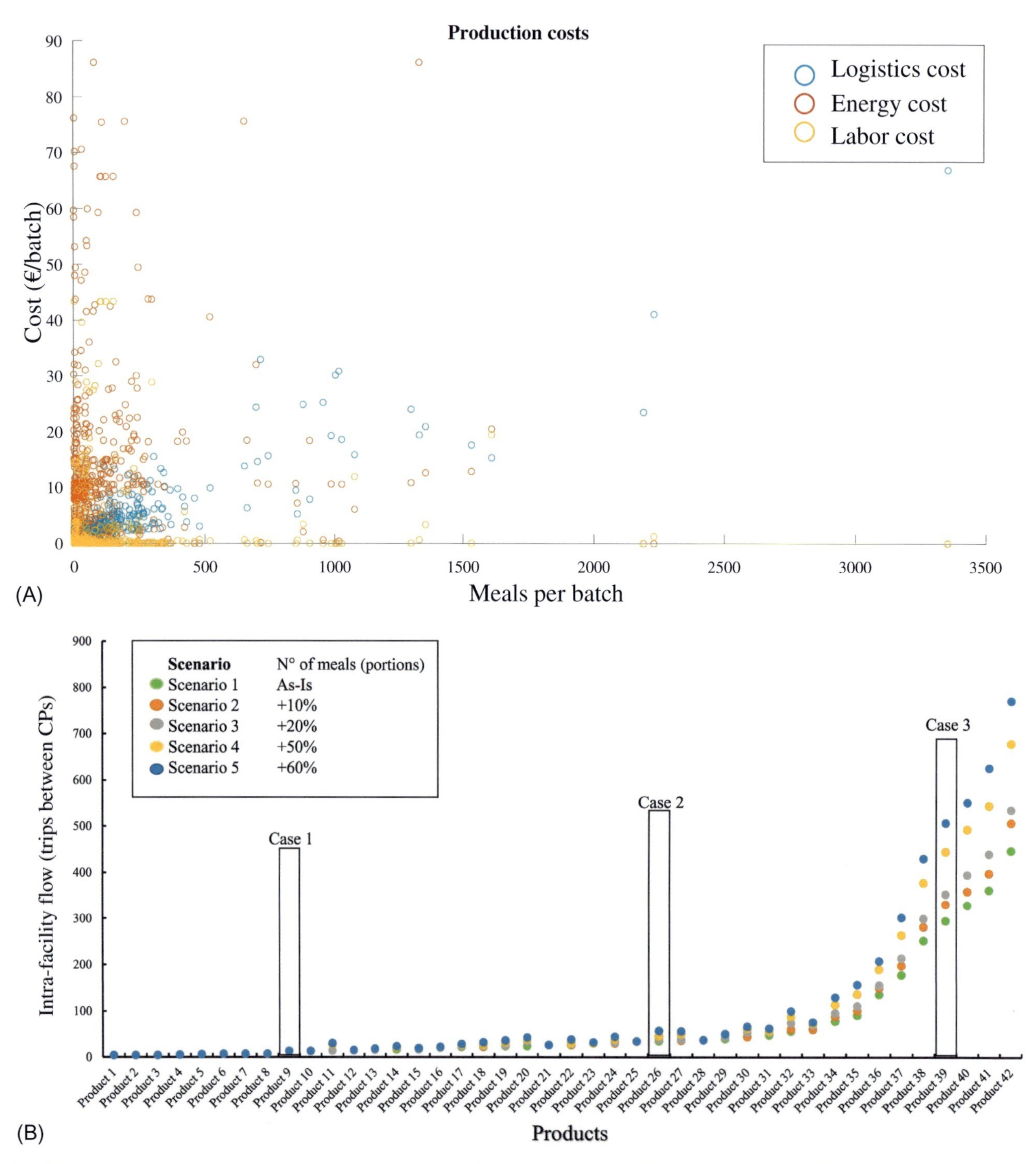

FIG. 7 Quantity-cost exploratory analysis and multiscenario sensitivity analysis: (A) Cost sensitivity to the size of the production batch; (B) intrafacility trip sensitivity to the number of meals (portions) served per day per recipe (product).

The multiscenario sensitivity analysis, in Fig. 7B, compares the performance of five scenarios in terms of number of trips traveled within the production system to complete the working cycles of a given menu. The aim of this analysis is to study the sensitivity of the processing system to the varying number of meals, raised by 10%, 20%, 50%, and 70%. The chart showcases three cases, each depicting a different product's behavior. Case 1 relates to those products whose number of trips does not vary with the growth of the demanded volume. As a consequence, the current machines and equipment are capable of processing higher demand without an evident increase in the number of trips. In Case 2, the distances between the as-is scenario and the simulated scenarios becomes evident from an increase of 50% of the produced number of meals. Finally, Case 3 relates to those products that suffer from volume growth. Machines and handling equipment reach a saturation point. Consequently, the system reacts by increasing the number of trips necessary to complete the working cycle.

Where the analyses proposed in Figs. 6 and 7 contribute by specifying and describing the role of the facility layout in reducing costs and material flows, Fig. 8 investigates the level of machine utilization and focuses on the production bottlenecks, assuming that the number of resources is to be designed according to the machine-based policy. Fig. 8 shows results with histograms of the level of utilization of the main processing resources. The workload for each machine is defined and calculated in agreement with the indicators proposed in Table 1 of the methodology section.

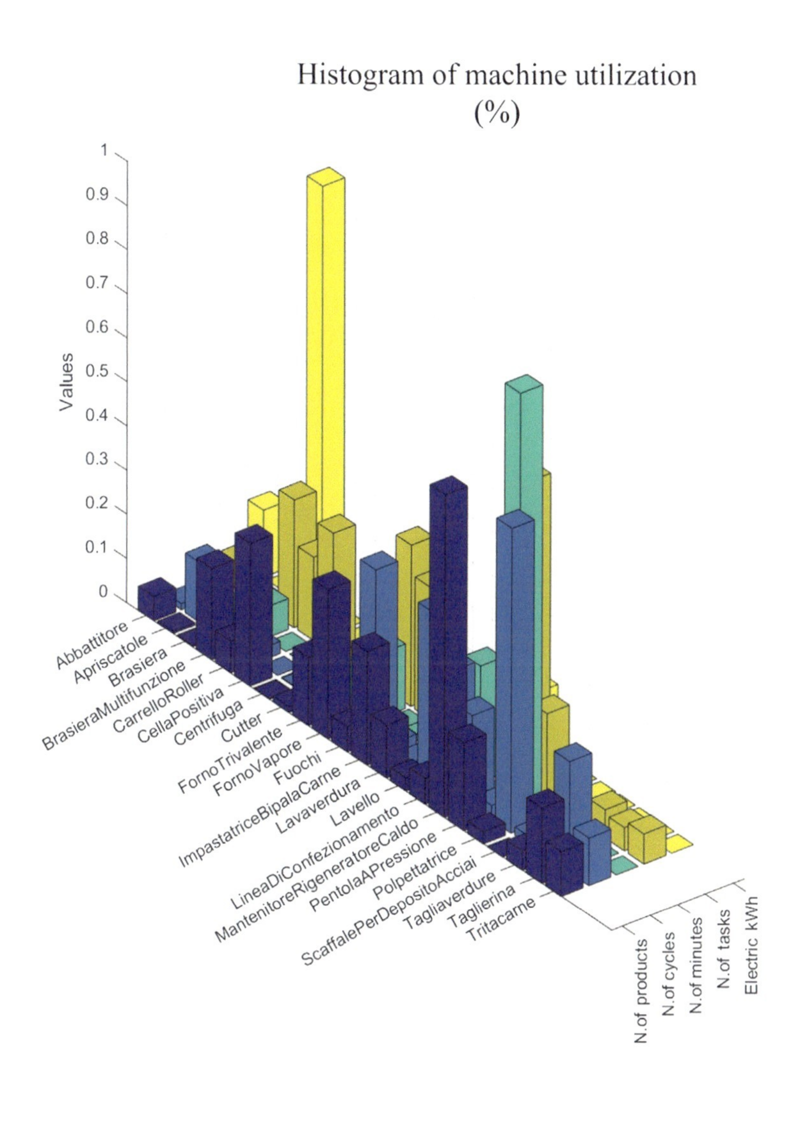

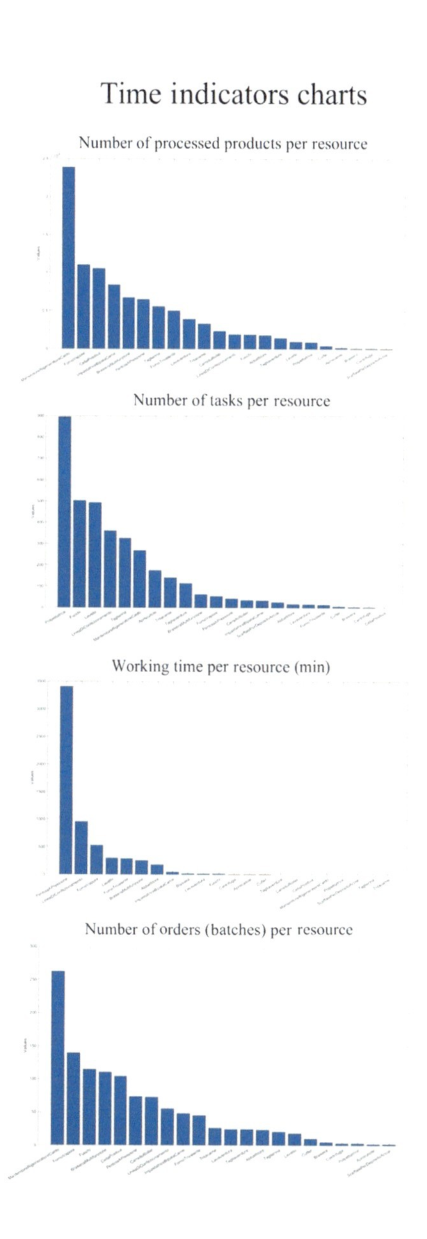

FIG. 8 Histograms of the workload per type of machine.

In addition to the previous analyses, the three resource design policies (i.e., machine-, layout-, and recipe-based) are implemented and compared, to identify which type of machine presents more criticalities.

Fig. 9 shows the results of such comparison, reporting on the *X*-axis the machine types and on the *Y*-axis the number of machines established according to the three policies in logarithmic scale. The logarithmic scale is necessary due to the high number of machines established by the recipe-based policy for those resources processing one piece at a time (e.g., cutting, slicing, and packing machines). Conversely, the machine-based policy set the number of all resources equal to 1, which represents the lower bound of the optimal number of resources.

When the number of resources from the machine-based is equal to that from the recipe-based policy, one resource of that type is enough to avoid bottlenecks and queues. Conversely, when a difference exists, bottlenecks and queues can occur depending on the size of the production batch. In such cases, a higher number of machines contributes to reduce bottlenecks and related inefficiencies and to avoid products waiting during the process, with possible implications for their quality and conservation.

Nevertheless, the cost of each machine type determines whether or not the number of resources to establish is affordable. In the case of cheap resources (e.g., tables or sinks) an additional resource does not affect the investment significantly, while more packing or cooking machines would result in higher investment cost and consequently a higher production cost due to the depreciation of such resources.

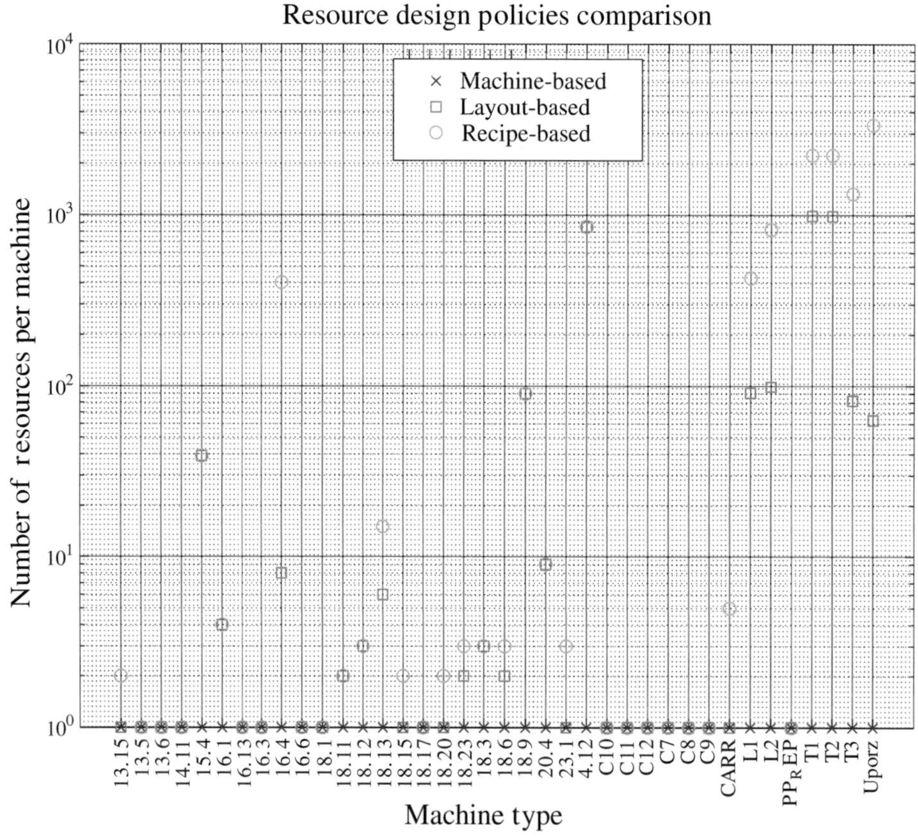

FIG. 9 Resource design policies comparison.

Table 4 illustrates the full dashboard of KPIs as quantified from the previous analyses and according to the proposed methodology. Specifically, the environmental impact indicator that quantifies the carbon emissions resulting from energy consumption is obtained, considering the impact of the energy production mix supplied by the power grid.

TABLE 4 Production system KPIs

	Value	Unit of measure
Cost metrics		
Total energy cost	2099	€/week
Total handling cost	6993	€/week
Total depreciation cost	4553	€/week
Total direct labor cost	1224	€/week
Time metrics		
Total handling time	1558	min/week
Total working time	10,593	min/week
Total employees time	2341	min/week
Energy metrics		
Electric consumption	40,923	kWh/week
Gas consumption	9356	kWh/week
Environmental metrics		
Equivalent CO_2	1450	kg/week

5 Discussion and lessons learned

The results of the case study have been discussed with the company managers and, based on these results, three strategic levers have been used to redesign the production system according to the expected meals demand. These levers are as follows:

1. *Balancing the workload* (i.e., the menu). While the menu is typically the result of consumer requests, the contribution and the impact of each recipe on the processing and handling operations should be carefully considered to set a production mix able to avoid bottlenecks and a balanced workload of the processing resources.
2. *Improving the processing resources.* If the menu review does not lead to significant improvements, investments in equipment and machines can be necessary. The quantified KPIs enable identification of products (i.e., recipes) and working cycles that most strongly impact the production system and would require new or more machines or handling equipment.
3. *Periodically redesign the layout.* The facility relayout planning can reduce the daily traveling considered as a no-value-added activity. As an example, reducing the distance between those departments/CPs accumulating most of the trips may significantly decrease the total traveling distance and improve the production efficiency.

On the other hand, the comparison of the three resource design policies shows that some resources are more sensitive to workload than others, since they behave as production bottlenecks. Future applications should be based on new instances and simulate the effects of different workload distributions and production scheduling strategies on the efficiency of the production system.

In addition to this, since the KPI dashboard illustrates the behavior and the performance of the production plant, no attention is paid to the storage and distribution phases or to the impact of the coordination between production and other supply chain operations (Accorsi et al., 2018a,b). For this reason, future research developments are expected on the integration of production planning with shipping, storage, and distribution activities (Accorsi et al., 2017b) in the food service industry.

6 Summary and conclusions

A set of tools for the analysis and relayout of a catering facility is illustrated. A comprehensive dashboard of KPIs leads the decision makers to study the impact of their choices on the performance of the layout of a food service facility. Furthermore, a multiscenario analysis is conducted to test the effectiveness of the current layout in the case of demand growth. A resource design methodology is implemented to identify a suitable number of resources in the food batch production industry.

A case study is introduced and explained with regards to the methods presented; it addresses the redesign of an existing food service facility. The results and their managerial and academic implications are discussed, remarking on where future research efforts should be directed.

References

Accorsi, R., Bortolini, M., Baruffaldi, G., Pilati, F., Ferrari, E., 2017a. Internet-of-things paradigm in food supply chains control and management. Procedia Manuf. 11, 889–895. https://doi.org/10.1016/j.promfg.2017.07.192.

Accorsi, R., Gallo, A., Manzini, R., 2017b. A climate driven decision-support model for the distribution of perishable products. J. Clean. Prod. 165, 917–929. https://doi.org/10.1016/j.jclepro.2017.07.170.

Accorsi, R., Baruffaldi, G., Manzini, R., 2018a. Picking efficiency and stock safety: a bi-objective storage assignment policy for temperature-sensitive products. Comput. Ind. Eng. 115, 240–252. https://doi.org/10.1016/j.cie.2017.11.009.

Accorsi, R., Baruffaldi, G., Manzini, R., Tufano, A., 2018b. On the design of cooperative vendors' networks in retail food supply chains: a logistics-driven approach. Int. J. Logist. Res. Appl. 21, 35–52. https://doi.org/10.1080/13675567.2017.1354978.

Amit, N., Suhadak, N., Johari, N., Kassim, I., 2012. Using simulation to solve facility layout for food industry at XYZ Company. In: SHUSER 2012–2012 IEEE Symposium on Humanities, Science and Engineering Research, pp. 647–652. https://doi.org/10.1109/SHUSER.2012.6268900.

Amorim, P., Almada-Lobo, B., 2014. The impact of food perishability issues in the vehicle routing problem. Comput. Ind. Eng. 67, 223–233. https://doi.org/10.1016/j.cie.2013.11.006.

Betoret, E., Betoret, N., Vidal, D., Fito, P., 2011. Functional foods development: trends and technologies. Trends Food Sci. Technol. 22, 498–508. https://doi.org/10.1016/j.tifs.2011.05.004.

CIBSE, 2009. TM-50 Energy Efficiency in Commercial Kitchens. https://doi.org/10.1108/EUM0000000002227.

Clark, J.A., 2009. Solving kitchen ventilation problems. ASHRAE J. 51, 20–24.

Connell, B.C., Adam, E.E., Moore, A.N., 1984. Aggregate planning in health care foodservice systems with varying technologies. J. Oper. Manag. 5, 41–55. https://doi.org/10.1016/0272-6963(84)90006-8.

Curt, C., Hossenlopp, J., Trystram, G., 2007. Control of food batch processes based on human knowledge. J. Food Eng. 79, 1221–1232. https://doi.org/10.1016/j.jfoodeng.2006.04.052.

de Keizer, M., Akkerman, R., Grunow, M., Bloemhof, J.M., Haijema, R., van der Vorst, J.G.A.J., 2017. Logistics network design for perishable products with heterogeneous quality decay. Eur. J. Oper. Res. 262, 535–549. https://doi.org/10.1016/j.ejor.2017.03.049.

Devapriya, P., Ferrell, W., Geismar, N., 2017. Integrated production and distribution scheduling with a perishable product. Eur. J. Oper. Res. https://doi.org/10.1016/j.ejor.2016.09.019.

Drira, A., Pierreval, H., Hajri-Gabouj, S., 2007. Facility layout problems: a survey. Annu. Rev. Control. 31, 255–267. https://doi.org/10.1016/j.arcontrol.2007.04.001.

Europe Food Service, 2017. European Industry Overview (WWW Document). http://www.ferco-catering.org/en/european-industry-overview. (Accessed 23 December 2017).

Farahani, P., Grunow, M., Günther, H.O., 2009. A heuristic approach for short-term operations planning in a catering company. In: IEEM 2009—IEEE International Conference on Industrial Engineering and Engineering Management, pp. 1131–1135. https://doi.org/10.1109/IEEM.2009.5372966.

Farahani, P., Grunow, M., Günther, H.-O., 2012. Integrated production and distribution planning for perishable food products. Flex. Serv. Manuf. J. 24, 28–51. https://doi.org/10.1007/s10696-011-9125-0.

Fusi, A., Guidetti, R., Azapagic, A., 2015. Evaluation of environmental impacts in the catering sector: the case of pasta. J. Clean. Prod. 132, 146–160. https://doi.org/10.1016/j.jclepro.2015.07.074.

Garayoa, R., Vitas, A.I., Díez-Leturia, M., García-Jalón, I., 2011. Food safety and the contract catering companies: food handlers, facilities and HACCP evaluation. Food Control 22, 2006–2012. https://doi.org/10.1016/j.foodcont.2011.05.021.

Henderson, S.M., Perry, R.L., Young, J.H., 1997. Principles of Process Engineering. American Society of Agricultural Engineers.

Heragu, S.S., 2008. Facilities Design, third ed. CRC Press.

Hiassat, A., Diabat, A., Rahwan, I., 2017. A genetic algorithm approach for location-inventory-routing problem with perishable products. J. Manuf. Syst. https://doi.org/10.1016/j.jmsy.2016.10.004.

Ji, G., Tan, K., 2017. A big data decision-making mechanism for food supply chain. In: Global Congress in Manufacturing and Management. pp. 1–10. https://doi.org/10.1051/matecconf/201710002048.

Minner, S., Transchel, S., 2017. Order variability in perishable product supply chains. Eur. J. Oper. Res. 260, 93–107. https://doi.org/10.1016/j.ejor.2016.12.016.

Mudie, S., Essah, E.A., Grandison, A., Felgate, R., 2014. Electricity use in the commercial kitchen. Int. J. Low-Carbon Technol. 11, 66–74. https://doi.org/10.1093/ijlct/ctt068.

Penazzi, S., Accorsi, R., Ferrari, E., Manzini, R., Dunstall, S., 2017. On the design and control of food job-shop processing systems. A simulation analysis in catering industry. Int. J. Logist. Manag. 28, 782–797.

Tufano, A., Accorsi, R., Gallo, A., Manzini, R., 2018a. Simulation in food catering industry. A dashboard of performance. In: International Food Operations and Processing Simulation Workshop, pp. 20–27.

Tufano, A., Accorsi, R., Garbellini, F., Manzini, R., 2018b. Plant design and control in food service industry. A multi-disciplinary decision-support system. Comput. Ind. 103, 72–85. https://doi.org/10.1016/j.compind.2018.09.007.

Urbani, D., Delhom, M., Garredu, S., 2010. A multi-agent touristic catering market modeling methodology toward a DSS: a new approach based on multi-agents and geographic information systems. In: European Simulation and Modelling Conference.

Van Kampen, T., Van Donk, D.P., 2013. Coping with product variety in the food processing industry: the effect of form postponement. Int. J. Prod. Res. 7543, 1–15. https://doi.org/10.1080/00207543.2013.825741.

Williams, P.G., 1996. Vitamin retention in cook/chill and cook/hot-hold hospital foodservices. J. Am. Diet. Assoc. https://doi.org/10.1016/S0002-8223(96)00135-6.

Chapter 9

The storage of perishable products: A decision-support tool to manage temperature-sensitive products warehouses

Giulia Baruffaldi, Riccardo Accorsi, Daniele Santi, Riccardo Manzini and Francesco Pilati
Department of Industrial Engineering, Alma Mater Studiorum—University of Bologna, Bologna, Italy

Abstract

With the growth of customer awareness and regulations in force on quality and safety of perishable products, the handling of environmental conditions experienced by products along the supply chain assumes increasing relevance. The storage temperature is one of the most important drivers in controlling the physicochemical properties of a perishable product throughout the distribution chain. Therefore, a fair strategy for ensuring the quality and safety of products to the final consumer is to monitor the temperature conditions experienced by products during their storage time spent in warehouses, through dedicated location-mapping activities. However, the warehousing literature reveals a lack of methods and tools to lead managers through temperature warehouse mapping. In view of this, this chapter proposes a tool tailored for the calculation and visualization of temperature profiles of each storage location over a given time period. The proposed tool is validated through a multiple case study provided by two third-party logistics (3PL) companies that deal with the storage of biomedical products and beverages, respectively. Then, a support-decision model devoted to the storage assignment problem for a perishable item warehouse is formulated. This model, properly fueled by the proposed tool, can be used to handle operational decisions and the put-away process when stored items can be affected by thermal stresses.

1 Introduction

In the last several decades, the global food trade has dealt with consumer concerns about the credence attributes of products (i.e., quality, security, sustainability, and organic and fair-trade issues) (Bernués et al., 2003). Critical events, such as food safety scandals, can damage brand reputation and increase the amount of information requested by consumers. Growing consumer awareness of the characteristics of the products they purchase has led to a higher demand for compliance with quality and safety standards (Marshall et al., 2016; Trienekens and Zuurbier, 2008).

The environmental conditions (i.e., temperature, humidity, vibrations) experienced by perishable products, such as food products and pharmaceutical items, over their life cycle strongly affect the expected shelf life and other product properties (i.e., taste, flavor, freshness) (Hertog et al., 2014). The effective management of such conditions is important, especially in warehouses and in other nodes along the supply chain where products pause for a significant time (Meneghetti and Monti, 2014). Both in food and pharmaceutical distribution chains, temperature is considered the most important driver to control the safety and quality of products. Researchers have widely demonstrated how uncontrolled storage temperature conditions affect the physicochemical characteristics of products, leading to loss of quality and spoilage (Stoecker, 1998). Regulators have focused on the development of standards of quality and safety to guide companies during distribution operations in order to ensure proper storage conditions and preserve the safety of stocks (European Commission, 2013; Council Regulation, 2002).

Tracking of the temperature conditions experienced by products during storage in warehouses is a good practice to follow. Climate, weather seasonality, and infrastructural characteristics related to the layout of the warehouse, such as proximity to the docks, fans, and skylights, may affect the uniformity of the temperature within the storage facility, despite the presence of heating, ventilation and air conditioning (HVAC) systems. For this reason, quality managers are often asked to schedule temperature-mapping activities in the facilities in order to record data on the temperature fluctuation over time.

Sustainable Food Supply Chains. https://doi.org/10.1016/B978-0-12-813411-5.00009-0

Despite temperature mapping being a common practice in warehousing, there is no unique sharing protocol guiding managers through the mapping process. Both regulatory agencies and private organizations provide several nonbinding guidance documents. However, the literature reveals a lack of publications on tools and methods for the temperature mapping of warehouses.

For these reasons, we propose a tailored tool aimed at supporting warehouse managers during temperature-mapping processes. The tool can elaborate on data from a temperature/humidity tracking campaign performed using a limited number of data loggers and calculate the temperature/humidity conditions experienced by each location of the warehouse. The tool provides a set of graphical user interfaces (GUIs) enabling the user to quickly obtain statistics and assess the results, highlighting how an effective mapping of the indoor conditions can lead to improving the safety of the inventory. Furthermore, it provides reliable temperature spatial data to the warehouse manager that support strategic and operational decisions on the warehouse infrastructure or the storage operations, respectively. Infrastructural aspects deal with the design of roofs, the choice of insulating materials, the location of windows and docks, and the height of racks, while operations decisions may deal with the storage assignment problem, namely location assignments, with temperature-sensitive products. The remainder of this chapter is organized as follows: Section 2 explores the main criticalities affecting temperature distribution in warehouses; Section 3 illustrates the temperature-mapping activity with respect to the regulatory framework; while Section 4 describes the proposed tool; Section 5 illustrates two case studies provided by two third-party logistics (3PL) companies located in north-central Italy; Section 6 formulates a decision-support model for the storage-assignment problem for perishable items that can receive input data by the proposed tool and provides operational suggestions to practitioners and warehouse managers; finally, Section 7 describes the obtained results and gives topics for future developments.

2 Warehouse temperature distribution

The development of strategies to measure and control the indoor temperatures in warehouses have caught the attention of both researchers and practitioners in order to prevent product quality and safety loss caused by inadequate storage conditions. Particularly in the food sector, uncontrolled temperature and humidity conditions increase the growth rate of microorganisms, which may produce off-flavoring, texture modifications, slime production, and finally spoilage (Vaikousi et al., 2008). Moreover, evidence from the food industry reveals how indoor environmental conditions are strongly affecting the health and productivity of workers (Rohdin and Moshfegh, 2011). Technology progress has led to more efficient HVAC system designs. Nevertheless, the problem of deciding where to locate these systems within the storage layout is being widely explored by architects and engineers due to the nonhomogeneous temperature distribution inside buildings. In particular, the phenomenon of so-called air stratification (Armstrong et al., 2009) generates distinct air temperature values at different heights in the facility. This phenomenon is caused by different densities between hot and cold air, so that warmer air rises while cooler air falls. Without other airflow ventilation mechanisms, the stratification of air remains stable over time. Buildings characterized by high ceilings, such as warehouses, are the most affected by air stratification (Li, 2016). In addition, during the winter months, high temperatures beneath the roof lead to a significant reduction of the heating energy efficiency (Aynsley, 2005; Accorsi et al., 2017a). Porras-Amores et al. (2014) provide a detailed investigation of the air stratification phenomenon in five different types of warehouses in Spain. Given the vertical temperature variation, products stored in a warehouse for a long time may experience different thermal stresses according to the given height at which they are stored. In addition, the structural characteristics of the facility, such as the position of docks, doors, and windows that generate air infiltration (Brinks et al., 2015), as well as the operational practices (e.g., the ventilation, the position of the control points, and worker operations) also affect the temperature distribution among the locations at the same height. Fig. 1 summarizes the main issues affecting the temperature distribution inside warehouses.

The extant literature highlights how an efficient design of the layout of the cooling units may increase the uniformity of air temperature. In particular, some techniques and systems, such as air circulation and mechanical air ventilation devices, contribute to reducing the air stratification phenomenon. The set of these induced mechanisms can be defined as a thermal destratification process. Despite several studies that explore the calculation of air stratification effects and investigate the setting of thermal destratification mechanisms in warehouses, industrial practitioners usually adopt rough and approximate design approaches (Wang and Li, 2017). A few studies suggest the effective adoption of computational fluid dynamics (CFD) techniques. Ho et al. (2010) use CFD to analyze the temperature distribution in a large refrigerated warehouse with several product stacks and cooling units. This study demonstrates that numerical modeling can be used to predict the air temperature distribution in refrigerated warehouses, but the presence of a high number of design parameters and their interactions extensively increase the complexity of computation, increasing the costs and time to perform the analysis and the inaccuracy of the results. As a consequence, the on-field monitoring of the temperature conditions experienced by the storage locations is still the most common strategy when analyzing the temperature distribution within a storage area

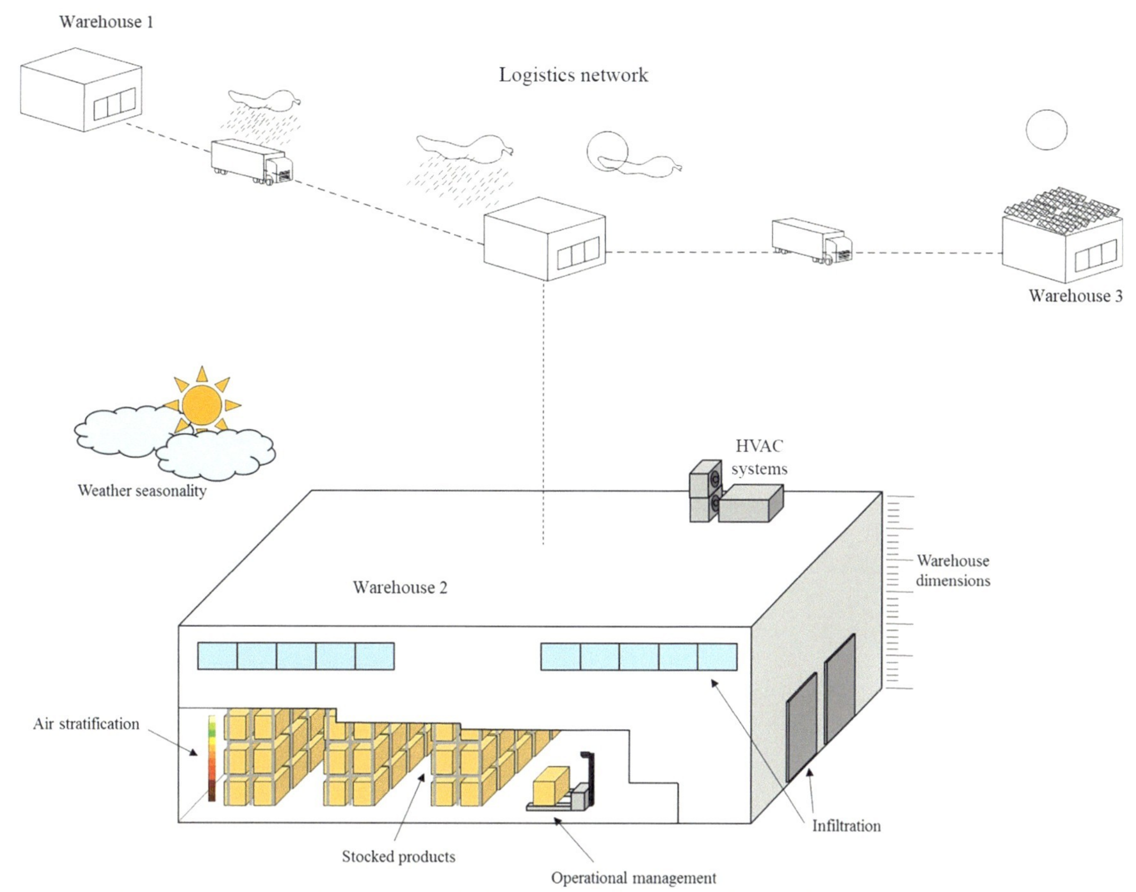

FIG. 1 Summary of main issues affecting the temperature distribution in warehouses as nodes of food supply chain.

(Zhu and Wang, 2013). The regulatory framework concerning the temperature mapping of warehouses suggests that the length of time to monitor relies on the type of stored products and whether the locations experience temperature fluctuations. Especially in those warehouses without HVAC systems, the temperature fluctuation is affected by weather seasonality, which becomes increasingly critical due to climate change (Chapman, 2007).

3 Warehouse temperature mapping

The mapping of temperatures and the indoor environmental conditions play key roles for those companies that deal with the storage of temperature-sensitive products, such as pharmaceutical products and fresh food like fruits and vegetables, frozen foods, and meat and dairy products. The temperature-mapping process involves the tracking of the temperature profile variation in storage areas, such as temperature-controlled chambers, industrial refrigerators and warehouses, over time. The aim of such a process is to verify whether the environmental conditions experienced by the stored products are in accordance with their labeled conservation conditions. The temperature mapping allows identification of temperature fluctuation due to architectural features, or to the properties of the observed facility and worker operations. Moreover, in warehouses the seasonal weather influences the indoor temperature and this has to be taken into account. The results obtained from the temperature mapping are applied to set the proper location to install monitoring systems, thus verifying the compliance of the temperature values with the quality conservation standards. Mapping is also used to identify potential improvement actions to reduce the so-called hot and cold spots within a storage area (WHO, 2014).

4 Methodology

In this section, the main activities developed during a typical warehouse mapping campaign are described. Among the aforementioned set of guidelines based on good manufacturing practices (GMP), we illustrate a nine-step protocol proposed by the company Vaisala (Vaisala, 2017). It is worth noting that, similarly to the other existing guidelines, the development of detailed documentation of the activities performed is fundamental in order to validate the obtained results. Therefore, the first step (Step 1) involves the creation of a validation plan, with the aim of establishing the company's goal and the adopted approach. Step 2 is the identification of the areas of potential risks within the facility, such as those locations where unacceptable temperatures and humidity values are expected, given the architectural characteristics of the facility. A fair practice is to realize a preliminary thermal inspection of the storage system in order to identify cold and hot spots. As an example, Fig. 2 presents two pictures obtained by a thermal camera in the warehouse used as a testbed. Fig. 2A shows a view of an aisle between two racks while in Fig. 2B, a gate in the inbound area is shown. It is worth noting how the temperatures within the truck achieve values up to 50°C, and this is the stress the load is expected to suffer. The brighter the color of the locations in the thermal pictures, the higher the intensity of the emitted infrared radiations will be.

In addition to the validation plan, in the third step (Step 3) a protocol for the mapping test that defines the data format, the number and the layout of the sensors, the calibration process description, and the acceptable range of temperature is developed. The sensor distribution is defined during Step 4. Currently, there are no specific formulas to determine the adequate sensor distribution, which is usually decided according to good practice. Reasonably, a higher number of sensors leads to a more accurate data collection. With respect to humidity monitoring, it should be driven by the sensitivity of the products to the effects of moisture. The selection of the best fitting tracking technology (Step 5) is important to validate the temperature mapping, and a phase involving control, calibration, set-up, and validation of the tracking equipment is recommended (Step 6). Step 7 involves the testing and data gathering phase. Results from this last phase are then used to propose possible adjustments, i.e., to the HVAC systems, or to decide on the location of particular products (Step 8). Finally, Step 9 involves the final temperature-mapping documentation and dissemination.

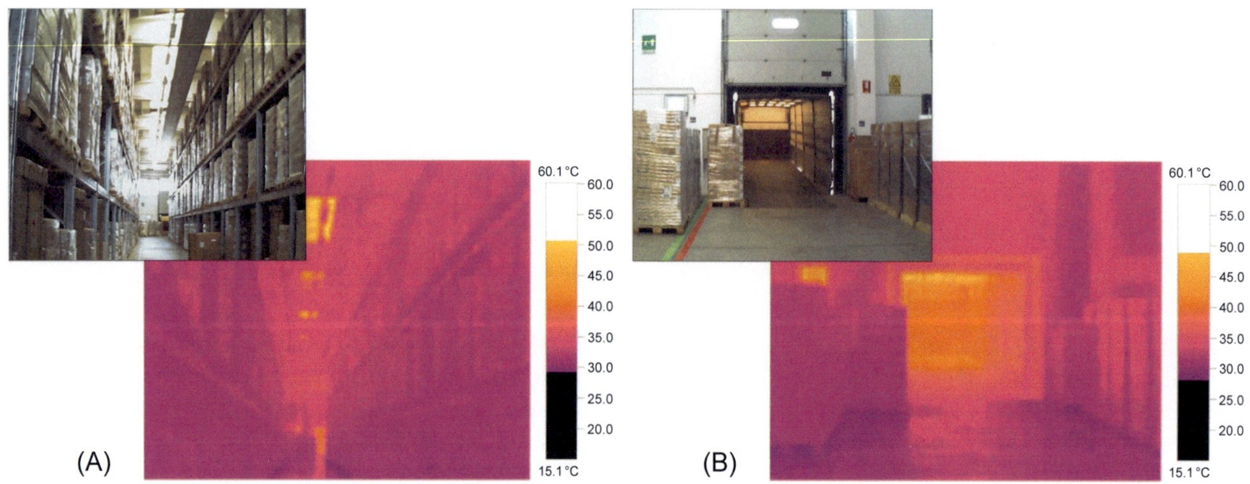

FIG. 2 (A) Thermal photo of a warehouse corridor; (B) thermal photo of an I/O dock.

5 A diagnostic-support tool

We illustrate an original diagnostic-support tool, developed in C#.Net, that manipulates a historical dataset representing the temperature and relative humidity profile recorded by the data loggers located in a warehouse. Giving a specific sample rate (e.g. 1 hour, 30 minutes), the tool is able to calculate the temperature and relative humidity values experienced by each location, even those not monitored, exploiting the data extracted by the data loggers located nearby. The results obtained aid the warehouse managers during the temperature mapping and, particularly, may support the activities performed in the aforementioned Step 8 and Step 9. It is worth noting that the reliability of the tool outputs is strictly related to the accuracy of the input data. Therefore, a high number of data loggers and their adequate distribution over the storage area are fundamental to produce valid data input. On the other hand, the cost of the equipment would increase. For these reason, managers are asked to carefully manage Step 4 in order to decide on the trade-off between costs and data accuracy.

The tool incorporates a relational database able to store and provide data and elaborate the results. Moreover, the main user-friendly graphical user interface (GUI) enables interactions between the tool and the final user and allows the graphic representation of the analysis. The database contains a set of tables that gather and organize the input and output data. The records of the monitoring campaign are appended in a table containing information on the code of the sensors (1), the recorded temperature and relative humidity values (2), the sample rate (3), and sensor positions inside the warehouse (4). The tool is organized around the main GUI that provides all the functionalities that aid implementing the six-phases procedure reported in Fig. 3.

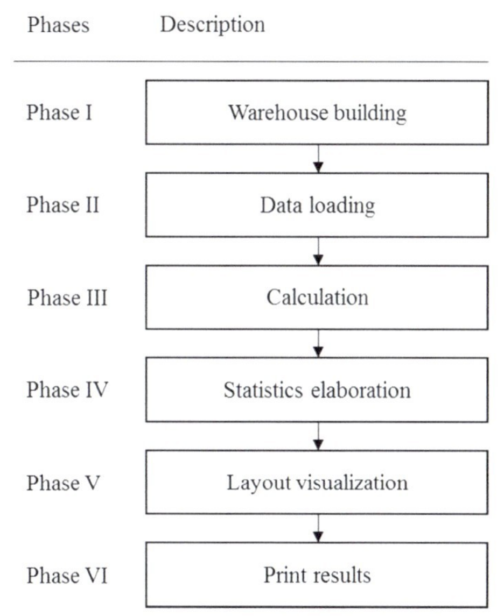

FIG. 3 The six-step procedure implemented by the proposed tool.

Phases I, II, and III are performed through the first module (i.e., warehouse (WH) design) of the GUI (see Fig. 4). The first phase involves the virtualization of the physical characteristics and the layout of the observed warehouse. A set of controls allows the user to define the size and the characteristics of the bays, the aisles, and the storage levels of each homogeneous area (i.e., zone) identified within the storage facility (Accorsi et al., 2014).

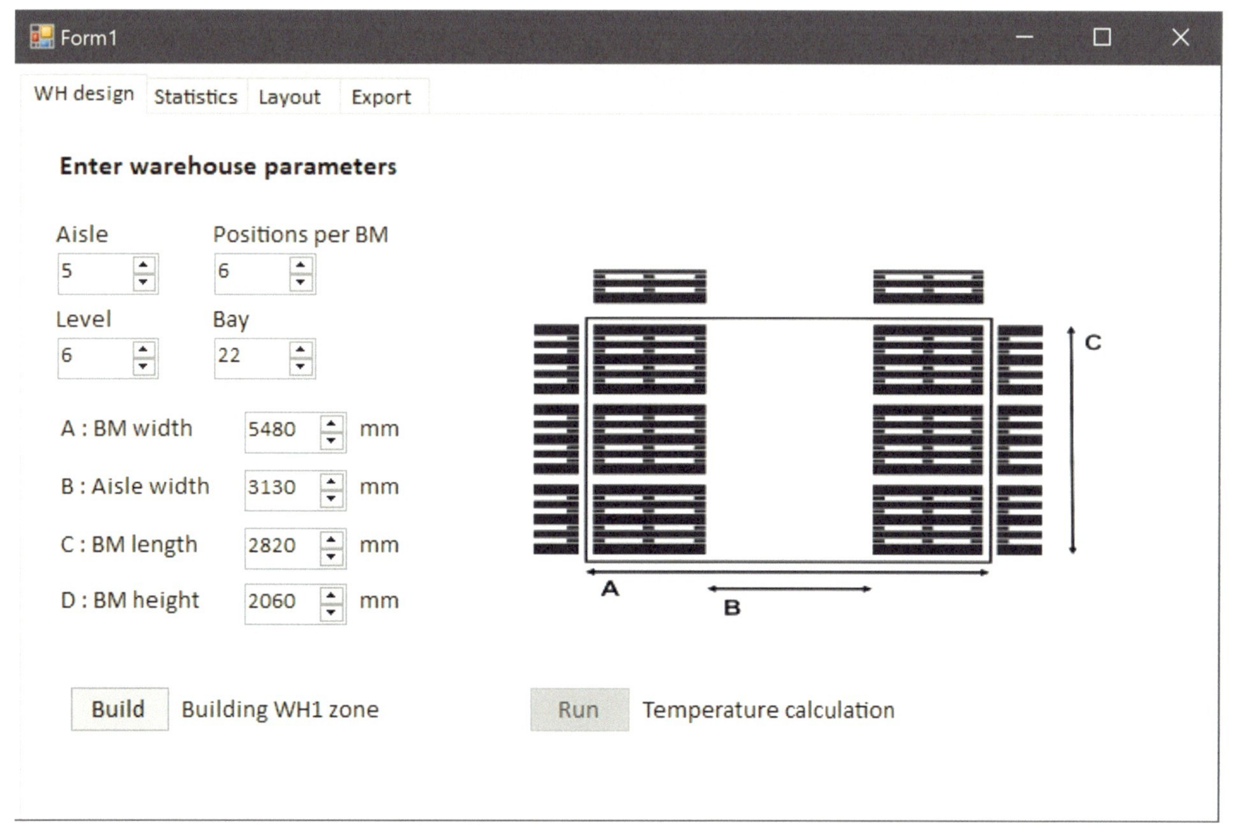

FIG. 4 Module "warehouse design" of the main GUI.

After the input loading (Phase II), Phase III calculates the temperature and humidity values expected at each non-monitored location based on the historical dataset. Since the calculation algorithm is the same for temperature and humidity, the following reports the exemplification of the temperature values only. Fig. 5 displays the 3D representation of a rack, where the pallet represents the physical location of the data loggers. Assume that the temperature experienced by a generic location A (i.e., not tracked) is calculated during a day d at time t. Given the sample rate of the sensors, this calculation is performed for each sample (e.g., at $t = 13{:}00$, $t = 14{:}00$, $t = 15{:}00$) over the length of time of monitoring.

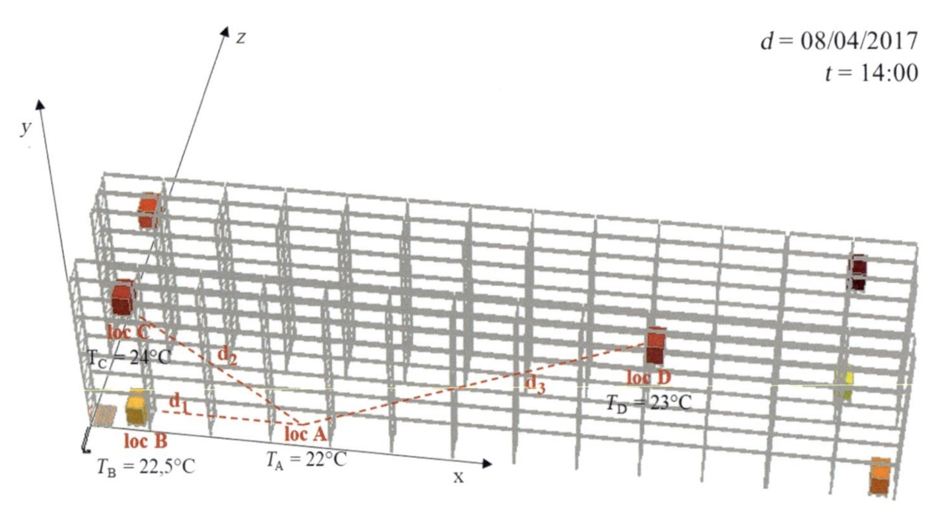

FIG. 5 3D representation of the rack within monitored and nonmonitored locations.

The calculation of the temperature expected at location A requires the identification of the three nearest locations to A through the three-dimensional Euclidean distances. Let the locations B, C, and D be the nearest available, as illustrated in Fig. 5. As an example, the distance between A and B (i.e., d_1) is calculated according to Eq. (1) as follows:

$$d_1 = \sqrt{(B_x - A_x)^2 + \left(B_y - A_y\right)^2 + (B_z - A_z)^2} \tag{1}$$

where B_x, B_y, and B_z are the coordinates of location B and A_x, A_y, and A_z are the coordinates of location A.

Then, the temperature expected at location A (i.e., T_A) is calculated using Eq. (2) as follows:

$$T_A = \frac{\frac{T_B}{d_1} + \frac{T_C}{d_2} + \frac{T_D}{d_3}}{\frac{1}{d_1} + \frac{1}{d_2} + \frac{1}{d_3}} \tag{2}$$

where T_B, T_C, and T_D are the temperature values, respectively, of B, C, and D while d_1, d_2, and d_3 are the Euclidean distance of B, C, and D from A.

Once Phase III is completed, the module "Statistics" of the tool allows visualization of the daily average temperature and humidity values through a line graph (Phase IV). This enables the user to identify and detect the most critical days or periods experienced by the warehouse overall and in detail, by each location, during the observed time horizon. During Phase V, the user can further analyze the temperature and humidity distribution over the selected days through the module "Layout," where each location is colored according to a given temperature range (see Fig. 6).

Finally, the last phase allows the user to store the results in the database and manipulate them through SQL queries.

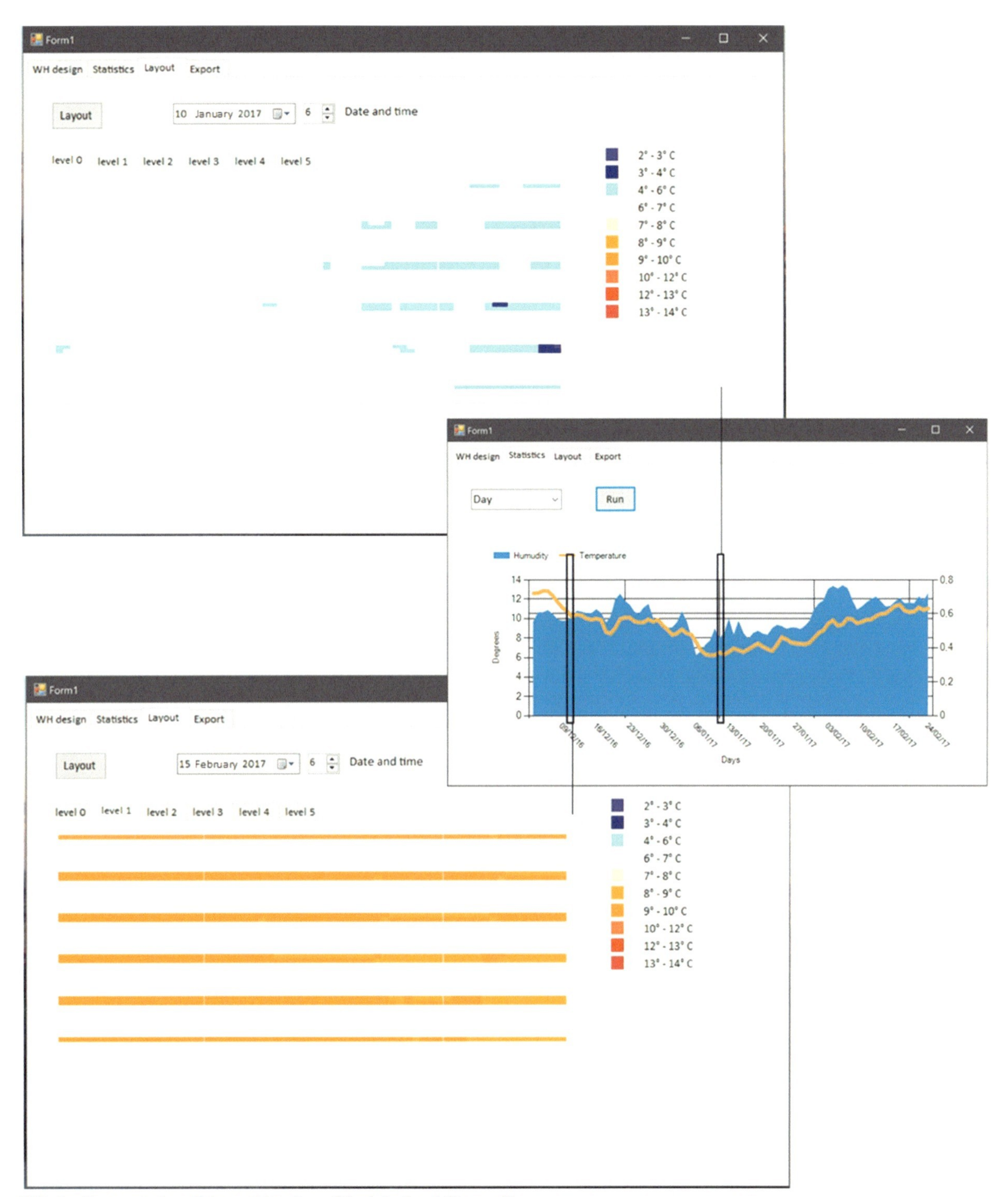

FIG. 6 Representation of the tool interfaces "Statistics" and "Layout."

6 Tool validation with industrial case studies

In order to assess the performance of the proposed tool, two testbeds from two different supply chains are illustrated. Both cases illustrate the application of the tool to support the temperature mapping in a real-world warehouse. The first warehouse stores biomedical products, i.e., products and devices for dialysis, and the second stores beverage items, such as wine, spirits, and other alcoholic drinks. The first warehouse belongs to a 3PL company that mostly deals with the storage and distribution of biomedical products. Due to the regulations in force on good distribution practices for pharmaceutical products for humans (European Commission, 2013) and the standards imposed by the clients, the main performance driver for the 3PL managers is product safety. On the other hand, due to the type of stored products, the second warehouse is not required to apply any severe procedures related to temperature monitoring. However, whether uncontrolled storage temperature conditions even slightly affect product safety, the perception of quality of some products, i.e., wine, for the final user is influenced by the temperature variations experienced by the products during storage. Consequently, this 3PL adopts a quality-driven approach in the storage operations, in order to guarantee a high level of service to its clients. Therefore, the warehouse mapping activity represents a strategic lever to enhance a competitive advantage. Both warehouses are located in the North of Italy, where weather conditions are usually characterized by a continental climate, with humidity and high temperature in summer and cold in winter. Since the warehouses are not dedicated to the aforementioned product categories only, they are not climate-controlled, and the weather seasonality highly affects the temperature conditions experienced by products. The two following subsections illustrate the case studies and the results obtained. Our analysis focuses on temperature, since the packaging in use protects the inventory from the effects of indoor moisture.

6.1 A 3PL warehouse for the storage of biomedical products

The warehouse is four levels high and contains more than 5000 locations mainly devoted to the storage of products and devices for dialysis. The warehouse handles approximately 700 different stock-keeping units (SKUs) every year. SKUs are picked at each rack level using counterbalance lift trucks and no reserve area is utilized. According to the safety-driven approach, during the temperature mapping a high level of accuracy is required in order to guarantee compliance with the imposed standard. For this reason, this case study represents a suitable testbed to validate the effectiveness of the tool. The inventory presents five different required conservation temperature ranges, but most belong to the range of 4–30°C. A preliminary thermometric inspection was performed in order to identify the hot and cold spots as well as the most adequate position to locate the sensors. According to the inspection, the most critical locations were those at the highest levels or near the skylights. In order to increase the accuracy, a total number of 82 data loggers were located inside the warehouse. Each aisle presents the same sensor distribution: six sensors distributed respectively in the front, in the center, and in the back of the rack. The same bay contains two sensors on the ground level and on the fourth level, respectively. We used self-sufficient digital thermometers and hygrometers with a resolution of 8 bits. We conducted a tracking campaign of almost 1 year (from March 2015 to February 2016).

All the records were analyzed using the proposed tool. The overall temperature profile showed a maximum variation of about 20°C from summer to winter. It is worth remembering that the weather conditions were particularly severe in summer of 2015. This period indeed was the most critical with respect to the compliance with the safety standards. For example, Fig. 7 shows some 3D views of the warehouse on August 4, 2015 at 6 p.m.

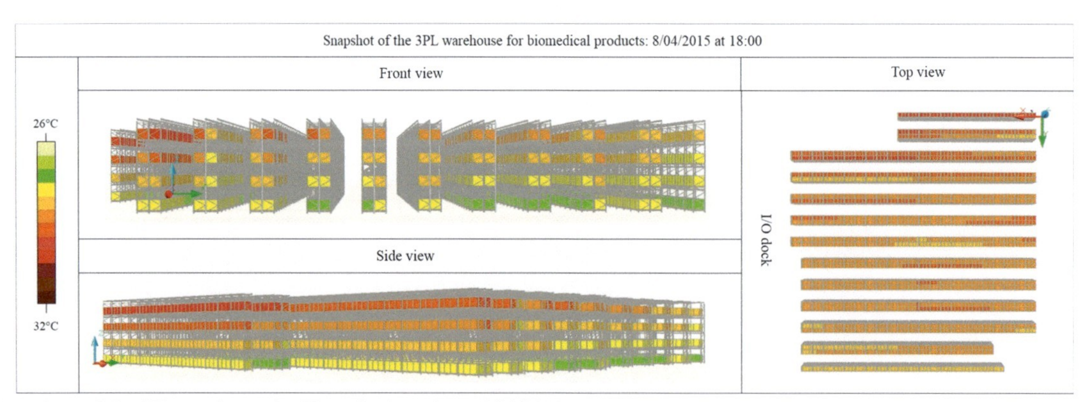

FIG. 7 Snapshots of the 3PL warehouse for biomedical products: 8/04/2018.

The storage locations are colored according to a temperature scale (from 26°C to 32°C). It is worth noting how the temperature values vary significantly all over the warehouse. Particularly, Fig. 7 reveals the impact of the aforementioned phenomenon of air stratification, showing a maximum difference of 4°C between the ground and the fourth rack level.

6.2 A 3PL warehouse for the storage of wine, spirits, and alcoholic items

This warehouse is five levels high and includes nine shelves and five aisles. Over a surface of 2400 m^2, a total number of 4000 locations is completely devoted to the storage of beverage products, such as water, beer, spirits, alcoholic drinks, and fine wine. The ground level is entirely dedicated to the picking tasks (i.e., cases from pallets and even bottles), which represent 98% of the retrieving, while the other levels belong to the reserve area where unit loads are stored according to a random assignment policy. Two or three operators per shift perform the picking activities over 12 hours a day. Due to its chemical characteristics, the most temperature-sensitive product over the inventory mix is wine. The safe storage temperature conditions vary from 6°C to 12.5°C for white wine (i.e., sparkling, light-bodied, and full-bodied white wine) and from 10°C to 18°C for red wine (i.e., light-bodied, medium-bodied, and full-bodied red wine). Similarly to the previous case study, an initial thermometric inspection was conducted in order to identify the hot and cold spots in the warehouse. To this purpose, 39 data loggers were located. Two temperature-tracking campaigns of 3 months each were conducted. The winter monitoring lasted from December 2016 to February 2017, while the summer monitoring took place from July 2016 to September 2016. Since the warehouse is developed along its length, in addition to the vertical air temperature distribution, warehouse managers were particularly interested in identifying the temperature variation among the same rack given the inbound doors placed over the front. The tool highlights how the temperature varies significantly from the front to the back of the rack for the ground level, while the upper levels present values that are more similar. This is evident in Fig. 8, which represents a warehouse view on January 10, 2015.

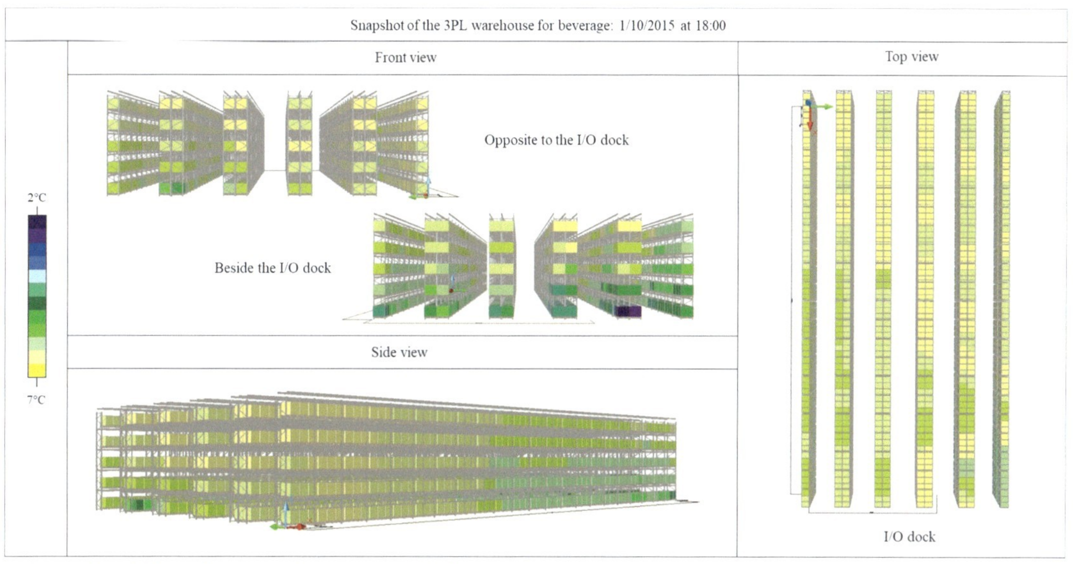

FIG. 8 Snapshots of the 3PL warehouse for beverages: 1/10/2016.

6.3 Outcomes from the case studies

The results from the case studies highlight the role of the tool in supporting the warehouse mapping activity both through providing reliable data and facilitating the reporting step. In addition, the results provide the manager with a chance to use a daily spatial temperature picture to identify those locations or storage zones characterized by a similar thermal profile. Such details may support decision making about the storage assignment problem for incoming temperature-sensitive goods (Accorsi et al., 2018), or the redesign of an energy-effective storage facility (Accorsi et al., 2017a) alike.

Moreover, the joint study of the temperature profile of a generic location over a time horizon and the inventory profile (i.e., the location turnover) of that location enables managers to monitor the compliance with the safe temperature conditions of the inventory. Hence, this tool can be used for the early diagnosis of the state, safety, and quality of conservation of the inventory and the assessment of the overall indoor conditions of the storage facility.

7 Modeling storage assignment for temperature-sensitive products

While the previous sections illustrate *ex-post* analysis of the environmental conditions experienced by the inventory, this section focuses on the *ex-ante* and prevention functionality of the tool, which incorporates an optimization model aimed at optimizing the storage assignment problem for the incoming unit loads, given the obtained temperature map of the storage locations. The storage assignment problem is widely debated in the warehousing literature (Gu et al., 2007; Petersen and Aase, 2004). It deals with decisions on the proper storage location for an incoming pallet of a given SKU, in order to maximize a generic function of convenience. Given its impact on the performance of the warehousing operations (i.e., picking and put-away), the storage assignment is usually aided by enterprise resource planning (ERP) software (i.e., warehouse management systems) and other support decision tools (Dotoli et al., 2015; Manzini et al., 2011; Accorsi et al., 2014). The storage assignment problem is typically formalized through mathematical programming techniques (De Koster et al., 2007; Heragu et al., 2007; Manzini et al., 2015) and can be driven by several functions of convenience. These vary from minimization of the traveling to complete a picking tour (Accorsi et al., 2012) to the stabilization of the loads on the racks to guarantee safety targets (Bortolini et al., 2015). Nevertheless, the storage conditions for the quality conservation of temperature-sensitive products represent a still underrated driver for the storage assignment problem.

Because the tool elaborates thermal logger records and provides a map of the historical and current indoor storage conditions, it offers the opportunity to implement storage assignment models aimed at minimizing the thermal stresses experienced by the stocks along their turnover. The remainder of this section is devoted to formalizing a generic storage assignment model that uses the quality of storage conservation conditions as the driver for the optimization problem. The following sets, parameters, and variables are assumed, according to Accorsi et al. (2018):

Sets and indices:

$j \in J$	Storage locations
$w \in W$	SKUs
$i_w \in I$	Unit load i of SKU w
T	Set of periods
$t \in T$	Period (e.g., a day)
Δs	Periods within an average inventory turnover
$T_j(t)$	Temperature measured at location j in period t
$T^{stress}_{i_w,j,\Delta s}(t)$	Expected stress temperature for unit load i_w measured at the location j during Δs starting from t
$[T^{safe-}_{i_w}, T^{safe+}_{i_w}]$	Safe temperature range for unit load i_w of SKU w

The solution defines the set of put-away tasks to assign to the incoming pallet i of SKU w to location j in period t.

The model optimizes the overall convenience that results from the daily assignment of the incoming loads to the available storage locations. The definition of convenience stems from the storage conditions experienced by each location. It pushes the assignment of the incoming unit-load i_w of SKU w to location j in period t according to the highest expected temperature stress that is measured in that location along a time horizon from t to $t+\Delta s$. The term Δs is the average period that a pallet is stored, which is calculated in the last $t-\Delta t$ periods. Assuming a location that is occupied until the load is completely retrieved, Δs is the time horizon within which a load may experience stresses or critical temperatures that affect its safe conservation.

Parameters:

$$T^{stress}_{i_w,j,\Delta s}(t^*) = \begin{cases} \max\left\{T_j(t) : t^* < t \le t^* + \Delta s\right\} & \text{If } \left|\max\left\{T_j(t) : t^* < t \le t^* + \Delta s\right\} - T^{safe+}_{i_w}\right| \ge \left|\max\left\{T_j(t) : t^* < t \le t^* + \Delta s\right\} - T^{safe-}_{i_w}\right| \\ \min\left\{T_j(t) : t^* < t \le t^* + \Delta s\right\} & \text{Otherwise} \end{cases} \tag{3}$$

$\delta_{jt} = \begin{cases} 1 \\ 0 \end{cases}$	If the location j is empty during period t Otherwise
$\alpha_{i_w t} = \begin{cases} 1 \\ 0 \end{cases}$	If the unit load i_w arrives between the periods $(t-1)$ and t Otherwise
$d^T_{i_w j} = \left\| T^{safe}_{i_w} - T^{stress}_{i_w, j, \Delta s}(t) \right\|$	Distance between the location j and the optimal location for load i_w according to the safe temperature $T^{safe}_{i_w}$

This highest stress temperature during Δs is measured per load i_w and location j using Eq. (3). A smaller difference (i.e., degrees) between this temperature $T^{stress}_{i_w, j, \Delta s}(t^*)$ and the temperature range of safe conservation for load i_w $[T^{safe-}_{i_w}, T^{safe+}_{i_w}]$ corresponds to a smaller difference from the optimal storage configuration.

The following list of assumptions introduces the model sets, objectives, and constraints:

- The number and type of incoming pallets are known at each period t.
- The incoming unit load i_w is assigned to a generic location j at the end of the daily operations.
- The unit loads are single SKU and each storage location holds one unit load.
- The unit loads have standard sizes and can fill every storage location.

Decision variables:

$x_{i_w jt} = \begin{cases} 1 \\ 0 \end{cases}$	If the unit load i_w is assigned to the location j during the time – window t Otherwise

Then, the model will be formulated as follows and solved per each period t (i.e., a day):

$$\min \sum_{i_w \in I} \sum_{j \in J} \left(x_{i_w jt} \cdot d^T_{i_w j} \right) \tag{4}$$

$$\text{s.t.} \quad \sum_{j \in J} x_{i_w jt} = 1 \quad \forall i_w \in I : \alpha_{i_w t} = 1 \tag{5}$$

$$\sum_{i_w \in I} x_{i_w jt} \leq \delta_{jt} \quad \forall j \in J \tag{6}$$

$$x_{i_w jt}, \in \{0, 1\} \quad \forall i_w \in I, j \in J \tag{7}$$

The quality conservation objective of Eq. (4) quantifies the difference between the as-is and the storage configuration that minimizes the temperature stress of the inventory.

Constraint (5) imposes the assignment of load i_w to a single storage location in period t. Constraint (6) ensures that each storage location j is filled by a single load only if j is empty in t. Constraint (7) denotes the binary nature of the decision variables.

As a combinatory problem, its complexity grows significantly with the instance, but it is solvable at the optimum in reasonable time for small- and medium-size warehousing receiving up to 5–10 trucks per day.

For an extended version of the model, see the multiobjective storage assignment formulation by Accorsi et al. (2018) aimed at mutually managing efficiency and stock safety goals. In this paper, the influence and the hidden impact of climate and weather, as well as outdoor environmental conditions, on the performance of the logistics operations for perishable items have been investigated for the storage systems, as was previously done by the authors for the food distribution and delivery process (Manzini and Accorsi, 2013; Manzini et al., 2014; Valli et al., 2013; Ayyad et al., 2017; Accorsi et al., 2016), and for the design of energy-effective cold chains (Gallo et al., 2017; Accorsi et al., 2017a,b,c).

8 Conclusion and future outlooks

An original tool is illustrated to support managers during the mapping of the indoor environmental conditions of a warehouse. In particular, the tool manipulates data collected from a set of temperature loggers located inside a warehouse and calculates the temperature/humidity profiles experienced by each storage location over a time horizon. The study of the temperature variation of a given location not only allows verification of the compliance of the storage conditions with the imposed quality and safety standards, but also enables managers to design proper storage assignment strategies for temperature-sensitive products. This tool could indeed fuel decision-support models aimed at minimizing the thermal stresses experienced by the stock along their turnover. A short formulation of the storage assignment problem incorporating the temperature conservation aspect is then given in this chapter. For an extended formulation of the problem the authors remand to Accorsi et al. (2018).

Furthermore, understanding the relationship between the weather conditions and the indoor temperatures measured at each location allows managers to estimate the expected stresses suffered by the inventory and thereby design storage assignment policies accordingly. In order to validate the tool, the experiences of two real-world 3PL warehouses are illustrated. Wide opportunities exist for future research in the management of the impacts that indoor temperatures have on the working conditions of labor and the resulting worker health effects. Despite the fact that further case studies could increase the accuracy of the analysis and the results, this tool still aids managers during the environmental data manipulation as well as in the detection of stresses. For this reason, further research may explore the development of a tool functionality aid to the automatic release of documentation in accordance with the regulations in force for specific products and perishable goods categories. The tool could be embedded into WMS platforms to lead managers through the real-time mapping of warehouse conditions for both inventory and labor (Baruffaldi et al., 2019). Furthermore, as the proposed tool and model have been intended for selective rack storage systems, new formulations are expected in the case of deep lane storage systems, which are typical in the food and beverage industry (Accorsi et al., 2017c).

To take another step forward, the tool could also be embedded within a support-decision web-based platform leading managers through the warehouse conditions mapping. Similar ready-to-use software applications are widespread online and can support managers in facing several practical industrial issues. As an example, in 2015 a Web-based database called *Bank of Solutions* (Banca delle Soluzioni, 2017) has been released to organize and showcase tools and informative reports that enhance the awareness of companies about safety issues in work places (Botti et al., 2015). These include technologies for confined spaces and ready-to-use solutions to improve ergonomics handling and working tasks. Similarly, a tailored decision-support platform for the mapping of storage conditions might valuably aid the companies involved in the perishable products supply chain.

References

Accorsi, R., Manzini, R., Bortolini, M., 2012. A hierarchical procedure for storage allocation and assignment within an order-picking system. A case study. Int. J. Log. Res. Appl. 15 (6), 351–364.

Accorsi, R., Manzini, R., Maranesi, F., 2014. A decision-support system for the design and management of warehousing systems. Comput. Ind. 65 (1), 175–186.

Accorsi, R., Ferrari, E., Gamberi, M., Manzini, R., Regattieri, A., 2016. A closed-loop traceability system to improve logistics decisions in food supply chains. A case study on dairy products. In: Montserrat Espiñeira, M., Santaclara, F.J. (Eds.), Advances in Food Traceability Techniques and Technologies: Improving Quality Throughout the Food Chain. Woodhead Publishing, Elsevier, pp. 337–351.

Accorsi, R., Bortolini, M., Gamberi, M., Manzini, R., Pilati, F., 2017a. Multi-objective warehouse building design to optimize the travel time, total cost and carbon footprint. Int. J. Adv. Manuf. Technol. 92 (1–4), 839–854.

Accorsi, R., Gallo, A., Manzini, R., 2017b. A climate-driven decision-support model for the distribution of perishable products. J. Clean. Prod. 165, 917–929.

Accorsi, R., Baruffaldi, G., Manzini, R., 2017c. Design and manage deep lane storage system layout. An iterative decision-support model. Int. J. Adv. Manuf. Technol. 92 (1–4), 57–67.

Accorsi, R., Baruffaldi, G., Manzini, R., 2018. Picking efficiency and stock safety: a bi-objective storage assignment policy for temperature-sensitive products. Comput. Ind. Eng. 115, 240–252.

Armstrong, M., Chihata, B., MacDonald, R., 2009. Cold weather destratification energy savings of a warehousing facility. ASHRAE Trans. 115 (2), 513–518.

Aynsley, R., 2005. Saving heating costs in warehouses. ASHRAE J. 47 (12), 47–51.

Ayyad, Z., Valli, E., Bendini, A., Accorsi, R., Manzini, R., Bortolini, M., Gamberi, M., Gallina Toschi, T., 2017. Simulating international shipments of vegetable oils: focus on quality changes. Ital. J. Food Sci. 29, 38–49.

Banca delle Soluzioni, 2017. Available at: http://safetyengineering.din.unibo.it/banca-delle-soluzioni (Accessed December 2017).

Baruffaldi, G., Accorsi, R., Manzini, R., 2019. Warehouse management system customization and information availability in 3pl companies: a decision-support tool. Ind. Manag. Data Syst. 119 (2), 251–273.

Bernués, A., Olaizola, A., Corcoran, K., 2003. Extrinsic attributes of red meat as indicators of quality in Europe: an application for market segmentation. Food Qual. Prefer. 14 (4), 265–276.

Bortolini, M., Botti, L., Cascini, A., Gamberi, M., Mora, C., Pilati, F., 2015. Unit-load storage assignment strategy for warehouses in seismic areas. Comput. Ind. Eng. 87, 481–490.

Botti, L., Duraccio, V., Gnoni, M.G., Mora, C., 2015. A framework for preventing and managing risks in confined spaces through IOT technologies. In: Safety and Reliability of Complex Engineered Systems. Proceedings of the 25th European Safety and Reliability Conference (ESREL 2015). CRC Press, Taylor & Francis Group, London, pp. 3209–3217.

Brinks, P., Kornadt, O., Oly, R., 2015. Air infiltration assessment for industrial buildings. Energy Build 86, 663–676.

Chapman, L., 2007. Transport and climate change: a review. J. Transp. Geogr. 15, 354–367.

Council Regulation (EC) No 178/2002, 2002. European Parliament and of the Council of 28 January 2002 laying down the general principles and requirements of food law, establishing the European Food Safety Authority and laying down procedures in matters of food safety. Off. J. Eur. Communities. 1.2.2002, 2001–2024.

De Koster, R.B., Le-Duc, T., Roodbergen, K.J., 2007. Design and control of warehouse order picking: a literature review. Eur. J. Oper. Res. 182, 481–501.

Dotoli, M., Epicoco, N., Falagario, M., Costantino, N., Turchiano, B., 2015. An integrated approach for warehouse analysis and optimization: a case study. Comput. Ind. 70, 56–69.

European Commission, C 343/1, 2013. Guidelines of 5 November 2013 on good distribution practice of medicinal products for human use. Off. J. Eur. Union. 23.11.2013, 2001–2024.

Gallo, A., Accorsi, R., Baruffaldi, G., Manzini, R., 2017. Designing sustainable cold chains for long-range food distribution: energy-effective corridors on the silk road belt. Sustainability 9 (11), 2044. https://doi.org/10.3390/su9112044.

Gu, J., Goetschalckx, M., McGinnis, L.F., 2007. Research on warehouse operation: a comprehensive review. Eur. J. Oper. Res. 177, 1–21.

Heragu, S., Du, L., Mantel, R.L., Schuur, P.C., 2007. Mathematical model for warehouse design and product allocation. Int. J. Prod. Res. 43 (2), 327–338.

Hertog, M.L.A.T.M., Uysal, I., Verlinden, B.M., Nicolaï, B.M., 2014. Shelf life modelling for first-expired-first-out warehouse management. Philos. Trans. R. Soc. A Math. Phys. Eng. Sci. 372.

Ho, S.H., Rosario, L., Rahman, M.M., 2010. Numerical simulation of temperature and velocity in a refrigerated warehouse. Int. J. Refrig. 33, 1015–1025.

Li, W., 2016. Numerical and Experimental Study of Thermal Stratification in Large Warehouses. (April) 121 p.

Manzini, R., Accorsi, R., 2013. The new conceptual framework for food supply chain assessment. J. Food Eng. 115 (2), 251–263.

Manzini, R., Accorsi, A., Pattitoni, L., Regattieri, A., 2011. A supporting decisions platform for the design and optimization of a storage industrial system. In: Jao, C. (Ed.), Efficient Decision Support Systems: Practice and Challenges – From Current to Future. InTech Publisher, Rijeka, ISBN: 978-953-307-441-2, pp. 437–458.

Manzini, R., Accorsi, R., Ziad, A., Bendini, A., Bortolini, M., Gamberi, M., Valli, E., Gallina Toschi, T., 2014. Sustainability and quality in the food supply chain. A case study of shipment of edible oils. Br. Food J. 116 (12), 2069–2090.

Manzini, R., Accorsi, R., Gamberi, M., Penazzi, S., 2015. Modeling class-based storage assignment over life cycle picking patterns. Int. J. Prod. Econ. 170 (2015), 790–800.

Marshall, D., McCarthy, L., McGrath, P., Harrigan, F., 2016. What's your strategy for supply chain disclosure? MIT Sloan Manag. Rev. 57 (2), 37–45.

Meneghetti, A., Monti, L., 2014. Greening the food supply chain: an optimisation model for sustainable design of refrigerated automated warehouses. Int. J. Prod. Res. 7543, 1–21.

Petersen, C.G., Aase, G., 2004. A comparison of picking, storage, and routing policies in manual order picking. Int. J. Prod. Econ. 92, 11–19.

Porras-Amores, C., Mazarrón, F., Cañas, I., 2014. Study of the vertical distribution of air temperature in warehouses. Energies 7 (3), 1193–1206.

Rohdin, P., Moshfegh, B., 2011. Numerical modelling of industrial indoor environments: a comparison between different turbulence models and supply systems supported by field measurements. Build. Environ. 46, 2365–2374.

Stoecker, W.F., 1998. Industrial Refrigeration Handbook. McGraw-HIll Book Co.

Trienekens, J., Zuurbier, P., 2008. Quality and safety standards in the food industry, developments and challenges. Int. J. Prod. Econ. 113, 107–122.

Vaikousi, H., Biliaderis, C.G., Koutsoumanis, K.P., 2008. Development of a microbial time/temperature indicator prototype for monitoring the microbiological quality of chilled foods. Appl. Environ. Microbiol. 74 (10), 3242–3250.

Vaisala, 2017. GMP warehouse mapping: step-by-step guidelines for validating life science storage facilities. Available at:https://www.vaisala.com/sites/default/files/documents/CEN-LSC-AMER-GMP-Warehouse-Mapping-White-Paper-B211170EN-A.pdf.

Valli, E., Manzini, R., Accorsi, R., Bortolini, M., Gamberi, M., Bendini, A., Lercker, G., Gallina Toschi, T., 2013. Quality at destination: simulating shipment of three bottled edible oils from Italy to Taiwan. Riv. Ital. Sostanze Grasse 90, 163–169.

Wang, L.L., Li, W., 2017. A study of thermal destratification for large warehouse energy savings. Energy Build. 153, 126–135.

WHO, 2014. Temperature mapping of storage areas. In: Annex 9: Model guidance for the storage and transport of time and temperature–sensitive pharmaceutical products. Technical supplement to WHO technical report series, no. 961, 2011.

Zhu, Y.C., Wang, S.S., 2013. The design of warehouse temperature and humidity monitoring system. Adv. Mater. Res. 709, 453–457.

Further reading

WHO, 2003. Annex 9: Guide to good storage practices for pharmaceuticals. Technical supplement to WHO technical report series, No. 908, 2003.

Chapter 10

Mathematical modeling of food and agriculture distribution

Christine Nguyen*, Zhané Goff* and Riccardo Accorsi†

*Department of Industrial and Systems Engineering, Northern Illinois University, DeKalb, IL, United States †Department of Industrial Engineering, Alma Mater Studiorum—University of Bologna, Bologna, Italy

Abstract

The food industry has grown from a local industry into a global industry with the birth of technology. Products are delivered via trucking, water, rail, air, or some combination. The distribution of products in a supply chain is a difficult problem to solve, and perishable products bring additional complexity. While technology exists to slow down the deterioration rate of foods, products need to be delivered urgently to fulfill customer demands satisfactorily, with the expected quality. Most of the literature focuses on minimizing the time these products spend in the supply chain, as well as considering the measurement of perishability in various ways. This chapter covers the mathematical models that have been developed in the areas of routing, network design, and distribution with respect to the food distribution network.

1 Introduction

All products are transported using some combination of rail, water, air, and truck. In 2015, "the U.S. freight transportation system moved 49.5 million tons of goods valued at more than $52.7 billion each day" (2016 Transportation Statistics Annual Report). However, final deliveries, or last-mile deliveries, are most commonly transported using truck. According to the 2015 Freight Facts and Figures report from the United States Department of Transportation, truck transportation was responsible for 77% of all domestic shipments (by weight) in 2013 (U.S. DOT, 2015). The food and agriculture supply chain is more difficult to manage, because most of the products deteriorate and require temperature-controlled systems. But even with better technology, according to the Food and Agriculture Organization of the United Nations, "roughly one third of the food produced in the world for human consumption – approx. 1.3 billion tons – gets lost or wasted" (FAO, 2011).

Many challenges in the food supply chain (FSC) are still unaddressed. FSCs are complex and continuously change. With an efficient supply chain network design, product flows and transportation costs can decrease and food safety can increase. Trienekens et al. (2012) argued that transparency in the FSC to government, consumers, and food product companies is necessary to maximize food quality and reduce waste.

The perishability of the product can be modeled directly or indirectly. Several works discussed in this chapter demonstrate ways perishability can be handled. Many researchers model perishability directly by calculating quality and deterioration. Typically, quality and deterioration are measured as linear or exponential functions with respect to time and/or temperature. Refrigeration, temperature-controlled environments, or similar efforts can be modeled as factors that slow down the deterioration of the product. In this chapter, there are also papers that have strict delivery dates (i.e., time windows) to handle the perishable products.

The distribution of products and supplies in a local or global network has extensive existing research. It ranges from vehicle routing, inventory routing, and production routing, to transportation distribution and scheduling. The complexity of these problems are not only due to the size of the network, products, origins, and destinations. The specific characteristics of the shipments contribute greatly to the difficulty of solving these problems. The existing research literature includes works in a variety of product types, such as oil (Ye et al., 2017), petroleum (Cornillier et al., 2012), blood (Stanger et al., 2012), drugs (Wu and Mao, 2017), food, and dairy (Adenso-Diaz et al., 1998), as well as shipping constraints like time windows, split delivery, consolidation, and transportation modes. Much research also exists in the areas of replenishment, inventory management, and production. Most of these problems are formulated as optimization problems and solved using techniques ranging from column generation to metaheuristics. Modeling any transportation problem with perishable products requires

Sustainable Food Supply Chains. https://doi.org/10.1016/B978-0-12-813411-5.00010-7

the consideration of quality. As products deteriorate, the quality decreases. The deterioration occurs with respect to time, so most distribution models will tackle this problem by controlling for time. Sometimes this is done directly by calculating a quality factor either in the objective function or constraint. In other models, it is accomplished by minimizing the time the product spends in transportation, distribution, inventory, and handling. Ideally, the sooner the product gets from an origin to a destination, the better the quality of the product. This chapter covers the different ways perishable products have been handled in the literature. The focus is primarily on routing problems, network design, and distribution problems, as illustrated in Fig. 1 and Table 1.

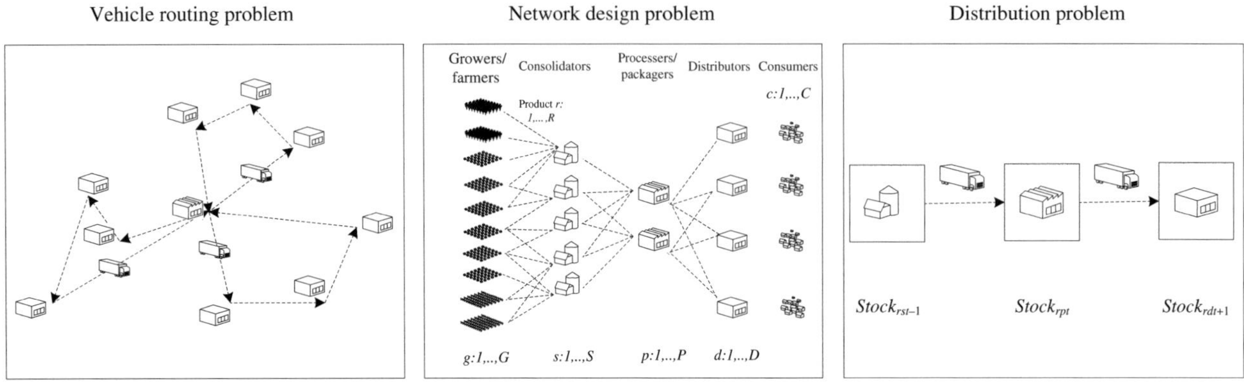

FIG. 1 Schematic representation of network problems (assume variable $stock_{rst-1}$ is the inventory of product r, in consolidation node s, at period $t-1$).

TABLE 1 Decision problems description

Decision problem	Demand time horizon	Decision perspective	Entities	Decisions
Network design	Long-term	Strategic	Network nodes (e.g., growers, packagers, warehouses)	(1) Which node to open; (2) how to allocate distribution flows throughout the network
Distribution problem	Mid-term	Tactical/ operational	Inventory	(3) Food inventory management; (4) when to deliver food
Vehicle routing	Short-term	Operational	Vehicles, routes	(5) Which vehicle to load; (6) which route to travel.

Food network typical issues

- Inventory management with shelf life and food quality decay constraints;
- Temperature control within nodes and along the delivery tours;
- Special handling required for transportation;
- Strong demand seasonality;
- High number of actors involved both upstream (growers) and downstream (distributors, retailer, markets).

2 Routing problems

The first area of research to consider is routing problems. These problems consider the locations that vehicles will visit and the order they should be visited. The most general problem is the vehicle routing problem, the purpose of which is to determine the routes for multiple vehicles from a single depot such that demand at a given set of locations will be satisfied. Variations of the problem include, but are not limited to, capacitated and incapacitated vehicles, homogeneous and heterogeneous vehicles, multiple depots, inventory decisions, delivery and pick-up decisions, and time windows for delivery and/or pick-up.

2.1 Vehicle routing problems

The area of vehicle routing problems (VRPs) is an extremely popular research topic. The VRP dates back to 1959 in a paper that generalized the traveling salesman problem to the optimum routing of a fleet of delivery trucks and was shown to be NP-hard (Dantzig and Ramser, 1959). The VRP formulation can be modeled in three different ways:

- vehicle flow formulation,
- commodity flow formulation, and
- the set partitioning formulation.

The integer variables in the vehicle flow model represent whether or not a vehicle uses an arc. The commodity flow formulation has additional continuous variables that represent the demand flow for each arc. The last formulation, set-partitioning, contains an exponential number of integer variables that represent whether a feasible route is used and additional variables that indicate whether a node is visited by the optimal routes.

The classic capacitated vehicle routing problem and its common variants can be found in Toth and Vigo (2002). Various exact methods (i.e., branch-and-bound, branch-and-cut), relaxations (i.e., Lagrangian relaxation), and metaheuristics (i.e., tabu search, genetic algorithms) have been developed and perform well in solving VRPs and their variants (Tarantilis et al., 2005; Toth and Vigo, 2002). However, the consideration of transporting perishable products dates to more recent research.

Specific products, like milk and dairy, have been modeled using VRP. Tarantilis and Kiranoudis (2001) formulate a VRP with a heterogeneous fleet for the distribution of fresh milk in Greece. The vehicles have different capacities and are part of a multidistribution center network. A threshold accepting algorithm was used to satisfy the demand while the objective was to minimize transportation time and distribution time.

Ambrosino and Sciomachen (2007) consider a food distribution problem for dry/fresh products and frozen products. Dry or fresh products require special handling and the frozen foods fight against time as they are transported. The authors formulated a mixed integer programming problem that includes special handling for foods as well as limits the number of stops a vehicle can make if it is delivering frozen foods. The application focuses on a highway network and takes into consideration the temperature effects on the frozen foods during transport. These factors were captured in the capacitated vehicle routing problem, and the authors developed a two-step heuristic algorithm approach to solve the problem.

Song and Ko (2016) addressed the distribution of multicommodity perishable food products in a network with delivery capacity, delivery time, and number of refrigerated and general-type vehicles. They developed a nonlinear mathematical model and a heuristic algorithm to solve it. The objective function was to maximize total customer satisfaction, which was measured by the freshness of the product at the time of delivery.

The vehicle routing problem is also appropriate for solving food waste and disposal problems. For example, it was effective in solving a reverse logistics problem to reduce disposed food waste by transporting waste to designated treatment facilities (Kim et al., 2014). The issues and concerns with the increasing amounts of food waste and disposal in Korea was the primary motivation for the project. The authors formulated a dynamic capacitated vehicle routing problem to determine the optimal route for daily food waste collection.

Ceselli et al. (2009) developed a column generation and dynamic programming algorithm approach to solve a real-world vehicle routing problem. While this work does not directly consider perishable products, many aspects of it allow for such an application. For example, the authors consider an upper limit on the number of consecutive driving hours, the ability to fulfill some orders using express courier services, maximum route length, and maximum route duration. These characteristics can be used to ensure certain foods do not spend too much time in transport.

The multitrip VRP for deliveries and pick-ups modeled a road freight transportation service for air cargo. The purpose of this research was to minimize the number of trucks in a large multimodel transportation network for cargo. This serves as a supplement to the cargo transported by air to and from certain cities. The problem aimed to minimize total cost and minimize the total number of multiple trips. The authors developed two approaches: one that simultaneously solves both objectives and the other that decomposes the problem. Both approaches use a metaheuristic guided neighborhood search (Derigs et al., 2011a,b).

These are only a few examples of classic VRP problems that considered food distribution. Many of these problems include additional quality or deterioration terms in the objective function or characteristics of the vehicles. However, another popular VRP approach for food transport uses the vehicle routing problem with time windows (VRPTW) model, where delivery and/or pick-up of products can be more definitively defined.

A generic formulation of the routing problem, suitable for both food products with expiration time limit and other commodities follows:

Sets and indices:

$i \in \aleph$	Set of delivery nodes i
$a \in M$	Set of arc in the food distribution network
$v \in V$	Set of vehicles in the network
g	Hub of the supply chain from where delivery tours depart
$l \in L$	Set of food load to be delivered

Parameters:

d_{il}	Demand of food load l from delivery node i
tc_{va}	Travel cost of arc a traveled by vehicle v
T_{va}	Travel time of arc a traveled by vehicle v
w^l	Weight of load l
v^l	Volume of load l
wc^v	Weight capacity of vehicle v
vc^v	Volume capacity of vehicle v
T_v^{max}	Maximum time for the traveling of vehicle v before food product expiration

Decision variables:

x_{va}	Binary variable that is 1 if vehicle v travels on arc a; 0 otherwise
y_{vi}	Binary variable that is 1 if vehicle v visits delivery node i; 0 otherwise

$$\min \sum_{v \in V} \sum_{a \in M} tc_{va} \cdot x_{va} \tag{1}$$

$$\text{s.t.} \sum_{i \in \aleph} \sum_{l \in L} d_{il} \cdot w^l \cdot y_{vi} \leq wc^v \quad \forall v \in V \tag{2}$$

$$\sum_{i \in \aleph} \sum_{l \in L} d_{il} \cdot v^l \cdot y_{vi} \leq vc^v \quad \forall v \in V \tag{3}$$

$$\sum_{a \in M} T_{va} \cdot x_{va} \leq T_v^{max} \quad \forall v \in V \tag{4}$$

$$\sum_{a \in M^-(i)} x_{va} = y_{vi} \quad \forall v \in V, i \in N \tag{5}$$

$$\sum_{a \in M^+(i)} x_{va} = y_{vi} \quad \forall v \in V, i \in N \tag{6}$$

$$\sum_{v \in V} y_{vi} = 1 \quad \forall i \in N \tag{7}$$

$$\sum_{a \in M^-(g)} x_{va} = 1 \quad \forall v \in V \tag{8}$$

$$\sum_{a \in M^+(g)} x_{va} = 1 \quad \forall v \in V \tag{9}$$

$$\sum_{a \in M^+(g)} x_{va} \geq y_{vi} \quad \forall v \in V, i \in N \tag{10}$$

$$y_{vi}, x_{va} \in \{0, 1\} \tag{11}$$

Whereas the objective function (1) minimizes the transportation costs of the delivery tours, constraints (2)–(4) guarantee the consideration of the vehicle weight and volume capacities as well as the product expiration time limit. Constraints (5) and (6) ensure the flow balancing at each node, and constraint (7) imposes that a delivery node is visited just one time by one vehicle. Constraints (8)–(10) set the flow balancing at the Hub from which the delivery tour departs, while constraint (11) denotes the nature and domain of the decision variables. With respect to this formulation of the routing problem, the definition of adequate solving algorithms or heuristics able to find the optimal solution in reasonable time when the size of the instance increases is not an aim of this chapter and is left to other sources to provide further details (e.g., Toth and Vigo, 2002; Manzini et al., 2014).

2.2 Vehicle routing problem with time windows

While there are many variants of the vehicle routing problem, the VRPTW was adopted quickly for the transportation of perishable products, agriculture, and food. In this mathematical formulation, time windows for pick-up and/or delivery of goods are incorporated into the formulation for the vehicle routing problem. This provides more accurate planning and scheduling of vehicles from the depot. A major difference in the formulation is the addition of a time window for each node. Let $[a_i, b_i]$ be the time window for delivery node i, where a vehicle must arrive before time b_i. If the delivery is made before a_i, then it will wait until a_i to be received.

Sets and indices:

$i \in \aleph$	Set of delivery nodes i
$a \in M$	Set of arcs in the food distribution network, where i and j are nodes and $a = (i, j)$
$v \in V$	Set of vehicles in the network
g	Hub of the supply chain from which delivery tours depart
$l \in L$	Set of food load to be delivered

Parameters:

d_{il}	Demand of food load l from delivery node i
tc_{va}	Travel cost of arc a traveled by vehicle v
T_{va}	Travel time of arc a traveled by vehicle v
ts_{vi}	Service time of node i by vehicle v
w^l	Weight of load l
v^l	Volume of load l
wc^v	Weight capacity of vehicle v
vc^v	Volume capacity of vehicle v
T_v^{max}	Maximum time for the traveling of vehicle v before food product expiration
$[a_i, b_i]$	Time window for node i, where a vehicle must arrive before b_i and after a_i

Decision variables:

x_{va}	Binary variable that is 1 if vehicle v travels on arc a; 0 otherwise.
y_{vi}	Binary variable that is 1 if vehicle v visits delivery node i; 0 otherwise
s_{vi}	Time that vehicle v begins service at node i

Most of the parameters and decision variables remain the same as the VRP formulation. New information is in the time windows $[a_i, b_i]$ that are defined for every node, and the service time ts_{vi} required by a vehicle at every node. The new decision variable s_{vi} is added to track the time period that vehicle v begins its service at delivery node i. This time must be within the time windows provided. It is also important to point out that if the vehicle arrives before a_i, the vehicle must wait until a_i before service will begin.

$$\min \sum_{v \in V} \sum_{a \in M} tc_{va} \cdot x_{va} \tag{12}$$

$$\text{s.t. Constraints } (2)-(11) \text{ of VRP formulation}$$

$$a_i \leq s_{vi} \leq b_i \quad \forall v \in V, i \in N \tag{13}$$

$$x_{va} \cdot \left(s_{vi} + T_{va} + ts_{vi} - s_{vj}\right) \leq 0 \quad \forall v \in V, i \in N, j \in N, i \neq j, a = (i,j) \tag{14}$$

$$s_{vi} \geq 0 \quad \forall v \in V, i \in N \tag{15}$$

The objective function will minimize the total transportation cost, just like previously. Constraints (2) through (11) of the VRP formulation is still true in this formulation. The difference is the addition of constraints (13) through (15). Constraint (13) ensures the relationship between the vehicle departure time from a node i and its immediate successor node j. If an arc $a = (i, j)$ is traveled by a vehicle, then it will reach node j after it reaches node i. The time it reaches node j is the time it reached node i plus the service time at node i and the time it takes to travel the arc. Constraint (15) is the domain conditions for the new decision variable.

While the preceding formulation is primarily standard, the vehicle routing problem with time windows has been extensively used to handle perishable food delivery. The time window constraints allow for scheduling at a more detailed level. Time windows can be defined as days or even hours that better fit the real operations at distribution centers. This formulation also fits well with modern grocery delivery services such as Amazon Fresh and meal preparation delivery services like Blue Apron.

Food transport can better be modeled to reflect the real world's constraints on retailer demand and expectation of the quality of a product. For example, using a vehicle routing problem with soft time windows formulation, a computer program was developed to determine delivery routes for a meat company in Spain (Belenguer et al., 2005). The computer program served as a decision support system for a company's operations. It used a variety of heuristic algorithms to solve a multiobjective cost function to minimize total lateness and total distance traveled.

In Hsu et al. (2007), the researchers used a stochastic VRPTW in a study of perishable, temperature-sensitive food being transported from a distribution center to various destinations. The amount of food remaining in a vehicle decreased based on a function of food deterioration, travel time, and temperature. The model allowed soft time window constraints, which could be violated at the cost of a large penalty in the objective function. The authors used the time-oriented nearest-neighbor heuristic to solve the problem.

Quite a bit of research in food transport considers travel time along with the use of time windows for deliveries. Osvald and Stirn (2008) extended the VRPTW to include time-dependent travel times to model the distribution of fresh vegetables. The travel time between two locations was a function of the time of day and distance. A mathematical model was formulated considering the perishability costs and a tabu search–based heuristic approach solved the problem efficiently. Similarly, Li et al. (2015) modeled the VRP with soft time windows for fruits and vegetables. Due to the sensitivity of the product, the authors included deterioration due to the bruising of the produce during transportation (i.e., vibration, shock).

While most of the papers deal with a single depot, Mahdavi et al. (2017) proposed a new formulation for the VRPTW with perishable products and intermediate depots. This unique problem dealt with the uncertainty in the demand from customers. In addition, the problem included a heterogeneous fleet of vehicles, an asymmetric graph network, and the capability for vehicles to make multiple trips. The objective function focused on minimizing total distance traveled, vehicle usage cost, and earliness and tardiness penalty costs of services.

The core VRPTW formulation can be easily modified to include freshness-maintaining efforts like refrigeration and to consider performance metrics other than minimizing total cost or distance. In particular, quality measurements with respect to the product itself are useful in determining how to transport the product. This is particularly used in problems where inventory management is important.

2.3 Inventory routing problems

The preceding two types of problems only consider the routing of vehicles. Inventory routing problems (IRP) are an extension of vehicle routing problems, where the difference lies in the consideration of customer usage of product instead of their orders. It combines two complex problems: inventory management and vehicle routing. With information on the current inventory at each customer's location, the distributor can determine when to ship product and to which customers, so there are no shortages and demand is met. This allows for more flexibility in shipping strategy for the distributor, but is also more difficult to solve. Unlike the vehicle routing problem, inventory management decisions are made over a time horizon of several days, weeks, or months. In the classic vehicle routing problem, customers place orders and the distributor has no knowledge of the consumption and inventory at each customer's location.

Inventory routing problems consider multiple decisions and events across multiple time periods. At each customer and in each time period, product is received, consumed, and the inventory level changes. At the depot, product is made available, shipped, and the inventory level changes. The amount consumed and made available in each time period can change over time. Therefore, the routing for each vehicle can also be different in different time periods, which is shown in Fig. 2. The formulation now contains a time index with respect to when deliveries are made and the route for each vehicle. Much like the VRP, there are multiple types of formulation structure:

- vehicle-indexed formulations, and
- flow formulations

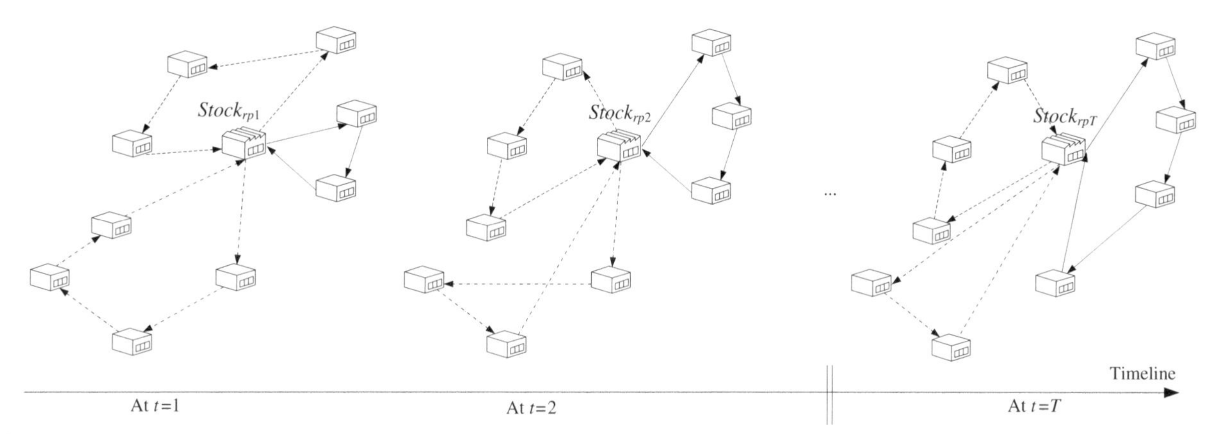

FIG. 2 Schematic representation of inventory routing problems.

The following formulation is based on Archetti et al. (2014) and illustrates a vehicle-indexed formulation. Since inventory will be tracked over time, the sequence of events in a time period must be defined. In other words, the inventory level at time t is at the end of the time period where delivery or consumption has occurred in time t. The inventory level at time t is equal to the inventory level at time $t-1$ plus the amount delivered in time t and minus the amount consumed in time t.

Sets and indices:

$i \in \aleph$	Set of delivery nodes i
$a \in M$	Set of arcs in the food distribution network
$v \in V$	Set of vehicles in the network
$\mathcal{g}$	Depot; Hub of the supply chain from which delivery tours depart
$l \in L$	Set of food load to be delivered
$t \in T$	Set of time periods

Parameters:	
d_{ilt}	Demand of food load l from delivery node i at time t
s_{it}	Supply of food load l made available at time t at the depot
h_i	Holding or inventory cost at node i
tc_{va}	Travel cost of arc a traveled by vehicle v
w^l	Weight of load l
v^l	Volume of load l
wc^v	Weight capacity of vehicle v
vc^v	Volume capacity of vehicle v

Decision variables:	
x_{at}^v	Integer variable representing the number of times vehicle v travels on arc a at time t
y_{it}^v	Binary variable that is 1 if vehicle v visits delivery node i at time t; 0 otherwise
z_{it}^v	Continuous variable representing the volume delivered to customer i at time t by vehicle k
I_{it}	Continuous variable representing the inventory level of node i at time t
I_{gt}	Continuous variable representing the inventory level of the depot at time t

$$\min \sum_{t\in T}\sum_{v\in V}\sum_{a\in M} tc_{va}\cdot x_{at}^v + \sum_{t\in T}\sum_{i\in \aleph} h_i \cdot I_{it} \tag{16}$$

$$\text{s.t.} \sum_{i\in \aleph}\sum_{l\in L} d_{ilt}\cdot w^l \cdot z_{it}^v \le wc^v \quad \forall v\in V, t\in T \tag{17}$$

$$\sum_{i\in \aleph}\sum_{l\in L} d_{ilt}\cdot v^l \cdot z_{it}^v \le vc^v \quad \forall v\in V, t\in T \tag{18}$$

$$\sum_{a\in M^-(i)} x_{at}^v = y_{it}^v \quad \forall v\in V, i\in N, t\in T \tag{19}$$

$$\sum_{a\in M^+(i)} x_{at}^v = y_{it}^v \quad \forall v\in V, i\in N, t\in T \tag{20}$$

$$\sum_{v\in V} y_{it}^v = 1 \quad \forall i\in N, t\in T \tag{21}$$

$$\sum_{a\in M^-(g)} x_{at}^v = 1 \quad \forall v\in V, t\in T \tag{22}$$

$$\sum_{a\in M^+(g)} x_{at}^v = 1 \quad \forall v\in V, t\in T \tag{23}$$

$$\sum_{a\in M^+(g)} x_{at}^v \ge y_{it}^v \quad \forall v\in V, i\in N, t\in T \tag{24}$$

$$I_{gt} = I_{gt-1} + \sum_{l\in L} s_{lt} - \sum_{v\in V}\sum_{i\in N} z_{it}^v \quad \forall t\in T \tag{25}$$

$$I_{it} = I_{it-1} - \sum_{l\in L} d_{lit} + \sum_{v\in V} z_{it}^v \quad \forall i\in N, t\in T \tag{26}$$

$$y_{it}^{v} \in \{0, 1\} \quad \forall v \in V, i \in N, t \in T \tag{27}$$

$$x_{at}^{v} \text{ are integer} \quad \forall v \in V, a \in M, t \in T \tag{28}$$

$$z_{it}^{v}, I_{it} \geq 0 \quad \forall v \in V, i \in N, t \in T \tag{29}$$

The objective function for this formulation is to minimize the total transportation and inventory costs, as shown in constraint (16). The travel cost can include the decrease in quality over time for the product. Constraints (17)–(18) are the vehicle capacity constraints for each time period and each vehicle. Constraints (19)–(20) ensure flow balancing at each node and time period. Constraint (21) ensures that a delivery node is visited once by only one vehicle for all time periods. Constraints (22)–(24) are the flow balancing constraints at the depot or Hub. The inventory flow balancing at the depot or Hub and for each delivery node are handled in constraints (25)–(26). Constraints (27)–(29) are the domain constraints for all the decision variables.

With respect to food transport, deterioration cannot solely be considered in the traveling and handling time. The additional visibility at each customer location could allude to issues with on-time delivery. Product delivered too early can result in large penalties due to insufficient inventory space at the customer's location. If product is delivered too late, then the customer incurs lost sales and potentially missed opportunities. Laganà et al. (2015) studied the inventory routing problem for a supermarket distribution industry. An optimal delivery strategy must be determined for a set of customers with multiple product categories that have different inventory policies and constraints. Penalties were incurred when product was delivered too early or too late. The decomposition solution approach had two phases: phase one focused on determining the best replenishment quantity, and phase two solved a VRPTW using a heuristic approach.

Many inventory routing problems can be considered a vendor-managed inventory system. This is where inventory levels at a vendor are managed by the distributor or supplier. Diabat et al. (2016) considered an inventory routing problem of perishable products that minimizes total costs while ensuring that customers do not run out of stock. The products are delivered from a centralized distribution center whenever products at the customers are close to running out or reaching maximum shelf life. Using an arc-based formulation, the authors applied a tabu search algorithm to solve it.

When determining the best route to take, the locations of the depot with respect to the customers can also be an influencing factor. The research by Hiassat et al. (2017) considered a location-inventory-routing problem that combines location decisions for warehouses with an inventory routing problem. It calculated the required number and location of warehouses as well as made routing and inventory decisions. Using a genetic algorithm approach, they demonstrated that the algorithm finds high-quality near-optimal solutions in a reasonable time. For this problem, perishable products had a shelf life defined by a number of time periods. Inventory levels at each customer location were limited by the physical capacity and the shelf life of the products.

Perishable products already contribute to the complex nature of these problems, but perishability also occurs within the production process. Vahdani et al. (2017) proposed a production-inventory-routing problem with time windows that contains production and distribution decisions. The authors pointed out that since perishable products have shorter shelf life, supply chain decisions would be more effective if production decisions were also considered. Specifically, the production decisions involved a multistate, multisite, multiperiod production scheduling problem that included inventory levels at each stage of the production process. The distribution decisions were determined by solving closed VRPTWs with a heterogeneous fleet of vehicles. The problem also included a decay rate and value to motivate the delivery of perishable products. In general, solution approaches to solving the IRP include branch-and-cut, decomposition techniques, and heuristics, to name a few. Solution approaches are discussed in a variety of other sources (e.g., Toth and Vigo, 2002; Archetti et al., 2014).

3 Network design and distribution problems

For dealing with the design of the food distribution network, a generic formulation is hereby given (see Accorsi et al., 2016 for further details).

Sets and indices:

$p \in P$	Set of processing/packaging plants
$r \in R$	Set of agro-food products
$d \in D$	Set of storage/consolidation facility
$c \in C$	Set of demand points (e.g., retailers, gross markets)

Parameters:

cf^n	Cost to establish a facility *n*
cv^t	Distribution costs per distance unit (e.g., kilometers) by mode *t*
$cv^{p(c)}$	Production (consolidation) costs per unit
cv^s	Storage costs per unit

Decision variables:

$y_{p(s)r}$	Binary variable that is 1 if facility *p*(*s*) is opened and devoted to product *r*; 0 otherwise.
y_d	Binary variable that is 1 if facility *d* is opened; 0 otherwise.
x_{ijrt}	Flow of product *r* distributed from node *i* to *j* by mode *t*

$$\begin{aligned}\min \; & \sum_{p\in P}\sum_{r\in R} y_{pr}\cdot cf^p + \sum_{d\in D} y_d \cdot cf^d + \sum_{s\in P}\sum_{r\in R} y_{sr}\cdot cf^s \\ & + \sum_{p\in P}\sum_{s\in S}\sum_{t\in T}\sum_{r\in R}\left(x_{sprt}\cdot cv^t \cdot d_{sp} + x_{sprt}\cdot cv^c\right) \\ & + \sum_{p\in P}\sum_{d\in D}\sum_{t\in T}\sum_{r\in R}\left(x_{pdrt}\cdot cv^t \cdot d_{pd} + x_{pdrt}\cdot cv^p\right) \\ & + \sum_{d,d'\in D}\sum_{r\in R}\sum_{t\in T}\left(x_{dd'rt}\cdot cv^t\cdot d_{dd'} + x_{dd'rt}\cdot cv^s\right) + \sum_{d\in D}\sum_{c\in C}\left(x_{dcrt}\cdot cv^t + x_{dcrt}\cdot cv^s\right)\end{aligned} \tag{30}$$

$$\text{s.t.} \quad \sum_{d\in D}\sum_{t\in T} x_{dcrt} = d_{cr} \quad \forall r,c \tag{31}$$

$$\sum_{p\in P}\sum_{t\in T} x_{sprt} \le C_{sr}\cdot y_{sr} \quad \forall r,s \tag{32}$$

$$\sum_{d\in D}\sum_{t\in T} x_{pdrt} \le C_{pr}\cdot y_{pr} \quad \forall r,p \tag{33}$$

$$\sum_{p\in P}\sum_{t\in T}\sum_{r\in R} x_{pdrt} + \sum_{d'\in D}\sum_{t\in T}\sum_{r\in R} x_{d'drt} \le C_d \cdot y_d \quad \forall d \tag{34}$$

$$\sum_{s\in S}\sum_{t\in T} x_{sprt} \ge \sum_{d\in D}\sum_{t\in T} x_{pdrt} \quad \forall p,r \tag{35}$$

$$\sum_{p\in P}\sum_{t\in T} x_{pdrt} \ge \sum_{d'\in D}\sum_{t\in T} x_{dd'rt} + \sum_{c\in C}\sum_{t\in T} x_{dcrt} \quad \forall d,r \tag{36}$$

$$y_{pr}, y_{sr} \in \{0,1\} \quad \forall p,s,r \tag{37}$$

$$x_{ijrt} \ge 0 \quad \forall i,j \in P\cup D\cup C, r, t \tag{38}$$

The objective function minimizes the network costs in terms of node establishment, flow management at nodes, and flow distribution, while the set of linear constraints ensures the demand fulfillment (31) and the respect of capacities (32–34) and guarantees the flow balancing (35 and 36). The schematic representation of the entities and the decisions involved in this problem is found in Fig. 3.

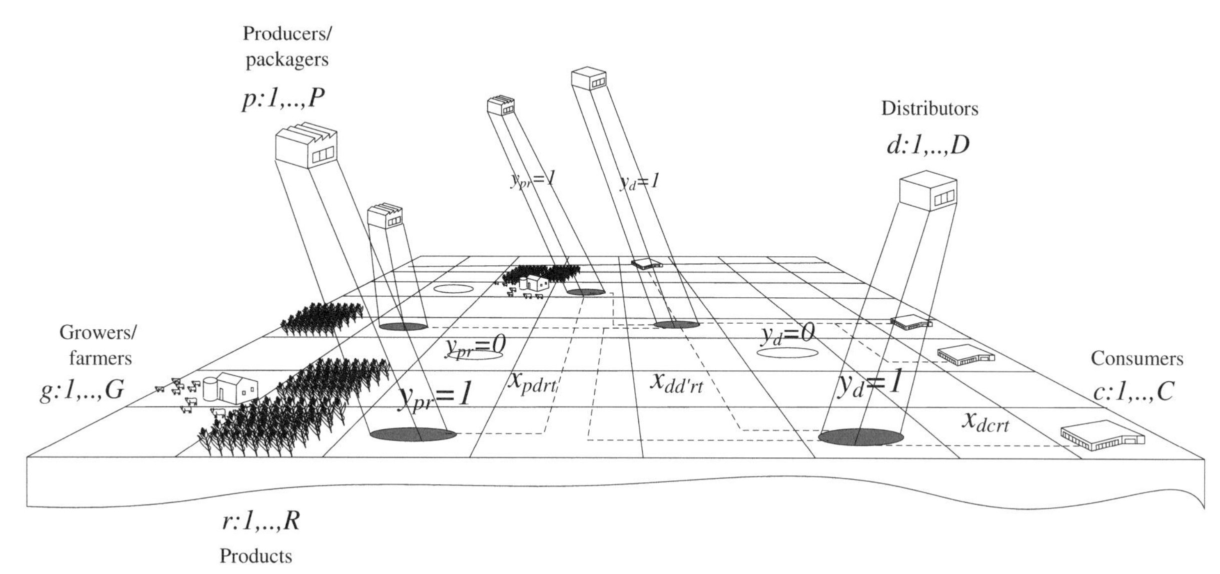

FIG. 3 Schematic representation of the location-allocation problem in FSC design.

As we've seen so far, routing problems have been combined with production problems and location problems. To reduce waste and more efficiently transport food, the supply chain design is an effective and important approach to explore. Any perishable product deteriorates as it travels through the supply chain network from supplier to customer. Leat et al. (2013) examined a pork supply chain facing business risks in the areas of production, market, and institution, covering quality management and food safety issues in the supply chain and presenting solutions for the development of efficient risk management and resilience through collaboration and coordination among different levels of the supply chain. The cantaloupe market of the United States also motivated the study of an FSC management problem that considers product differentiation based on product freshness and food safety. A network-based problem was formulated. Product deterioration throughout the network was included by using arc multipliers, which determined a closed-form equation for fresh product flows (Yu and Nagurney, 2013).

Agricultural produce marking corporations (APMCs) are active for many products in certain countries. They are very complex systems to study and model, and they serve as a link between the farmers and markets. Tsao (2013) formulated anonlinear programming model for a fresh food product supply network. They focused on APMC optimal service area, replenishment cycle times, and freshness-maintenance efforts. The objective of solving the integrated facility location and inventory allocation problems was to maximize profit. However, the author uniquely incorporated the freshness-maintenance efforts into the design of the supply chain network. As these efforts increased, the perishable rate of the fresh food decreased.

Supply chains typically contain mixed product flow since consolidating product flow can decrease transportation costs. Heterogeneous product decay could be incorporated into a network design as shown in the work by de Keizer et al. (2015). A new mixed integer linear programming formulation determined which hubs in the network to open and how to allocate flows to the hubs, between hubs, and from hubs. Inventory and process decisions were also included in the formulation. Agricultural test data demonstrated how different levels of decay resulted in different network structures. High and low product quality were also split in the design.

Galal and El-Kilany (2016) addressed the problem of harmful emissions from storage and transportation of a two-echelon agriculture FSC. Specifically, they studied the effects of order quantity on cost, emissions, and service level using a discrete-event simulation model. Their analysis included stochastic demand and lead time, emissions along the supply chain, service levels, and product lifetime.

In addition to network design, distribution problems focus primarily on shipping decisions of an existing network or supply chain. These networks contain supplier to distributor or consolidation center, distributor to retailer, and, in some cases, all three. Viergutz and Knust (2014) extended the work by Armstrong et al. (2008), who studied an integrated production and distribution problem for a perishable product with a limited short lifespan. The problem considered a made-to-order scenario, one production facility with limited production rate, and one vehicle responsible for delivery. The extended problem allowed variable distribution and production sequences while the original problem assumed this was fixed. Customers from multiple locations requested the product to be delivered within a time window. A single vehicle with nonnegligible travel times performed the deliveries. The focus of the problem was to determine how to select the customers to ship to such that the total satisfied demand was maximized. Because of the limited shelf life and time windows, it was possible that not all customers would receive their demand.

Nguyen et al. (2014) imposed a hard time constraint on the number of time periods products can stay in inventory. By defining this time constraint, products with a perishable nature will reach the destination quickly. The network contained shipments from suppliers to a consolidation center and from the consolidation center to customer destinations. Travel from supplier to the hub was negligible, and transport from the consolidation center was only a few days. Combined with freshness-maintenance efforts, the few days in transit would not impact the quality of the product greatly. Three transportation modes were considered: full truckload, less-than-truckload, and courier. The authors applied this concept in the California cut flower industry, where California-grown flowers are shipped to customers within only a few days of harvest (Nguyen et al., 2013). The large savings were due to maximizing the economies of scale. Lamsal et al. (2016) also studied the distribution of agricultural products. They considered the transport of products from multiple, independent producers to one location for storage or processing. This model addressed the logistics decisions within a 24-hour period because the product had to be shipped as soon as it was harvested. Each producer identified the number of loads that could be harvested on a certain day. Harvest times were held constant at each location but start times varied. The objective was to minimize the cumulative deviation of actual deliveries to the facility while determining the number of trucks required to retrieve the harvested loads.

The planning of perishable product storage and distribution decisions has also been studied while including the interactions with weather conditions and the effect on temperature (Accorsi et al., 2017). The authors analyzed how climate influences the distribution along a supply chain. They modeled a more detailed objective function that included variable and fixed refrigeration costs corresponding to both transportation costs and storage costs as well as disposal costs.

4 Similar research problems

There are many research problems and techniques in the literature that have unique characteristics that can be applied to food transportation. This section contains several works that, while not directly applied to food and agriculture transport, focus on perishable and time-sensitive products that experience decay or quality deterioration.

While not food or agriculture, the transport of blood products also faces deterioration risks. Wen et al. (2016) proposed a multiobjective optimization model for solving a capacitated vehicle routing problem that considers blood deterioration over time. The vehicles in the model were unmanned aerial vehicles (UAVs) to be used in emergency situations. The authors had to maintain the blood's temperature during transport to multiple sites, and each UAV had a very limited capacity. The problem considered all factors affecting the product: external environment, travel time, and refrigerant or heating agent. A decomposition-based multiobjective evolutionary algorithm and a local search method was implemented. The work in Nagurney et al. (2015) aimed to optimize the complex operations of a medical nuclear supply chain. The perishable product under consideration was radioisotope molybdenum, which experiences radioactive decay over time. The model was a generalized network model focusing on minimizing the total operational cost, waste cost, and risk.

Derigs et al. (2011a,b) focused on modeling the VRP with compartments. This work allows for the transportation of heterogeneous products on the same vehicle but in separate compartments. While this work does not directly consider perishable products, it is a useful way to consider food transport. The different compartments can be used for different food and agricultural products and can include the different storage, transport, and handling conditions. This type of model has potential for consolidation centers in helping to minimize transportation costs and quality measurements.

5 Conclusions

As supply chain operations grow and become more visible, more data will become available; however, that may impact the research and complexity of these problems. As stated before, the amount of food waste is drastically high. Time becomes more valuable as products increase in perishability. While technology exists to decrease the decay rate, delays can still impact the overall quality of the food in terms of taste, texture, etc. Increased visibility of the supply chain and improvements to technology can provide additional and more accurate data to determine more efficient routes and network designs. Some research has been accomplished in the areas of routing, network design, and distribution for food and agricultural transport. However, there are still many problems that have not yet been solved but are crucial to food transportation. This area of research is rich with research opportunities and potential to develop methods that will impact practical operations.

References

Accorsi, R., Cholette, S., Manzini, R., Pini, C., Penazzi, S., 2016. The land-network problem: ecosystem carbon balance in planning sustainable agro-food supply chains. J. Clean. Prod. 112 (1), 158–171.

Accorsi, R., Gallo, A., Manzini, R., 2017. A climate driven decision-support model for the distribution of perishable products. J. Clean. Prod. 165, 917–929.

Adenso-Diaz, B., Gonzales, M., Garcia, E., 1998. A hierarchical approach to managing dairy routing. Interfaces 28, 21–31.

Ambrosino, D., Sciomachen, A., 2007. A food distribution network problem: a case study. IMA J. Manag. Math. 18 (1), 33–53.

Archetti, C., Bianchess, N., Irnich, S., Speranza, M.G., 2014. Formulations for an inventory routing problem. Int. J. Oper. Res. 21, 353–374.

Armstrong, R., Gao, S., Lei, L., 2008. A zero-inventory production and distribution problem with a fixed customer sequence. Ann. Oper. Res. 159 (1), 395–414.

Belenguer, J.M., Benavent, E., Martínez, M.C., 2005. RutaRep: a computer package to design dispatching routes in the meat industry. J. Food Eng. 70 (3), 435–445.

Ceselli, A., Righini, G., Salani, M., 2009. A column generation algorithm for a rich vehicle-routing problem. Transp. Sci. 43 (1), 56–69.

Cornillier, F., Boctor, F., Renaud, J., 2012. Heuristics for the multi-depot petrol station replenishment problem with time windows. Eur. J. Oper. Res. 220 (2), 361–369.

Dantzig, G., Ramser, J., 1959. The truck dispatching problem. Manag. Sci. 6, 80–91.

de Keizer, M., et al., 2015. Logistics network design for perishable products with heterogeneous quality decay. Eur. J. Oper. Res. 262, 535–549.

Derigs, U., Gottlieb, J., et al., 2011a. Vehicle routing with compartments: applications, modelling and heuristics. OR Spectr. 33 (4), 885–914.

Derigs, U., Kurowsky, R., Vogel, U., 2011b. Solving a real-world vehicle routing problem with multiple use of tractors and trailers and EU-regulations for drivers arising in air cargo road feeder services. Eur. J. Oper. Res. 213 (1), 309–319.

Diabat, A., Abdallah, T., Le, T., 2016. A hybrid tabu search based heuristic for the periodic distribution inventory problem with perishable goods. Ann. Oper. Res. 242 (2), 373–398.

Food and Agriculture Organization of the United Nations (FAO), 2011. Key facts on food loss and waste you should know. In: Save food: global initiative on food loss and waste reduction.http://www.fao.org/save-food/resources/keyfindings/en/. [(Accessed 10 December 2017)].

Galal, N.M., El-Kilany, K.S., 2016. Sustainable agri-food supply chain with uncertain demand and lead time. Int. J. Simul. Model. 15 (3), 485–496.

Hiassat, A., Diabat, A., Rahwan, I., 2017. A genetic algorithm approach for location-inventory-routing problem with perishable products. J. Manuf. Syst. 42, 93–103.

Hsu, C., Hung, S., Li, H., 2007. Vehicle routing problem with time-windows for perishable food delivery. J. Food Eng. 80, 265–475.

Kim, H., Kang, J., Kim, W., 2014. An application of capacitated vehicle routing problem to reverse logistics of disposed food waste. Int. J. Ind. Eng. 21 (1), 46–52.

Laganà, D., Longo, F., Santoro, F., 2015. Multi-product inventory-routing problem in the supermarket distribution industry. Int. J. Food Eng. 11 (6), 747–766.

Lamsal, K., Jones, P.C., Thomas, B.W., 2016. Harvest logistics in agricultural systems with multiple, independent producers and no on-farm storage. Comput. Ind. Eng. 91, 129–138.

Leat, P., Revoredo-Giha, C., Revoredo-Giha, C., 2013. Risk and resilience in agri-food supply chains: the case of the ASDA PorkLink supply chain in Scotland. Supply Chain Manag. 18 (2), 219–231.

Li, P., et al., 2015. Vehicle routing problem with soft time windows based on improved genetic algorithm for fruits and vegetables distribution. Discret. Dyn. Nat. Soc. 2015.

Mahdavi, I., et al., 2017. A robust multi-trip vehicle routing problem of perishable products with intermediate depots and time windows. Numer. Algebra Control Optim. 7 (4), 417–433.

Manzini, R., Accorsi, R., Bortolini, M., 2014. Operational planning models for distribution networks. Int. J. Prod. Res. 52 (1), 89–116.

Nagurney, A., Nagurney, L.S., Li, D., 2015. Securing the sustainability of global medical nuclear supply chains through economic cost recovery, risk management, and optimization. Int. J. Sustain. Transp. 9 (6), 405–418.

Nguyen, C., et al., 2013. Evaluation of transportation practices in the California cut flower industry. Interfaces 43 (2), 182–193.

Nguyen, C., Dessouky, M., Toriello, A., 2014. Consolidation strategies for the delivery of perishable products. Transp. Res. E Logist. Transp. Rev. 69, 108–121.

Osvald, A., Stirn, L.Z., 2008. A vehicle routing algorithm for the distribution of fresh vegetables and similar perishable food. J. Food Eng. 85 (2), 285–295.

Song, B.D., Ko, Y.D., 2016. A vehicle routing problem of both refrigerated- and general-type vehicles for perishable food products delivery. J. Food Eng. 169, 61–71.

Stanger, S.H.W., et al., 2012. What drives perishable inventory management performance? Lessons learnt from the UK blood supply chain. Supply Chain Manag. 17 (2), 107–123.

Tarantilis, C.D., Kiranoudis, C., 2001. A meta-heuristic algorithm for the efficient distribution of perishable foods. J. Food Eng. 50, 1–9.

Tarantilis, C.D., Ioannou, G., Prastacos, G., 2005. Advanced vehicle routing algorithms for complex operations management problems. J. Food Eng. 70 (3), 455–471.

Toth, P., Vigo, D., 2002. The Vehicle Routing Problem. Society for Industrial and Applied Mathematics.

Trienekens, J.H., et al., 2012. Transparency in complex dynamic food supply chains. Adv. Eng. Inform. 26 (1), 55–65.

Tsao, Y.-C., 2013. Designing a fresh food supply chain network: an application of nonlinear programming. J. Appl. Math. 2013, 1–8.

United States Department of Transportation (U.S. DOT), 2015. 2015 freight facts and figures report. Accessed February 2017.

Vahdani, B., Niaki, S.T.A., Aslanzade, S., 2017. Production-inventory-routing coordination with capacity and time window constraints for perishable products: heuristic and meta-heuristic algorithms. J. Clean. Prod. 161, 598–618.
Viergutz, C., Knust, S., 2014. Integrated production and distribution scheduling with lifespan constraints. Ann. Oper. Res. 213 (1), 293–318.
Wen, T., Zhang, Z., Wong, K.K.L., 2016. Multi-objective algorithm for blood supply via unmanned aerial vehicles to the wounded in an emergency situation. PLoS ONE 11 (5), 1–23.
Wu, D., Mao, H., 2017. Research on optimization of pooling system and its application in drug supply chain based on big data analysis. Int. J. Telemed. Appl. 2017, 1.
Ye, Y., Liang, S., Zhu, Y., 2017. A mixed-integer linear programming-based scheduling model for refined-oil shipping. Comput. Chem. Eng. 99, 106–116.
Yu, M., Nagurney, A., 2013. Competitive food supply chain networks with application to fresh produce. Eur. J. Oper. Res. 224 (2), 273–282.

Chapter 11

Using vehicle routing models to improve sustainability of temperature-controlled food chains

Helena M. Stellingwerf, Argyris Kanellopoulos and Jacqueline M. Bloemhof
Operations Research and Logistics, Wageningen University, Wageningen, The Netherlands

Abstract
In this chapter we discuss the importance of vehicle routing problems in improving sustainability of temperature-controlled transportation of perishable products. Transportation of these products requires substantial amounts of energy to optimize the product environment and delay the decay process. Because of this, minimizing traveling distance is not a sufficient objective to improve sustainability of this type of transportation systems. To achieve a more sustainable distribution of perishable products, other objectives related to spoilage and greenhouse gas emissions should also be evaluated. Moreover, it is important that the potential of new technologies and alternative logistical concepts (e.g., collaboration) are explored and analyzed quantitatively.

1 Introduction: Sustainability in temperature-controlled chains

In addition to the costs that transportation inflicts on companies for delivering and collecting products, transportation also causes externalities related to accidents, noise, air quality deterioration, congestion, and greenhouse gas emissions. These externalities affect society as a whole, and they are not paid for by the companies who cause them (Ricci, 2003). In urban areas, these externalities are more acute than in rural areas, because urban areas have more traffic, congestion, and pollution (Palmer, 2007).

Because of these externalities, it is important that logistics and transportation be designed to be as sustainable as possible. However, current transportation systems are very inefficient. During the period of 2001–10 it is estimated that between 18.0% and 20.4% of freight kilometers driven in the European Union were by empty vehicles. Apart from that, the average loading rate (in terms of weight) of trucks driving in the EU is only about 56% (Cruijssen, 2012). In temperature-controlled chains, the problem is even more severe. Temperature-controlled chains, also called "cold chains," are supply chains in which the products are kept at low temperature to preserve their quality (Zanoni and Zavanella, 2012). Examples of products that need to be transported in a cold chain are fresh foods, frozen foods, flowers, but also medical products such as pharmaceuticals, blood, and vaccines. The temperature control needed during transportation requires additional energy to regulate temperature and ensure quality, product safety, and shelf life (Adekomaya et al., 2016; Ketzenberg et al., 2015). Therefore, temperature-controlled transportation causes more emissions than ambient transportation; greenhouse gas emissions from conventional diesel engine vapor compression refrigeration systems can be as high as 40% of the vehicle's emissions (Tassou et al., 2009). A UK survey among 2300 trucks showed that the thermal energy requirement is between 15% and 25% of the motive energy requirement of trucks (McKinnon et al., 2003; Tassou et al., 2009).

Research has shown that 40% of all foods require refrigeration and 15% of all electricity worldwide is used for refrigeration (Adekomaya et al., 2016; James and James, 2010; Meneghetti and Monti, 2015). These data show that the efficiency and consequently the sustainability of food logistics can be improved substantially. Improving sustainability of logistics requires looking beyond economic objectives and targeting structures that also enhance social and environmental performance (Soysal et al., 2012).

Finding ways to create sustainable food logistics calls for the use of decision support tools. For example, traditional vehicle routing problem (VRP) models have been used to minimize the traveling distance given a specific demand. Extended variants of VRP models, such as the green VRP models, have also been used to account for environmental

Sustainable Food Supply Chains. https://doi.org/10.1016/B978-0-12-813411-5.00011-9

considerations. Multicriteria decision making techniques can be used to deal with multiple conflicting objectives of sustainability by calculating trade-offs and identifying efficient logistical designs that improve efficiency. Such decision support tools provide the opportunity to evaluate the potential of different sustainability improvement options, such as alternative vehicle technologies, but also new logistical structures, such as joint route planning.

The remainder of this chapter is organized as follows: in Section 2 we present how VRP models have been used to improve efficiency and sustainability of transportation systems; in Section 3 we review multiple-criteria decision making (MCDM) methods that have been used to deal with multiple objectives involved in sustainability evaluation of supply chain management; Section 4 discusses how collaboration can be used to improve sustainability of cold chains; Section 5 presents concluding remarks.

2 Vehicle routing problem for temperature-controlled food transportation

Models based on the VRP focus on identifying the optimal routes for a fleet of vehicles in order to serve a given set of costumers. The objective is generally to satisfy all known product demand at various locations while minimizing the costs of the route (Santillán et al., 2012). Recently, green VRP models have been developed. These variants of VRP models use environmental objectives, such as fuel consumption (Bektaş and Laporte, 2011), or they consider a cost for CO_2 emissions in their objective (cost minimization) function (Cheng et al., 2017).

2.1 Green vehicle routing

Bektaş et al. (2016) defines green vehicle routing as "a branch of green logistics which refers to vehicle routing problems where externalities of using vehicles, such as carbon dioxide-equivalents emissions, are explicitly taken into account so that they are reduced through better planning." Table 1 shows recent green vehicle routing literature, organized by objective and solution method. The table does not give a complete review but it provides an overview of different types of green VRP methods in the literature. A very common approach to account for environmental objectives is to convert environmental indicators into costs (e.g., including fuel costs or penalties for greenhouse gas emissions) and include them in the objective function of a cost-minimizing VRP model. It is interesting to note that three authors mentioned in Table 1 put special emphasis on the perishability of food in their green VRP articles (Hsu et al., 2007; Song and Ko, 2016; Soysal et al., 2015b).

TABLE 1 Recent Green VRP literature, categorized by objective function and solution method

Objective	Solution approach	Reference
Combined cost function (e.g., including fuel or emission costs)	Exact	Hsu et al. (2007)
		Bektaş and Laporte (2011)
		Soysal et al. (2015b)
	Tabu search procedure	Jabali et al. (2012)
	Greedy heuristic (GRASP: Greedy Randomized Adaptive Search Concept)	Figliozzi (2010)
	DSOP (Departure Time and Speed Optimization Problem)	Franceschetti et al. (2013)
	Convex programming approach	Fukasawa et al. (2016)
Weighted distance	Exact	Kara et al. (2007)
	Comparison of local search metaheuristic, tabu search, and branch-and-cut algorithm	Zachariadis et al. (2015)
Fuel	Exact	Kopfer and Kopfer (2013)
	New heuristic approach	Qian and Eglese (2016)
	Simulated annealing based algorithm	Xiao et al. (2012)

TABLE 1 Recent Green VRP literature, categorized by objective function and solution method—cont'd

Objective	Solution approach	Reference
Multiobjective	Exact	Soysal et al. (2015a)
		Stellingwerf et al. (2018)
	Block recombination algorithm	Tiwari and Chang (2015)
	Tabu search	Bing et al. (2014)
Consumer satisfaction	Priority based	Song and Ko (2016)

2.2 Solution approaches and solution methods

Table 1 shows that some green VRP instances are solved exactly, while others use heuristics. In general, small instances (i.e., less than 10 nodes) of green VRPs are solved exactly. Green VRPs that aim to solve larger-scale problems, or that aim to solve many variables simultaneously, are generally solved using a heuristic. Different heuristics are used; authors generally take a "standard" heuristic and adapt it to fit their specific problem description. The Adaptive Large Neighborhood Search (ALNS) algorithm (Demir et al., 2012) is an example of a heuristic that was developed for solving a green VRP, namely the pollution routing problem (Bektaş and Laporte, 2011).

2.3 An application: Green vehicle routing for temperature-controlled transportation

To illustrate how green VRPs have been developed, we describe one of the articles in Table 1 in more detail. Stellingwerf et al. (2018) have used a VRP-based fuel and emission model to confirm what survey research has suggested: temperature control can cause an increase in fuel use and emissions of up to 40%. In this article, the Load Dependent Vehicle Routing Problem (LDVRP) is extended to also account for the energy use and emissions caused by temperature control. This is done by considering how much heat enters the truck walls and increases the temperature of the loaded products during the trip and how much heat enters during each stop when the door of the truck is opened. These two factors result in an estimate of how much heat the diesel-driven vapor compression system should remove. The efficiency of the vapor compression system is also considered such that the amount of the heat to be removed can be translated into (additional) fuel consumption. Total CO_2 emissions are normally calculated based on fuel consumption, but this article also considers the CO_2 emissions caused by leakage of refrigerant (Stellingwerf et al., 2018). Other variables that influence the performance of temperature-controlled transport are the inside temperature of the truck, the temperature of the surrounding air, the slope, the quality of the isolation of the truck, the efficiency of the cooling engine, and the speed driven. To illustrate the type of results from this research, Figs. 1 and 2 show the effect of average driving speed on fuel consumption and emissions in ambient and temperature-controlled transportation.

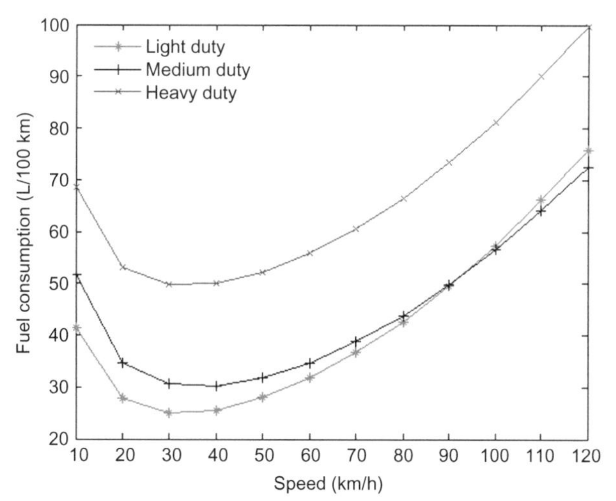

FIG. 1 Fuel consumption as function of speed for ambient temperature vehicles (so linearly related to emissions) (Cheng et al., 2017).

Fig. 1 shows fuel consumption for ambient trucks, where emissions are linearly related to fuel consumption. This figure shows that the emission-minimizing speed is around 40 km/h, and both lower and higher speeds cause an increase of fuel consumption and consequent emissions.

Fig. 2 shows emissions of temperature-controlled transportation. It shows that the optimal speed is around 60 km/h. This is caused by temperature control–related emissions: as these are mainly dependent on the time driven, a faster route causes less thermal emissions. This study shows that considering temperature control in a VRP can result in different optimal routes and speeds compared to ambient transport.

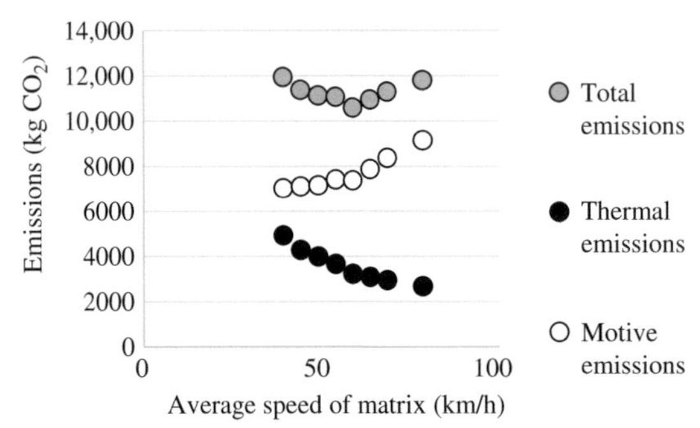

FIG. 2 Emissions as function of speed for temperature-controlled transportation (Stellingwerf et al., 2018).

3 Multicriteria problems for cold chains

Sustainability has been defined as meeting the needs of the current generation without compromising the needs of the future generations (Hurlem, 1987). Sustainability performance can be expressed in economic, environmental, and social measures: for example, costs, emissions, and employees' health (Akkerman et al., 2010; Kleindorfer et al., 2005). As the sustainability objectives can be conflicting, MCDM models can be used to capture the complexity of designing sustainable cold chains. These models are designed to comply with multiple criteria simultaneously, and they are also useful to capture the inherent complexity that characterizes temperature-controlled food supply chains (Banasik et al., 2016). The aim of cold chains is to avoid decrease in product value and to preserve product quality over the entire supply chain network from farm to consumers (Zanoni and Zavanella, 2012). Managing and planning of food supply chains is complicated due to perishable, seasonable, and fresh products (Geramian, 2018; Tsolakis et al., 2014). In food logistics, decision makers must take into account food safety support via transparency and traceability and efficient demand planning and forecasting. Complexity, existence of randomness in parameters, numerous variables and constraints, and conflicting objectives make this problem more challenging and in need of a sophisticated decision-making process and tools. The literature on sustainable supply chain network models that cover all three pillars of sustainability (i.e., economic, environmental, and social criteria) is scarce (Geramian, 2018).

Demir et al. (2014) did a study in which they described the biobjective Pollution Routing Problem. This study serves as a good example that demonstrates how multicriteria models can be used for sustainability-related issues (Demir et al., 2014). In this model, both minimization of driving time and minimization of fuel consumption are treated as objectives. These objectives are conflicting, as in general a higher driving speed (above 40 km/h) results in higher fuel consumption. In this study the trade-off between the two objectives was calculated. It was shown clearly that improving one objective can only be done by worsening the other (Demir et al., 2014).

Geramian (2018) also showed the trade-offs that can be found in supply chains, and they based their research on a case study done in a frozen food supply chain in Canada and the United States. In this study, the authors evaluated economic, environmental, and social sustainability by including them as objective functions in a multiobjective supply chain model. Total emissions were used as an environmental indicator, employment was used as a social indicator, and total cost was used as an economic indicator of sustainability. In their optimization, they also explored alternative energy sources (coal, hydroelectric, natural gas, and nuclear) for production. They showed that, within an environmentally friendly scenario, distribution centers should be located closer to environmentally friendly energy sources, which would reduce emissions from transportation activities substantially (Geramian, 2018).

4 Using a VRP to quantify sustainability effects of collaboration: An illustrative example

Previous sections showed the types of models that can help to improve the routing of a cold supply chain and make it more sustainable. However, when multiple companies work together on their logistics and, for example, merge part of their transportation needs, more opportunities for improvement arise. Literature confirms that both food companies and logistics service providers see logistics collaboration as a way to improve their sustainability performance (Bloemhof et al., 2015). Other authors confirm that logistics collaboration and coordination can create opportunities to improve the sustainability of food chains (Ramanathan et al., 2014; Vanovermeire et al., 2014). This section will introduce logistics collaboration and show how it can improve supply chains, and specifically how it can improve supply chain sustainability for cold chains.

4.1 Logistics collaboration

There are different types of logistics collaboration: vertical, horizontal, and lateral collaboration. Vertical logistics cooperation occurs between two different echelons in a distribution system, such as a supplier of raw material and a manufacturer (Krajewska et al., 2008). Horizontal logistics cooperation is defined as cooperation between partners at the same echelon of the distribution system. Lateral supply chain collaboration is defined as coordination between logistics and production (Leitner et al., 2011) and can thus be seen as a combination of horizontal and vertical cooperation (Simatupang and Sridharan, 2002).

4.2 Opportunities and impediments to logistics collaboration

Firms can decide to collaborate for different reasons: for example to decrease costs, to decrease emissions, or to make investments (Vanovermeire and Sörensen, 2014). Research has suggested that cooperation can yield larger efficiency and sustainability improvements than individual companies (Ramanathan et al., 2014; Vanovermeire and Sörensen, 2014). Apart from saving costs and reducing emissions, cooperative transportation also has other advantages: improved utilization rate of vehicles and increased transportation capacity (Cruijssen and Salomon, 2004). Despite the advantages of collaboration, companies often are hesitant to implement collaborative strategies because they fear losing competitive advantage to their collaborative partners. One of the most important impediments to cooperation is dividing the resulting gains (Cruijssen et al., 2007b).

4.3 Joint route planning

Joint route planning is a form of horizontal collaboration in which companies work jointly to make their routing more efficient (Cruijssen et al., 2007a). We use this concept in our example to show how multicriteria VRP models can be used to assess the potential collaborative sustainability improvements for the supply chain. We consider three supermarkets (A, B, C) that all get one delivery of fresh products per week. Supermarket A gets deliveries on Mondays, supermarket B on Tuesdays, and C on Wednesdays. The total demand is 600 kg, 250 kg, and 100 kg of product for supermarkets A, B, and C, respectively. When the products arrive at the supermarkets, the remaining shelf life (RSL) is 7 days. The supermarkets have perfect information on the uniformly distributed demand; the stock sells out just before the new delivery enters.

The supplier and the supermarkets consider two collaborative efficiency improvement options. In the first option, they decide to save costs by combining all orders and delivering every Tuesday; the total demand of A, B, and C fits in one truck. The second option they consider is to have two weekly collaborative routes. This will not give the same savings in costs and emissions as the first option, but with this option the supermarkets can increase the average RSL of the products on their shelves.

Fig. 3 shows the deliveries in the week of the current situation without collaboration. Fig. 4 shows the collaborative route (for both improvement options 1 and 2), which can be found with a traditional distance-minimizing VRP.

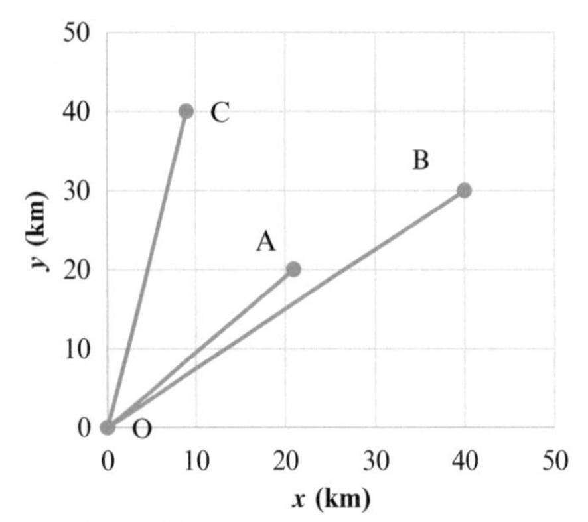

FIG. 3 A, B, and C all get delivered separately from the supplier.

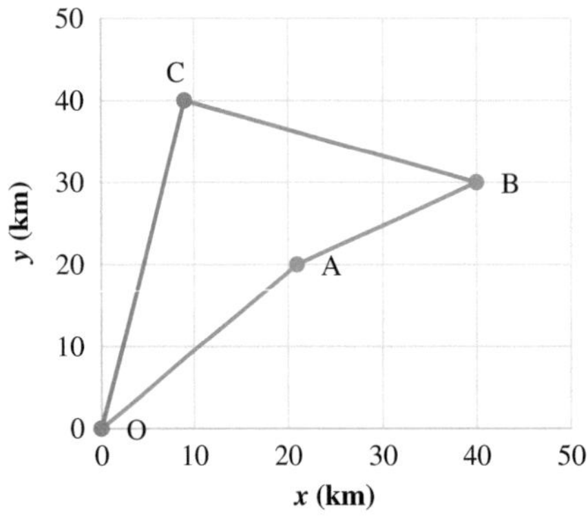

FIG. 4 A, B, and C get deliveries via one collaborative route from the supplier.

Changing from the three individual routes to one collaborative route results in a decrease in total distance from 240 to 106 km. When considering total distance, it does not matter if the truck takes route O-A-B-C-O or route O-C-B-A-O. However, if we use one of the green VRPs, it does matter. The energy minimizing VRP, for example, minimizes the weighted distance on a route (Kara et al., 2007). Assuming the truck weighs 2000 kg, the weighted distance of route O-A-B-C-O is 286 ton km, and the weighted distance of route O-C-B-A-O is 328 ton km, which means that the first route is 13% less in terms of weighted distance. In ambient temperature transportation, the weighted distance is linearly related to fuel use, which is linearly related to the CO_2 emissions if the speed is constant. Also, saving fuel helps to save costs. Using a simple green VRP can thus help to choose a route that helps to save costs and emissions between two equal-distance routes. Table 2 summarizes the current situation and the two improvement options in terms of remaining shelf life, distance driven, and weighted distance.

TABLE 2 RSL, distance, and weighted distance of a separate routing and two options of collaborative routing

Deliveries/week	Remaining shelf life (days)	Distance (km)	Weighted distance (ton km)
Current situation	4	240	514
Option 1: 1 collaborative route	4	124	286
Option 2: 2 collaborative routes	5.7	248	534

Table 2 shows that in terms of distance driven, and in terms of the weighted distance (which relates to fuel and emissions and indirectly to costs), it is best to take a once-a-week collaborative route. However, in terms of RSL, it is best to take two collaborative routes. This small example shows that there can be trade-offs in collaborative routing. The decision makers involved in collaboration should decide together on which indicator they find most important. The fact that decision makers can have differences in opinion on what is most important can hamper collaboration.

Another issue that the collaborative partners need to agree upon is how to share the cost savings. Assume that driver costs are the main costs factor, and that the logistics service provider charges the three supermarkets €0.10 per kilometer driven. The total costs of changing from three separate routes per week to one joint route per week would thus decrease from €24 to €12.4. One way to divide these costs is to split them equally. Another way is to divide the costs as a ratio of the total costs that the partners would have if they would not collaborate. However, these methods do not always correctly reflect each participant's contribution to the savings. Therefore, gain allocation methods have been developed. An overview of different types of gain allocation methods can be found in the literature (Frisk et al., 2010).

5 Concluding remarks

To improve sustainability of fresh transportation systems, it is important to deal not only with the complexity involved and the perishability of the product, but also the multidimensional nature of sustainability. In this chapter, we have showed how decision support tools can deal with this complexity, and we focused on VRP and MCDM models. These models can be used to test sustainability improvement options in cold transportation, not only alternative technologies like electric vehicles but also logistical concepts like horizontal collaboration. The studies discussed in this chapter clearly show that there are many possibilities for sustainability improvement and decision support models can help to assess the effects of the different improvement options. The specific aspects of cold chains such as perishability make the potential supply chain improvements even bigger compared to ambient chains.

References

Adekomaya, O., Jamiru, T., Sadiku, R., Huan, Z., 2016. Sustaining the shelf life of fresh food in cold chain–a burden on the environment. Alex. Eng. J. 55 (2), 1359–1365.

Akkerman, R., Farahani, P., Grunow, M., 2010. Quality, safety and sustainability in food distribution: a review of quantitative operations management approaches and challenges. OR Spectr. 32 (4), 863–904.

Banasik, A., Bloemhof-Ruwaard, J.M., Kanellopoulos, A., Claassen, G.D.H., van der Vorst, J.G.A.J., 2016. Multi-criteria decision making approaches for green supply chains: a review. Flex. Serv. Manuf. J., 1–31.

Bektaş, T., Laporte, G., 2011. The pollution-routing problem. Transp. Res. B Methodol. 45 (8), 1232–1250.

Bektaş, T., Demir, E., Laporte, G., 2016. Green vehicle routing. In: Green Transportation Logistics. Springer, pp. 243–265.

Bing, X., de Keizer, M., Bloemhof-Ruwaard, J.M., van der Vorst, J.G., 2014. Vehicle routing for the eco-efficient collection of household plastic waste. Waste Manag. 34 (4), 719–729.

Bloemhof, J.M., van der Vorst, J.G.A.J., Bastl, M., Allaoui, H., 2015. Sustainability assessment of food chain logistics. Int. J. Logist. Res. Appl. 18 (2), 101–117.

Cheng, C., Yang, P., Qi, M., Rousseau, L.-M., 2017. Modeling a green inventory routing problem with a heterogeneous fleet. Transp. Res. E Logist. Transp. Rev. 97, 97–112.

Cruijssen, F., 2012. Framework for collaboration: a CO3 position paper. Collaboration concepts for CO-modality. Report for the European Union's Seventh Framework Program.

Cruijssen, F., Salomon, M., 2004. Empirical study: order sharing between transportation companies may result in cost reductions between 5 to 15 percent. CentER Discussion Paper No. 2004-80.

Cruijssen, F., Bräysy, O., Dullaert, W., Fleuren, H., Salomon, M., 2007a. Joint route planning under varying market conditions. Int. J. Phys. Distrib. Logist. Manag. 37 (4), 287–304.

Cruijssen, F., Cools, M., Dullaert, W., 2007b. Horizontal cooperation in logistics: opportunities and impediments. Transp. Res. E Logist. Transp. Rev. 43 (2), 129–142.

Demir, E., Bektaş, T., Laporte, G., 2012. An adaptive large neighborhood search heuristic for the pollution-routing problem. Eur. J. Oper. Res. 223 (2), 346–359.

Demir, E., Bektaş, T., Laporte, G., 2014. The bi-objective pollution-routing problem. Eur. J. Oper. Res. 232 (3), 464–478.

Figliozzi, M., 2010. Vehicle routing problem for emissions minimization. Transp. Res. Rec. 2197, 1–7.

Franceschetti, A., Honhon, D., Van Woensel, T., Bektaş, T., Laporte, G., 2013. The time-dependent pollution-routing problem. Transp. Res. B Methodol. 56, 265–293.

Frisk, M., Göthe-Lundgren, M., Jörnsten, K., Rönnqvist, M., 2010. Cost allocation in collaborative forest transportation. Eur. J. Oper. Res. 205 (2), 448–458.

Fukasawa, R., He, Q., Song, Y., 2016. A disjunctive convex programming approach to the pollution-routing problem. Transp. Res. B Methodol. 94, 61–79.

Geramian, R., 2018. Tactical and Operational Planning of Sustainable Supply Chains for the Agrifood Industry. Ecole de technologie supérieure (ETS), Montréal, Quebec.

Hsu, C.-I., Hung, S.-F., Li, H.-C., 2007. Vehicle routing problem with time-windows for perishable food delivery. J. Food Eng. 80 (2), 465–475.

Hurlem, B.G., 1987. Our Common Future: World Commission on Environment and Development. Oxford University Press.

Jabali, O., Van Woensel, T., de Kok, A.G., 2012. Analysis of travel times and CO2 emissions in time-dependent vehicle routing. Prod. Oper. Manag. 21 (6), 1060–1074.

James, S.J., James, C., 2010. The food cold-chain and climate change. Food Res. Int. 43 (7), 1944–1956.

Kara, I., Kara, B.Y., Yetis, M.K., 2007. Energy minimizing vehicle routing problem. In: International Conference on Combinatorial Optimization and Applications. Springer, pp. 62–71.

Ketzenberg, M., Bloemhof, J., Gaukler, G., 2015. Managing perishables with time and temperature history. Prod. Oper. Manag. 24 (1), 54–70.

Kleindorfer, P.R., Singhal, K., Wassenhove, L.N., 2005. Sustainable operations management. Prod. Oper. Manag. 14 (4), 482–492.

Kopfer, H.W., Kopfer, H., 2013. Emissions minimization vehicle routing problem in dependence of different vehicle classes. In: Dynamics in Logistics. Springer, pp. 49–58.

Krajewska, M.A., Kopfer, H., Laporte, G., Ropke, S., Zaccour, G., 2008. Horizontal cooperation among freight carriers: request allocation and profit sharing. J. Oper. Res. Soc. 59 (11), 1483–1491.

Leitner, R., Meizer, F., Prochazka, M., Sihn, W., 2011. Structural concepts for horizontal cooperation to increase efficiency in logistics. CIRP J. Manuf. Sci. Technol. 4 (3), 332–337.

McKinnon, A., Ge, Y., Leuchars, D., 2003. Analysis of Transport Efficiency in the UK Food Supply Chain. Logistics Research Centre Heriot-Watt University, Edinburgh.

Meneghetti, A., Monti, L., 2015. Greening the food supply chain: an optimisation model for sustainable design of refrigerated automated warehouses. Int. J. Prod. Res. 53 (21), 6567–6587.

Palmer, A., 2007. The Development of an Integrated Routing and Carbon Dioxide Emissions Model for Goods Vehicles (PhD thesis). Cranfield University.

Qian, J., Eglese, R., 2016. Fuel emissions optimization in vehicle routing problems with time-varying speeds. Eur. J. Oper. Res. 248 (3), 840–848.

Ramanathan, U., Bentley, Y., Pang, G., 2014. The role of collaboration in the UK green supply chains: an exploratory study of the perspectives of suppliers, logistics and retailers. J. Clean. Prod. 70, 231–241.

Ricci, A., 2003. Valuation of externalities for sustainable development. In: International Conference on the Sustainable Development of the Mediterranean an Black Sea Environment, May, Thessaloniki. Extended abstract available at: www.iasonnet.gr/past_conf/abstracts/ricci.html.

Santillán, C.G., Reyes, L.C., Rodríguez, M.L.M., Barbosa, J.J.G., López, O.C., Zarate, G.R., Hernández, P., 2012. Variants of VRP to optimize logistics management problems. In: Logistics Management and Optimization Through Hybrid Artificial Intelligence Systems. IGI Global, pp. 207–237.

Simatupang, T.M., Sridharan, R., 2002. The collaborative supply chain. Int. J. Logist. Manag. 13 (1), 15–30.

Song, B.D., Ko, Y.D., 2016. A vehicle routing problem of both refrigerated- and general-type vehicles for perishable food products delivery. J. Food Eng. 169, 61–71.

Soysal, M., Bloemhof-Ruwaard, J.M., Meuwissen, M.P., van der Vorst, J.G., 2012. A review on quantitative models for sustainable food logistics management. Int. J. Food Syst. Dyn. 3 (2), 136–155.

Soysal, M., Bloemhof-Ruwaard, J.M., Bektaş, T., 2015a. The time-dependent two-echelon capacitated vehicle routing problem with environmental considerations. Int. J. Prod. Econ. 164, 366–378.

Soysal, M., Bloemhof-Ruwaard, J.M., Haijema, R., van der Vorst, J.G., 2015b. Modeling an inventory routing problem for perishable products with environmental considerations and demand uncertainty. Int. J. Prod. Econ. 164, 118–133.

Stellingwerf, H.M., Kanellopoulos, A., van der Vorst, J.G.A.J., Bloemhof, J.M., 2018. Reducing CO2 emissions in temperature-controlled road transportation using the LDVRP model. Transp. Res. Part D: Transp. Environ. 58, 80–93.

Tassou, S., De-Lille, G., Ge, Y., 2009. Food transport refrigeration–approaches to reduce energy consumption and environmental impacts of road transport. Appl. Therm. Eng. 29 (8), 1467–1477.

Tiwari, A., Chang, P.-C., 2015. A block recombination approach to solve green vehicle routing problem. Int. J. Prod. Econ. 164, 379–387.

Tsolakis, N.K., Keramydas, C.A., Toka, A.K., Aidonis, D.A., Iakovou, E.T., 2014. Agrifood supply chain management: a comprehensive hierarchical decision-making framework and a critical taxonomy. Biosyst. Eng. 120, 47–64.

Vanovermeire, C., Sörensen, K., 2014. Integration of the cost allocation in the optimization of collaborative bundling. Transp. Res. E Logist. Transp. Rev. 72, 125–143.

Vanovermeire, C., Sörensen, K., Van Breedam, A., Vannieuwenhuyse, B., Verstrepen, S., 2014. Horizontal logistics collaboration: decreasing costs through flexibility and an adequate cost allocation strategy. Int. J. Logist. Res. Appl. 17 (4), 339–355.

Xiao, Y., Zhao, Q., Kaku, I., Xu, Y., 2012. Development of a fuel consumption optimization model for the capacitated vehicle routing problem. Comput. Oper. Res. 39 (7), 1419–1431.

Zachariadis, E.E., Tarantilis, C.D., Kiranoudis, C.T., 2015. The load-dependent vehicle routing problem and its pick-up and delivery extension. Transp. Res. B Methodol. 71, 158–181.

Zanoni, S., Zavanella, L., 2012. Chilled or frozen? Decision strategies for sustainable food supply chains. Int. J. Prod. Econ. 140 (2), 731–736.

Chapter 12

Cool chain and temperature-controlled transport: An overview of concepts, challenges, and technologies

Behzad Behdani, Yun Fan and Jacqueline M. Bloemhof

Operations Research and Logistics, Wageningen University, Wageningen, The Netherlands

Abstract

Compared with dry cargo, temperature-sensitive products have some distinctive features that make managing the logistics process much more challenging. The importance of product quality, the need for specific transport fleets, the energy requirements for storage, handling, and transportation of perishable cargo are some of these specific characteristics. Meanwhile, as cool chains keep getting longer and more complex, transportation options are also changing. This chapter provides an overview of different stages of a cool chain and the changes in managing transportation processes in each stage. We also discuss the concepts and technologies for achieving high-quality products as well as sustainability in the agri-food supply chains.

1 Introduction

During the last couple of decades, the food industry and food retailing have been challenged by several trends and forces. Consumers increasingly expect high-quality foods in retail stores (van der Vorst et al., 2011). They also demand greater product variety and more convenience (e.g., prepared, frozen, or ready-to-eat products). Also, they have become more informed about sustainability issues and (more and more) consider them in their lifestyle choices. The other trend is consumer demand for year-round availability of fresh produce (Ahumada and Villalobos, 2009). Fruits and vegetables produced in Spain, Turkey, and North African countries are sold in North Europe or Asia for fulfilling the year-round consumer demands, which makes the supply chain more uncertain and complex. This has also been a driver for globalization and the need for long-haul intercontinental transportation in agri-food supply chains. As a result, maintaining the quality of agri-food products for a global market has increasingly become a challenge for most supply chains.

Agri-food supply chains include a wide range of products (Fig. 1). Different types of products have different characteristics and, accordingly, different logistics requirements. Perishable agricultural products—both crops and livestock—are considered to be the most dynamic sector in the food industry (Kumar and Nigmatullin, 2011). The demand for these products is continually changing as it is influenced by population trends, wage drift, and public awareness of diet (Ahumada and Villalobos, 2009). On the supply side, the availability of these products is also uncertain and influenced by environmental factors—like weather conditions—and continuous changes in regulations and laws.

A specific challenge for managing a perishable supply chain is the deterioration in quality of products over time throughout different stages of the chain (Rong et al., 2011). Agri-food products are mostly living products and their life processes remain active after harvest (Martin-Belloso and Fortuny, 2010). An important life process, which must be controlled to maintain good quality of fruit, is respiration. This is a complicated induction of chemical reactions that convert starch into sugars and those sugars into energy. To minimize respiration and avoid quality deterioration, perishable products need to be stored and transported in a controlled atmosphere environment. Among the many factors influencing the respiration process and quality degradation of fresh produce, temperature plays a dominant role (Caleb et al., 2012). The higher the temperature of the fruit's environment, the higher the temperature of the fruit itself, which results in a faster respiration rate and earlier spoilage of products (Kader, 1997). Also, the composition of the air that the fruit is surrounded by has a significant influence on the respiration rate. For example, a low amount of O_2 reduces the chance of respiration of agri-food products. On the other hand, a low amount of O_2 will stimulate anaerobic respiration, which may also adversely influence the quality of the agro products.

Sustainable Food Supply Chains. https://doi.org/10.1016/B978-0-12-813411-5.00012-0

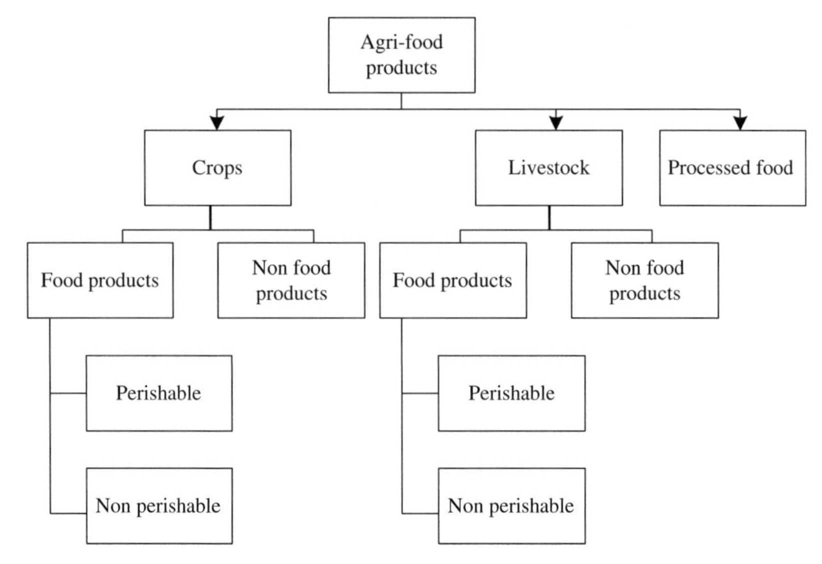

FIG. 1 Classification of agri-food supply chains. *(Based on Shukla, M., Jharkharia, S., 2013. Agri-fresh produce supply chain management: a state-of-the-art literature review. Int. J. Oper. Prod. Manag. 33 (2), 114–158 and Behzadi, G., O'Sullivan, M.J., Olsen, T. L., Zhang, A., 2017. Agribusiness supply chain risk management: a review of quantitative decision models. Omega.)*

Beside respiration, the ripening" process—especially through an accumulation of ethylene in packs—is another concern in the transportation of agri-food products (Abeles et al., 2012). Ethylene is a plant hormone that regulates many physiological reactions and is generally known as an influential factor in ripening and aging processes in fresh produce. It may cause changes in the texture, flavor, softening, or color of fresh products (e.g., yellowing of vegetables). The impact is more prominent on climacteric products such as avocado, banana, mango, papaya, and passion fruit (Saltveit, 2003). Exposure of these products to exogenous ethylene results in an acceleration of maturation, which reduces their shelf life and may lead to product spoilage. Therefore in order to extend the shelf life of ethylene-sensitive products, it is mandatory to extract the produced ethylene during the transportation and maintain the ethylene concentration at a low level. Exposure to ethylene may also happen in cases where multiple products are stored or transported in the same unit. For example, apples and pears produce ethylene with ripening. Ethylene is also used commercially to ripen some agro products, like tomatoes, pears, and bananas (Thompson, 2010).

The third physiological process that may influence the quality of agri-food products is transpiration, whereby water actively evaporates due to differences in temperature and pressure (Montanaro et al., 2010). To control transpiration, it is important to monitor and manage the relative humidity (RH) along the transportation chain. RH affects the physiological process of transpiration and the water activity of foods. At the same time, it may also increase the rate of growth of microorganisms. As such, inappropriate RH affects surface spoilage due to molds, yeast, and certain bacteria, and therefore the product should be stored at a relatively high RH (Dodd and Bouwer, 2014).

Considering all these aspects, to achieve an optimum quality and shelf life, fresh products may need additional conditions for storage and transportation. Table 1 shows a summary of the atmospheric conditions needed for a group of agri-food products. Providing such a tailored atmosphere for perishable products can be costly and may also add to the carbon footprint of agri-food chains.

TABLE 1 Recommended temperature and storage conditions for fruits and vegetables

Crop	Temperature (°C)	Relative humidity (%)	Storage life (days)	Ethylene sensitive
Apples	−1 to 4	90–95	90–240	Yes
Apricots	−0.5 to 0	90–95	7–21	Yes
Avocados	3–13	85–90	14–28	Yes
Bananas	13–15	90–95	7–28	Yes
Blackberries	−0.5 to 0	90–95	2–3	–
Blueberries	−0.5 to 0	90–95	14	–
Cherries	−1 to 0.5	90–95	14–21	–
Grapefruit	10–15	85–90	42–56	–
Grapes	−0.5 to 0	85	14–56	–

TABLE 1 Recommended temperature and storage conditions for fruits and vegetables—cont'd

Crop	Temperature (°C)	Relative humidity (%)	Storage life (days)	Ethylene sensitive
Kiwifruit	−0.5 to 0	95–100	90–150	Yes
Lemons	10–13	85–90	30–180	–
Nectarines	−0.5 to 0	95	14–28	Yes
Oranges	0–9	85–90	56–84	–
Peaches	−0.5 to 0	90–95	14–28	Yes
Pears	−1.5 to 0.5	90–95	60–210	Yes
Pineapple	7–13	90–95	14–28	–
Raspberries	−0.5 to 0	90–95	2	Yes
Strawberries	0–0.5	90–95	5–7	–
Watermelon	10–15	90	14–21	–

Based on Bachmann, J., Earles, R., 2000. Postharvest Handling of Fruits and Vegetables, ATTRA and El-Ramady, H.R., Domokos-Szabolcsy, É., Abdalla, N.A., Taha, H.S., Fári, M., 2015. Postharvest management of fruits and vegetables storage. In: Sustainable Agriculture Reviews. Springer, Cham, pp. 65–152.

2 Cool chain management

A cool chain includes all steps and facilities for storing, handling, and transportation of perishable products, for which controlled temperature conditions must be maintained from the point of production to the point of sale (Hundy et al., 2008). In principle, every cool chain is designed to maintain a specific temperature range for one or some specific products. Accordingly, cool chain management aims at planning and controlling processes and technologies to provide temperature integrity and avoid temperature violations for perishable products along the journey from origin to destination (Kuo and Chen, 2010). In other words, cool chain management is supply chain management of perishable items (Bishara, 2006). This implies that, besides responsiveness and cost efficiency, which are the primary objectives in managing every supply chain, cool chain management aims at preserving the quality of products along the chain (Fig. 2). This is mainly achieved by controlling the temperature along the chain (Aung and Chang, 2014). However, as discussed in the previous section, for many products several other factors like relative humidity or the level of oxygen, ethylene, or carbon dioxide may also significantly influence the quality of fresh produce. Therefore, in a broader sense, cool chain management is beyond temperature-controlled chain management and should be understood as controlled-atmosphere chain management. Furthermore, for a cool chain, the sustainability and carbon footprint need to be understood and addressed differently. This is because, unlike dry cargoes, the energy needs for fresh logistics are not limited to the direct fuel consumption in the transportation processes, but also include the cooling energy that is needed to preserve the product quality along the chain.

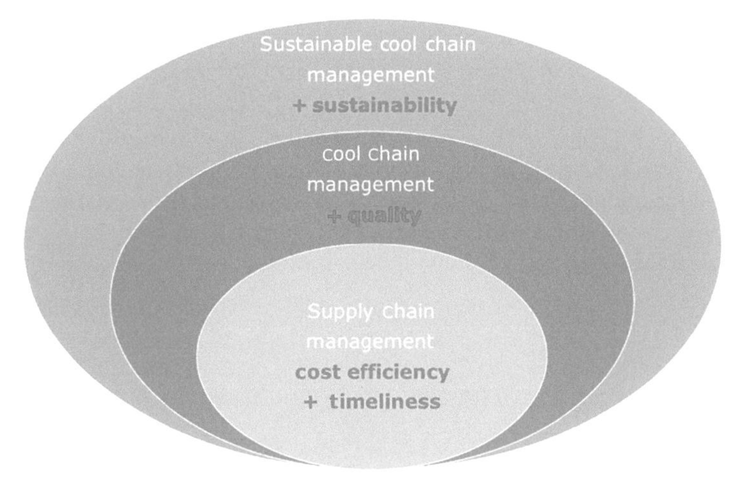

FIG. 2 Objectives and requirements in sustainable cool chain management.

2.1 Steps in a cool chain

A cool chain, in general, starts with the harvesting of products at the grower and ends at the table of the final consumer. There are several steps in between that highly influence the quality and sustainability of agri-food chains. Fig. 3 shows the main steps in a global cool chain. The first step is harvesting of agro products and precooling. Precooling aims at removal of field heat of the harvested produce shortly after the harvest of a crop (Brosnan and Sun, 2001). After precooling, the products need to be shipped to the market or put in cold storage facilities. Sometimes an additional processing or packaging process is needed to prepare the cargo for long-haul shipping. It is also possible that the pallets of products of different traders are consolidated in one truck or reefer container at the warehouse by a third party logistics company. At the destination country, the products need to be stored in the cold storage area after unloading from a vessel or airplane until they are shipped to the warehouse or a distribution center of a shipper. The final step is the last-mile distribution of products to the final shops or customers.

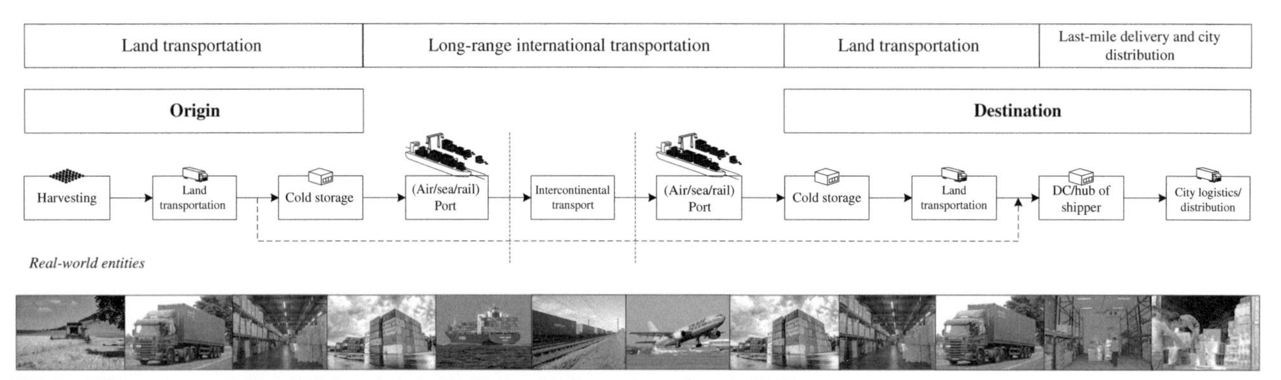

FIG. 3 Different segments in a global cool chain (for further details see Accorsi et al., 2018).

In general, three transportation steps are involved in each cool chain, which is discussed further in the following text.

2.1.1 International transport

For many decades, air transport has been the dominant modality for long and intercontinental transportation of agri-food products. This is especially because it provides high speed of delivery as well as accessibility to distant destinations, which are critical factors in preserving the quality and shelf life of perishable products in a global supply chain (Kemp et al., 2010). A recent trend in global transportation of agri-food products is a modal shift from air transport to sea transport. This is mainly driven by increased containerization and usage of reefer containers to transport perishable products (James and James, 2010). Reefer containers are insulated standard ISO containers that can be utilized in multiple transport modes, such as vessels, trains, and trucks and over long distances (Bömer and Tadeu, 2014). There are two main types of reefer containers: (1) porthole containers (also called insulated or CONAIR containers), which do not have their own refrigeration unit and are supplied with cold air; and (2) integrated reefers, which have refrigeration units built into their structure (Accorsi et al., 2014). These refrigeration units operate electrically, either from an external power supply during transportation or at a container yard (Rodrigue and Notteboom, 2015). Additionally, a reefer is equipped with a control system for temperature monitoring and alarm devices to ensure the safety of agri-food products. However, it should be emphasized that the existing reefer technologies are mainly designed to maintain the temperature of the load (and not reduce the temperature of products inside the reefer). Therefore, precooling cargos to the required (or optimum) temperature is very important before loading them into a reefer container (Ryan, 2017). The modular and standard structure of reefer containers provides handling flexibility and efficiency for multimodal transport and has been the main reason for a big shift from specialized reefer vessels to reefer containers in global transportation of perishable goods in the last two decades (Fig. 4).

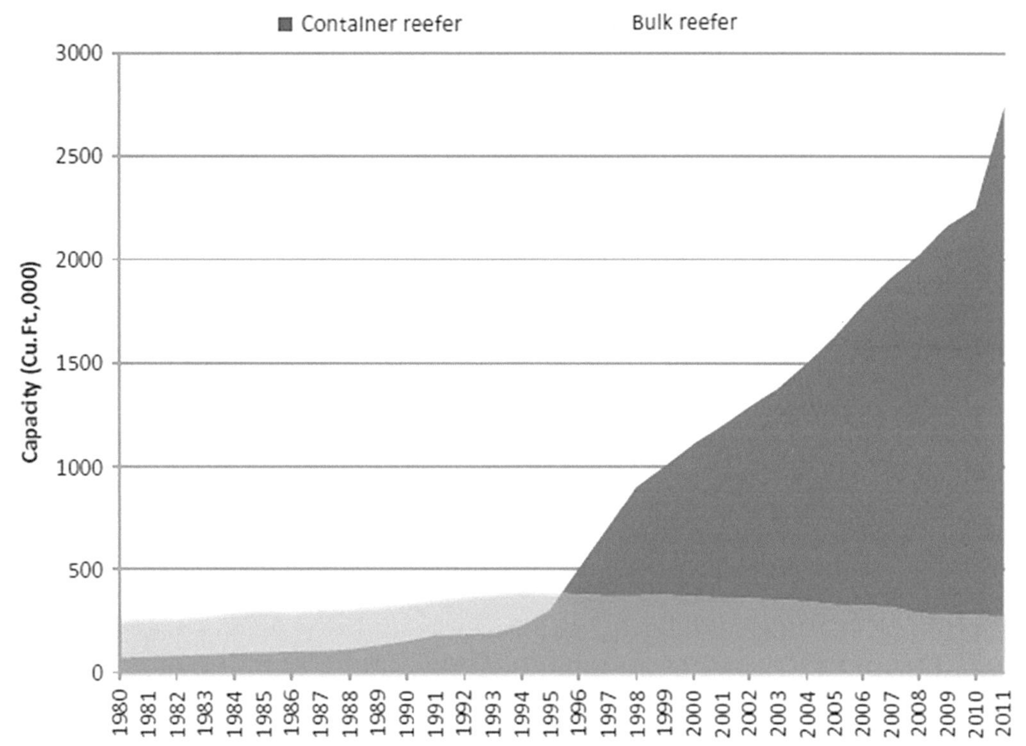

FIG. 4 Transition from reefer vessels to reefer containers (Arduino et al., 2015).

TABLE 2 The carbon emission and cost of different modalities

Modality	Emission (g CO_2/ton·km)	Cost ($/ton·km)
Sea transport	3 (container)[a]; 0.6 (bulk)[a]	0.01 (dry cargo) to 0.025 (reefer)
Air transport	218 (short-haul)[a]; 159 (long-haul)[a]	0.8–1.5
Inland waterways	50.62[b]	0.09
Truck	139.8[b]	0.2
Rail	15.6[b]	0.12

[a]*Based on data from BEIS (2017).*
[b]*Based on data from European Environment Agency (2015).*

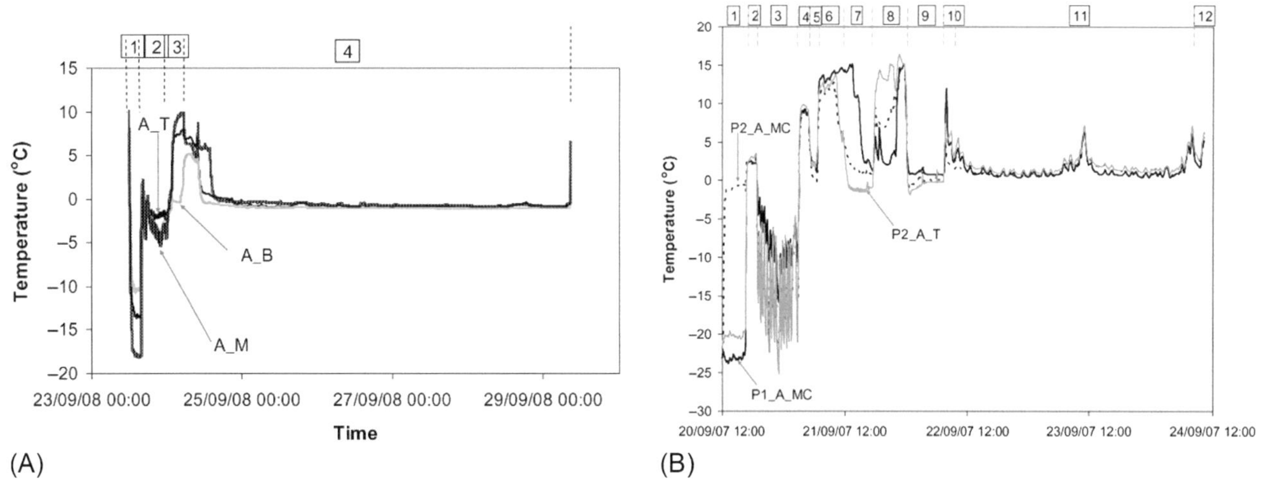

FIG. 5 Temperature profile of transportation of fish cargo using (A) maritime transportation, (B) air transportation (Mai et al., 2012).

Table 2 compares the cost and environmental impacts of different modalities. For international transport of perishable cargo, the cost and pollution of air freight are much higher than the cost and pollution of sea transport. Yet, the main reason for using air transport is shorter delivery time and consequently preserving the shelf life of perishable products. However, the developments in reefer containers as well as advanced packaging technologies to ensure the quality of cargo along the transportation chain (such as controlled atmosphere technologies; see Section 3 for further details) increasingly foster a modal shift from air to sea transport. Besides, a critical challenge with air transport is the temperature fluctuations and the possibility of a break in the cool chain during the flight or while cargo is loaded or unloaded from the aircraft (Brecht et al., 2003; James et al., 2006). The chance of breakage in the cool chain is much lower for sea transport, because as soon as cargo is loaded into a reefer, it is closed and sealed. The reefer is kept closed throughout the voyage until it reaches the DC of a shipper, where the reefer is opened and the cargo is unloaded into the warehouse. Therefore, there is no interface with ambient temperature along the chain and—unless the cooling is off for a long period of time—the temperature of perishable cargo remains stable (or with small oscillations) and in the expected range (Fig. 5). Similar temperature profiles during air and sea transport are reported by Rodríguez-Bermejo et al. (2007) and Jedermann et al. (2009). Insulated packaging, polystyrene slabs, insulation blankets/lining, together with cooling media such as dry ice and ice packs, can be used to protect temperature-sensitive products from thermal variations during air transport (James et al., 2006).

2.1.2 Inland and continental shipping

As soon as cargo arrives at a port (a seaport or an airport), it must be unloaded and stored in the temporary storage facilities before it is sent to a location inland (which is usually a warehouse or a shipping distribution center). In general, two main solutions are available for land transportation. The first solution is unimodal road transport. For most countries, road transport is the dominant mode of inland freight transport (Behdani et al., 2016). For instance, at the EU-28 level, almost three-quarters of the total freight inland movement (in tonne-kilometers) was performed by road in 2013 (Eurostat Database, 2014). Agri-food product transportation has been even more dominated by road transport (around 95%). This share has remained more or less stable over the last 5 years. Although road transport has several advantages in terms of speed and flexibility for inland transport, the huge demand for road freight transport leads to undesirable environmental and social impacts (Roso, 2007). A key issue is traffic congestion, particularly in regions around the main seaport. A primary reason for this congestion is that the capacity of road infrastructure is relatively small and, in most cases, less developed compared to the expected freight growth. Accordingly, it can be easily influenced by external conditions such as commuting peaks and working time regulations for truck drivers (Maloni and Jackson, 2005). This means that, for a port such as the Port of Rotterdam, over 40% of vehicles (including container trucks) going to/from the Maasvlakte area suffer heavy delays (Port of Rotterdam, 2013). The growing container road traffic can also cause many negative environmental impacts. It is estimated that an export/import container emits 50.6 g CO_2/tonne-km if carried by a barge and 139.8 g CO_2/tonne-km if carried by truck (European Environment Agency, 2015). In addition to GHG emissions, container trucking also contributes to the emissions of air pollutants such as nitrogen oxide (NO_x) and particulate matter (PM). Taking the Port of Los Angeles as an example, in 2010 the emissions of NO_x and PM from heavy-duty trucks was twice that from cargo-handling equipment (Starcrest Consulting Group, LLC, 2011).

The alternative to single-mode road transport is intermodal-freight transport (IFT) services. Intermodal transport is the use of a combination of two or more transport modes in a single transport chain, in the same unit (TEU container), without handling of the goods as the modalities are changed, and with a single tariff for whole transport service (SteadieSeifi et al., 2014; Tavasszy et al., 2017). Based on the type of modality, we can distinguish between three kinds of IFT services: land based (railway-road transport), water based (ocean-railway transport, ocean-inland waterway, inland waterway-railway transport, etc.) and air based (airline-road transport). Railway-road transport and inland waterway-road transport are substitutes for long-distance single-mode road transport that could reduce highway congestion, lead to less labor intensity, and conserve resources (Caris et al., 2013). IFT also leads to greater cost and operating efficiencies, especially for large flow over long distances (Bhattacharya et al., 2014).

Based on origin-destination, we can also distinguish between two types of IFT systems (Wiegmans, 2013). The first one is the continental IFT service, with an origin and destination on the same mainland (such as the European mainland). Pre- and end-haulage are performed by road transport to an inland terminal and from there barge or train can be chosen to execute the main haulage. The second type of IFT service is the hinterland transport service, in which a train or barge service transports containers from a seaport to an inland container terminal or a dry port where containers are unloaded and end-haulage is performed by truck, for a short trip of maximum 50 km, to deliver the cargo to the distribution center or shipping warehouse. It is possible for a barge to visit and collect containers from several terminals in the seaport—

especially in cases where the volumes of one container terminal are too small to offer a point-to-point service between a maritime and inland container terminal (Wiegmans and Konings, 2013).

As stated, trucking has been the preferred mode of transport in the agri-food sector for decades and the share of rail or inland waterway transport is quite marginal. Several initiatives in the last few years have promoted the use of intermodal freight transport. One of these projects is called Green Barge, which is aimed at using inland waterway transportation for delivering ornamental plants (Bestfact, 2014a). The project involved a pilot study from the North-East polder to the Amsterdam region (FloraHolland) via existing and IFT networks and inland terminals. A similar project, GreenRail, was initiated in 2009 by FloraHolland and the VGB (the trade association of flower and plant wholesalers) to transport floriculture products by rail (Bestfact, 2014b). In this project, a weekly rail transport service of ornamental plant products from the Netherlands to Italy and Romania was put into operation. The results of the pilot study showed 95% service reliability. Additionally, a built-in GPS/GPRS system was used to obtain continuous insight into the location of the container and check the temperature, relative humidity, and ethylene inside the conditioned container. The analysis of results showed no difference in the quality of products compared to the conventional road transport refrigerated trailers.

A recent initiative on using intermodal services for perishable products is the Cool Rail project. Cool Rail was launched by Bakker Barendrecht, a Dutch wholesaler of fruits and vegetables, and Euro Pool System to create a new rail link between Rotterdam and Valencia (Railfreight, 2016). The project was extended with a twice-weekly reefer service between Valencia and Cologne in Germany. The first results of the pilot project showed that, although the service takes 3 days, 1 day more than by road, each container travels 83% fewer truck kilometers, resulting in a 70% reduction in CO_2 emissions. The aim is to achieve a CO_2 reduction of 70%–90% and a cost savings of 20%–30% (CoolRail, 2016).

FIG. 6 Store deliveries by Franprix supermarket using inland waterways in Paris.

2.1.3 City distribution and last mile delivery

The final stage in the transportation of agri-food products is delivering the products from a distribution center or centralized warehouse to the retail stores. This is called city distribution and last mile delivery. It also includes the increasingly popular B2C models of product delivery to the final consumer (Wohlrab et al., 2012), which is especially stimulated by online shopping and e-commerce business models for agri-food products (Zeng et al., 2017). Although some articles distinguish between B2B processes (such as city distribution) and B2C processes (such as last mile delivery), there is not always a clear-cut distinction between these two terms in the literature. City distribution and last mile delivery face specific challenges as the size of loads are smaller, the shipment frequency is relatively high, and the product type is heterogeneous (Park and Regan, 2004). Furthermore, delivery inside cities is constrained by local or regional laws and regulations (e.g., low-emissions or clean air zones). Regulations enforce time access restrictions on the delivery of products as well as the type of vehicles that can be used for delivery inside a city (Anderson et al., 2005). The objective is managing the trade-offs between consumer convenience and the cost of product delivery. In the operations research literature, this trade-off is usually modeled by the vehicle routing problem (VRP) or different variants of it, such as VRP with Time Windows (VRPTW), with additional constraints to reflect the features of specific cases. The body of literature in this domain is very rich. An overview of applications to the agri-food domain can be found in Akkerman et al. (2010).

City distribution and last mile delivery is mostly done by trucks, semitrucks, and vans (Wohlrab et al., 2012). There are also some recent initiatives to arrange city distribution through multimodal transport. A successful case story is the case of Franprix supermarket stores in Paris, where the last transport leg between regional distribution centers and retail shops is handled via inland waterways (Fig. 6). The shipment is transported in a special container and sent from the warehouse to a river port (Port of Bonneuil) where it is loaded onto a barge (Bestfact, 2016a). From Bonneuil, containers are carried 20km on the Marne and then on the Seine River to the *quai de la Bourdonnais* in downtown Paris. From there another truck transports the containers to the shop on a very short trip. A similar initiative is the Zero-Emission Beer Boat in Utrecht, which is implemented by the Municipality of Utrecht. An electric beer boat is used to supply food and drinks to local businesses based along the canals of Utrecht (Bestfact, 2016b).

FIG. 7 Palliflex pallet storage system (Van Amerongen, 2017).

3 Technologies to maintain quality in perishable transport

To maintain the quality of fresh produce in a cool chain, several aspects should be considered. An overview of these aspects and the existing technologies are discussed in the following section.

3.1 Packaging of products

Increasing the shelf life of fresh products can be achieved by decreasing the speed of the ongoing respiration processes. Modified atmosphere packaging (MAP) is a widely used method to extend the quality and transit life of agri-food products by adjusting the environment of the fresh produce. This includes four primary environmental variables: relative humidity, and concentration of O_2, CO_2, and ethylene (Saltveit, 2003). To maximize the lifespan of agri-food products, an acceptable range for these variables is determined, monitored, and regulated (Saltveit, 2003). The most critical factor is the amount of oxygen in a package, because when oxygen is reduced, the respiration and metabolism of the fruit will slow down. The levels of O_2 and CO_2 in a package of a fresh produce depends on the interaction between the permeability properties of the packaging film and the commodity respiration (Kader, 2002). In fact, the atmosphere inside the package is achieved by a natural interplay between respiration rates of the products and the permeability of packaging films (Mangaraj et al., 2009). The amount of oxygen consumed by fresh produce in the packaging depends on the weight of the commodity, the temperature, the respiration rate, and the rate of carbon dioxide and oxygen movement in and out of the bag (Kader, 2002). There are two kinds of packaging films: continuous (or nonperforated) and perforated films (Beaudry, 2000). Perforated films have small holes (i.e., microperforations) as the routes through which gas exchange takes place. Perforated films allow a much higher gas exchange. Perforations enable the application of MAP for products that are sensitive even to small changes in the concentrations of O_2, CO_2, or ethylene (Dash, 2014). These films have been scaled up from small packages to bigger bags that can enclose pallets of fresh products as well, for instance, the Palliflex pallet storage system (Fig. 7). The cables in the figure add nitrogen and CO_2 to the bag to create the needed environment for the fruit and retard the degradation of bioactive compounds (Selcuk and Erkan, 2015).

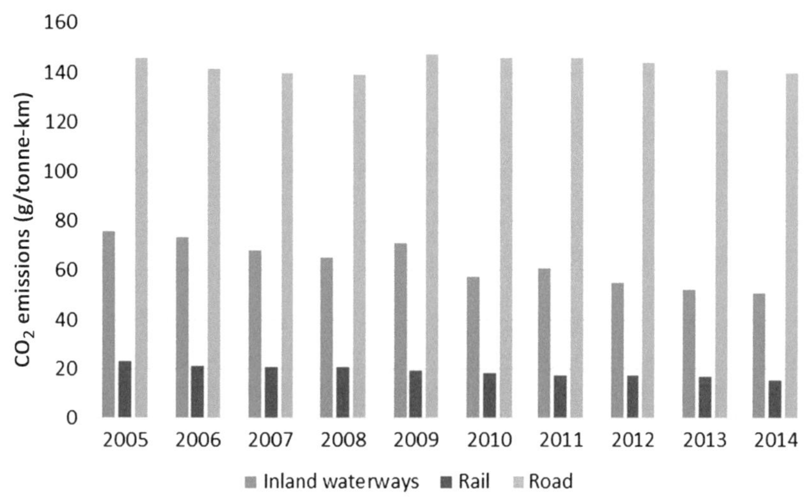

FIG. 8 CO_2 emissions per mode of transport in Europe. *(Based on European Environment Agency, 2015. Energy Efficiency and Specific CO2 Emissions. https://www.eea.europa.eu/data-and-maps/indicators/energy-efficiency-and-specific-co2-emissions/energy-efficiency-and-specific-co2-9. Accessed December 2017.)*

Packages should also protect the fruit against any large vibrations of transport and from physical damage. Chonhenchob and Singh (2005) performed a study on the physical protective packaging methods for papayas against vibration during shipment. In this research, papayas in different packaging systems were transported from the orchard to the laboratory. The packaging methods used consisted of foam nets and paper-based materials. The research showed that for papayas at an optimum maturity stage, single-layer packaging with cushioning (foam net or paper-based material) provides the best product quality at the delivery point.

3.2 Temperature control in cool chain

For most perishable cargoes, temperature control is the primary factor influencing the quality and shelf life of products. Precooling is the first step in a cool chain for temperature management of fresh produce. It aims at removing field/respiratory heat of fruits and vegetables immediately after harvesting (Brosnan and Sun, 2001). Different horticultural products need to be cooled to different storage temperatures. Table 1 gives an overview of preferred temperatures for different products. Several methods for precooling are discussed in the literature and are used in practice. An overview of the main methods and their characteristics are presented in Table 3.

TABLE 3 An overview of different precooling methods for cool chains

Precooling method	Definition	Characteristics
Air cooling or room cooling	Horticultural produce is placed into a cold insulated room where it is exposed to cold air.	The cost of this method is relatively low, but it is a slow method.
		Applicable to all fruits and vegetables, but it is satisfactory only for commodities with slow respiration rates.
Hydrocooling	Removing heat by hydrocooling is done by putting the produce into chilled water or by flowing the water over the fruit	Heat transfer is faster than air-cooling, but it is a more expensive method than room cooling.
		Commodities that are often hydrocooled include asparagus, snap beans, carrots, sweet corn, cantaloupes, celery, snow peas, radishes, tart cherries, and peaches. Cucumbers, peppers, melons, and early crop potatoes are sometimes hydrocooled (ASHRAE, 2010).
Forced-air cooling or pressure cooling	In forced-air cooling, cold air is forced through the stacked product and it allows more rapid heat removal.	This method can cool as much as four times faster than room cooling (Sargent, 1998). It causes dehydration of fresh produce; to avoid this, humidified air is used. It is most appropriate in dry climates and for chilling sensitive produce.
		Forced-air cooling is suggested for many products including avocado, banana, berries, Brussels sprouts, cassava, coconut, cucumber, eggplant, ginger, grapefruit, guava, kiwi fruit, mango, melons, mushrooms, pineapple, pumpkin, strawberry, and tomato (El-Ramady et al., 2015).
Vacuum cooling	Vacuum cooling is based on the principle of latent heat of vaporization of water and rapid evaporation of water from the product in a flash chamber under low pressure.	It is a very rapid and energy efficient precooling method. Vacuum is created in an airtight chamber by evacuation of air. At reduced pressure, water on the surface of the produce rapidly evaporates, which removes the field heat.
		Vacuum cooling is a very rapid method of cooling and is most efficient for commodities with a high surface-to-volume ratio such as leafy crops.
Packaged icing	Package icing involves direct crushed or flaked ice on top of the produce. Ice melts and cold water runs down through the produce, which cools the produce.	Produce is cooled faster than forced air, but product and packaging must tolerate the contact with water and ice.
		There should be appropriate holes in packages for water drainage.
		This method can result in uneven cooling because the ice generally remains where it was placed until it has melted.
		It can be used for products that are not harmed by contact with ice. Spinach, collards, kale, Brussels sprouts, broccoli, radishes, carrots, and green onions are commonly packaged with ice (ASHRAE, 2010).

Based on ASHRAE (2010). Methods of precooling fruits, vegetables and cut flowers. Refrigeration Systems and Applications Handbook. Atlanta, GA: American Society of Heating, Refrigerating and Air-Conditioning Engineers, Cap, 10, 1-10.. Methods of precooling fruits, vegetables and cut flowers. In: Refrigeration Systems and Applications Handbook. American Society of Heating, Refrigerating and Air-Conditioning Engineers, Atlanta, GA, Cap, 10, pp. 1–10 and Brosnan, T., Sun, D.W., 2001. Precooling techniques and applications for horticultural products—a review. Int. J. Refrig. 24 (2), 154–170.

After precooling, the next step is keeping the temperature of fresh products in the acceptable range during the entire trip. This is usually achieved by an adequate airflow and circulation of cooled air within the refrigerated container on a truck or in a storage room. There are also other technologies available to keep products cool during transport (Rodrigue and Notteboom, 2013). The main technologies include dry ice, gel packs, eutectic plates, liquid nitrogen, quilts, and reefers. Dry ice and liquid nitrogen are not very suitable for transportation of agricultural products, as the achieved temperature is far too low and may cause product freezing and texture damage. Gel packs are mostly used in the pharmaceutical sector. Eutectic plates are suitable for fruit transport, especially for short product deliveries. Quilts act as a temperature buffer and are common for transport of agri-food products. For long and intercontinental transport, the reefer is often the preferred option to control the temperature and avoid temperature breaks in the cool chain. When being transported by ship, the integral reefer refrigeration units must be connected to the on-board power supply system (Jedermann et al., 2009). Similarly, when reefers are unloaded from the vessels and stored at the terminal, the containers need to be connected to the terminal's power supply system as soon as possible. For transport by road and rail, most integral unit refrigeration units are operated by a diesel power generator set (or genset). This may either be a component of the refrigeration unit or connected to the refrigeration unit. In addition to temperature regulation, reefer integral units allow controlled fresh air exchange, e.g., for the removal of metabolic products such as CO_2 and ethylene in the case of the transport of fruits.

3.3 Atmosphere control in cool chains

In addition to refrigeration and temperature control, to extend the shelf life of fresh products we need to minimize the respiration and ripening processes. These processes change product quality, color, texture, flavor, and water content and can influence the shelf life of fresh produce (Caleb et al., 2012). By controlling oxygen, carbon dioxide, humidity, and ethylene levels, controlled atmosphere (CA) systems for refrigerated containers aim at slowing down the respiration and ripening processes (Thompson, 2010). For this purpose, sensors are installed in a reefer, or in storage rooms, to measure and adjust the level of gases (especially CO_2 and O_2) to predetermined levels through the fresh air intake and removal of excess CO_2. In cases of highly sensitive and low-respiring produce, sometimes nitrogen injection is also needed to establish a desirable balance of gases inside the reefer (Toivonen et al., 2010).

CA technologies have been a cornerstone of the globalization of agri-food supply chains. They allow fresh produce to travel longer distances, enabling delivery of a greater variety of products to consumers around the world. CA systems have been applied to both reefer vessels and reefer containers. However, they are potentially more effective in the case of reefer containers, as they help to arrange door-to-door "controlled conditions" for fresh produce. Still, the effectiveness of CA for reefer containers is dependent on the type of commodity—i.e., how fast the commodity respires and the level of maturity of products—and the maintenance of reefers (Bessemans et al., 2016). Potential air leakage in a reefer, especially through the doors and plastic curtains that fit inside the door, can reduce the degree of control over the gases in CA containers (Thompson, 2010).

A CA system works best for high-respiring products like bananas, avocados, mangos, asparagus, and nectarines (Kader, 2002). It is also suggested for fresh-cut products like mango cubes, sliced kiwifruit, sliced strawberry, and cubed cantaloupe (Gorny, 2001). For low-respiring products that may require a specific O_2 or CO_2 level that differs from normal atmospheric air—e.g., grapes, blueberries, apples, and papaya—a standard reefer can still be used by injecting nitrogen or CO_2 into the container prior to departure to establish the desired levels (Kader, 2002).

CA technology can considerably increase the shelf life of fruits and vegetables in a global supply chain. It is also useful to actively control the in-transit quality of products in different stages of a chain. By continuous monitoring and adjusting of the atmospheric situation of fresh cargo, reefer containers can be turned into an atmospheric conductive for ripening fruits (or an in-transit ripening chamber). Therefore, we can regulate the expected quality of products in different stages of the chain and minimize the operational cost, by minimizing the storage or avoiding unnecessary transport. In other words, we can integrate quality control and logistics functions for agri-food supply chains and achieve "quality-driven logistics" for fresh products.

4 Achieving sustainability in cool chains

Transportation plays an important role in the carbon footprint and sustainability of agri-food supply chains (Bloemhof et al., 2015). To improve the sustainability of cool chains, two main options are available. The first option is improving the efficiency and minimizing energy consumption of transport modalities. The second option concerns the efficiency of logistics processes. This also includes promoting a modal shift for fresh produce from less-sustainable transport modalities (i.e., road or air transport) to more sustainable modalities (i.e., sea, rail, and inland waterways). These two options are further discussed in the following sections.

4.1 Achieving sustainability by improvement in the transport modalities

As shown in Fig. 8, the overall sustainability of all transport modalities has been considerably improved in the last two decades. Since trucking is the major modality used for inland transport, many advancements have been focused on improving the fuel efficiency and reducing the emissions of trucks. Examples of improvements include engine down-speeding, friction and parasitic loss reduction, engine and after-treatment controls, and waste energy recovery (Stanton, 2013). Besides improving the engine, reducing the weight of a truck and reducing the aerodynamic drag are discussed as methods to increase the energy efficiency of transport vehicles (Chu and Majumdar, 2012). Additionally, there has been a lot of research on using alternative fuels (e.g., methanol, biodiesel, hydrogen, fuel cells, and electricity) for freight transport. An overview of these studies can be found in Salvi et al. (2013). An exceptional trend is the development of electric trucks, especially for city distribution and last mile delivery. Small trucks (<20 tonnes) with an electric engine have already been used for city distribution. An overview of existing vehicle electric technologies and the challenges in using them for distribution of products is provided in Pelletier et al. (2016). In general, the electric vehicles are classified into battery electric vehicles (which have an electric motor with the battery pack that can be charged from the electricity grid); hybrid electric vehicles (which have an internal combustion engine and an electric motor as well as a battery pack for electrical energy storage); and fuel cell electric vehicles (in which a fuel cell stack generates electricity from the chemical reaction of hydrogen, which then powers the motor or charges a battery). Several recent studies have showed the value of using electric trucks in achieving more sustainable transport. Lee et al. (2013) studied the sustainability of urban delivery trucks and showed that for a drive cycle with frequent stops and low average speed, which is typical for the city distribution of medium to big cities, an electric truck emits 42%–61% less GHGs and consume 32%–54% less energy than a diesel truck. For cases with fewer frequent stops and high average speed, such as city–suburban transportation, electric trucks emit 19%–43% less GHGs and consume 5%–34% less energy. Still, there are some major challenges for using electric vehicles for freight transport. The main issues involve the lifespan of batteries, the availability of charging infrastructure, and the cost of existing technologies, which may influence the competitiveness of commercial electric trucks (Davis and Figliozzi, 2013). Besides land transport, some recent initiatives have aimed at developing electric engines and batteries for a container ship—especially for transportation of goods within zero emission zones (Kongsberg, 2017).

Besides the transportation fleet, there have been many developments in improving the cooling technologies for fresh transportation. About 40% of the total energy need for distribution of perishable products is due to cooling and refrigeration needs (James and James, 2010). Accordingly, improvements in refrigeration technologies have a great impact on the environmental performance of a cool chain. An overview of refrigeration technologies is provided in Tassou et al. (2009). The most common refrigeration system is mechanical refrigeration with a vapor compression system. The drive systems for refrigerated transport vapor compression are different based on the size of the vehicle and the total cooling load (Tassou et al., 2009, 2010). For small delivery vans, the vehicle alternator unit is commonly used as a drive system (in which the vehicle engine crankshaft drives an upgraded single alternator and the alternator charges the vehicle battery, which feeds a small refrigeration system with 12 V DC supply). For van-size vehicles, the compressor of the refrigeration unit is directly driven from the vehicle engine through a belt (which is called direct belt drive). For medium to large vehicles, an auxiliary diesel engine built into the refrigeration unit drives the mechanical compression. An emerging research domain is using solar energy for cooling systems. The developments are mainly in three directions (Kim and Ferreira, 2008; Mathiesen et al., 2015). The first technology is solar electric refrigeration systems that use photovoltaic panels and an electrical refrigeration device to provide the cooling requirements. The two other sets of technologies are both solar thermal refrigeration technologies: thermo-mechanical refrigeration and sorption refrigeration (Kim and Ferreira, 2008). Thermo-mechanical refrigeration systems use a heat engine to convert solar heat to mechanical energy. This mechanical energy drives the compressor of refrigeration machine. Sorption refrigeration uses chemical or physical attraction between a pair of substances to produce the refrigeration effect (Kim and Ferreira, 2008; Abdulateef et al., 2009).

4.2 Achieving sustainability by improving the efficiency of transportation processes

A main factor causing inefficiency in freight transportation is asset repositioning and empty running. For instance, one out of four heavy goods vehicles in the EU runs empty (European Commission, 2012). A similar problem can be observed in container transportation, which is mainly caused by geographical trade imbalances in the global market. It is estimated that around 20.5% of container turnover around the world is empty container handling (Rodrigue and Notteboom, 2015). A second problem is underutilization of transportation assets. At the EU level, the average load factor for loaded trucks is about 56% (European Environment Agency, 2010). Léonardi and Baumgartner (2004) also performed a study on road freight transport among 50 German haulage companies and showed that, for heavy-duty trucks (>40 tons), the mean weight and volume load factors are only 44.7% and 63.6%, respectively. This indicates that the transportation fleet has been less than efficiently used. This is also caused by decreasing load sizes due to just-in-time operations and tight time windows for product delivery, especially for city distribution and last mile delivery.

TABLE 4 Application of ICT technologies to improve the efficiency of freight transportation

Software components	Transportation Management Systems (TMS)	Planning and execution of transportation of goods (Gonzalez-Feliu and Salanova, 2012)
	Port Community Systems (PCS)	Platform to connect the ICT systems operated by multiple stakeholders of a port or inland port (Carlan et al., 2016)
	Truck Appointment Systems	Improving yard efficiency in a container terminal (Zhao and Goodchild, 2013)
		Minimizing congestion at the port terminal gates (Zehendner and Feillet, 2014)
	Advanced Fleet Management Systems (AFMS)	Real-time management, planning and replanning of vehicles (Crainic et al., 2009)
Hardware components	Global Positioning System (GPS)	Tracking and monitoring of transportation fleet (Heywood et al., 2009)
		Anticipating delays and dynamic scheduling of deliveries (Klumpp et al., 2012)
		Container positioning and monitoring (Jing and Siyuan, 2010)
	Radio Frequency Identification (RFID)	Vehicle tracking in the container terminal (Ting et al., 2012)
		Port gate automation (Ro and Kim, 2006)
		Assuring container security and monitoring electronic seals (Rizzo et al., 2011)
		Container tracking and operations systems (Shi et al., 2011)
		Minimizing cargo loading and misplacement (Prasanna and Hemalatha, 2012)
		Temperature control and traceability in agri-food chains (Abad et al., 2009)
		Train tracking and automatic identification of border crossing in railway systems (Mašek et al., 2016)
	Wireless Sensor Networks (WSN)	Container tracking and monitoring systems (Mahlknecht and Madani, 2007)
		Monitor the environmental parameters in the (refrigerated) container (Wang et al., 2009; Ruiz-Garcia et al., 2008)

To cope with these challenges and minimize the inefficiencies in global transportation networks, several ICT technologies and solutions are suggested and implemented in practice. Table 4 gives an overview of these technologies and their application to the transportation domain. ICT technologies help to have a better visibility along the chain and avoid unnecessary movements. In addition to technical solutions, promoting collaboration among shippers and service providers may result in reducing empty running and improving the utilization of transportation assets. As an example, Cruijssen et al. (2007) discuss a case of collaboration between eight Dutch producers of sweets and candy, called Zoetwaren Distributie Nederland (ZDN: Dutch Sweets Distribution), in which they consolidate the shipments and delivery to the customers. Consolidated shipments reduced the total number of deliveries and, accordingly, the cost of operation for each company and also resulted in higher service level and increased product quality to the customers. Matching the drop-off locations for cargo and backhaul loads is the other solution to minimizing the empty running of trucks (Ubeda et al., 2011). A similar concept, called a triangulation strategy, is applied to container transportation. Triangulation is matching the nearby origins of container journeys and sending empty containers directly from an importer to the exporter, without returning/stacking at a port or inland depot (Braekers et al., 2011; Van den Berg and De Langen, 2015). This strategy reduces empty container movements and consequently the operating cost of container logistics. However, it is difficult to apply to the reefer domain because each reefer container needs to pass a pretrip inspection (PTI) to check if it is operating properly prior to loading a new cargo (Sørensen et al., 2014). In most cases, the facilities for testing reefers are located in the port area, which implies that each reefer must come back to the port for the PTI test before it can be sent back to the hinterland for loading cargo by another shipper.

5 Concluding remarks

This chapter discusses the concepts of cool chain management and temperature-controlled transport. Cool chains are some of the most difficult types of supply chains to manage, as the quality of products as well as the cost efficiency and timeliness must be addressed, which are the primary objectives in all logistics processes. Additionally, for a cool chain, the sustainability is a critical and distinct issue, because, for perishable supply chains, the environmental impacts are not limited to the emissions or carbon/water footprints but also the waste of products due to quality degradation and product spoilage. Additionally, to address sustainability in a cool chain, we need to minimize the direct fuel consumption for transportation of cargo as well as additional energy needs for keeping products cool along the chain.

In this chapter, we discuss the concepts and technologies to achieve high-quality products as well as sustainability in the agri-food supply chains. Considering existing technologies, an interesting area for future studies is the integration of quality control and logistics management for perishable supply chains. There have been some studies—like the intelligent containers concept (Jedermann et al., 2014)—that initiated such an integration. But still, a supply chain perspective on quality-driven logistics is lacking in the literature. In fact, we need to systematically study which quality levels are needed at each stage of the chain, and how we can regulate the factors influencing quality—e.g., by changing the atmospheric conditions along the voyage—to achieve the expected quality. The quality-driven logistics for fresh products is expected to reduce the operational cost for the whole chain, by minimizing the storage of products, reducing investment costs for ripening facilities, and avoiding unnecessary transport along the chain.

References

Abad, E., Palacio, F., Nuin, M., De Zarate, A.G., Juarros, A., Gómez, J.M., Marco, S., 2009. RFID smart tag for traceability and cold chain monitoring of foods: demonstration in an intercontinental fresh fish logistic chain. J. Food Eng. 93 (4), 394–399.

Abdulateef, J.M., Sopian, K., Alghoul, M.A., Sulaiman, M.Y., 2009. Review on solar-driven ejector refrigeration technologies. Renew. Sust. Energ. Rev. 13 (6–7), 1338–1349.

Abeles, F.B., Morgan, P.W., Saltveit Jr., M.E., 2012. Ethylene in Plant Biology. Academic Press.

Accorsi, R., Manzini, R., Ferrari, E., 2014. A comparison of shipping containers from technical, economic and environmental perspectives. Transp. Res. Part D: Transp. Environ. 26, 52–59.

Accorsi, R., Cholette, S., Manzini, R., Tufano, A., 2018. A hierarchical data architecture for sustainable food supply chain management and planning. J. Clean. Prod. 203, 1039–1054.

Ahumada, O., Villalobos, J.R., 2009. Application of planning models in the agri-food supply chain: a review. Eur. J. Oper. Res. 196 (1), 1–20.

Akkerman, R., Farahani, P., Grunow, M., 2010. Quality, safety and sustainability in food distribution: a review of quantitative operations management approaches and challenges. OR Spectr. 32 (4), 863–904.

Anderson, S., Allen, J., Browne, M., 2005. Urban logistics—how can it meet policy makers' sustainability objectives? J. Transp. Geogr. 13 (1), 71–81.

Arduino, G., Carrillo Murillo, D., Parola, F., 2015. Refrigerated container versus bulk: evidence from the banana cold chain. Marit. Policy Manag. 42 (3), 228–245.

ASHRAE, 2010. Methods of precooling fruits, vegetables and cut flowers. In: Refrigeration Systems and Applications Handbook. American Society of Heating, Refrigerating and Air-Conditioning Engineers, Atlanta, GA, pp. 1–10. Cap, 10.

Aung, M.M., Chang, Y.S., 2014. Temperature management for the quality assurance of a perishable food supply chain. Food Control 40, 198–207.

Beaudry, R.M., 2000. Responses of horticultural commodities to low oxygen: limits to the expanded use of modified atmosphere packaging. HortTechnology 10 (3), 491–500.

Behdani, B., Fan, Y., Wiegmans, B., Zuidwijk, R., 2016. Multimodal schedule design for synchromodal freight transport systems. Eur. J. Transp. Infrastruct. Res. 16 (3), 424–444.

BEIS, 2017. Greenhouse Gas Reporting: Conversion Factors 2017. UK Department for Business, Energy & Industrial Strategy. https://www.gov.uk/government/publications/greenhouse-gas-reporting-conversion-factors-2017. (Accessed December 2017).

Bessemans, N., Verboven, P., Verlinden, B.E., Nicolaï, B.M., 2016. A novel type of dynamic controlled atmosphere storage based on the respiratory quotient (RQ-DCA). Postharvest Biol. Technol. 115, 91–102.

Bestfact, 2014a. Green Barge—FloraHolland. http://www.bestfact.net/green-barge-floraholland/. (Accessed December 2017).

Bestfact, 2014b. Green Rail—FloraHolland. http://www.bestfact.net/green-rail-floraholland/. (Accessed December 2017).

Bestfact, 2016a. Supermarket Stores Deliveries Using Waterways in Paris. http://www.bestfact.net/wp-content/uploads/2016/01/CL1_051_QuickInfo_Franprix-en-Seine-16Dec2015.pdf. (Accessed December 2017).

Bestfact, 2016b. Zero-Emission Beer Boat in Utrecht. http://www.bestfact.net/wp-content/uploads/2016/01/CL1_151_QuickInfo_ZeroEmissionBoat-16Dec2015.pdf. (Accessed December 2017).

Bhattacharya, A., Kumar, S.A., Tiwari, M.K., Talluri, S., 2014. An intermodal freight transport system for optimal supply chain logistics. Transp. Res. C Emerg. Technol. 38, 73–84.

Bishara, R., 2006. Cold chain management-an essential component of the global pharmaceutical supply chain. Am. Pharm. Rev. 1, 105–109.

Bloemhof, J.M., van der Vorst, J.G., Bastl, M., Allaoui, H., 2015. Sustainability assessment of food chain logistics. Int. J. Logist. Res. Appl. 18 (2), 101–117.

Bömer, G.C., Tadeu, R.L., 2014. The South America East Coast reefer cargo: a diagnosis of a competitive market. IBIMA Bus. Rev. 2014, 1–14.

Braekers, K., Janssens, G.K., Caris, A., 2011. Challenges in managing empty container movements at multiple planning levels. Transp. Rev. 31 (6), 681–708.

Brecht, J.K., Chau, K.V., Fonseca, S.C., Oliveira, F.A.R., Silva, F.M., Nunes, M.C.N., Bender, R.J., 2003. Maintaining optimal atmosphere conditions for fruits and vegetables throughout the postharvest handling chain. Postharvest Biol. Technol. 27 (1), 87–101.

Brosnan, T., Sun, D.W., 2001. Precooling techniques and applications for horticultural products—a review. Int. J. Refrig. 24 (2), 154–170.

Caleb, O.J., Mahajan, P.V., Opara, U.L., Witthuhn, C.R., 2012. Modelling the respiration rates of pomegranate fruit and arils. Postharvest Biol. Technol. 64 (1), 49–54.

Caris, A., Macharis, C., Janssens, G.K., 2013. Decision support in intermodal transport: a new research agenda. Comput. Ind. 64 (2), 105–112.

Carlan, V., Sys, C., Vanelslander, T., 2016. How port community systems can contribute to port competitiveness: developing a cost–benefit framework. Res. Transp. Bus. Manag. 19, 51–64.

Chonhenchob, V., Singh, S.P., 2005. Packaging performance comparison for distribution and export of papaya fruit. Packag. Technol. Sci. 18 (3), 125–131.

Chu, S., Majumdar, A., 2012. Opportunities and challenges for a sustainable energy future. Nature 488 (7411), 294.

CoolRail, 2016. Launch of new rail link for fresh produce. Available from: https://www.coolraileurope.com/en/launch-of-new-rail-link-for-fresh-produce (Accessed December 2017).

Crainic, T.G., Gendreau, M., Potvin, J.Y., 2009. Intelligent freight-transportation systems: assessment and the contribution of operations research. Transp. Res. C Emerg. Technol. 17 (6), 541–557.

Cruijssen, F., Dullaert, W., Fleuren, H., 2007. Horizontal cooperation in transport and logistics: a literature review. Transp. J., 22–39.

Dash, S.K., 2014. Modified atmosphere Packaging of food. In: Alavi, S., et al. (Eds.), Polymers for Packaging Applications. CRC Press.

Davis, B.A., Figliozzi, M.A., 2013. A methodology to evaluate the competitiveness of electric delivery trucks. Transp. Res. E Logist. Transp. Rev. 49 (1), 8–23.

Dodd, M.C., Bouwer, J.J., 2014. The supply value chain of fresh produce from field to home: refrigeration and other supporting technologies. In: Postharvest Handling, third ed, Academic Press, pp. 449–483.

El-Ramady, H.R., Domokos-Szabolcsy, É., Abdalla, N.A., Taha, H.S., Fári, M., 2015. Postharvest management of fruits and vegetables storage. In: Sustainable Agriculture Reviews. Springer, Cham, pp. 65–152.

European Commission, 2012. Road Transport: A Change of Gear. https://ec.europa.eu/transport/sites/transport/files/modes/road/doc/broch-road-transport_en.pdf.

European Environment Agency, 2010. Load factors for freight transport (TERM 030)—Assessment report. https://www.eea.europa.eu/data-and-maps/indicators/load-factors-for-freight-transport/load-factors-for-freight-transport-1. (Accessed December 2017).

European Environment Agency, 2015. Energy Efficiency and Specific CO2 Emissions. https://www.eea.europa.eu/data-and-maps/indicators/energy-efficiency-and-specific-co2-emissions/energy-efficiency-and-specific-co2-9. (Accessed December 2017).

Eurostat Database, 2014. Freight Transport Statistics—Modal Split. http://ec.europa.eu/eurostat/statistics-explained/index.php/Freight_transport_statistics_-_modal_split. (Accessed December 2017).

Gonzalez-Feliu, J., Salanova, J.M., 2012. Defining and evaluating collaborative urban freight transportation systems. Procedia Soc. Behav. Sci. 39, 172–183.

Gorny, J.R., 2001. A summary of CA and MA requirements and recommendations for fresh-cut (minimally processed) fruits and vegetables. In: Postharvest Horticulture Series No. 22A, University of California, Davis, CA, pp. 95–145.

Heywood, C., Connor, C., Browning, D., Smith, M.C., Wang, J., 2009. GPS tracking of intermodal transportation: system integration with delivery order system. In: Systems and Information Engineering Design Symposium, 2009. SIEDS'09. IEEE, pp. 191–196.

Hundy, G.H., Trott, A.R., Welch, T.C., 2008. Refrigeration and Air-Conditioning. Butterworth-Heinemann, pp. 214–225.

James, S.J., James, C., 2010. The food cold-chain and climate change. Food Res. Int. 43 (7), 1944–1956.

James, S.J., James, C., Evans, J.A., 2006. Modelling of food transportation systems—a review. Int. J. Refrig. 29 (6), 947–957.

Jedermann, R., Ruiz-Garcia, L., Lang, W., 2009. Spatial temperature profiling by semi-passive RFID loggers for perishable food transportation. Comput. Electron. Agric. 65 (2), 145–154.

Jedermann, R., Nicometo, M., Uysal, I., Lang, W., 2014. Reducing food losses by intelligent food logistics. Philos. Transact. Ser. A Math. Phys. Eng. Sci. 372, 1–20. 20130302.

Jing, M., Siyuan, T., 2010. Container real-time positioning on the terminal. In: 2010 International Conference on Logistics Systems and Intelligent Management. vol. 2. IEEE, pp. 1069–1071.

Kader, A.A., 1997. A summary of CA requirements and recommendations for fruits other than apples and pears. In: Kader, A.A. (Ed.), Proceedings of the 7th International Controlled Atmosphere Research Conference, Davis, CA, USA. vol. 3, pp. 1–34.

Kader, A.A., 2002. Postharvest Technology of Horticultural Crops. Vol. 3311. University of California Agriculture and Natural Resources.

Kemp, K., Insch, A., Holdsworth, D.K., Knight, J.G., 2010. Food miles: do UK consumers actually care? Food Policy 35 (6), 504–513.

Kim, D.S., Ferreira, C.I., 2008. Solar refrigeration options—a state-of-the-art review. Int. J. Refrig. 31 (1), 3–15.

Klumpp, M., Sandhaus, G., ild Essen, F.O.M., 2012. Dynamic scheduling in logistics with agent-based simulation. In: The 2012 European Simulation and Modelling Conference, Conference Proceedings October, pp. 22–24.

Kongsberg, 2017. YARA and KONGSBERG Enter into Partnership to Build World's First Autonomous and Zero Emissions Ship. https://www.km.kongsberg.com/ks/web/nokbg0238.nsf/AllWeb/98A8C576AEFC85AFC125811A0037F6C4?OpenDocument. (Accessed December 2017).

Kumar, S., Nigmatullin, A., 2011. A system dynamics analysis of food supply chains–case study with non-perishable products. Simul. Model. Pract. Theory 19 (10), 2151–2168.

Kuo, J.C., Chen, M.C., 2010. Developing an advanced multi-temperature joint distribution system for the food cold chain. Food Control 21 (4), 559–566.

Lee, D.Y., Thomas, V.M., Brown, M.A., 2013. Electric urban delivery trucks: energy use, greenhouse gas emissions, and cost-effectiveness. Environ. Sci. Technol. 47 (14), 8022–8030.

Léonardi, J., Baumgartner, M., 2004. CO2 efficiency in road freight transportation: status quo, measures and potential. Transp. Res. Part D: Transp. Environ. 9 (6), 451–464.

Mahlknecht, S., Madani, S.A., 2007. On architecture of low power wireless sensor networks for container tracking and monitoring applications. In: 5th IEEE International Conference on Industrial Informatics, 2007. vol. 1, pp. 353–358.

Mai, N.T.T., Margeirsson, B., Margeirsson, S., Bogason, S.G., Sigurgísladóttir, S., Arason, S., 2012. Temperature mapping of fresh fish supply chains–air and sea transport. J. Food Process Eng. 35 (4), 622–656.

Maloni, M., Jackson, E.C., 2005. North American container port capacity: an exploratory analysis. Transp. J. 44 (3), 1–22.

Mangaraj, S., Goswami, T.K., Mahajan, P.V., 2009. Applications of plastic films for modified atmosphere packaging of fruits and vegetables: a review. Food Eng. Rev. 1 (2), 133.

Martin-Belloso, O., Fortuny, R.S., 2010. Advances in Fresh-Cut Fruits and Vegetables Processing. CRC Press.

Mašek, J., Kolarovszki, P., Čamaj, J., 2016. Application of RFID technology in railway transport services and logistics chains. Procedia Eng. 134, 231–236.

Mathiesen, B.V., Lund, H., Connolly, D., Wenzel, H., Østergaard, P.A., Möller, B., Hvelplund, F.K., 2015. Smart Energy Systems for coherent 100% renewable energy and transport solutions. Appl. Energy 145, 139–154.

Montanaro, G., Dichio, B., Xiloyannis, C., 2010. Significance of fruit transpiration on calcium nutrition in developing apricot fruit. J. Plant Nutr. Soil Sci. 173 (4), 618–622.

Park, M., Regan, A., 2004. Issues in Emerging Home Delivery Operations. University of California Transportation Center.

Pelletier, S., Jabali, O., Laporte, G., 2016. 50th anniversary invited article—goods distribution with electric vehicles: review and research perspectives. Transp. Sci. 50 (1), 3–22.

Port of Rotterdam, 2013. Annual report 2012. www.portofrotterdam.com/annualreport. (Accessed December 2017).

Prasanna, K.R., Hemalatha, M., 2012. RFID GPS and GSM based logistics vehicle load balancing and tracking mechanism. Procedia Eng. 30, 726–729.

Railfreight, 2016. Cool Rail's Fresh Produce Service Lowers Emissions. https://www.railfreight.com/business/2016/12/13/cool-rails-fresh-produce-freight-cuts-emissions/. (Accessed December 2017).

Rizzo, F., Barboni, M., Faggion, L., Azzalin, G., Sironi, M., 2011. Improved security for commercial container transports using an innovative active RFID system. J. Netw. Comput. Appl. 34 (3), 846–852.

Ro, C.W., Kim, K.M., 2006. Design and implementation of port container management system using RFID. J. Korea Contents Assoc. 6 (2), 1–8.

Rodrigue, J.-P., Notteboom, T., 2013. The cold chain and its logistics. In: The Geography of Transport Systems. Routledge, New York.

Rodrigue, J.P., Notteboom, T., 2015. Looking inside the box: evidence from the containerization of commodities and the cold chain. Marit. Policy Manag. 42 (3), 207–227.

Rodríguez-Bermejo, J., Barreiro, P., Robla, J.I., Ruiz-García, L., 2007. Thermal study of a transport container. J. Food Eng. 80 (2), 517–527.

Rong, A., Akkerman, R., Grunow, M., 2011. An optimization approach for managing fresh food quality throughout the supply chain. Int. J. Prod. Econ. 131 (1), 421–429.

Roso, V., 2007. Evaluation of the dry port concept from an environmental perspective. Transp. Res. Part D: Transp. Environ. 12 (7), 523–527.

Ruiz-Garcia, L., Barreiro, P., Robla, J.I., 2008. Performance of ZigBee-based wireless sensor nodes for real-time monitoring of fruit logistics. J. Food Eng. 87 (3), 405–415.

Ryan, J.M., 2017. Guide to Food Safety and Quality During Transportation: Controls, Standards and Practices. Academic Press.

Saltveit, M.E., 2003. Is it possible to find an optimal controlled atmosphere? Postharvest Biol. Technol. 27 (1), 3–13.

Salvi, B.L., Subramanian, K.A., Panwar, N.L., 2013. Alternative fuels for transportation vehicles: a technical review. Renew. Sust. Energ. Rev. 25, 404–419.

Sargent, S.A., 1998. Handling and cooling techniques for maintaining postharvest quality. In: Vegetable Production Guide for Florida. University of Florida, Gainesville, FL.

Selcuk, N., Erkan, M., 2015. The effects of modified and palliflex controlled atmosphere storage on postharvest quality and composition of 'Istanbul'-medlar fruit. Postharvest Biol. Technol. 99, 9–19.

Shi, X., Tao, D., Voß, S., 2011. RFID technology and its application to port-based container logistics. J. Organ. Comput. Electron. Commer. 21 (4), 332–347.

Sørensen, K.K., Nielsen, J.D., Stoustrup, J., 2014. Modular simulation of reefer container dynamics. Simulation 90 (3), 249–264.

Stanton, D.W., 2013. Systematic development of highly efficient and clean engines to meet future commercial vehicle greenhouse gas regulations. SAE Int. J. Engines 6, 1395–1480 (2013-01-2421).

Starcrest Consulting Group, LLC, 2011. The Port of Los Angeles Inventory of Air Emissions for Calendar Year 2010. Port of Los Angeles, Los Angeles, CA.

SteadieSeifi, M., Dellaert, N.P., Nuijten, W., Van Woensel, T., Raoufi, R., 2014. Multimodal freight transportation planning: a literature review. Eur. J. Oper. Res. 233 (1), 1–15.

Tassou, S.A., De-Lille, G., Ge, Y.T., 2009. Food transport refrigeration–approaches to reduce energy consumption and environmental impacts of road transport. Appl. Therm. Eng. 29 (8–9), 1467–1477.

Tassou, S.A., Lewis, J.S., Ge, Y.T., Hadawey, A., Chaer, I., 2010. A review of emerging technologies for food refrigeration applications. Appl. Therm. Eng. 30 (4), 263–276.

Tavasszy, L.A., Behdani, B., Konings, R., 2017. Intermodality and synchromodality. In: Geerlings, H., Kuipers, B., Zuidwijk, R. (Eds.), Ports and Networks: Strategies, Operations and Perspectives. Routledge, London, UK.

Thompson, A.K., 2010. Controlled Atmosphere Storage of Fruits and Vegetables. CABI.

Ting, S.L., Wang, L.X., Ip, W.H., 2012. A study on RFID adoption for vehicle tracking in container terminal. J. Ind. Eng. Manag. 5 (1), 22.

Toivonen, P., Wiersma, P.A., Hampson, C., Lannard, B., 2010. Effect of short-term air storage after removal from controlled-atmosphere storage on apple and fresh-cut apple quality. J. Sci. Food Agric. 90 (4), 580–585.

Ubeda, S., Arcelus, F.J., Faulin, J., 2011. Green logistics at Eroski: a case study. Int. J. Prod. Econ. 131 (1), 44–51.

Van Amerongen, 2017. Storing Fruit on Pallets. https://www.van-amerongen.com/en/pallet. (Accessed December 2017).

Van den Berg, R., De Langen, P.W., 2015. Towards an "inland terminal centred" value proposition. Marit. Policy Manag. 42 (5), 499–515.

van der Vorst, J.G., van Kooten, O., Luning, P.A., 2011. Towards a diagnostic instrument to identify improvement opportunities for quality controlled logistics in agrifood supply chain networks. Int. J. Food Syst. Dyn. 2 (1), 94–105.

Wang, X., Bischoff, O., Laur, R., Paul, S., 2009. Localization in wireless ad-hoc sensor networks using multilateration with RSSI for logistic applications. Procedia Chem. 1 (1), 461–464.

Wiegmans, B., 2013. Intermodalism: competition or cooperation? In: Prokop, D. (Ed.), The Business of Transportation. Praeger Publishers.

Wiegmans, B., Konings, R., 2013. The performance of intermodal inland waterway transport: modeling conditions influencing its competitiveness. In: WCTR 2013: 13th World Conference on Transport Research, Rio de Janeiro, Brazil, 15–18 July 2013.

Wohlrab, J., Harrington, T.S., Srai, J.S., 2012. Last Mile Logistics Evaluation—Customer, Industrial and Institutional Perspectives. Cambridge University Press, Cambridge, pp. 5–20.

Zehendner, E., Feillet, D., 2014. Benefits of a truck appointment system on the service quality of inland transport modes at a multimodal container terminal. Eur. J. Oper. Res. 235 (2), 461–469.

Zeng, Y., Jia, F., Wan, L., Guo, H., 2017. E-commerce in agri-food sector: a systematic literature review. Int. Food Agribus. Manag. Rev. 20 (4), 439–460.

Zhao, W., Goodchild, A.V., 2013. Using the truck appointment system to improve yard efficiency in container terminals. Marit. Econ. Logist. 15 (1), 101–119.

Further reading

Arvidsson, N., 2013. The milk run revisited: a load factor paradox with economic and environmental implications for urban freight transport. Transp. Res. A Policy Pract. 51, 56–62.

Bachmann, J., Earles, R., 2000. Postharvest Handling of Fruits and Vegetables. ATTRA.

Behzadi, G., O'Sullivan, M.J., Olsen, T.L., Zhang, A., 2017. Agribusiness supply chain risk management: a review of quantitative decision models. Omega, 21–42.

Christiansen, M., Fagerholt, K., 2014. Ship routing and scheduling in industrial and tramp shipping. In: Vehicle Routing: Problems, Methods, and Applications, second ed. Society for Industrial and Applied Mathematics, pp. 381–408 (Chapter 13).

Leprogres, 2012. Franprix approvisionne Paris par fleuve··· Lyon en question. http://www.leprogres.fr/france-monde/2012/10/01/franprix-approvisionne-paris-par-fleuve-lyon-en-questiojn. (Accessed December 2017).

Shukla, M., Jharkharia, S., 2013. Agri-fresh produce supply chain management: a state-of-the-art literature review. Int. J. Oper. Prod. Manag. 33 (2), 114–158.

Wiegmans, B., Behdani, B., 2018. A review and analysis of the investment in, and cost structure of, intermodal rail terminals. Transp. Rev. 38 (1), 33–51.

Chapter 13

Food transportation and refrigeration technologies—Design and optimization

Christian James

Food Refrigeration and Process Engineering Research Centre (FRPERC), Grimsby Institute, Grimsby, United Kingdom

Abstract

Millions of unrefrigerated and refrigerated transport systems are used to distribute dry, chilled, and frozen foods throughout the world. In general, these transportation systems are expected to maintain the temperature of the food within close limits to ensure its optimum safety and high-quality shelf-life. Increasingly the sustainability of such systems is also being considered. When considering the sustainability of transport systems, a holistic approach is needed. For example, lower temperatures may require greater energy consumption but may significantly extend storage life, thus reducing waste and leading to a more sustainable system. Models and other tools are increasingly being used to design and optimize food transportation.

This chapter covers the available technology and the modeling approaches that can be used to aid the design and optimization of food transportation technologies.

1 Introduction

Millions of unrefrigerated and refrigerated transport systems are used to distribute dry, chilled, and frozen foods throughout the world. In general, these transportation systems are expected to maintain the temperature of the food within close limits to ensure its optimum safety and high-quality shelf-life. Increasingly, the sustainability of such systems is also being considered. When considering the sustainability of transport systems, a holistic approach is needed. For example, lower temperatures may require greater energy consumption but may significantly extend storage life, thus reducing waste and leading to a more sustainable system. Models and other tools are increasingly being used to design and optimize food transportation.

This chapter covers the available technology and the modeling approaches that can be used to aid the design and optimization of food transportation technologies.

2 Food transportation methods

There are four modes of transport: road, water, rail, and air. Transportation methods range from 12-m containers for long-distance water, road, or rail movement of bulk ambient, chilled, or frozen products to small, uninsulated vans supplying food to local retail outlets or even directly to the consumer. A considerable amount of international transportation of foods is intermodal, i.e., the food is transported in a single container unit using two or more modes of transport.

2.1 Road

Most road vehicles for long-distance transport are mechanically refrigerated. However, in many countries perishable foods are still transported without any temperature or environmental control.

There are substantial difficulties in maintaining the temperature of perishable foods transported in small refrigerated vehicles that conduct multidrop deliveries to retail stores and caterers. In a UK survey it was found that during a delivery run, the refrigerated product can be subjected to as many as 50 door openings, where there is heat ingress directly from outside and from personnel entering to select and remove product (James et al., 2006). The design of the refrigeration system has to allow for extensive differences in load distribution on different delivery rounds and days of the week, and for the removal of product.

Sustainable Food Supply Chains. https://doi.org/10.1016/B978-0-12-813411-5.00013-2

The rise in supermarket home delivery services where there are requirements for mixed loads of products that may each require different storage temperatures is introducing a new complexity to local land delivery (Cairns, 1996). At the same time, by replacing consumer journeys, this trend can significantly reduce the carbon footprint (Wakeland et al., 2012).

The role of the consumer in food transport can be significant. It has been estimated that about one in ten car journeys in the United Kingdom are for food shopping (Department for Transport, 2007).

2.2 Water

Historically it was the need to preserve food (particularly meat) during sea transport that led to the development of mechanical refrigeration and the modern international trade in foods. Developments in frozen transport in the 19th century established the international frozen meat trade.

There are two standard options for the marine transport of refrigerated foods: (1) container ships and (2) refrigerated ships. Refrigerated ships are particularly suited to bulk cargoes and can provide tight temperature control for the cargo being transported (Tanner, 2016). Refrigerated vessels may also transport refrigerated containers above deck. Containerization has revolutionized world trade. Container ships can carry a range of refrigerated and nonrefrigerated products and can be rapidly loaded and unloaded in a few hours compared with days for a traditional cargo vessel. About 90% of nonbulk cargo is transported by container globally (Tanner, 2016).

In order to reduce fuel consumption and cost, "slow steaming" is increasingly being introduced. This practice involves reducing the ship's speed from about 27–24 knots to 21–18 knots (or even lower, in the case of "super slow steaming") which requires four times less power (and therefore fuel). However, this practice can add an additional week's sailing time on Asia-Europe routes (Wiesmann, 2010).

2.3 Rail

While the refrigerated transport of foods by rail has a long history, in many countries it has been displaced by road transport (Tanner, 2016). However, with the carbon footprint of modes of transport receiving more scrutiny, rail transport of bulk products is receiving renewed interest. While dedicated refrigerated railcars/boxcars continue to be used for some routes, the integrated refrigerated shipping container offers a lot more flexibility, as it can be transported by road and water, as well as rail (Tanner, 2016).

2.4 Air

Air-freighting is generally used for high-value, short–shelf life products, such as strawberries, asparagus, Wagyu beef, and live lobsters (Sharp, 1988; Stera, 1999; Tanner, 2016). It is considered to be the most expensive mode of transportation, and often is that with the largest carbon footprint (however, this depends on the product and markets being supplied).

Air-freight can be carried on board passenger, combi (passenger and cargo), or cargo aircraft. The ambient temperature of the cargo hold of the aircraft will depend in part on the type of aircraft, the location, and the type of cargo being transported. Many aircraft use standardized unit load devices (ULD) for easy loading, storage, and unloading of freight. These units come in a number of standard sizes and can be refrigerated.

Although air freighting of foods offers a rapid method of serving distant markets, there has been a problem in that the product may be unprotected by refrigeration for much of its journey. Up to 80% of the total journey time is made up of waiting on the tarmac and transport to and from the airport (James et al., 2006). During flight the hold may be between 15°C and 20°C and the perishable cargo is usually carried in standard containers, sometimes with an insulating lining and/or dry ice, but often unprotected. Refrigerated ULDs, using standard mechanical refrigeration systems or passive PCMs, are now common for the transport of perishable foods (Baxter and Kourousis, 2015).

3 Food transportation technologies

A range of technologies exists for maintaining the temperature and quality of chilled, frozen, and ambient foods during transport.

3.1 Containers

Most International Standard Organisation (ISO) containers for food transport are either 6 or 12 m long, hold up to 26 tonnes of product and can be insulated or refrigerated (Heap, 1986). The refrigerated containers incorporate insulation and have refrigeration units built into their structure. The units usually operate electrically, either from an external power supply on board the ship or dock, or from a generator on a road vehicle. Insulated containers for refrigerated foods may utilize either plug type refrigeration units or may be connected directly to an air-handling system in a ship's hold or at the docks. Close temperature control is most easily achieved in containers that are placed in insulated holds and connected to the ship's refrigeration system. However, suitable refrigeration facilities must be available for any overland sections of the journey. When the containers are fully loaded and the cooled air is forced uniformly through the spaces between cartons, the maximum difference between delivery and return air can be less than 0.8°C. The entire product in a container can be maintained to within ±1.0°C of the set point (Heap, 1986).

Refrigerated containers are easier to transport by road than the insulated types, but, when transported by water, they may have to be carried on deck because of problems in operating the refrigeration units within closed holds. Therefore, they may be subjected to much higher ambient temperatures on board ship and consequently larger heat gains that make it far more difficult to control product temperatures. There may also be problems on docks where often there are not enough power supply plug-in points, because it is difficult to predict the maximum number of refrigerated containers to be managed at any one time.

3.2 Controlled atmosphere transport

Products such as fruits and vegetables that produce heat by respiration, or products that have to be cooled during transit, require circulation of air through the product. Control of the oxygen and carbon dioxide levels in containers has allowed fruits and vegetables, such as apples, pears, avocados, melons, mangoes, nectarines, blueberries, and asparagus, to be shipped (typically 40 days in the container) from Australia and New Zealand to markets in the United States, Europe, Middle East, and Japan (Adams, 1988). If the correct varieties are selected and rapidly cooled immediately after harvest, the product arrives in good condition and has a long subsequent shelf life.

3.3 Insulation

Vehicles and containers for the transport of refrigerated foods depend on good thermal insulation to minimize the transfer of heat between the inside and the external ambient. Providing the product is fully cooled prior to loading and the loading is carried out in a refrigerated loading dock, the only heat load of consequence is infiltration through the structure. Over time, insulation materials deteriorate (an overall mean deterioration rate of 3.1% per annum has been reported; Heap, 1990) and containers are periodically tested to see if they are within thermal specifications. The insulation of choice has typically been expanded polyurethane, but other materials being considered include vacuum insulated panels (VIPs), aerogel insulation, and mineral wool (Lawton and Marshall, 2007; James and James, 2010). In practice VIPs can be five times more efficient than insulated foam panels, so wall thickness can be thinner and load capacity increased (James and James, 2010). However, currently they are expensive, and problems occur at corners and junctions. Currently, most transport containers use internal and external sheet metal to provide support and protection. Replacing these with lighter, less conductive materials could potentially improve energy consumption and effectiveness (Adekomaya et al., 2016).

3.4 Types of refrigeration system

Most refrigerated containers and vehicles use conventional mechanical refrigeration systems, but alternatives are being developed. The use of photovoltaics (solar power) to power mechanical units has been pioneered in some countries.

Despite the different refrigeration technologies and the different sources of energy used to power refrigeration units, a quantitative metric is used to compare alternative solutions in agreement with the efficiency of the energy use. The energy efficiency ratio (EER), also known as coefficient of performance (COP), is used to indicate the efficiency of a thermal machine when it is used to absorb heat (i.e., cooling process). The COP can be defined as follows:

$$\mathrm{COP} = \mathrm{EER} = \frac{Q_{\mathrm{Cool}}}{|Pw|} \leq \frac{T_{\mathrm{Cold}}}{T_{\mathrm{Hot}} - T_{\mathrm{Cold}}} \tag{1}$$

where Q_{Cool} is the joules of heat absorbed from the room to be cooled, and Pw is the power absorbed to reduce the temperature of the room. By considering the ideal Carnot cycle, the ratio can also be expressed as a function of the temperatures. The COP is used to compare the energy efficiency of different refrigeration technology and calculate the associated costs and environmental impacts.

3.4.1 Mechanical systems

Many types of independent engine- and/or electric motor-driven mechanical refrigeration units are available for trucks or trailers (see Fig. 1). One of the most common is a self-contained "plug" unit that mounts in an opening provided in the front wall of the vehicle. The condensing section is on the outside and the evaporator is on the inside of the unit. Units have one or two compressors, depending upon their capacity; these can be belt driven from the vehicle but are usually driven directly from an auxiliary engine. This engine may use petrol from the vehicle's supply, an independent tank, or liquid petroleum gas. Many are equipped with an additional electric motor for standby use or for quiet running, e.g., when parked or on a ferry.

Irrespective of the type of refrigeration equipment used, the product will not be maintained at its desired temperature during transportation unless it is surrounded by air or surfaces at or below that temperature. This is usually achieved by a system that circulates air, either forced, or by gravity, around the load. Inadequate air distribution is probably the principal cause of product deterioration and loss of shelf life during transport. Conventional forced air units usually discharge air over the stacked or suspended products either directly from the evaporator or through ducts towards the rear cargo doors. If the products have been cooled to the correct temperature before loading and do not generate heat, then they only need to be isolated from external heat ingress. Surrounding them with a blanket of cooled air achieves this purpose. Care must be taken during loading to avoid any product contact with the inner surfaces of the vehicle, because this would allow heat ingress during transport. Many vehicle containers are now being constructed with an inner skin that forms a return air duct along the side walls and floor, with the refrigerated air being supplied via a ceiling duct.

The application of photovoltaics (PV) to refrigeration for the distribution of chilled supermarket produce has been pioneered in some countries, such as the United Kingdom (James and James, 2010). For example, in 1997 Sainsbury's, a major UK supermarket chain, commissioned the world's first solar-powered refrigerated trailer. The high capital cost of PV systems at the time apparently limited adoption. It is anticipated that with time, the costs should come down and payback times shorten.

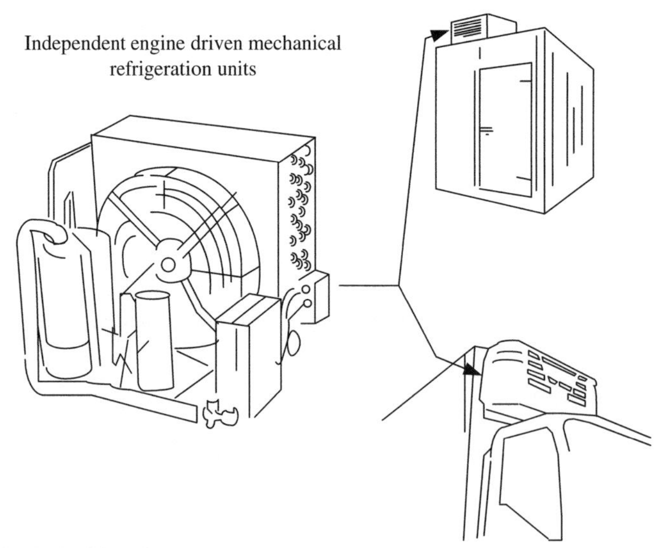

FIG. 1 Electric motor-driven mechanical refrigeration.

3.4.2 Adsorption systems

Adsorption systems (see Fig. 2) have been demonstrated in principle for food transportation. Though these systems have low COPs and cost more than a conventional mechanical system to produce, they can utilize waste heat (e.g., from a vehicle exhaust), which improves the overall consumption of such systems (Tassou et al., 2009; Gao et al., 2016). Gao et al. (2016) estimated the payback on a sorption system for a refrigerated truck, compared to a mechanical system, to be less than a year.

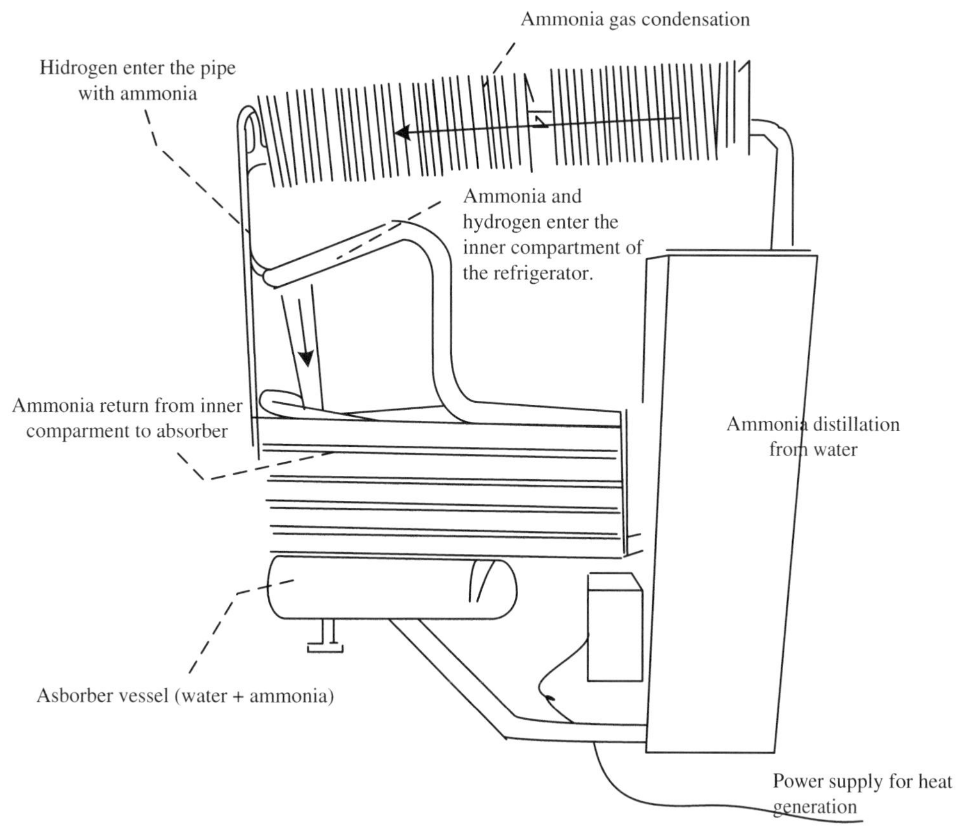

FIG. 2 Absorption refrigeration unit.

3.4.3 Air cycle

Historically some of the first mechanical refrigeration systems used for the transport of refrigerated foods used air cycle systems. While air cycle systems can be less energy efficient than chlorofluorocarbon (CFC) based systems, the impact on global warming potential (GWP) is substantially lower. Thus, there has been a reappraisal of air cycle systems for food transport in recent years, with studies by Engelking and Kruse (1996), Spence et al. (2005), and Li et al. (2017). The main limits to air cycle systems at present are their low COP and the unavailability of off-the-shelf systems (Tassou et al., 2009).

3.4.4 Phase change materials

The first refrigeration systems utilized a naturally produced phase change material (PCM), i.e., ice. Ice was once commonly used to provide cooling in road and rail transport. Phase change cooling systems are still used for some refrigerated vehicles serving local distribution chains (James and James, 2014). The phase change system consists of a coil, through which a primary refrigerant can be passed, mounted inside a thin tank filled with a PCM. The PCM is frozen by the refrigeration system in a static, or overnight, situation. During the working day no power is required, the cooling being provided by the melting of the frozen PCM. Standard PCMs freeze at temperatures between −3°C and −50°C. A number of these plates are mounted on the walls and ceilings or used as shelves or compartment dividers in the vehicles. Two methods are commonly used for charging up the plates: [1] when the vehicle is in the depot the solutions are frozen by coupling the plates to stationary refrigeration plants via flexible pipes; [2] a condensing unit on the vehicle is driven by an auxiliary drive when the vehicle is in use and an electric motor when stationary. Ideally the PCMs are charged with renewable energy. PCM systems are chosen for the simplicity, low maintenance, and quietness of their operation but can suffer from poor temperature control (James and James, 2014).

3.4.5 Cryogenic systems

Many advantages are claimed for liquid nitrogen transport systems, including minimal maintenance requirements, uniform cargo temperatures, silent operation, low capital costs, environmental acceptability, rapid temperature reduction, and increased shelf life due to the modified atmosphere (Smith, 1986). Overall costs are claimed to be comparable with mechanical systems (Smith, 1986; Pedolsky et al., 2004). However, published trials on the distribution of milk have shown that the operating costs using liquid nitrogen, per 100L of milk transported, may be 2.2 times that of a mechanically refrigerated transport system (Nieboer, 1988).

A typical liquid nitrogen system consists of an insulated liquid nitrogen storage tank (see Fig. 3) connected to a spray bar that runs along the ceiling of the transport vehicle. Liquid nitrogen is released into the spray bar via a thermostatically controlled valve and vaporizes instantly as it enters the body of the vehicle. The air is then cooled directly, utilizing the change in the latent and sensible heat of the liquid nitrogen. Once the required air temperature has been reached, the valve shuts off the flow of liquid nitrogen and the temperature is subsequently controlled by intermittent injections of liquid nitrogen.

It is claimed that long hauls can be carried out, with vehicles available that will maintain a chilled cargo at 3°C for 50h after a single charge of liquid nitrogen, and with overall costs comparable to mechanical systems (James and James, 2014). However, uncertainty regarding cryogen prices, availability, and a lack of charging infrastructure is currently hindering the use of cryogenic transport systems (Rai and Tassou, 2017).

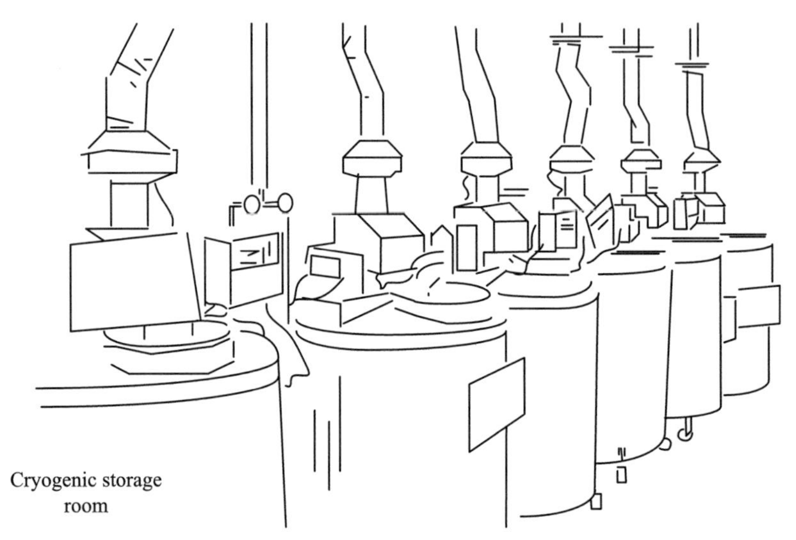

FIG. 3 Cryogenic storage room.

4 Food transportation requirements

It is particularly important that the food is at the correct temperature before loading, since the refrigeration systems used in most transport containers are generally not designed to extract heat from the load but to maintain the temperature of the load. The potential for cooling during transport has been explored for a range of foods, such as citrus fruit (Defraeye et al., 2015, 2016) and banana (Jedermann et al., 2013, 2014b). In the EU red meat generally cannot be transported until the meat has reached a temperature throughout the meat of not more than 7°C. However, recent changes allow transport if the competent authority authorizes it for specific products, provided that the meat leaves the slaughterhouse, or a cutting room on the same site, immediately and transport takes no more than 2h. While studies have shown that foods can be cooled during transport, the cooling is much slower than in dedicated cooling systems that have been designed to cool.

Studies have shown a notable range of temperatures in foods within a container. For example, a 5°C difference in temperature in a load of strawberries (Pelletier et al., 2011), and 10.3°C difference within a load of chicken portions (Raab et al., 2008). Such a range can have a significant impact on product quality. To quote Jedermann et al. (2014a) "a temperature-controlled container's set point is an extremely poor approximation of the actual product temperature."

Publications such as the International Institute of Refrigeration (IIR) *Recommendations for the Chilled Storage of Perishable Produce* (2000) and *Recommendations for the Processing and Handling of Frozen Foods* (2006) provide data on the storage life of many foods at different temperatures. The Agreement on the International Carriage of Perishable Foodstuffs (ATP Agreement) specifies maximum temperatures for the transportation of chilled and frozen foods (Table 1).

TABLE 1 Maximum temperatures for the transportation of chilled and frozen foods specified in the Agreement on the International Carriage of Perishable Foodstuffs (ATP Agreement)

	Maximum temperature (°C)	Description
Chilled foods	7	Red meat and large game (other than red offal)
	6	Raw milk
	6[a]	Meat products, pasteurized milk, fresh dairy products (yogurt, kefir, cream, and fresh cheese), ready cooked foodstuffs (meat, fish, vegetables), ready to eat (RTE) prepared raw vegetables and vegetable products, concentrated fruit juice and fish products not listed
	4	Poultry, game (other than large game) and rabbits
	3	Red offal
	2[a]	Minced meat
	0[b]	Untreated fish, molluscs, and crustaceans
Frozen foods	−20	Ice cream
	−18	Frozen or quick (deep)-frozen fish, fish products, molluscs and crustaceans and all other quick (deep)-frozen foodstuffs
	−12	All other frozen foods (except butter)
	−10	Butter

[a]At temperature indicated on the label and/or on the transport documents.
[b]At temperature of melting ice.

Foods have different optimal transport temperatures and environmental conditions. As well as temperature, gaseous atmosphere can be particularly important with fruits and vegetables. Ethylene production/sensitivity, chilling or freezing sensitivity, and off-odor or colors due to cross-contaminations will all affect the quality and shelf life of the food. Although in general the lower the storage temperature, the longer the shelf life, many fruits and vegetables are sensitive to low temperatures (chill injury), examples of which are listed in Table 2. Aung and Chang (2014) addressed methods used to improve the ability to define an optimal target temperature for multicommodity refrigerated storage. Their simulations suggest that sensor-based methods for real-time quality monitoring and assessment may be superior to the traditional visual assessment method.

TABLE 2 Fruits and vegetables susceptible to chill injury

Commodity	Lowest safe temperature (°C)	Damage
Apples		
certain varieties	2–3	Internal browning, brown core
Avocados		
West Indian	11	Pitting, internal browning
Other varieties	5–7	Pitting, internal browning
Bananas	13–12	Dull color, blackening of skin
Beans	7–10	Pitting and russeting
Cucumbers	7–10	Pitting, water soaked spots, decay
Grapefruit	7	Scald, pitting, watery breakdown, internal browning
Lemons	13–14	Internal discoloration, pitting
Mangoes	10–5	Internal discoloration, abnormal ripening

Continued

TABLE 2 Fruits and vegetables susceptible to chill injury—cont'd

Commodity	Lowest safe temperature (°C)	Damage
Melons		
Cantaloupe	7	Pitting, surface decay
Honeydew	4–10	Pitting, surface decay
Watermelons	2–4	Pitting, objectionable flavor
Oranges	3	Pitting, brown stains
Papaya	6	Pitting, water soaking of flesh, abnormal ripening
Peppers, sweet	7	Sheet pitting, alternaria rot on pods and calyxes, darkening of seeds
Pineapples	7–10	Dull green when ripe, internal browning
Potatoes	3–4	Mahogany browning, sweetening
Sweet potatoes	13	Decay, pitting, internal discoloration, Hard core when cooked
Tomatoes	7–10	Watersoaking and softening

McGlasson, W.B., Scott, K.J., Mendoza, Jr, D.B., 1979. The refrigerated storage of tropical and subtropical products. Int. J. Refrig. 2(6), 199–206; Hardenburg, R.E., Alley, E., Watada, C.Y.W., 1986. The Commercial Storage of Fruits, Vegetables, Florist and Nursery Stocks, United States Department of Agriculture Handbook No. 66; McGregor, B.M., 1989. Tropical Products Transport Handbook, USDA OT Agricultural Handbook No. 688; International Institute of Refrigeration, 2000. Recommendations for Chilled Storage of Perishable Produce, International Institute of Refrigeration (IIR), Paris; Wang, C.Y., 2016. Chilling and Freezing Injury. In: Gross, K. (Ed.), In: The Commercial Storage of Fruits, Vegetables, and Florist and Nursery Stocks. USDA-ARS. Agriculture Handbook Number 66, In: Gross, K., Wang, C.Y., Saltveit, M. (Ed.).

5 Traceability and temperature monitoring

Traceability during transport is very important. Transport is often carried out by a third party, but the supplier is still responsible for the product. Thus, being able to monitor the state of the food is important, in order to maintain the safety, security, and integrity of the product. Temperature is usually the prime parameter, but other parameters, such as relative humidity and environmental gases, may also be useful. Traditional data loggers are increasingly being supplemented with the use of radio frequency identification (RFID) and wireless sensor networks (WSNs).

Telematics, the Internet of Things (IoT), cloud computing capabilities, and other innovative Internet technologies are creating better visibility of cargo through the global food supply chain, providing more and better intelligence, which ultimately should lead to more responsive and efficient logistics decision making. Such technologies enable the user to make decisions in real time, rather than based on historical datasets. This is particularly useful in the case of long global supply chains. The development of such technologies has been discussed by Kaloxylos et al. (2013), Badia-Melis et al., 2016, Verdouw et al. (2016), Shih and Wang (2016), and Zou et al. (2014), among others.

Monitoring devices are used to ensure the temperature integrity in the cold chain. Ideally a high number of temperature data loggers are required at first to identify any localized temperature hot spots. Once these are identified, a smaller number are subsequently required (Jedermann and Lang, 2009). Jedermann and Lang (2009) recommends one sensor per meter in a delivery truck to adequately monitor temperatures. However, resource limitations and cost factors often prohibit their use to one device per pallet or even perhaps one device per container, but this may be ineffective (Badia-Melis et al., 2016). A number of studies have validated such systems for the transport of fruit (Ruiz-García et al., 2007), fresh fish (Abad et al., 2009; Hayes et al., 2005), beef (Zhang et al., 2015), pineapple (Amador et al., 2009), etc. Fully integrated returnable packaging and transport units with active RFID have been developed that can be used throughout the supply chain, from production to retailing (Martínez-Sala et al., 2009).

Methods such as artificial neural networks (ANNs), kriging, and capacitive heat transfer have been used to interpret and predict temperatures in order to identify problems during transport (Badia-Melis et al., 2016).

5.1 Modeling of the impact of transport on carbon footprint and sustainability

Although important, full life-cycle analyses indicate that for most foods, transportation does not have the largest environmental impact, according to Wakeland et al. (2012). It has been estimated that in the United States transportation may account for 50% of the total carbon emissions for many fruits and vegetables, but less than 10% for red meat products

(Weber and Matthews, 2008). The type of transportation used will substantially affect the energy used. It has been estimated that the same amount of fuel can transport 5 kg of food only 1 km by personal car, 43 km by air, 740 km by truck, 2400 km by rail, and 3800 km by ship (Brodt et al., 2007). It is often stated that longer transportation distances lead to increased air pollution and greenhouse gas (GHG) emissions, which affect human health and contribute to climate change (Govindan et al., 2014), thus leading to the issue of "food miles." When taken in isolation this may be true, but transport is only one element of the carbon footprint of a food. A holistic approach needs to be taken that looks at the entire production and consumption process. The concept of "food miles" is clearly of concern to countries with well-established export markets, such as Australia and New Zealand. However, a comparison of dairy and sheep meat production by Saunders et al. (2006) concluded that New Zealand-produced products for the UK market were "by far more energy efficient" than those produced in the United Kingdom. This included the energy used in transportation, with production being twice as efficient in the case of dairy, and four times as efficient in the case of sheep meat.

Many studies have looked at sustainable supply chain management and numerous models proposed. Detailed reviews of these approaches have been written by Srivastava (2007), Carter and Rogers (2008), Seuring and Müller (2008), Ageron et al. (2012), Soysal et al. (2012), Konieczny et al. (2013), Ting et al. (2014), Govindan et al. (2014), and Stellingwerf et al. (2018), among others.

5.2 Models of the environment in refrigerated transport units

Computational fluid dynamics (CFD) has been used to investigate the optimization of air distribution in refrigerated vehicles in order to decrease the temperature variation within the load space (Zertal-Menia et al., 2002; Moureh et al., 2002; Moureh et al., 2009; Kayansayan et al., 2017). It has additionally been used to characterize the airflow generated by a wall jet within a long and empty slot-ventilated enclosure (Moureh and Flick, 2004), a design stated to be extensively used in refrigerated transport. The work was extended to look at the effect of air distribution with and without air ducts on temperature difference throughout the cargo (Moureh et al., 2002; Moureh and Flick, 2004). The predictions showed that air ducts would reduce the maximum air temperature from −16°C to −20°C and the overall temperature difference from 12°C to 8°C. Moureh et al.'s (2009) study on modeling temperature and air distributions in containers showed clearly that although air and product temperatures at either end of a load may be adequate (in one example, 9.7°C and 7.8°C for an orange transport), higher temperatures (11.7°C in the same example) can be encountered in the middle of a container due to local stagnant areas.

Tso et al. (2002) used CFD to model the effect of door openings on air temperature within a refrigerated truck, and the infiltration heat gain due to frequent door openings of an empty and midsize (12-ton) refrigerated truck has also been modeled by Lafaye de Micheaux et al. (2015).

Tapsoba et al. (2006) used CFD to model air velocities and flow patterns throughout a container loaded with two rows of slotted pallets. The flow rate through the last pallet was calculated to be about 35 times smaller than for the five first pallets. However, this work was carried out on empty pallets. Getahun et al. (2017a, b) used CFD to investigate the performance of commonly used ventilated packaging boxes (Getahun et al. (2017b). Adding vent-holes (3.5% vent area) on the bottom face of the package reduced vertical airflow resistance and reduced the seven-eighths cooling time by 37%, compared to a package with no bottom vent-holes.

5.3 Models of heat and mass transfer in foods and packages during transport

Rushbrook (1974, 1976) developed a simple one-dimensional model to represent heat flow into cartons of chilled meat in a standard mechanically refrigerated container. Although the author stated that the model was limited, he thought it useful in predicting that: (1) action would be improved if the temperature sensor measured the air off the carton stack instead of the return air; (2) proportional control on the refrigeration capacity was more stable and gave an improved response over on/off action; (3) temperature control was very sensitive to changes in system parameters.

Moureh and Derens (2000) used CFD to model temperature rises in pallet loads of frozen food during distribution. They specifically looked at the times during loading, unloading, and temporary storage when the pallets would be in an ambient temperature above 0°C. Experiments were carried out with pallets of frozen fish blocks in a shaded loading bay (4°C, 80% RH) and an open bay (22°C, 50 RH). The model took into account conductive and radiative heat transfer into the surface of the pallet, but ignored condensation. As would be expected, fish in the top corners of the pallet showed the largest temperature rise. In the shaded bay, the predicted temperature rise after 25 min in the corner was 2.7°C compared with an average of 2.5°C experimentally. In the exposed bay the corresponding figures were 6.4°C and 6°C. As the authors point out, under European quick-frozen food regulations, the fish must be distributed at −18°C or lower with brief upward fluctuation of no more than 3°C allowed within distribution. In the case of the open loading bay, the initial temperature of the fish would have to be below −25°C to keep it within the regulations.

There are stages in transportation where food is not in a refrigerated environment, i.e., in loading bays, in supermarkets before loading into retail displays, domestic transportation from shop to home, etc. The presence of an insulating cover on pallets can aid the delivery of thermosensitive food (Bennahmias et al., 1997). Studies showed that the presence of the cover increased the time taken for temperatures in the corner of the pallet to rise from 12°C to 24°C from 1.5 to 5.5 h. Ten millimeters into the load, the time was increased from approximately 2 h to over 8 h.

Insulation has a substantial effect on the temperature rise in food supplied direct to consumers by post. Direct supply is a growing market brought about by the popularity of Web-based shopping. Stubbs et al. (2004) developed a numerical model for the length of time a food packed in an expanded polystyrene box with a gel coolant could remain below 8°C or 5°C. As would be expected if the cold gel lined the top, sides, and base of the box, the time for the food to reach 5°C or 8°C was substantially longer than with gel at the sides and top or just the top. Assuming that the product would be delivered within 24 h of posting, this was the only configuration that would maintain the product below 8°C in temperatures of up to 30°C and below 5°C in temperatures of up to 25°C.

Simple numerical models have been used to identify the relative importance of different factors in the airfreight of perishable produce (Sharp, 1988). This showed clearly that some form of insulation was required around the produce, and that precooling of the produce before transportation was essential, while dry ice was unnecessary. Amos and Bollen (1998) developed a simple model to evaluate the effect of pallet wrapping on the quality of asparagus during air transport. Covering pallets with insulated blankets increased the shelf life by 0.5 to 0.7 days, while the use of a eutectic blanket increased shelf life by 2 to 3 days.

Commonly utilized methods for calculating average cooling load, including not only the product heat load but also heat transmission losses through the container walls, respiration heat, latent heat due to moisture evaporation, heat loss due to infiltration and ventilation, and heat produced by equipment (evaporator fans) can be found in the ASHRAE handbooks.

Hoang et al. (2012) developed a simplified heat transfer model of a refrigerated vehicle in which two loads (front and rear) were considered. This model allowed for the difference in front and rear temperatures that is often found due to the location of the supply air duct (and higher air speeds) that are found at the front of a container, compared to the rear load near the doors.

5.4 Models of refrigeration performance during transport

Jolly et al. (2000) developed a model to simulate the steady-state performance of a container refrigeration system. The model was shown to be within a ±10% agreement of experimentally measured data from cooling capacity tests conducted on a 2.2 m full-scale container housed in a temperature-controlled environmental test chamber. Such a model is useful for looking at the performance of different refrigerants in such systems, but, being steady state, cannot show the effect of dynamically changing external ambient conditions. Bagheri et al. (2017) developed a model for real-time performance evaluation of a refrigerated trailer in order to estimate potential energy savings and GHG reduction. This showed that significant savings could be obtained by replacing engine-driven refrigeration systems with battery-powered systems. Rai and Tassou (2017) developed a simple spreadsheet model to provide a comparative analysis of the energy consumption, production- and operation-related GHG emissions, and running costs of cryogenic liquid carbon dioxide and liquid nitrogen transport refrigeration systems with traditional conventional diesel-run systems. The operating costs and GHG emissions of the three technologies were shown to be fairly similar.

5.5 Combined models

A software model called Censor was developed by Cambridge Refrigeration Technology to estimate cargo temperatures in refrigerated containers during normal and abnormal operations (Frith, 2004). Censor was able to model the effect of different defrost intervals, the type of reefer unit used, power on and off times, ambient conditions, and up to 17 different types of food. The software was able to simulate two control modes, either modulated or on/off return air, and allow for varying effects of solar heat on the sides and roof.

One of the largest and most systematic attempts to predict the temperature of foods during multidrop deliveries has been the CoolVan program in the United Kingdom (James et al., 2006; Novaes et al., 2015). Three main types of refrigeration system were identified: a conventional diesel-driven unit, a hydraulic drive unit, and a eutectic system. Data on van performance in commercial operation were obtained during seven separate delivery trips with two major food companies. At the end of the program development, the complete model was found to be able to predict the mean temperature of the food in the vehicle with an accuracy better than 1°C at any time throughout the journey. The program was able to predict the effect of journeys with multiple stops under a wide range of environmental conditions on temperatures within the van. The model was utilized recently by Novaes et al., (2015) to investigate the regional distribution of ready-to-eat refrigerated meat products (ham, turkey and chicken breasts, salami, sausage) in Brazil.

5.6 Modeling of shelf life and microbial growth during transport

There are many microbial growth models that could be applied to modeling the growth of microorganisms in food during transport (Baranyi and Pin, 2001; McMeekin et al., 1993; Van Impe et al., 1992), such as the freely available Pathogen Modeling Program (PMP) (http://ars.usda.gov/Services/docs.htm) and Combase. The Seafood Spoilage and Safety Predictor (SSSP) software (DTU Aqua, Denmark) has also been used to predict remaining shelf life during transport (Mai et al., 2012).

Tijskens and Polderdijk (1996) developed a generic static and dynamic model for the storage and transport of about 60 different types of perishable fruits and vegetables that includes the effects of temperature, chilling injury, and different levels of initial quality. Specific models describing quality or remaining shelf life as a function of temperature and other environmental conditions have been developed for many perishable food products, including fresh-cut vegetables (Jacxsens et al., 2002), bananas (Jedermann et al., 2014b), pork and poultry (Bruckner et al., 2013), lamb (Mack et al., 2014), shrimp (Dabadé et al., 2015), and yogurt (Mataragas et al., 2011). According to Mercier et al. (2017) the application of such systems has been very limited thus far by "the inaccuracy of shelf-life estimates from temperature measurements."

The safety of the multitemperature small vans used for home deliveries has been investigated by Estrada-Flores and Tanner (2005). Recorded temperature histories were integrated with mathematical models to predict growth of pseudomonads and *Escherichia coli*. Their results showed that product temperatures were such that pseudomonads could grow, but that less than half the temperatures measured were suitable for the growth of *Escherichia coli*. The thermal behavior of the food products inside the van was strongly influenced by the loading period. A similar approach was adopted by James and Evans (1992) looking at domestic transport from the supermarket to the home. This work showed the importance of a cool box in transported refrigerated products to homes. Ambient temperatures around uninsulated products rapidly rose to approaching 40°C during a 1-h car journey, theoretically resulting in up to 1.8 generations in growth in bacterial numbers.

5.7 Other transport factors that have been modeled

Although the main modeling emphasis has been on routing, temperature control, and shelf life, other factors have also been modeled. Modeling has been used to investigate the performance of chambers developed to test systems for transporting perishable foods in accordance with the United Nations ATP agreement (Chatzidakis et al., 2004). Heat transmission through the structure of the holds of ships has also been modeled, taking into account the geometrical complexity of the structure (Magini et al., 1982).

Mechanical damage during transport is a major problem, responsible for immediate wastage and shortening shelf life. Vibration during the transportation of fresh fruit and vegetables is thought to be more important than impacts as a source of damage (Hinsch et al., 1993). It was found that for some frequencies, between 5 and 30 Hz, the top box of a stack vibrated considerably more than the middle and bottom boxes. Acceleration levels in trailers fitted with air-ride suspension systems were typically 60% of that with steel-spring suspension. Microbial growth on strawberries subjected to simulated transport has been shown to correlate with vibration levels (La Scalia et al., 2016); thus vibration levels could be used to predict shelf life.

Models have also been developed to study the effect of moving cargoes on vehicle stability. Mantriota (2002) developed a mathematical model for the dynamic study of articulated vehicles carrying suspended cargos. In addition, equations have also been developed (Ryska et al., 1999) to show that there can be considerable differences in the refrigeration performance of nominally similar transport refrigeration units when vehicle engines are idling, i.e., when the vehicles are moving slowly or stationary.

6 Conclusions and future trends

An increasing number of models and tools can be used to help design and optimize the transport of food. How many of these models/tools are used in actuality is difficult to assess. However, it is clear that increased connectivity and the growth of wireless sensors and networks are leading to greater traceability and monitoring during transport, which should strengthen this part of the food supply chain. Better control during transport should hopefully lead to improvements in sustainability, ideally reducing energy consumption, waste, and the overall carbon footprint of food supply.

References

Abad, E., Palacio, F., Nuin, M., De Zarate, A.G., Juarros, A., Gómez, J.M., Marco, S., 2009. RFID smart tag for traceability and cold chain monitoring of foods: Demonstration in an intercontinental fresh fish logistic chain. J. Food Eng. 93, 394–399.

Adams, G.R., 1988. In: Controlled atmosphere containers.Refrigeration for Food and People, Meeting of IIR Commissions C2, D1, D2/3, E1, Brisbane (Australia), pp. 244–248.

Adekomaya, O., Jamiru, T., Sadiku, R., Huan, Z., 2016. Sustaining the shelf life of fresh food in cold chain—a burden on the environment. Alexandria Eng. J. 55 (2), 1359–1365.

Ageron, B., Gunasekaran, A., Spalanzani, A., 2012. Sustainable supply management: An empirical study. Int. J. Prod. Econ. 140 (1), 168–182.

Amos, N.D., Bollen, A.F., 1998. Predicting the deterioration of asparagus quality during air transport. In: Refrigerated transport, storage and retail display, Meeting of IIR Commission D2/3 with D1, Cambridge (UK). International Institute of Refrigeration, Paris, pp. 163–170.

Amador, C., Emond, J.P., Nunes, M.C., 2009. Application of RFID technologies in the temperature mapping of the pineapple supply chain. Sens. Instrum. Food Qual. Saf. 26–33.

Aung, M.M., Chang, Y.S., 2014. Temperature management for the quality assurance of a perishable food supply chain. Food Control 40, 198–207.

Badia-Melis, R., Mc Carthy, U., Uysal, I., 2016. Data estimation methods for predicting temperatures of fruit in refrigerated containers. Biosyst. Eng. 151, 261–272.

Bagheri, F., Fayazbakhsh, M.A., Bahrami, M., 2017. Real-time performance evaluation and potential GHG reduction in refrigerated trailers. Int. J. Refrig. 73, 24–38.

Baranyi, J., Pin, C., 2001. Modelling microbiological safety. In: Tijskens, L.M.M., Hertog, M.L.A.T.M., Nicolai, B.M. (Eds.), Food Process Modelling. Woodhead Publishing Ltd, Cambridge, UK, pp. 383–401.

Baxter, G., Kourousis, K., 2015. Temperature controlled aircraft unit load devices: the technological response to growing global air cargo cool chain requirements. J. Technol. Manag. Innov. 10 (1), 157–172.

Bennahmias, R., Gaboriau, R., Moureh, J., 1997. The insulating cover, a particular logistic means for thermo- sensitive foodstuffs. Int. J. Refrig. 20 (5), 359–366.

Brodt, S., Chernoh, E., Feenstra, G., 2007. Assessment of Energy Use and Greenhouse Gas Emissions in the Food System: A Literature Review. Agricultural Sustainability Institute, University of California Davis.http://asi.ucdavis.edu/Research/Literature_Review__Assessment_of_Energy_Use_and_Greenhouse_Gas_Emissions_in_the_Food_system_Nov_2007.pdf.

Bruckner, S., Albrecht, A., Petersen, B., Kreyenschmidt, J., 2013. A predictive shelf life model as a tool for the improvement of quality management in pork and poultry chains. Food Control 29, 451–460.

Cairns, S., 1996. Delivering alternatives: Success and failures of home delivery services for food shopping. Transp. Policy (3), 155–176.

Carter, C.R., Rogers, D.S., 2008. A framework of sustainable supply chain management: moving toward new theory. Int. J. Phys. Distrib. Logist. Manag. 38 (5), 360–387.

Chatzidakis, S.K., Athienitis, A., Chatzidakis, K.S., 2004. Computational energy analysis of an innovative isothermal chamber for testing of the special equipment used in the transport of perishable products. Int. J. Energy Res. 28 (10), 899–916.

Dabadé, D.S., den Besten, H.M.W., Azokpota, P., Nout, M.J.R., Hounhouigan, D.J., Zwietering, M.H., 2015. Spoilage evaluation, shelf-life prediction, and potential spoilage organisms of tropical brackish water shrimp (Penaeus notialis) at different storage temperatures. Food Microbiol. 48, 8–16.

Defraeye, T., Nicolai, B., Kirkman, W., Moore, S., van Niekerk, S., Verboven, P., Cronjé, P., 2016. Integral performance evaluation of the fresh-produce cold chain: A case study for ambient loading of citrus in refrigerated containers. Postharvest Biol. Technol. 112, 1–13.

Defraeye, T., Verboven, P., Opara, U.L., Nicolai, B., Cronjé, P., 2015. Feasibility of ambient loading of citrus fruit into refrigerated containers for cooling during marine transport. Biosyst. Eng. 134, 20–30.

Department for Transport, 2007. Personal Travel Factsheet. Department for Transport, London.

Engelking, S., Kruse, H., 1996. Development of air cycle technology for transport refrigeration. In: International Refrigeration and Air Conditioning Conference. Paper 348.

Estrada-Flores, S., Tanner, D., 2005. Temperature variability and prediction of food spoilage during urban delivery of food products. Acta Hort. (ISHS) (674), 63–69.

Frith, J., 2004. Temperature prediction software for refrigerated container cargoes. In: Proceedings of the Institute of Refrigeration 2003-04, pp. 1–12.

Gao, P., Wang, L.W., Wang, R.Z., Li, D.P., Liang, Z.W., 2016. Optimization and performance experiments of a MnCl 2/CaCl 2–NH 3 two-stage solid sorption freezing system for a refrigerated truck. Int. J. Refrig. 71, 94–107.

Getahun, S., Ambaw, A., Delele, M., Meyer, C.J., Opara, U.L., 2017a. Analysis of airflow and heat transfer inside fruit packed refrigerated shipping container: Part I–Model development and validation. J. Food Eng. 203, 58–68.

Getahun, S., Ambaw, A., Delele, M., Meyer, C.J., Opara, U.L., 2017b. Analysis of airflow and heat transfer inside fruit packed refrigerated shipping container: Part II–Evaluation of apple packaging design and vertical flow resistance. J. Food Eng. 203, 83–94.

Govindan, K., Jafarian, A., Khodaverdi, R., Devika, K., 2014. Two-echelon multiple-vehicle location–routing problem with time windows for optimization of sustainable supply chain network of perishable food. Int. J. Prod. Econ. 152, 9–28.

Heap, R.D., 1986. Container transport of chilled meat. In: Recent Advances in the Refrigeration of Chilled Meat, Meeting of IIR Commissions C2 Bristol, UK, pp. 505–510.

Hayes, J., Crowley, K., Diamond, D., 2005. Simultaneous web-based real-time temperature monitoring using multiple wireless sensor networks. In: IEEE Sensors. IEEE pp. 4.

Heap, R.D., 1990. Developments in measurement of the thermal deterioration of insulated containers in service. In: IIF/IIR-Commissions B2, C2, D1, D2/3—Dresden (Germany) (Vol. 2). p. C2.

Hinsch, R.T., Slaughter, D.C., Craig, W.L., Thompson, J.F., 1993. Vibration of fresh fruits and vegetables during refrigerated truck transport. Trans. ASAE 36 (4), 1039–1042.

Hoang, M.H., Laguerre, O., Moureh, J., Flick, D., 2012. Heat transfer modelling in a ventilated cavity loaded with food product: Application to a refrigerated vehicle. J. Food Eng. 113 (3), 389–398.

International Institute of Refrigeration, 2000. Recommendations for Chilled Storage of Perishable Produce. International Institute of Refrigeration (IIR), Paris.

International Institute of Refrigeration, 2006. Recommendations for the Processing and Handling of Frozen Foods. International Institute of Refrigeration (IIR), Paris.

Jacxsens, L., Devlieghere, F., Debevere, J., 2002. Temperature dependence of shelf-life as affected by microbial proliferation and sensory quality of equilibrium modified atmosphere packaged fresh produce. Postharvest Biol. Technol. 26, 59–73.

James, S.J., Evans, J., 1992. Consumer handling of chilled foods: temperature performance. Int. J. Refrig. 15 (5), 299–306.

James, S.J., James, C., 2014. Meat marketing: Transport of meat and meat products. In: Devine, C., Dikeman, M. (Eds.), Encyclopedia of Meat Sciences, second ed. In: vol. 2. Academic Press, Elsevier Science Ltd, pp. 236–243.

James, S.J., James, C., 2010. The food cold-chain and climate change. Food Res. Int. 43, 1944–1956.

James, S.J., James, C., Evans, J.A., 2006. Modelling of food transportation systems – a review. Int. J. Refrig. 29, 947–957.

Jedermann, R., Lang, W., 2009. The minimum number of sensors–Interpolation of spatial temperature profiles in chilled transports. In: European Conference on Wireless Sensor Networks. Springer, Berlin, Heidelberg, pp. 232–246.

Jedermann, R., Geyer, M., Praeger, U., Lang, W., 2013. Sea transport of bananas in containers—parameter identification for a temperature model. J. Food Eng. 115, 330–338.

Jedermann, R., Nicometo, M., Uysal, I., Lang, W., 2014a. Reducing food losses by intelligent food logistics. Phil. Trans. R. Soc. A 372.

Jedermann, R., Praeger, U., Geyer, M., Lang, W., 2014b. Remote quality monitoring in the banana chain. Phil. Trans. R. Soc. A 372.

Jolly, P.G., Tso, C.P., Wong, Y.M., Ng, S.M., 2000. Simulation and measurement on the full-load performance of a refrigeration system in a shipping container. Int. J. Refrig. 23, 112–126.

Kaloxylos, A., Wolfert, J., Verwaart, T., Terolc, C.M., Brewster, C., Robbemond, R.M., Sundmaker, H., 2013. The use of Future Internet technologies in the agriculture and food sectors: integrating the supply chain. Procedia Technol. 8, 51–60.

Kayansayan, N., Alptekin, E., Ezan, M.A., 2017. Thermal analysis of airflow inside a refrigerated container. Int. J. Refrig. 84, 76–91.

Konieczny, P., Dobrucka, R., Mroczek, E., 2013. Using carbon footprint to evaluate environmental issues of food transportation. LogForum 9 (1), 3–10.

La Scalia, G., Aiello, G., Miceli, A., Nasca, A., Alfonzo, A., Settanni, L., 2016. Effect of vibration on the quality of strawberry fruits caused by simulated transport. J. Food Process Eng. 39 (2), 140–156.

Lafaye de Micheaux, T., Ducoulombier, M., Moureh, J., Sarte, V., Bonjour, J., 2015. Experimental and numerical investigation of the infiltration heat load during the opening of a refrigerated truck body. Int. J. Refrig. 54, 170–189.

Lawton, A.R., Marshall, R.E., 2007. Developments in refrigerated transport insulation since the phase out of CFC and HCFC refrigerants. International Congress of Refrigeration, Beijing.

Li, S., Wang, S., Ma, Z., Jiang, S., Zhang, T., 2017. Using an air cycle heat pump system with a turbocharger to supply heating for full electric vehicles. Int. J. Refrig. 77, 11–19.

Mack, M., Dittmer, P., Veigt, M., Kus, M., Nehmiz, U., Kreyenschmidt, J., 2014. Quality tracing in meat supply chains. Phil. Trans. R. Soc. A 372.

Magini, U., Milano, G., Pisoni, C., 1982. On the heat transmission through ships structures, Refrigeration of perishable products for distant markets, Meeting of IIR Commissions C2, D1, D2/3 Hamilton (New Zealand). International Institute of Refrigeration, Paris, pp. 275–280.

Mai, N.T.T., Margeirsson, B., Margeirsson, S., Bogason, S.G., Sigurgísladóttir, S., Arason, S., 2012. Temperature mapping of fresh fish supply chains–air and sea transport. J. Food Process Eng. 35 (4), 622–656.

Mantriota, G., 2002. Influence of suspended cargoes on dynamic behaviour of articulated vehicles. Heavy Veh. Syst. 9 (1), 52–75.

Martínez-Sala, A.S., Egea-López, E., García-Sánchez, F., García-Haro, J., 2009. Tracking of returnable packaging and transport units with active RFID in the grocery supply chain. Comput. Ind. 60 (3), 161–171.

Mataragas, M., Dimitriou, V., Skandamis, P.N., Drosinos, E.H., 2011. Quantifying the spoilage and shelf-life of yoghurt with fruits. Food Microbiol. 28, 611–616.

McMeekin, T.A., Olley, J.N., Ross, T., Ratkowsky, D.A., 1993. Predictive microbiology: theory and application. In: Research Studies Press Ltd. Taunton, UK.

Mercier, S., Villeneuve, S., Mondor, M., Uysal, I., 2017. Time–temperature management along the food cold chain: A review of recent developments. Compr. Rev. Food Sci. Food Saf. 16 (4), 647–667.

Moureh, J., Derens, E., 2000. Numerical modelling of the temperature increase in frozen food packaged in pallets in the distribution chain. Int. J. Refrig. 23, 540–552.

Moureh, J., Menia, N., Flick, D., 2002. Numerical and experimental study of airflow in a typical refrigerated truck configuration loaded with pallets. Comput. Electron. Agric. 34 (1-3), 25–42.

Moureh, J., Flick, D., 2004. Airflow pattern and temperature distribution in a typical refrigerated truck configuration loaded with pallets. Int. J. Refrig. 27 (5), 464–474.

Moureh, J., Tapsoba, S., Derens, E., Flick, D., 2009. Air velocity characteristics within vented pallets loaded in a refrigerated vehicle with and without air ducts. Int. J. Refrig. 32 (2), 220–234.

Nieboer, H., 1988. Distribution of dairy products. Cold-Chains in Economic Perspective, Meeting of IIR Commission C2. Wageningen (The Netherlands) 16, 1–16.9.

Pedolsky, H., Fedder, S., Gavrylov, K., 2004. In: Economic competitiveness of cryogenic refrigeration systems for transported product. International Refrigeration and Air Conditioning Conference at Purdue, July 12-15, 2004. R162.

Novaes, A.G., Lima Jr., O.F., Carvalho, C.C.D., Bez, E.T., 2015. Thermal performance of refrigerated vehicles in the distribution of perishable food. Pesquisa Operacional 35 (2), 251–284.

Pelletier, W., Brecht, J.K., do Nascimento Nunes, M.C., Émond, J.P., 2011. Quality of strawberries shipped by truck from California to Florida as influenced by postharvest temperature management practices. HortTechnology 21 (4), 482–493.

Raab, V., Bruckner, S., Beierle, E., Kampmann, Y., Petersen, B., Kreyenschmidt, J., 2008. Generic model for the prediction of remaining shelf life in support of cold chain management in pork and poultry supply chain. J. Chain Network Sci. 8, 59–73.

Rai, A., Tassou, S.A., 2017. Energy demand and environmental impacts of alternative food transport refrigeration systems. Energy Procedia 123, 113–120.

Ruiz-García, L., Elorza, P.B., Rodríguez-Bermejo, J., Robla, J.I., 2007. Monitoring the intermodal, refrigerated transport of fruit using sensor networks. Span. J. Agric. Res. (2), 142–156.

Rushbrook, A.J., 1974. Temperature control of chilled meat shipments in containers: Part 1, One-dimensional model studies. Meat Industry Research Institute of New Zealand Report No, p. 418.

Rushbrook, A.J., 1976. In: Temperature control simulation of chilled meat in containers.Proceedings of International Institute of Refrigeration Melbourne, pp. 277–283.

Ryska, A., Kral, F., Ota, J., 1999. Method of determination of the effective capacity of refrigeration and A/C units of variable speeds. Int. J. Refrig. 23 (5), 402–410.

Saunders, C., Barber, A., Taylor, G., 2006. Food Miles—Comparative Energy/Emissions Performance of New Zealand's Agriculture Industry, Research Report. Agribusiness & Economics Research Unit, Lincoln University, New Zealand, p. 285.

Seuring, S., Müller, M., 2008. From a literature review to a conceptual framework for sustainable supply chain management. J. Clean. Prod. 16 (15), 1699–1710.

Sharp, A.K., 1988. In: Air freight of perishable produce. Refrigeration for food and people, Meeting of IIR Commissions C2, D1, D2/3, E1, Brisbane (Australia). International Institute of Refrigeration, Paris, pp. 219–224.

Shih, C.W., Wang, C.H., 2016. Integrating wireless sensor networks with statistical quality control to develop a cold chain system in food industries. Comput. Stand. Interfaces 45, 62–78.

Smith, B.K., 1986. Liquid nitrogen in-transit refrigeration. In: Recent Advances and Developments in the Refrigeration of Meat Chilling. Proceedings of the Conference of: IIR Commission C2, Bristol, UK. International Institute of Refrigeration, Paris, France, pp. 383–390.

Soysal, M., Bloemhof-Ruwaard, J.M., Meuwissen, M.P., van der Vorst, J.G., 2012. A review on quantitative models for sustainable food logistics management. Int. J. Food Syst. Dyn. 3 (2), 136–155.

Spence, S.W., Doran, W.J., Artt, D.W., McCullough, G., 2005. Performance analysis of a feasible air-cycle refrigeration system for road transport. Int. J. Refrig. 28 (3), 381–388.

Srivastava, S.K., 2007. Green supply-chain management: a state-of-the-art literature review. Int. J. Manag. Rev. 9 (1), 53–80.

Stellingwerf, H.M., Kanellopoulos, A., van der Vorst, J.G., Bloemhof, J.M., 2018. Reducing CO2 emissions in temperature-controlled road transportation using the LDVRP model. Transp. Res. Part D: Transp. Environ. 58, 80–93.

Stera, A.C., 1999. Long distance refrigerated transport into the third millennium. In: 20th International Congress of Refrigeration, IIF/IIR Sydney, Australia. paper 736.

Stubbs, D.M., Pulko, S.H., Wilkinson, A.J., 2004. Wrapping strategies for temperature control of chilled foodstuffs during transport. Trans. Inst. Meas. Control. 26 (1), 69–80.

Tanner, D., 2016. Refrigerated transport. In: Reference Module in Food Science. Elsevier. https://doi.org/10.1016/B978-0-08-100596-5.03485-5.

Tapsoba, M., Moureh, J., Flick, D., 2006. Airflow patterns in an enclosure loaded with slotted pallets. Int. J. Refrig. 29 (6), 899–910.

Tassou, S.A., De-Lille, G., Ge, Y.T., 2009. Food transport refrigeration–Approaches to reduce energy consumption and environmental impacts of road transport. Appl. Therm. Eng. 29 (8-9), 1467–1477.

Tijskens, L.M.M., Polderdijk, J.J., 1996. A generic model for keeping quality of vegetable produce during storage and distribution. Agric. Syst. 51 (4), 431–452.

Ting, S.L., Tse, Y.K., Ho, G.T.S., Chung, S.H., Pang, G., 2014. Mining logistics data to assure the quality in a sustainable food supply chain: A case in the red wine industry. Int. J. Prod. Econ. 152, 200–209.

Tso, C.P., Yu, S.C.M., Poh, H.J., Jolly, P.G., 2002. Experimental study on the heat and mass transfer characteristics in a refrigerated truck. Int. J. Refrig. 25, 340–350.

Van Impe, J.F., Nicolaï, B.M., Martens, T., De Baerdemaeker, J., Vandewalle, J., 1992. Dynamic mathematical model to predict microbial growth and inactivation during food processing. Appl. Environ. Microbiol. 58 (9), 2901–2909.

Verdouw, C.N., Wolfert, J., Beulens, A.J.M., Rialland, A., 2016. Virtualization of food supply chains with the internet of things. J. Food Eng. 176, 128–136.

Wakeland, W., Cholette, S., Venkat, K., 2012. Food transportation issues and reducing carbon footprint. In: Boye, J.I., Arcand, Y. (Eds.), Green Technologies in Food Production and Processing. Springer, US, pp. 211–236.

Weber, C.L., Matthews, H.S., 2008. Food-miles and the relative climate impacts of food choices in the United States. Environ. Sci. Technol. 42 (10), 3508–3513.

Wiesmann, A., 2010. Slow steaming – a viable long-term option? Wärtsilä Tech. J. 2, 49–55.

Zertal-Menia, N., Moureh, J., Flick, D., 2002. Simplified modelling of air flows in refrigerated vehicles. Int. J. Refrig. 25 (5), 660–672.

Zhang, W., Yang, X., Song, Q., 2015. Construction of traceability system for maintenance of quality and safety of beef. Int. J. Smart Sens. Intelligent Syst. 8(1).

Zou, Z., Chen, Q., Uysal, I., Zheng, L., 2014. Radio frequency identification enabled wireless sensing for intelligent food logistics. Phil. Trans. R. Soc. A 372.

Further reading

Hardenburg, R.E., Alley, E., Watada, C.Y.W., 1986. The Commercial Storage of Fruits, Vegetables, Florist and Nursery Stocks. United States Department of Agriculture. Handbook No. 66.

McGlasson, W.B., Scott, K.J., Mendoza Jr., D.B., 1979. The refrigerated storage of tropical and subtropical products. Int. J. Refrig. 2 (6), 199–206.

McGregor, B.M., 1989. Tropical Products Transport Handbook. USDA OT Agricultural Handbook No. 688.

United Nations Economic Commission for Europe Working Party on the Transport of Perishable Foodstuffs, 2012. ATP Handbook. United Nations.

Wang, C.Y., 2016. Chilling and Freezing Injury. In: Gross, K. (Ed.), The Commercial Storage of Fruits, Vegetables, and Florist and Nursery Stocks. USDA-ARS. Agriculture Handbook Number 66, ed. K. Gross, C. Y. Wang, & M. Saltveit.

Chapter 14

Quality assessment of temperature-sensitive high-value food products: An application to Italian fine chocolate distribution

Riccardo Manzini, Riccardo Accorsi, Marco Bortolini and Andrea Gallo
Department of Industrial Engineering, Alma Mater Studiorum—University of Bologna, Bologna, Italy

Abstract
Modern supply chains collect and deliver products worldwide and link manufacturers and consumers over thousands of miles. In the food industry, the safety and quality of products is affected by manufacturing/processing and logistics activities, such as transportation and packaging. Transportation is likely the most critical step throughout the "food journey" from "farm-to-fork" because of the potential stresses that affect the products during shipment and storage activities. When the value and the perceived quality of products is high, the impact of such stresses can be significant and can reasonably affect the economic value of those items, as well as the experience of consumers.

Among high-value and temperature-sensitive food, chocolate is a product that is loved by many because of its desirable qualities, which include a shiny gloss on the surface, snap when the chocolate is broken, and a smooth texture that becomes apparent only when the chocolate melts in the mouth.

The aim of this chapter is the application of a two-step quality assessment methodology to the case of distribution of fine chocolate products, e.g., the well-known Italian Cremino and Gianduiotto. The first step consists of monitoring the temperature fluctuation during the distribution from the production plant to the point of demand, usually a candy shop. The second step deals with the simulation of the monitored shipment in a temperature- and humidity-controlled system. The simulation of the physical and environmental conditions generated during the real shipment is followed by chemical analysis and sensitivity analysis conducted separately on nonstressed products (the so-called "time zero products") and stressed products. The effects of temperature are evaluated by comparing stressed and time zero products belonging to the same production batch. The analyses are repeated after 6 months in order to evaluate the effects due to time and shelf-life erosion. The obtained results validate the proposed quality assessment methodology and demonstrate the importance and criticality of packaging and container solutions to face unexpected temperature variations during logistic operations.

1 Introduction and literature review

In the last decades, the food industry has experienced profound transformations. Companies are increasingly facing new challenges and issues: on one hand is the access to new and attractive markets, and on the other are the increasing numbers of products, competition in a global market, and consumer requirements in terms of quality and environmental sustainability of the products (Li et al., 2014).

To meet the growing trends of food demand, products intensively travel across continents and oceans, thereby shifting the focus from the management of local agriculture models to the optimization of the global food supply chain (Ahumada and Villalobos, 2009; Accorsi et al., 2017). These processes result in more complex supply networks, which emphasize the distance between the grower and the consumer and affect consumer awareness of the supply chain stages from farm-to-fork. Beyond the appealing shelves of the grocery store, food is grown, harvested, handled, stored, and shipped, and each of these steps may affect the product's quality and safety (Accorsi et al., 2018c).

In particular, due to biological variations, the product quality is determined by time and environmental conditions, which may be affected by the type of packaging, the loading method, and the availability of temperature-controlled packages, the vehicles, the warehouses, etc., and generally, by the set of entities and processes configuring the supply chain (Accorsi et al., 2018c).

Sustainable Food Supply Chains. https://doi.org/10.1016/B978-0-12-813411-5.00014-4

The physical and environmental conditions, i.e., temperature, humidity, light, vibrations, and mechanical shocks, etc., experienced through the whole supply chain can affect the safety, quality, and conservation conditions causing significant losses in the product value, undermining their safe and healthy final consumption (Manzini and Accorsi, 2013).

In this context the Food Supply Chain research center, at Alma Mater Studiorum, University of Bologna (http://foodsupplychain.din.unibo.it), developed and applied an original ex-post and closed-loop traceability system to predict and control the performance of a packaging system and the quality of a perishable product at the place of consumption (Manzini and Accorsi, 2013; Manzini et al., 2014; Ayyad et al., 2017).

The aim of this chapter is twofold: (1) to illustrate a closed-loop traceability and control methodology proposed and adopted at the Food Supply Chain Center and (2) to present an original case study and real-world application, which deals with the distribution of fine Italian chocolate products. This case study was selected because chocolates demonstrate poor melting resistance during storage and transportation in tropical regions and/or summer seasons (Lillah et al., 2017). In particular, this chapter presents an original comparative analysis of the performance of the chocolate supply chain conducted by the adoption of a two-step traceability and quality assessment approach. It is made up of a monitoring activity of real shipments (1) and a laboratory simulation process (2) to predict the expected quality levels of the product at the place of consumption.

The literature presents several studies and reviews on the food supply chain including different problems and issues, e.g., processing and manufacturing (Tufano et al., 2018), logistics and operations (Accorsi et al. 2017, 2018a; Gallo et al., 2017), storage (Accorsi et al., 2018b), packaging and containment (Manzini et al., 2017; Accorsi et al., 2014), traceability, and sustainability (Li et al., 2014; Dabbene et al., 2014; Govinand, 2018). On the other hand, many contributions focus on food science technologies, products, and processes, but only a few studies simultaneously deal with food science issues and expertise (focused on food processing, food quality, and food safety) mixed with engineering problems and decisions (focused on traceability, physical and environmental stresses, packaging and containment solutions, storage and warehousing, transportation, etc.).

A recent study presents a literature review of agri-fresh food supply chain quality (Siddh et al., 2017). It demonstrates that research publications on quality issues in the food supply chain are increasing, empirical investigation is speedily increasing, and about 13% of publications deal with logistic management. The interest in the integration of food quality and logistics is growing more and more.

Valli et al. (2013), Manzini et al. (2014), and Ayyad et al. (2017) present some significant applications of the proposed quality assessment system applied to the shipment of Italian edible oils in the presence of different containment and packaging solutions. Manzini et al. (2014) also conduct a life cycle assessment (LCA) on the selected and simulated containment solutions, moving attention simultaneously onto (1) environmental impacts, (2) logistic efficiency, and (3) quality/safety issues. The closed-loop simulation system developed by the Food Supply Chain Center and illustrated in this chapter was also adopted to conduct accelerated life test analyses (ALTA) of different food packaging systems (Manzini and Accorsi, 2013; Manzini et al., 2017), demonstrating the effectiveness of a multiscenario simulation approach to support the robust design of a packaging system when subject to different stresses during storage and transportation.

This paper presents an application to temperature-sensitive and high-value food items such as chocolate. A comprehensive and recent review on the state-of-the-art of chocolate supply chains is presented by Gutiérrez (2017). He affirms that "the ingredients added and the methods used in the various stages of postharvest processing and chocolate manufacture, all have an effect on the finished product." He deeply illustrates the factors influencing the quality of chocolate from the processing of dry cocoa beans to the final product manufacture (kneading, laminating, conching, intermediate storage, tempering, molding, cooling, shaping, and packaging) passing through ingredient definitions. He presents all processing and logistic activities in the chocolate supply chain from farm-to-fork, including storage and transportation issues of the final products. Only a brief discussion on packaging, storage, and distribution is presented by Gutiérrez (2017). The aim of a robust control system is to avoid fluctuations in temperature during storage of the final product, as these can accelerate the formation of an opaque appearance and fat bloom, and during transport in accordance with common practices of using refrigerated shipments.

The literature on packaging and containment solutions and technology to control the quality of final chocolate products during storage and shipment is still poor, and there are no discussions of distribution and logistics issues with the simultaneous investigation and analysis of the impacts on the product safety and quality (Armstrong, 2008; Ntiamoah and Afrane, 2009; Gutiérrez, 2017).

Buettner (2017) defines flavor as "the set of olfactory, tactile, and kinesthetic perceptions that allow a consumer to identify a food, and to establish the degree of pleasure or dislike that it brings" and it is determinated by parameters such as aroma, taste, shine, snap, melting, and mouth feel. Quality parameters of chocolate products are discussed by De Clercq et al. (2017) and Gutiérrez (2017).

In particular, De Clercq et al. (2017) analyze the cocoa butter equivalents (CBE) functionality when used in chocolate products. CBE was introduced in chocolate products as alternatives to the cocoa butter (CB) for the following main reasons: limited supply, high demand, and fluctuating prices. But another important issue that distinguishes CB from CBE is the melting behavior and properties. Chocolate products are solid at ambient temperature (20–25°C) and on ingestion (oral temperature is 37°C) melt to undergo dissolution in oral saliva (Afoakwa et al., 2007).

Afoakwa et al. (2007) presents a review of factors influencing the rheological qualities of chocolate. Moreover, some recent studies present models and investigations of chocolate rheology during processing activities, e.g., extrusion (Engmann and Mackley, 2006; Saputro et al., 2017).

All these studies demonstrate that ingredient composition is the most effective way to modify the rheological behavior. In addition, De Clercq et al. (2017) demonstrate that, from sensory analyses with final consumers, the differences between commercially available CB and CBE products are very hard for humans to detect. Several contributions on CBE-based products and other alternatives (e.g., hard palm midfraction (PMF) and/or other exotic fats) replacing the cocoa butter are presented in the literature (De Silva Souza and Block, 2018). Discussions on the environmental impacts and human health impacts when using substitutes for cocoa butter are still open (Khatun et al., 2017; Xiusong et al., 2017).

In addition, De Clercq et al. (2017), Afoakwa et al. (2007) and other recent and not so recent research contributions do not discuss the melting point and the quality effects due to shipments using and not using temperature-controlled containment solutions. Chocolate usually melts down at 33.8°C due to liquefaction of solid cocoa butter (Lillah et al., 2017).

As a result of this review, two basic questions have not yet been investigated:

1. What happens in storage and shipment activities of final chocolate products?
2. What about fine chocolate products, i.e., those not using CBE and other special ingredients?

This chapter deals with CB chocolate products and regional shipments using unrefrigerated (i.e., reefer) containers with a focus on the quality of the product on the table of the final consumer. In particular, reefer containers are used when the climate conditions seem favorable, i.e., not critical due to high temperatures.

A major discussion on the so-called “heat resistant chocolate (HRC)” animates the literature (Stortz and Marangoni, 2011, 2013; Lillah et al., 2017) in response to the necessity of producing chocolate that resists deformation at temperatures above 40°C, e.g., in tropical climates (“high critical climate”). Lillah et al. (2017) identifies three different approaches to HRCs: addition of an oil binding polymer (1), enhancement of network microstructure (2), and the use of fats having high-melting points (3).

In particular, the addition of organic compounds like ethycellulose (EC) represents an effective strategy for HRC mass production and distribution in tropical locations or during hot months of the year. The quality effects on chocolate products at such temperatures is still an open question.

A recent study on the chocolate supply chain is presented by Recanati et al. (2018). They illustrate the results of an LCA of an Italian chocolate from cocoa cultivation to packaging end-of-life where the distribution contribution is significant and quantified according to an average distance assumed equal to 150 km.

Another two additional basic questions that emerge and animate this chapter are the following:

3. What happens to the quality of a CB product not moving in a high critical climate (e.g., tropical over 40°C)?
4. What are the effects on the quality of a final product due to a regional, i.e., not international, distribution?

In response to these questions, the assessment methodology applied in this chapter is focused on the quality of products when subjected to “short regional shipments,” e.g., from Italy to Italy, but significantly larger than 150 km. The selected shipments discussed here are around 1000 km long. We demonstrate that the quality of fine chocolate at the customer point of demand is a very important issue even if the shipment is relatively short and does not experience tropical or arid climates. The effects generated by the uncontrolled environmental temperature can be unrepairable and can compromise the products on the table of the final consumer. For this reason, it is necessary to support purchasing, processing, manufacturing, logistics, packaging, and containment decisions, transportation, etc., in agreement with guidelines and best practices that result from past experience and simulation activities.

The remainder of this chapter is organized as follows: Section 2 illustrates the closed-loop control system and the adopted thermostatic chamber; Section 3 introduces the case study presenting selected products and shipments; Section 4 illustrates the sensory analysis; Section 5 presents and discusses the results obtained by the comparative and competitive analyses; and finally, Section 6 summarizes the conclusions and proposes suggestions for further research.

2 Closed-loop control system and thermostatic chamber

This section proposes a closed-loop control system intended for the simulation of environmental conditions experienced by food products and packages along the supply chain. It is made up of two steps that correspond to two subsystems: (1) the monitoring step, whose aim is to measure the shipment conditions, collecting data on-field, about the most critical parameters (e.g., temperature and humidity) potentially affecting the product safety and quality, and (2) the simulation step, to experimentally assess the combined impacts of such parameters on the product features during the life cycle.

Fig. 1 illustrates this ex-post traceability system, which differs from real-time control systems as are largely discussed in past research works (Accorsi et al., 2016). In fact, this monitoring system protocol is based on inserting data loggers within the product primary, secondary, and/or tertiary package, gathering the data loggers at the end of the selected logistic process (e.g., a storage and/or a distribution activity), and collecting the distribution profile data. An ex-post detailed analysis of the collected data identifies the most critical steps of the supply chain and aids operative decisions (e.g., packaging improvements, containment decisions, etc.) to improve the control of the food safety and quality. The simulation step ("virtual shipment" in Fig. 1) gives the decision maker the opportunity to compare the performance of different supply chain configurations in a what-if multiscenario environment. In particular, a selection of key performance indicators (KPIs) can be evaluated thanks to chemical and sensorial analyses conducted on simulated products.

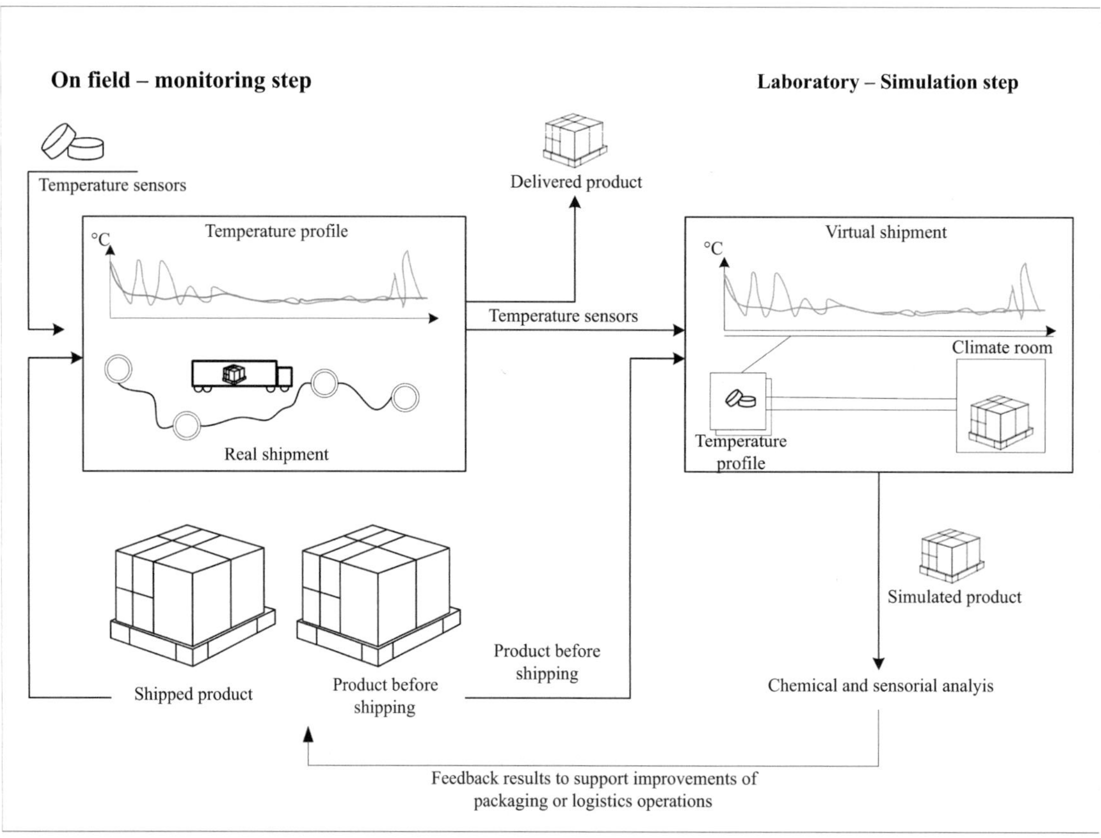

FIG. 1 Two-step closed-loop and ex-post traceability system. *(From Accorsi, R., Ferrari, E., Gamberi, M., Manzini, R., Regattieri, A., 2016, A closed-loop traceability system to improve logistics decisions in food supply chains: a case study on dairy products. In: Advances in Food Traceability Techniques and Technologies. Elsevier Ltd. doi:10.1016/B978-0-08-100310-7.00018-1.)*

The physical system developed at the Food Supply Chain Center and used to simulate in-lab the selected shipments is made of a 1800 × 1800 × 2000(h) mm steel climate chamber (part 1 in Fig. 2A) equipped with the auxiliary devices necessary to measure and control the temperature and relative humidity.

The actuators and components of the system are:

- heater, made of a 2kW power electric resistor;
- cooler, made of a 0.9kW vapor compression refrigeration cycle using R404A refrigerant fluid;

- ultrasonic humidifier with capacity 1.2 L/h. The electric power of this component is 0.08 W;
- deliquescence dehumidifier with calcium chloride salts. It is devoted to reducing significantly the humidity within the chamber in a few minutes;
- electric dehumidifier, 0.2 kW and capacity to reduce humidity equal to 10 L in 24 h;
- a tank of distilled water.

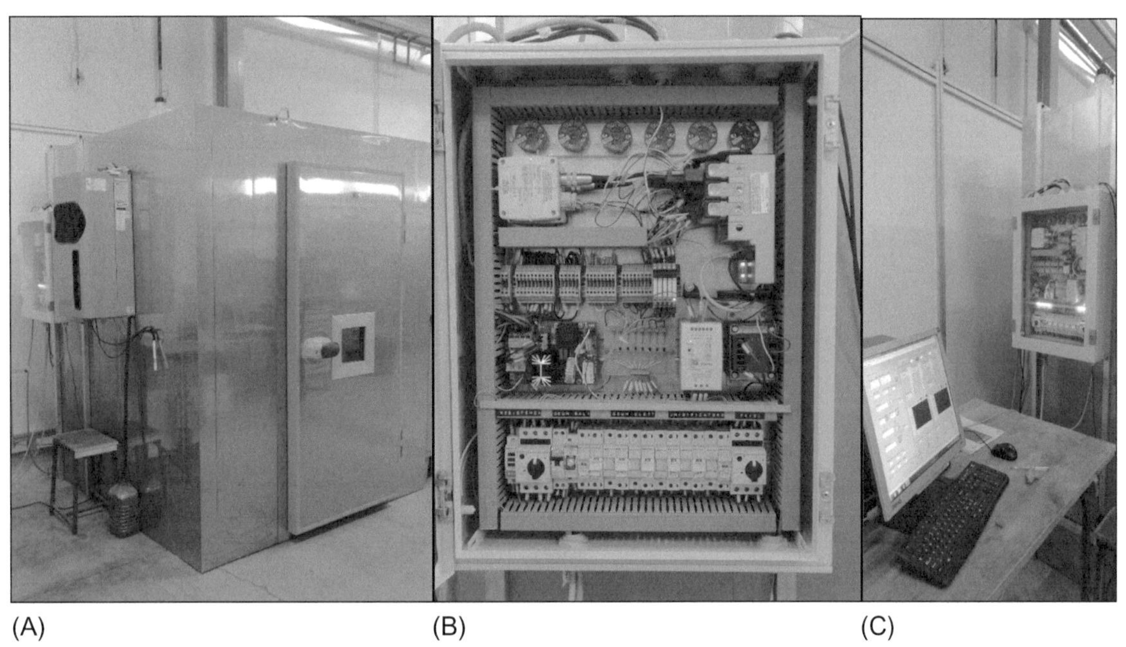

FIG. 2 Thermostatic chamber system. Components including the cooler (A), power panel (B), and control panel (C).

All devices are controllable both in on/off mode and by modulating their rated voltage through external power controllers according to a proportional-integral-derivative (PID) control system. Fig. 2A) shows the chamber with the cooler, the power system and the control panel. Air flow circulation inside the chamber is due to fans.

Fig. 2B and C shows the power and control panel devoted to controlling the system in agreement with an input target profile of temperature and humidity. The LabView hardware modules are at the top and the right of the picture, while the power units are at the bottom-right corner.

Finally, the temperature and relative humidity measurement is conducted through a commercial integrated sensor including a PT100 resistance transducer with $\pm 0.3°C$ accuracy and a capacitive humidity sensor with $\pm 2\%$ accuracy.

To simulate a generic profile of temperature and humidity in agreement with the previously illustrated closed-loop approach, a customized LabView virtual instrument (VI) has been designed and implemented. This tool allows the user to import the temperature and relative humidity profiles to track and to physically control the sensors and actuators installed into the thermostatic chamber. It further allows selecting of the control strategy. Fig. 4 presents a snapshot of the LabView control panel.

The key modules implemented in the interface are as follows:

- Temperature and relative humidity setpoint load and preliminary check (yellow section, named "set point profile" in Fig. 3);
- Real-time temperature and relative humidity field acquisition (blue section in Fig. 3);
- Strategy coding and implementation (red section in Fig. 3);
- Actuator control and command (fuchsia section in Fig. 3);
- Field data tracing and saving (green section in Fig. 3);
- Monitoring and danger detection.

The setpoint profiles for both variables are visible in the yellow box. The green box includes the input and output paths for data load and store. The control parameters and the run setting options are in the red boxes, while the monitored conditions

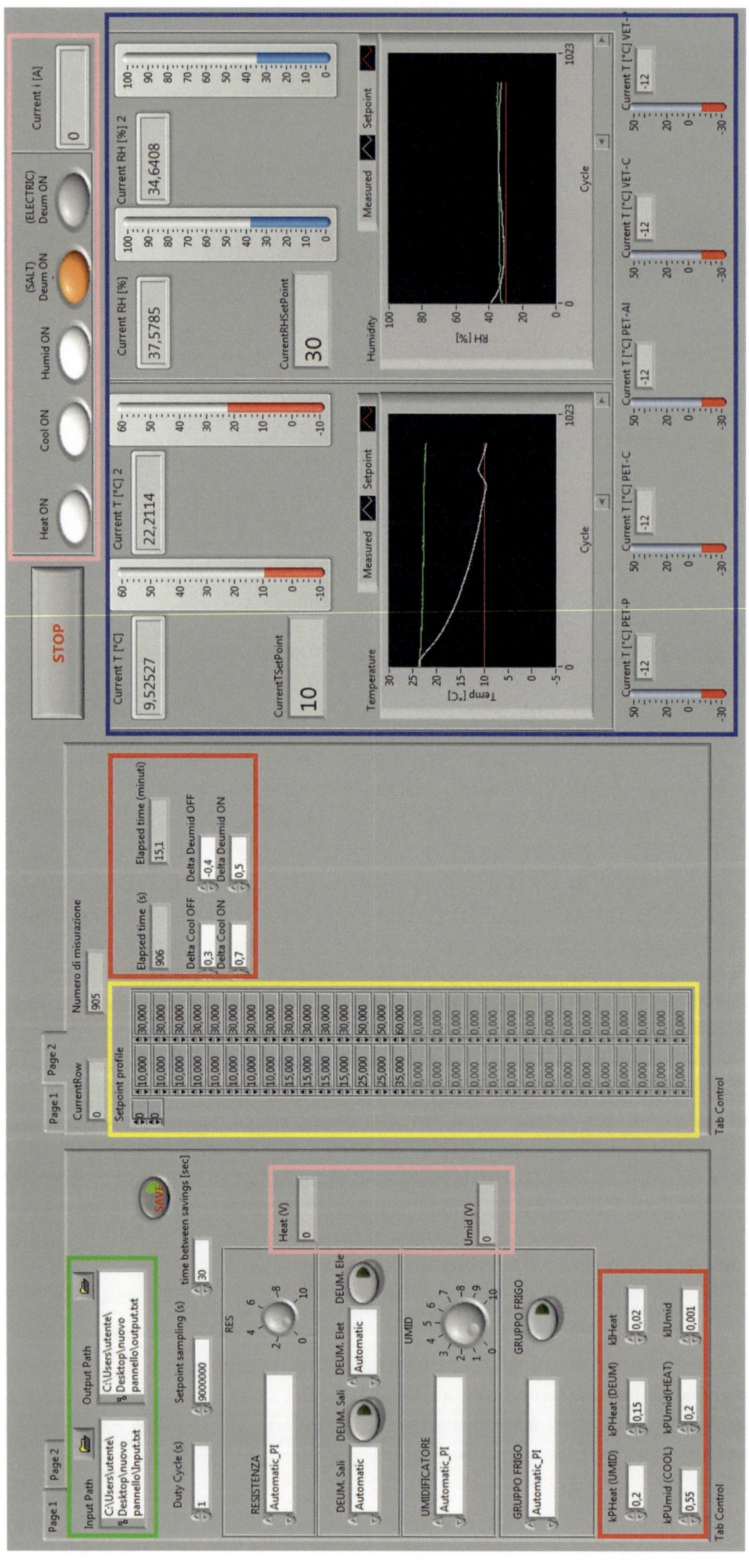

FIG. 3 LabView interface for food shipment laboratory simulation.

are visible in the blue boxes in both numeric and graphic representation. Finally, the fuchsia box includes data about the actuator state, i.e., on/off condition, and the control signals coming from the control strategy.

3 Case study: Products, monitoring step and shipments selection

This section illustrates the selected fine chocolate products and the selected trends of stress temperature simulated in the previously illustrated closed-loop system. The products are the well-known Gianduiotto and Cremino (see Fig. 4A and B, respectively) and a special product named Scorza (see Fig. 4C). This third selection comes from the wrinkled and irregular design of chocolate similar to tree bark. The first recipe dates back to 1832, when chocolate was known only as a drink. The careful selection and mixing of the finest cocoa powders give to this product a unique crumbly mouthfeel and vintage aroma. For this reason, the Scorza has been selected for this simulation process.

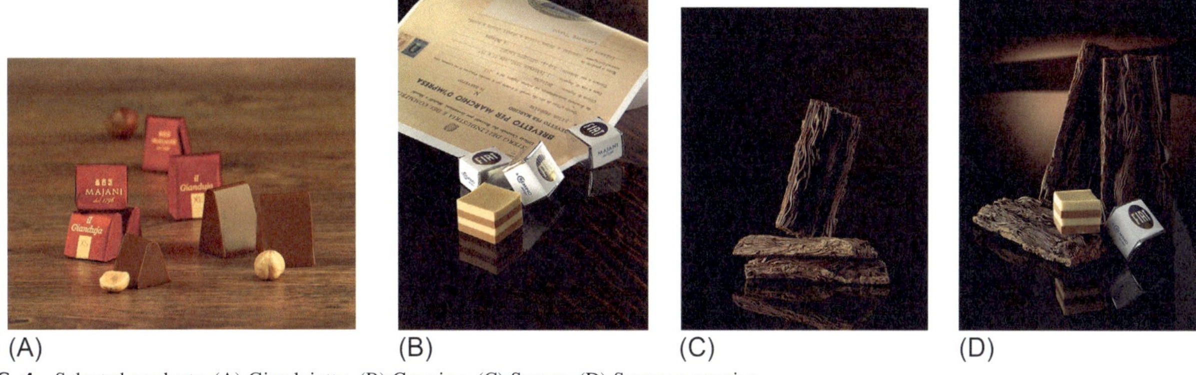

FIG. 4 Selected products. (A) Gianduiotto, (B) Cremino, (C) Scorza, (D) Scorza e cremino.

A campaign of monitoring regional and international shipments executed in different periods of the year from the Italian production plant to different points of demand has been conducted. Two critical shipments from a large sample of monitored orders were selected. They were executed in September and October from the north of Italy towards the south. They are critical because in this period of shipment the mode of transportation is a nonrefrigerated truck, i.e., one not equipped with a temperature control because it seems unnecessary at this time of year, unlike in other periods of the year such as summer. However, the climate was extremely and unexpectedly hot in the selected period of time, especially in the south of Italy. Fig. 5 (shipment from Bologna to Messina) and Fig. 6 (shipment from Bologna to Trapani) report the trend of the monitored temperature inside the truck containers.

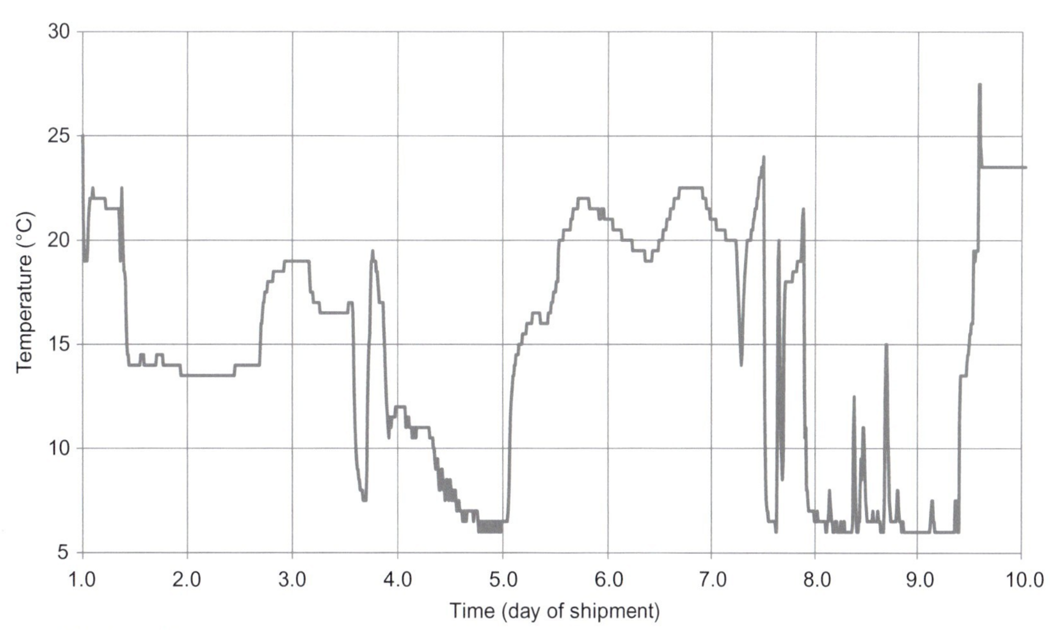

FIG. 5 Shipment 1, Bologna-Messina.

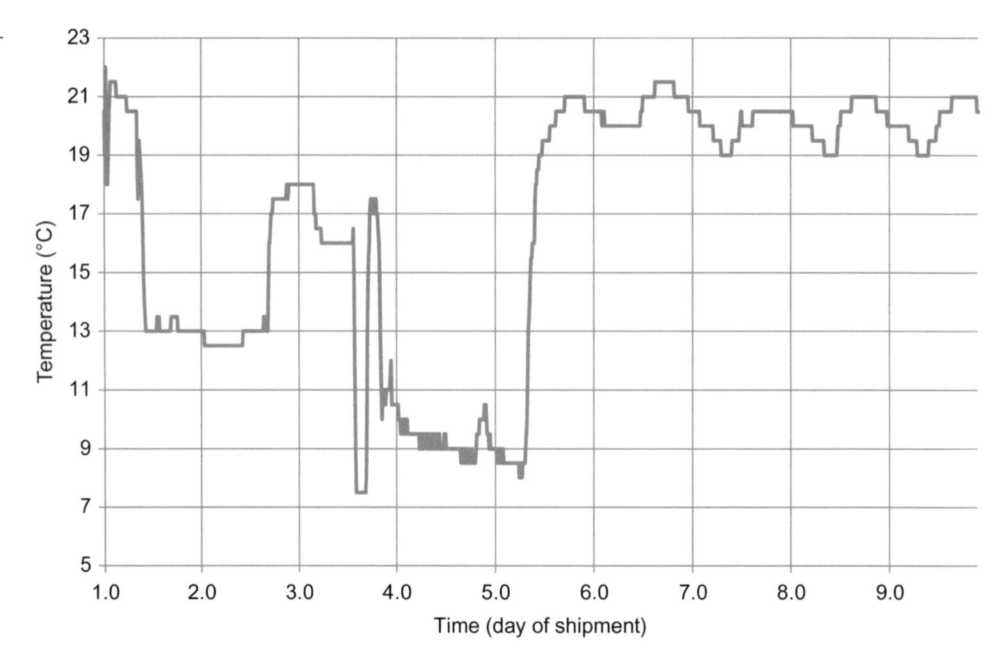

FIG. 6 Shipment 2. Bologna-Trapani.

4 Sensory analysis

A standardized sensory analysis procedure was adopted. Three pictures from this process are reported in Fig. 7. The sensory analysis is based on a panel of human assessors, named "experts." They measure several drivers grouped according to the five senses: sight, smell, taste, touch, and hearing. Each group is made of several descriptors subject to a quantitative evaluation. Table 1 reports the descriptors without any additional information about the weights adopted. Each descriptor has a finite set of options, i.e., alternatives. We decided not to detail further the structure of the test because that is not the object of this study. The test generates a summary score for each group, i.e., sense. The generic summary score is the weighted mean quantified on different "basic sensations." For example, considering the taste summary score, some of the basic sensations are acidity, bitterness, crunchiness, sweetness, primary sensations, etc.

FIG. 7 Some pictures of the sensory analysis conducted with experts.

TABLE 1 Descriptors and options in the standardized sensory analysis

Sense group	Descriptors	Options								
Sight	Homogeneity	Uniform	Striped	Cracked	Spotted	Patinated	With bubbles	With smear		
	Color	Blackish (tending to black)	Dark brown	Light brown	Tan (reddish brown)	Cinnamon	Butter	Ivory	White milk	Beige
	Shine	Bright	Highly polished	Polished	Translucent	Opaque				
Touch	Fluency	Smooth	Silky	Plush	Rough	Sandy	Granular			
	Ductility	Ductile	Somewhat ductile	Inflexible	Absent					
Hearing	Snap	Very clear	Clear	Quite dull	Obtuse	Absent				
Smell	Primary aromas	Aggressive	Penetrating	Evident	Delicate	Weak				
	Aroma intensity	Very intense	Intense	Somewhat intense	Not very intense					
	Secondary aromas (multiple choices)	Fresh fruit	Floral	Dried fruit	Milk	Licorice	Caramel	Toffee aroma	Ricotta	Smoked
		Toasted	Tobacco	Leather	Spices	Vanilla	Bread	Rancid	Wood	Different aromas
	Persistence smell	Very persistent	Persistent	Somewhat persistent	Little persistent					
	Balance flavors	Very balanced	Balanced	Somewhat balanced	Unbalanced					
Taste	Primary sensations	Aggressive	Penetrating	Full (solid)	Delicate	Weak				
	Intensity of taste	Very intense	Intense	Somewhat intense	Little intense					
	Lasting taste	Very persistent	Persistent	Somewhat persistent	Little persistent					
	Secondary aromas (multiple choices)	Fresh fruit	Floral	Dried fruit	Milk	Licorice	Caramel	Toffee aroma		
		Ricotta	Smoked	Toasted	Tobacco	Leather	Spices	Vanilla	Sugar	Bread
		Rancid	Wood	Burnt	Phenol	Honey	Different aromas			

	Sweetness	Exceptionally sweet	Very sweet	Sweet	Sweet enough	Little sweet	Absent	
	Bitterness	Exceptionally bitter	Very bitter	Bitter	Bitter enough	Little bitter	Absent	
	Acidity	Exceptional acid	Very acid	Acid	Acid enough	Little acid	Absent	
	Tactile sensation in the mouth	Grainy	Sandy	Round	Sharp	Buttery	Creamy	
	Astringency	Very astringent	Astringent	Astringent enough	Slightly astringent	Weakly astringent	Absent	
	Crunchiness	Hard	Very crunchy	Crunchy	Slightly crunchy	Absent		
	Fusibility	Very fusible	Fusible	Somewhat fusible	Not very meltable			
	Creaminess	Very creamy	Creamy	Not very creamy	Absent			
Final sensation	Overall balance	Harmonic	Somewhat harmonic	Somewhat discordant	Discordant			
	Global persistence	Extremely long	Very long	Long	Somewhat long	Short	Very short	Extremely short

In addition to the five groups there is another score named "final sensation." Finally, a "global score" summarizes all sensations and descriptors. The final sensation is a sort of summarized sensation perceived by the expert, while the global score is a weighted mean quantified on the five summary scores (assuming taste weight 73%, smell 20%, sight 4%, touch 1%, hearing 1%) and the final sensation score (assuming weight 1%). Consequently, the heaviest weights participating in this mean are taste and smell.

Given a generic score, the higher the value, the higher is the perceived level of quality.

This standardized approach can be applied to quickly compare different items of the same products subject to different stress profiles and in coherence with different logistic and packaging decisions. The comparative and competitive analysis is conducted using the evaluation of the time-zero stress profile and the comparison of performance of alternative scenarios. The time-zero usually represents the benchmark profile.

5 Results from simulation and comparative analyses

The previously selected shipments have been simulated in the controlled thermal system presented in Section 2. Given the selected chocolate products, three identical samples were compared: one subject to the stress run "simulation 1", i.e., the Shipment 1 (Bologna-Messina, Fig. 6); one subject to the stress run "simulation 2", i.e., the Shipment 2 (Bologna-Trapani, Fig. 7); and finally, one representing the so-called time-zero sample.

The time-zero sample, corresponding to nominal values of temperature of 14°C and RH (relative humidity) of 60%, was not subjected to any temperature stress and it represents a benchmark for the proposed quality assessment. The comparison consists of a chemical (microbiological) analysis (1) and a sensitivity analysis conducted by a team of five anonymous experts (2).

The microbiological analysis conducted on the time-zero sample and the simulated samples does not reveal any significant effects in terms of safety. Consequently, safe products are consumed by the customers at the place of consumption.

Regarding the sensitivity analysis, all the involved experts (21, 32, 43, 58, and 64 in the following summary reports and graphs) did not know anything about the nature of the samples. They executed the following three blind analysis runs:

- Analysis 1. Immediately after the conclusion of the simulation run. This corresponds to the virtual arrival time of the products at the destination (Messina and Trapani for Shipment 1 and 2, respectively);
- Analysis 2. After 2 weeks. This is a repetition of the first set of analyses in order to avoid censored data.
- Analysis 3. After 6 months in order to identify possible effects visible on mature, i.e., not fresh, products.

The comparative what-if analysis illustrated in this chapter involves four main factors: Product, Analysis, Expert, and Sample. Each factor counts multiple levels, i.e., options:

- Expert, i.e., different combinations of experts;
- Product, i.e., different chocolate products (Cremino, Gianduiotto, and Scorza);
- Sample, i.e., different shipments (Shipment 1, Shipment 2, and time-zero);
- Analysis, i.e., different times of analysis execution (1, 2, and 3).

The scores are reported in an interaction plot (Minitab Inc. Statistical Software) that is a double-entry analysis (i.e., 2 factors varying and 1 response analysis) that shows a performance index combining different factors and levels.

5.1 Taste and smell evaluations

Figs. 8 and 9 report the values of the summary score from the taste evaluation and the smell evaluation, respectively. The highest values correspond to the highest levels of quality in terms of taste and smell.

The following results can be reported:

- There are no significant variations comparing Analysis 1 and Analysis 2 in terms of taste and smell. Consequently, the expert behavior is coherent in the first and the second analyses;
- Considering the taste test, Cremino and Scorza products vary passing from the second to the third analysis, i.e., there is an effect due to the so-called "passing-time." In particular, the quality of the Scorza improves, while the quality of Cremino is reduced in terms of taste.
- Considering the smell test, both Cremino and Gianduiotto reduce their level of quality after 6 months. An insignificant effect is observed in the Scorza product.
- The effect on quality due to the stress of temperature is significant on the smell and insignificant on the taste.
- Time-zero analysis performs better for all products in terms of smell.
- The largest part of the group of experts (with the exception of 45) found the time-zero samples better. In other words, they identified the nontravelling samples, i.e., those not subject to any simulations of the stress of temperature.

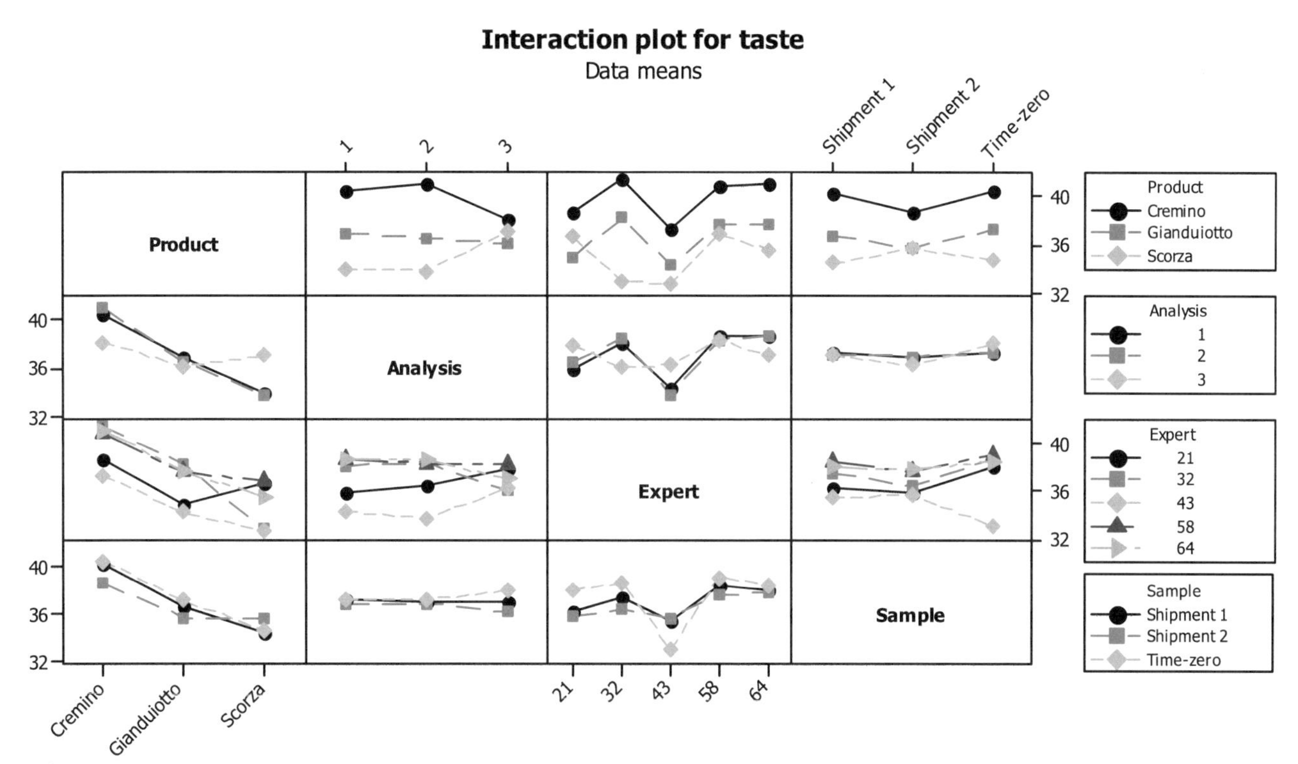

FIG. 8 Sensitivity analysis. Taste evaluation results, interaction plot (Minitab Inc. Statistical software).

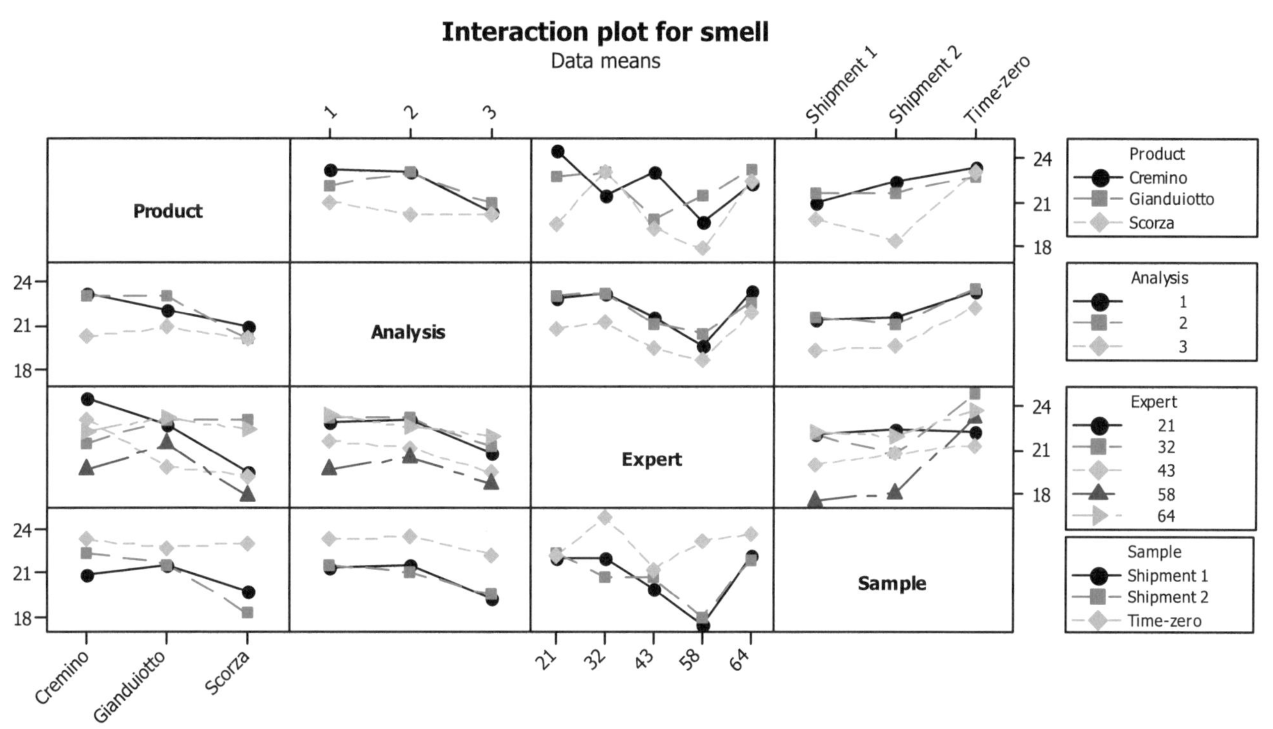

FIG. 9 Sensitivity analysis. Smell evaluation results, interaction plot (Minitab Inc. Statistical software).

5.2 Global score evaluation

Fig. 10 presents the previously defined global score quality summary result in an interactive plot and main effect plot limited to 3 of the 4 factors (Product, Analysis, and Sample), i.e., renouncing to the Expert factor. This performance index is measured in percentage, i.e., the highest quality level is when it is equal to 100%. The following considerations can be drawn:

- The time-zero sample performs better, i.e., the consumer prefers products that do not travel but are consumed at the place of origin/production;
- Gianduiotto does not change its level of global quality with passage of time;
- Scorza improves its global quality after 6 months, while Cremino loses quality.
- Insignificant variations between Shipment 1 and Shipment 2, but both perform worse than the time-zero sample.

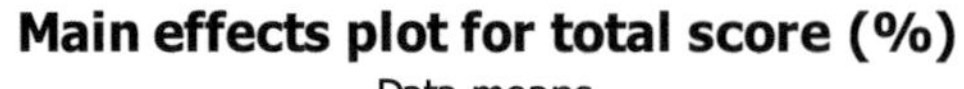

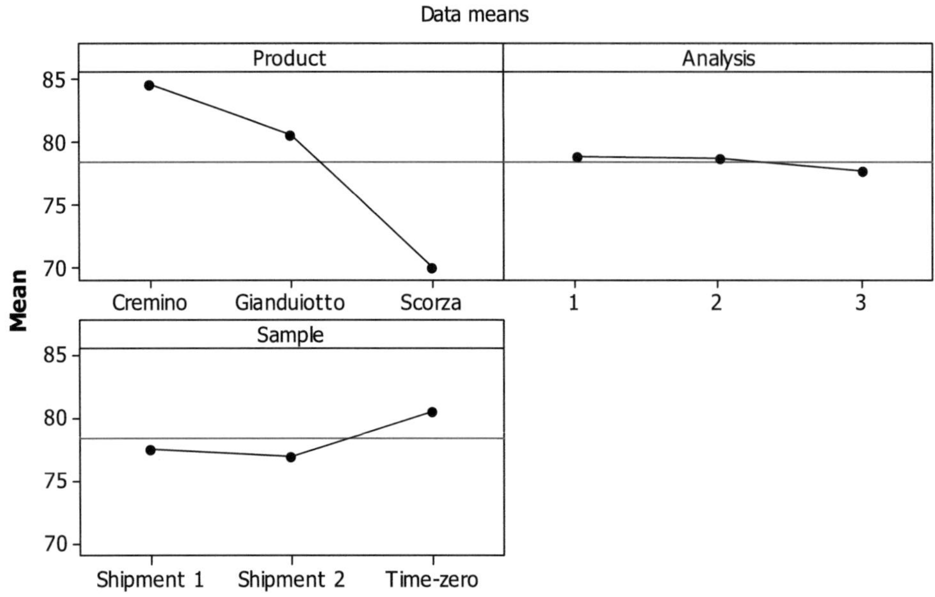

Interaction plot for total score (%)

Data means

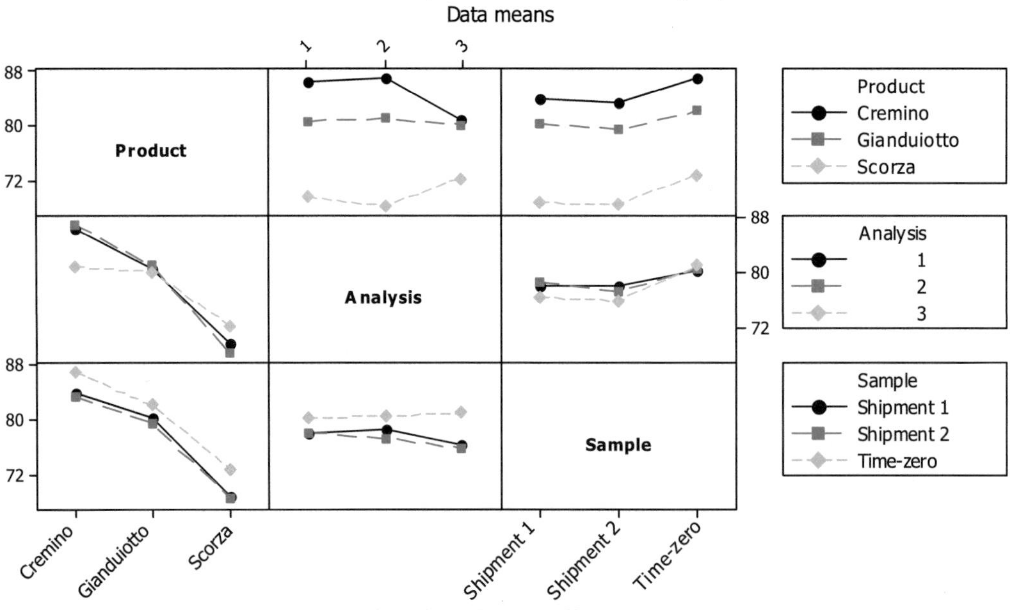

FIG. 10 Global score results [%], varying product, analysis, and sample. Main effects plot and interactive plot (Minitab Inc. Statistical Software).

5.3 Final sensation analysis

The effect due to time passing on Cremino when compared to Scorza is illustrated in Fig. 11 in terms of the previously defined "final sensation" score, whose range of values is [0–5]. Similar to all the other scores, the higher the score, the higher the perceived level of quality. These values confirm the improvement of the quality of Scorza in Analysis 3 if compared to Analysis 2 and Analysis 1. In particular, the time-zero sample performs better, especially in Analysis 3.

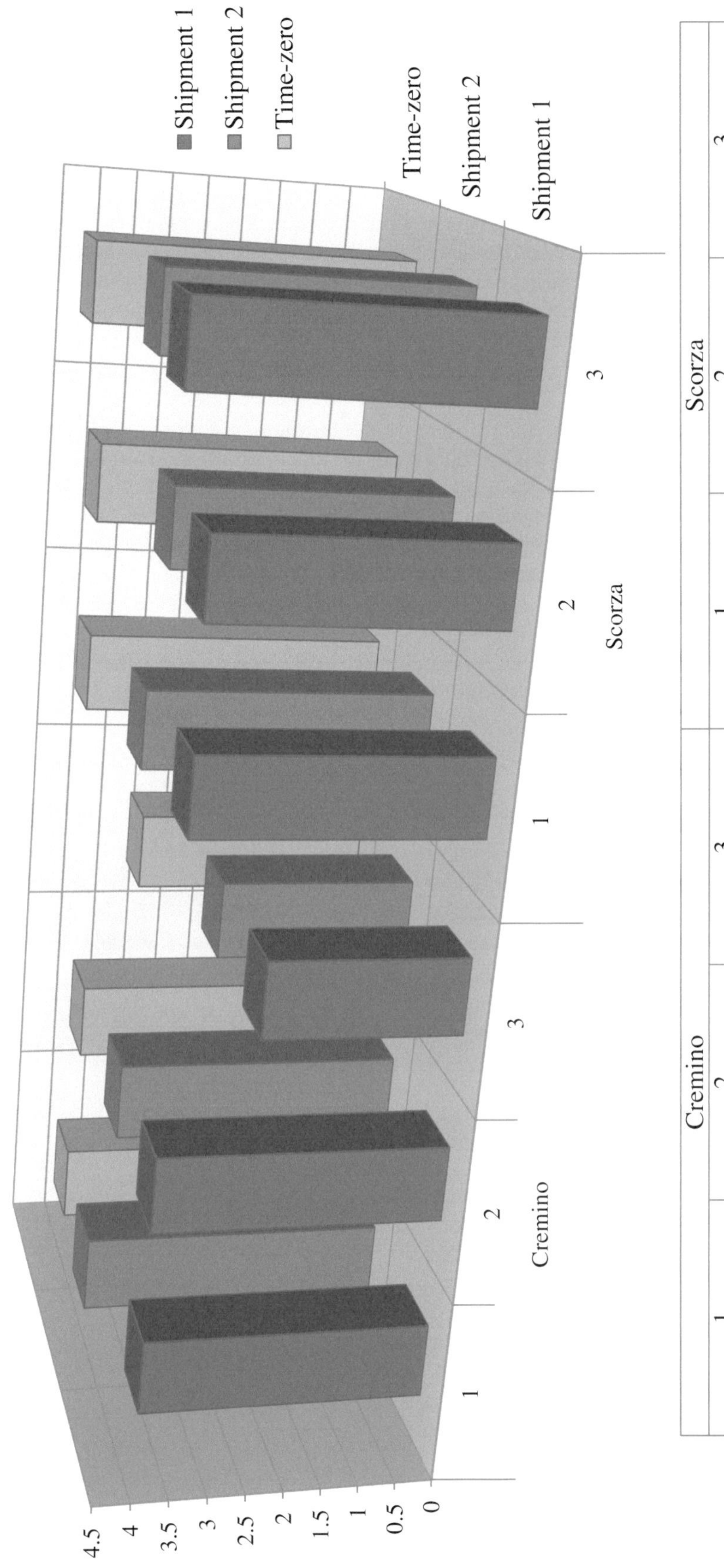

	Cremino			Scorza		
	1	2	3	1	2	3
Shipment 1	3.8	3.8	2.6	3.7	3.7	4.1
Shipment 2	4.1	3.8	2.6	3.8	3.6	3.9
Time-zero	4	3.9	3.2	4.1	4.1	4.3

FIG. 11 Final sensation score [0–5]. Comparison between cremino and scorza.

These quality perceptions might be generated by the different processes of oxidation, i.e., with different acceleration, involving the basic components (e.g., some fats, cocoa butter, etc.). The variation of temperature accelerates this process of deterioration of perceived quality differently.

6 Conclusions and further research

This chapter illustrates a two-step simulation-based and quality traceability and control system suitable for assessing the level of safety and quality of a perishable product and packaging system when subject to a portfolio of environmental and physical stresses, e.g., temperature and humidity. In particular, the monitoring activity gives the decision maker the opportunity to predict the performance of a product/package when it follows a specific target temperature profile.

Whether this profile has been collected on-field during critical storage and/or shipment activity, this system helps the decision maker to design a supply chain robust to unexpected conditions, e.g., variability and peaks of temperature. This chapter applies this quality assessment platform to a chocolate supply chain of an Italian producer of fine chocolate products.

In recent years, the literature has discussed the effectiveness of "heat resistant products" for production and/or distribution in tropical climates and summer seasons. But the environmental and nutritional effects of such modified products are still uncertain and critical.

The quality assessment illustrated in this chapter was conducted via simulation of the environmental stress of temperature monitored during two local shipments from Bologna to the south of Italy. The safety assessment demonstrates that there are no significant effects due to microbiological phenomena. But the quality assessment demonstrates that there is a significant effect on the perceived quality due to the unexpected hot temperatures monitored in the south of Italy in the month of September.

The quality of such products at the place of consumption can change. This change depends on the level of temperature stressing the product, i.e., on the climate conditions. The uncertainty of the climate conditions can be responsible for uncertainty regarding the quality of the product at the place of consumption.

The variability in quality perception also depends on the time the consumer waits, from the time of receiving to the time of consumption.

Further research is expected dealing with situations where international shipments are involved. The following questions should animate further research:

- What happens in cold (near 0°C) and very cold temperature (under −15°C)?
- What happens in the presence of significant variations of humidity?
- What about the role of packaging solutions, including the primary package and containment in multimodal transportation?

Other interesting issues are the assessment of the product life cycle including logistic and packaging decisions when the quality of the final product is controlled at the place of consumption. These are new issues of interest and research challenges for further studies in chocolate and other food supply chains.

Acknowledgments

The authors thank the Majani S.p.A Company (Bologna, Italy) for the fruitful collaboration on the development and application of the proposed closed-loop quality assessment. Special greetings to Dr. Maurizio Trombini, production manager of Majani S.p.A.

References

Accorsi, R., Manzini, R., Ferrari, E., 2014. A comparison of shipping containers from technical, economic and environmental perspectives. Transp. Res. Part D: Transp. Environ. 26, 52–59.

Accorsi, R., Ferrari, E., Gamberi, M., Manzini, R., Regattieri, A., 2016. A closed-loop traceability system to improve logistics decisions in food supply chains: a case study on dairy products. In: Advances in Food Traceability Techniques and Technologies. Elsevier Ltd. https://doi.org/10.1016/B978-0-08-100310-7.00018-1.

Accorsi, R., Gallo, A., Manzini, R., 2017. A climate driven decision-support model for the distribution of perishable products. J. Clean. Prod. 165, 917–929.

Accorsi, R., Baruffaldi, G., Manzini, R., Tufano, A., 2018a. On the design of cooperative vendors' networks in retail food supply chains: a logistics-driven approach. Int. J. Log. Res. Appl. 21 (1), 35–52.

Accorsi, R., Baruffaldi, G., Manzini, R., 2018b. Picking efficiency and stock safety: a bi-objective storage assignment policy for temperature-sensitive products. Comput. Ind. Eng. 115, 240–252.

Accorsi, R., Cholette, S., Manzini, R., Tufano, A., 2018c. A hierarchical data architecture for sustainable food supply chain management and planning. J. Clean. Prod. 203, 1039–1054.

Afoakwa, E.O., Peterson, A., Fowler, M., 2007. Factors influencing rheological and textural qualities in chocolate e a review. Trends Food Sci. Technol. 18, 290–298.

Ahumada, O., Villalobos, R., 2009. Application of planning models in the agro-food supply chain: a review. Eur. J. Oper. Res. 195, 1–20.

Armstrong, M.J., 2008. Inventory velocity in the Ottawa chocolate bar supply chain. In: IIE Annual Conference and Expo, pp. 1031–1036.

Ayyad, Z., Valli, E., Bendini, A., Accorsi, R., Manzini, R., Bortolini, M., Gamberi, M., Toschi, T.G., 2017. Simulating international shipments of vegetable oils: focus on quality changes. Ital. J. Food Sci. 29, 38–49.

Buettner, A., 2017. Springer Handbook of Odor. Springer, London.

Dabbene, F., Gay, P., Tortia, C., 2014. Traceability issues in food supply chain management: a review. Biosyst. Eng. 120, 65–80.

De Clercq, N., Kadivar, S., Van de Walle, D., De Pelsmaeker, S., Ghellynck, X., Dewettinck, K., 2017. Functionality of cocoa butter equivalents in chocolate products. Eur. Food Res. Technol. 243 (2), 309–321.

De Silva Souza, C., Block, J.M., 2018. Impact of the addition of cocoa butter equivalent on the volatile compounds profile of dark chocolate. J. Food Sci. Technol. 55 (2), 767–775.

Engmann, J., Mackley, M.R., 2006. Semi-solid processing of chocolate and cocoa butter modelling rheology and microstructure changes during extrusion. Food Bioprod. Process. 84 (2), 102–108.

Gallo, A., Accorsi, R., Baruffaldi, G., Manzini, R., 2017. Designing sustainable cold chains for long-range food distribution: Energy-effective corridors on the Silk Road belt. Sustainability 9 (11), 2044. https://doi.org/10.3390/su9112044.

Govinand, K., 2018. Sustainable consumption and production in the food supply chain: a conceptual framework. Int. J. Prod. Econ. 195, 419–431.

Gutiérrez, T.J., 2017. State-of-the-art chocolate manufacture: a review. Compr. Rev. Food Sci. Food Saf. 16, 1313–1344.

Khatun, R.K., Hasan Reza, M.I., Moniruzzaman, M., Yaakob, Z., 2017. Sustainable oil palm industry: the possibilities. Renew. Sust. Energ. Rev. 76, 608–619.

Li, D., Wang, X., Chan, H.K., Manzini, R., 2014. Sustainable food supply chain management. Int. J. Prod. Econ. 152, 1–8.

Lillah, M.S., Asghar, A., Pasha, I., Murtaza, G., Maratab, A., 2017. Improving heat stability along with quality of compound dark chocolate by adding optimized cocoa butter substitute (hydrogenated palm kernel stearin) emulsion. LWT Food Sci. Technol. 80, 531–536.

Manzini, R., Accorsi, R., 2013. The new conceptual framework for food supply chain assessment. J. Food Eng. 115, 251–263. ISSN 0260-8774.

Manzini, R., Accorsi, R., Ayyad, Z., Bendini, A., Bortolini, M., Gamberi, M., Valli, E., Gallina Toschi, T., 2014. Sustainability and quality in the food supply chain. A case study of shipment of edible oils. Br. Food J. 16 (12), 2069–2090.

Manzini, R., Accorsi, R., Piana, F., Regattieri, A., 2017. Accelerated life testing for packaging decisions in the edible oils distribution. Food Packaging Shelf 12, 114–127.

Ntiamoah, A., Afrane, G., 2009. Life cycle assessment of chocolate produced in Ghana. In: Appropriate Technologies for Environmental Protection in the Developing World—Selected Papers from ERTEP 2007, pp. 35–41.

Recanati, F., Marveggio, D., Dotelli, G., 2018. From beans to bar: a life cycle assessment towards sustainable chocolate supply chain. Sci. Total Environ. 613–614, 1013–1023.

Saputro, A.D., Van de Walle, D., Kadivar, S., et al., 2017. Investigating the rheological, microstructural and textural properties of chocolates sweetened with palm sap-based sugar by partial replacement. Eur. Food Res. Technol. 243, 1729. https://doi.org/10.1007/s00217-017-2877-3.

Siddh, M.M., Soni, G., Jain, R., Sharma, M.K., Yadav, V., 2017. Agri-fresh food supply chain quality (AFSCQ): a literature review. Ind. Manag. Data Syst. 117 (9), 2015–2044.

Stortz, T.A., Marangoni, A.G., 2011. Heat resistant chocolate. Trends Food Sci. Technol. 22 (5), 201–214.

Stortz, T.A., Marangoni, A.G., 2013. Ethylcellulose solvent substitution method of preparing heat resistant chocolate. Food Res. Int. 51 (2), 797–803.

Tufano, A., Accorsi, R., Garbellini, F., Manzini, R., 2018. Plant design and control in food service industry. A multidisciplinary decision-support system. Comput. Ind. 103, 72–85.

Valli, E., Manzini, R., Accorsi, R., Bortolini, M., Gamberi, M., Bendini, A., Lercker, G., Gallina Toschi, T., 2013. Quality at destination: simulating shipment of three bottled edible oils from Italy to Taiwan. Riv. Ital. Sostanze Grasse 90 (3), 163–169.

Xiusong, Y., Yahong, H., Chenjing, Y., Jiguo, H., 2017. The nutrition and food processing safety in palm oil. J. Chinese Inst. Food Sci. Technol. 17 (7), 191–198.

Chapter 15

Models and technologies for the enhancement of transparency and visibility in food supply chains

Fredrik Nilsson*, Malin Göransson* and Klara Båth[†]

*Packaging Logistics, Department of Design Sciences, Lund University, Lund, Sweden [†]Microbiology and Hygiene, RISE Research Institutes of Sweden, Gothenburg, Sweden

Abstract

With the increased pressure and challenges of economic, environmental, and social character, the need for innovations (including both the generation and adoption of innovations) that can be implemented in supply chains increases. A number of novel concepts focusing on intelligent logistics and packaging systems are being developed and tested in the food industry, all over the world. Several of these concepts predict quality and product safety of foods for use along the food supply chain (FSC) by the food industry, distributors, and retailers, as well as consumers. In this chapter, the focus is set on models and technologies related to increased transparency and visibility in FSCs for the purpose of lowering food waste, increasing food safety, and increasing overall resource efficiency. An overview of models and concepts for transparency with specific emphasis on food monitoring systems and technologies is presented, together with an in-depth field study of an industry case. The field study covers a whole supply chain in which all actors were provided with real-time data on time and temperature of a product from production until consumption. It is concluded that the use of new technologies holds great potential and huge value in collecting and sharing quality data. However, the main challenges are found in the business relationships where the risks and willingness to share information, i.e., of being more transparent, are the major hurdles in the development of sustainable food supply chains.

1 Introduction

A number of novel concepts focusing on intelligent logistics and packaging systems are being developed and tested in the food industry worldwide. Several of these concepts predict quality and product safety of foods for use along the food supply chain (FSC) by the food industry, distributors, ad retailers, as well as consumers. With such new concepts implemented in FSCs, real-time monitoring will enable the identification of abnormalities, e.g., excessive temperatures, as well as cooperative improvement efforts between supply chain actors. Based on such information, actions can be taken directly. Information can also be provided on documented cold chain data for changes of predicted quality and shelf life. In-house efficiency improvements are possible through the early relocation of products whose shelf life has changed during the management process. Food distribution controlled according to needs and actual quality opens up new business opportunities for both carriers and retail trade. This chapter focuses on the advancements and applications of new sensor technologies, including new insights into how actors in the supply chain can cooperate and utilize each other's information for a more sustainable society. Information is presented on how innovation, traceability, and transparency in the food industry can be used to reduce food waste, increase resource efficiency, and improve food security.

The rest of the chapter addresses the following areas:

- Introduction to the challenges related to food waste/resource efficiency and the need to monitor food quality in the supply chain.
- Overview of models and concepts for increased transparency and visibility.
- Overview of technologies.
- Industry case on cold chain monitoring.
- Value and challenges related to sharing information and providing increased transparency in FSCs.

Sustainable Food Supply Chains. https://doi.org/10.1016/B978-0-12-813411-5.00015-6

2 Resource efficiency and food waste challenges in supply chains

The challenge of sustainable development have been on the agenda for many years, but it is not until recently that they have become of strategic importance for companies. The interest and involvement both from stakeholders concerned about what a firm should do in terms of sustainable practices (Gonzalez-Benito et al., 2011; Gray, 2013), and consumers demanding more sustainable services and products as well as transparency in practices (Trienekens et al., 2012), have all influenced this change. In the food industry, a number of challenges exist in sustainable development, with food waste, food safety, and resource inefficiencies being the major ones. Globally, it is estimated that one-third of all the edible food for human consumption is wasted (Gustavssonet al., 2011), i.e., 1.3 billion tons of food per year. While households are found responsible for the largest amount of food waste (Jensen et al., 2013), supply chain practices cause large amounts of food waste, both directly and indirectly. The set-up of FSCs today, i.e., where retailers have the most power, often leads to cost pressure on suppliers as well as demands on implementing quality and sustainability standards without sufficient support. This leads to short-term solutions and causes unnecessary waste due to liability contracts and reimbursement agreements. For several of the food actors, the waste is handled financially with reimbursements, i.e., the cost of food waste is pushed upstream to the suppliers, while in terms of actual food waste the amounts are high but often unknown. However, with increased understanding of customer behaviors, more efficient and secured FSCs, and more effective communication and information between supply chain actors as well as with consumers, food waste both in the supply chain and among consumers could also decrease.

The segment of chilled food products includes many of the most important food products of the European daily diet, especially for people that live in the northern part of Europe. This segment covers the high value and high environmental impact of products like meat, fish, and poultry as well as bulk products like milk. In order to keep chilled food products safe and healthy, consistent low temperatures are required (Olsson and Skjöldebrand, 2008). The physical and microbial quality is highly dependent on the food product storage and handling conditions, i.e., an intact cold chain with minimal fluctuations of storage temperature. Hence, chilled food products are demanding to manage in supply chains due to their perishability, often short shelf life, and natural variety in quality (Aung and Chang, 2014; Mena et al., 2011). The cold supply chains' complexity increases with the many variants of products, different temperature levels on different products, the coloading of products, and consumer expectations regarding new product launches (Aung and Chang, 2014; Naturvårdsverket, 2013).

Most food products today are labeled with a barcode for identity and a production or use-by date for quality and traceability. While this method of providing product data is efficient, it also has a number of limitations due to the fact that it is static and does not take into consideration what happens to a product during its handling and use. The current date labeling system is also a major reason for returns and rejections at different points in the supply chain. Lindbom et al. (2014) estimate that over 50% of all food waste in the industry derives from expired "best before" dates. Furthermore, two-thirds of household food waste is still fit for consumption (WRAP, 2007). There are significant challenges in being able to detect any deficiencies and manage deviations safely without recalling larger batches, which is critical, especially for food producers and retailers, both for profitability and customer confidence. If the state of food products can be identified in a real-time feed, throughout the whole supply chain, including households, the quality of the food can be assured or appropriate actions taken when needed (e.g., targeted recalls).

Food producers state their products' shelf life, in terms of expiry date label, but nevertheless have no control over how the products are treated downstream in the supply chain. With an intact cold chain and correct handling, most products could in theory still be of high quality after the labeled best before date. Common examples of this are products like milk, which still can be drinkable up to 2 weeks after the labeled best before date. Also eggs, stored in the refrigerator, retain their high quality months after the expiry date has passed. In today's society, it seems that many consumers lack sufficient knowledge of food and food handling. Thus, many consumers seek their safety and reliability information from their purchased products, which some of them find in the date labeling system. The safety is, however, as already stated, somewhat misleading, since many consumers completely depend on the labeled date and consider it a health risk to eat food that has passed its "use by" date (Rahelu, 2009). A number of these consumers do not even trust the date labeling system, leading them to throw away food even before the labeled best before date has passed (WRAP, 2011).

3 Transparency and visibility in food supply chains

The need and demands for information from authorities as well as supply chain actors set the standard for information carriers in packaging. In several industries, barcodes have come to be obligatory in order to identify goods. However, even though barcodes are reliable and easy to use, they have a number of limitations, e.g., being passive, needing manual handling, and only providing limited information. Furthermore, barcodes can also easily be removed and replaced with new ones. The development of other solutions, i.e., auto-ID and other information-enriching solutions that can be integrated with the product and its packaging, has been ongoing during the past 20 years. A number of researchers and practitioners have proclaimed different potentials with radio frequency identification (RFID) and other auto-id technologies. Nonetheless, no widespread effects of RFID have yet been reported in grocery supply chains. However, with advances in sensor technologies and a growing attention to the Internet of Things (IoT), increased visibility and transparency is promising. Today many companies have implemented new solutions or are conducting pilot studies of different auto-ID solutions, e.g., Gillette, Volvo, and DHL. In these cases, packaging should be designed in order to function in any situation, whether the demand is for information in order to track and trace, to follow the temperature of a package in a cold chain, or to inform a customer of containment, origin, and handling tips. The latest advancements, both conceptually and technologically, provide clear opportunities for new ways to identify, measure, and follow in real time what happens to products in supply chains. From this information, necessary adjustments can be made to minimize unwanted losses of food or safety/quality issues that occur.

In terms of traceability, all types of food should be traceable through all stages of production, processing and distribution, where each party is responsible for tracing the food one step back and one step forward. Based on European legislation, traceability for the food industry is defined in Regulation (EC) No 178/2002 as "the ability to trace and follow food, feed, food producing animal or substance intended to be, or expected to be incorporated into a food or feed, through all stages of production, processing and distribution" (EU, 2002). The purpose of this is to secure that the complete history can be restored if needed, which is especially important if the food has been found contaminated. As a part of international standards, Moe (1998) and Olsen and Borit (2013) explain that the traceability definition is found in the International Standardization Organization (ISO) 8402 as "the ability to trace the history, application, or location of an entity by means of recorded identifications" (ISO, 1994).

A traceability system must support both tracking and tracing, where tracking is used to keep records of the product at each stage, and tracing is the process to identify the origin of a product, i.e., reconstructing the history of the data recorded by the tracking process (Pizzuti and Mirabelli, 2015). "Tracking is the informative process by which a product is followed along the supply chain keeping records at each stage [..]. Tracing is defined as the ability of reconstructing the history of a product, identifying its origin through the complexity of resources involved in its lifecycle." (Pizzuti and Mirabelli, 2015, pp. 17–18).

The identity of the product is crucial to conduct traceability within a company. The batch is the most common way to group a certain quantity of the same product, and it is identified by using some type of labeling with a batch number, unique and unrepeatable. This number is the clue to tracing during the processing and to connecting either upstream with the raw material or downstream with the finished product.

Related to traceability is transparency. Traceability for logistics firms is the track and trace services that allow a higher degree of visibility (Hultman and Axelsson, 2007). Doorey (2011) and Mol (2015) define transparency as disclosure of information. Besides information sharing within the supply chain, there is an increased demand for transparency from other stakeholders such as consumers and government (Carter and Rogers, 2008; Doorey, 2011). The potential benefits from transparency are that it can create business opportunities (Svensson, 2009), improve business processes (Carter and Rogers, 2008) and lead to a favorable reputation (Fombrun, 1996) for the firm. Another important aspect of transparency is information asymmetry, meaning that one party to a business transaction has more information than the other party, which makes it impossible to choose the product that is believed to yield greater value (Wognum et al., 2011). Further, in terms of corporate social responsibility (CSR), it is crucial to implement transparency in order to obtain a CSR policy that is sustainable, since a company that perform well in CSR cannot distinguish itself from other competitors without transparency (Dubbink et al., 2008).

One way to ensure food quality and at the same time enable transparency is the use of different types of sensors in the FSC. These sensors can function as adaptive shelf-life indicators in consideration of temperature changes, microbiological growth, quality of raw materials, spoilage indicators, metabolites, harmful/food poisoning bacteria, and the quality of the food handling depending on the sensor level of intelligence. In addition to the food quality assurance brought by the sensors, the introduction of sensors will also increase the traceability and information flow within the supply chain.

This type of technology has advantages compared to barcodes and date labels, especially for microbiologically sensitive products such as chilled food. If the actual state of products can be communicated in an easy way, the "safety scare" among consumers might decrease or maybe even disappear. A new and deeper trust can be built between the producers and the consumers that could result in a decrease of unnecessary food waste on the consumer side. However, in order to implement sensor solutions in FSCs, it is extremely important to find incentives for all the actors involved (Hellström et al., 2011). Recognized and acknowledged actor incentives will enable sustainable development for both the food industry and the environment.

4 Overview of concepts and models for increased visibility and transparency

In FSCs, traceability has become an integrating concept covering several industrial objectives (Coff et al., 2008) and business perspectives (van Dorp, 2002). Some of the central objectives for improved traceability are: risk management and food safety, control and verification, supply chain management and efficiency, provenance and quality assurance of products, and information and communication to the customer (Coff et al., 2008). Traceability has also become a risk management tool that enables governmental authorities to improve safety control over FSCs (Popper, 2007; Banterle and Stranieri, 2008; Thakur and Hurburgh, 2009). A functioning system for traceability is also necessary in order to gain increased visibility and transparency, both being concepts that are essential for competitive advantage and sustainable development. Transparency builds on a willingness to share, to be open, and to communicate. Carter and Rogers (2008), in their concept of sustainable supply chain management, put forward transparency as one of the four key facets. With reference to Hart (1995, p. 1000) who states that "Increasingly, local communities and external stakeholders are demanding that corporate practices become more visible and transparent [...] To maintain legitimacy and build reputation, therefore, companies may need to open their operations to greater public scrutiny," transparency is emphasized as a way to engage stakeholders to build confidence. This is especially central for FSCs due to the many food scandals, which have led to decreased consumer confidence and trust. However, transparency also includes risks related to visibility in the sharing of information in business relationships (Aung and Chang, 2014; Bosona and Gebresenbet, 2013). According to Storøy et al. (2013) the current status of traceability in FSCs is that many food producers have good, often electronic traceability systems internally, but the sharing of information (especially electronic exchange) among actors in the supply chain is insufficient. The diversity and proprietary nature of the actor's internal systems, together with an unwillingness to share information due to business risks, are the main reasons. Nonetheless, the need and demand for increased traceability as well as transparency have triggered a number of projects from which concepts and models been developed and published. In the following sections, some of the frameworks, concepts, and models are presented along with the main findings from these.

4.1 Traceability systems

A number of researchers have provided different aspects of traceability, which have evolved into different frameworks of traceability systems. Bendaoud et al. 2007, (p. 2) define a traceability system as a "system structured in such a way that it allows to totally or partially reconstruct the lifecycle of a given set of physical products." Moe (1998) provides in Fig. 1A system description that allows tracing of both the product and the activities (core entities) that should be included in a traceability system. Both of them have essential descriptors to be traced, such as type of product, type of activity, amount of product, time and/or duration of the activity. The information related to these must be collected or measured and stored.

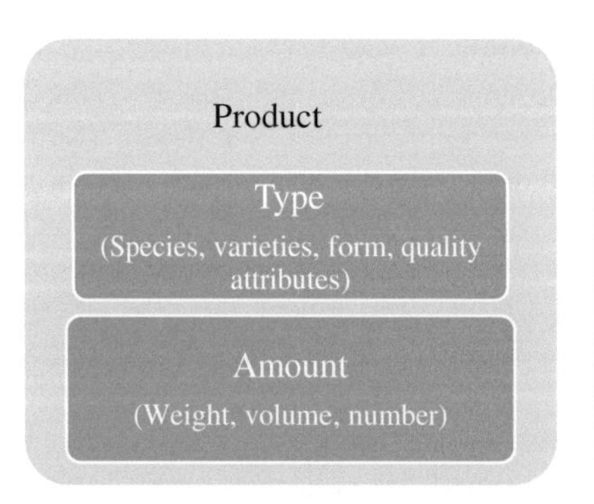

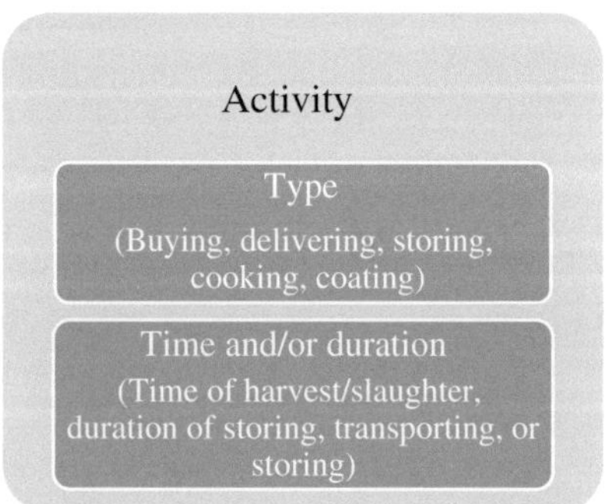

FIG. 1 Fundamental structure of a traceability system (Moe, 1998). Specific examples of the descriptors are presented in this figure including quality attributes and/or environmental factors.

The four pillars of the traceability framework proposed by Regattieri et al. (2007) coincide with Moe's point of view regarding product identification and data to trace the product's routing or processes (activities). In addition to these shared dimensions between the frameworks, the framework by Regattieri et al. (2007) is broader since it also includes tools to achieve the traceability. Technical solutions and operative resources, such as alphanumerical codes, bar codes, or RFID, are included for the proper functioning of a traceability system.

Another framework that focuses on the way traceability systems can be designed, assessed, and managed is presented by Bendaoud et al. (2012). Unlike the previous frameworks, this takes on a generic approach with specific emphasis on the surrounding environment for a traceability system. The essential aspects of the surrounding environment to understand and include are the supply chain actors, regulations, standards, products, government bodies, and internal beneficiaries. The framework is exhaustive and based on known systematic frameworks, such as the FAST (Function Analysis System Technique) (Bytheway, 2005). The surroundings are evaluated from three complementary points of view:

1. the functional point of view, by focusing on what the traceability system is expected to perform. The generic primary function is "to provide the beneficiaries of the traceability system with data on product traceability," where the beneficiaries are the entities that surround the traceability system and the data comprises the upstream (origin of input material), internal, and downstream traceability (destination(s) of output products).
2. the technical perspective, by focusing on how the traceability system should work, the technical functions needed to do so, and which technical criteria allow them to be assessed.
3. the informational point of view, which is considered in order to build a generic traceability data model; thus it is defined by the different data to be taken into account for the good performance of this system.

Nevertheless, a traceability system alone is not sufficient to achieve safety requirements in the supply chain; it should be seen as a complementary tool to quality safety activities (Bosona and Gebresenbet, 2013).

4.2 Traceability concepts and projects

The CATRENE PASTEUR project, a European research consortium of academic and industrial partners (2009–12) set out with the aim of tracking and displaying food quality and the actual remaining shelf life (Pasteur Project, 2012). With empirical studies on meat and fruit products, a wireless sensor platform and tags were developed during the project. The Pasteur sensor tag combined RFID with sensors measuring, e.g., temperature, humidity, and quality. According to the Netherlands Packaging Centre, NVC, which coordinated the project, all firms along the supply chain would gain benefits from the Pasteur sensor technology. For the producer it would be easier to be certain about the product quality. For freight forwarders the evaluation of correct temperatures during transport would be enabled. The retailers would be able to check and control the product quality on the shelves. Lastly, if sensor tags were placed on each package, the consumer would be able to see the product's quality at home with their cell phone (NVC, 2012).

The DynahMat project (dynamic shelf life for minimized food waste) set out to develop and provide road maps and implementation strategies for an open system solution that enables the integration and interconnection of different sensor solutions, food quality prediction, communication solutions, databases, business systems, and user applications. With specific focus on the static date-labeling system, the project has explored ways to minimize food waste and supply chain inefficiencies, and to contribute to a more sustainable food industry.

The DynahMat concept (see Fig. 2 for an illustration) entails an infrastructure for reliable, safe, and certified sensors attached to a food product (primary or secondary packaging) that provide data (position, time, identity, temperature, mechanical impact, etc.) to a cloud-based information system. The data is processed with, for example, prediction models for dynamic shelf life and other information that can be gained from the data and then communicated to the supply chain actors involved in the product flow, as well as the end customer/consumer through different digital interfaces (Web pages, mobile apps, business systems).

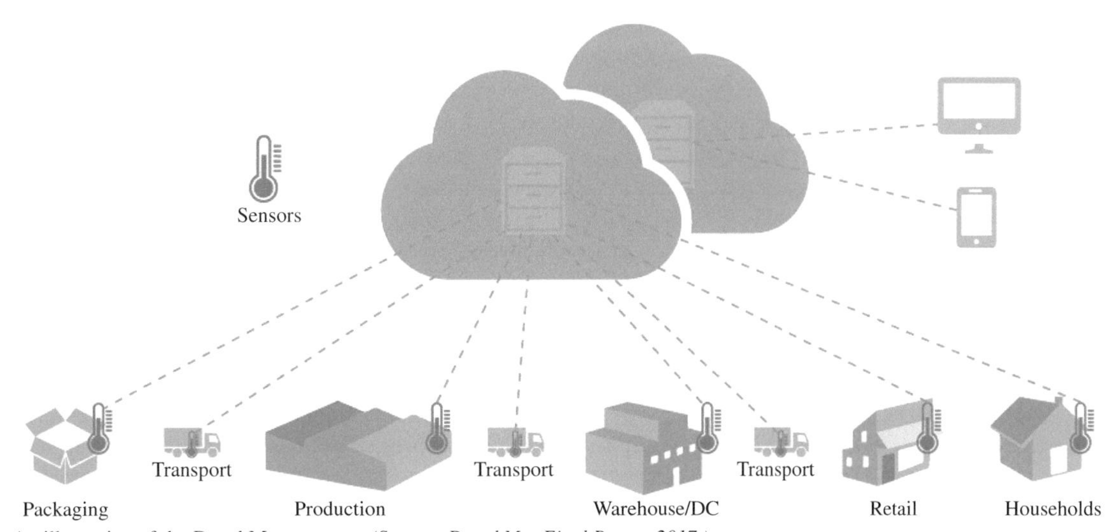

FIG. 2 An illustration of the DynahMat concept. *(Source: DynahMat Final Report 2017.)*

The main learning outcomes from the DynahMat project were that a secured, stable, and cold food supply chain enables longer shelf life for the products investigated as most product flows are below recommended temperatures. However, it would need real-time monitoring in order to identify abnormalities; i.e., if too high a temperature is registered it would enable actions to be taken directly as well as provide registration of documented cold chain data for updates on predicted quality and shelf life. A second key outcome relates to a supply chain vs. consumer solution, where it was found that in terms of implementation potential and food waste reduction a supply chain solution on secondary packages shows the greatest potential. Furthermore, the actual measuring of goods temperatures at every step of the supply chain for every delivery takes quite a lot of time. A central area of efficiency improvements was found from having a unified and automatic way of measuring temperature in the supply chain in terms of time reduction related to temperature controls.

5 Overview of technologies

5.1 Time and temperature indicators

The examples of time and temperature indicators (TTIs) have increased since 2000. Several companies provide TTI solutions used to track chill chains worldwide, e.g., ColdStream by SensiTech, SmartTrace by Smart Trace Online Monitoring, and MonitorMark by 3M, among others. TTIs are most commonly used by food producers and distributors who perform random or specific tests to monitor the quality of their cold chain distribution. These types of quality controls are mainly performed using puck-sized data loggers, such as the TempTale4 USB (TT4 USB), which contains microprocessor, battery, and memory with easily accessible data that is transferred through an integrated 2.0 USB. This can be connected to a server at the final destination (SensiTech, 2018). The increasing interest in rigid cold chain quality control has increased demand on the service and technology levels of TTI solutions. Smart Trace Online Monitoring has developed a transportation monitoring system that notifies time, temperature, and location of the products distributed with only minutes of online delay. The Smart Trace indicator is a small nonreturnable tag placed on the pallet level. After the tags are activated, they continuously upload time and temperature information to a gateway placed in the transportation unit. The data analyzed is customized to suit the needs of the sender (SmartTrace, 2018). This type of monitoring system can be valuable for food producers supplying perishable food products sensitive to temperature changes, such as meat, fish, seafood, and ready-to-eat products. Time and temperature indicators are mainly used between producer and wholesaler, i.e., during transportation, and few of these indicators can be seen or used by the retail staff or the end consumers.

5.2 Biosensors

Biosensors are physiochemical detectors that measure and/or categorize biological material (Rodriguez-Mozaz et al., 2004). Biosensors can detect specific microorganisms by containing active sites that bind to target molecules such as specific enzymes, DNA sequences, antibodies, or proteins (Zhou and Dong, 2011). Up to now, biosensors that target specific microorganisms only exist in the laboratory. However, upcoming biosensors within the food industry are promising, especially those that have the possibility of determining pathogenic microorganisms in food products (Rodriguez-Mozaz et al., 2004). Biosensors with more general biochemical mechanisms have come further in research development and there are several research projects around the world, developing biosensor concepts for food quality control (Pasteur Project, 2012; Fraunhofer Press, 2011). General biosensors detect and categorize chemical changes in food, such as conductance, pH, or gas composition. These chemical changes in food are due to biochemical breakdown or changes or to an increasing amount of microbial growth (Loessner et al., 2005). Examples of biosensors for food products are RipeSense (RipeSense, 2018), and the biosensor film developed by the Fraunhofer Research Institution for Modular Solid State Technologies EMFT (Fraunhofer Press, 2011).

5.3 Internet of things

The Internet of things (IoT) is a collective term for the development of connected devices, i.e., machinery, vehicles, goods, household appliances, packages, and other things equipped with small built-in sensors and computers. These devices can perceive their surroundings and communicate with other devices or humans, thus creating a situational behavior, and they can help to create smarter, more attractive, and more helpful environments, goods, and services. It is proclaimed that the IoT can create prosperity through efficiencies and innovations in various industries, as well as novel usages for consumers, e.g., health, facility management, and personal security and convenience. Verdouw et al. (2016, p. 129) state that, based on

IoT, "food supply chains can be monitored, controlled, planned and optimized remotely and in real-time via the Internet based on virtual objects instead of observation on site."

With sensor technologies connected to the Internet that register, monitor, and transfer food quality data as well as logistics data, advanced solutions that not only can track and trace but also reallocate, optimize, or manage deviations in ongoing product flows can be implemented. The latest advancements in IoT include the release of new standards. The Narrowband Internet of Things (NB-IoT) standards introduced in 2017 represent one example of technological development that enables new ways to implement traceability systems. The NB-IoT is connected to 4G and the coming 5G system for telecommunication and enables small, battery-efficient sensors that directly communicate through the base stations instead of being dependent on different gateways to send data to cloud servers. The communication advancements with NB-IoT implies that the substantial costs of investing in and installing gateways in each supply chain facility can be reduced.

5.4 Block chain technologies

One technology that has been given much attention during the last few years, which can offer both traceability and transparency, is blockchain technology (Yli-Huumo et al., 2016), which initially was invented to support the digital currency of Bitcoin (Nakamoto, 2008). The blockchain technology is designed to store data in blocks; placed in chronological order and based on a mathematical trapdoor (Brennan and Lunn, 2016) the data stored in the blocks is impossible to alter or remove (Nakamoto, 2008; Fanning and Centers, 2016). Copies of the chain of blocks (hence the term blockchain), and thereby the information, are distributed among the participants in the network (Tsai et al., 2016). The copies of the blockchain are then updated when a new block of information is added to the chain (Swan, 2016). While initially research conducted related to blockchains has focused on digital currencies (e.g., Bitcoin) (Yli-Huumo et al., 2016), the increasing usage in other industries to handle transactional information is growing, especially as a recordkeeping technology (Lemieux, 2016). The irreversible data-storing technology that blockchain enables has made the food supply chain industry an interesting application area (Tian, 2016), where the technology could support traceability and thereby achieve transparency (Hancock and Vaizey, 2016).

From a consumer perspective, blockchain technology could open up new possibilities for the food supply chain actors. By reading a simple QR code with a smartphone, data such as an animal's date of birth, use of antibiotics, vaccinations, and location where the livestock was raised can be presented and conveyed to the consumer in a reliable way. For food suppliers a traceability system based on blockchain technology would enable source information about the origin, condition, and movement of food, and contaminated produce could be quickly traced. According to IBM, a major opportunity with blockchain technology is to provide transparency across entire business ecosystems. They have announced blockchain collaboration projects with major food actors, including Nestle, Unilever, and Walmart. Tian, 2017 argues that the centralized traceability systems of today are monopolistic, opaque, and asymmetric, which could cause trust problems. The author (Tian, 2017) argues that the decentralized set-up of blockchain technology together with IoT applications shows great potential for increased transparency and openness among supply chain actors, as well as in consumer relations. However, it is also proclaimed that the technology and its applications are new and further developments are needed.

6 Industrial field studies—Putting concept and models into reality

In order to increase understanding and explore the "real" challenges as well as opportunities with new concepts and models for increased tractability and transparency, a number of field studies have been performed (Göransson et al. 2018a). In this section, a specific study within the DynahMat project is presented, based on two field tests in which all actors had real-time information about the time and temperature of a product from production until storing and consumption in a household. The field studies were performed during the summer of 2016 (field test 1) and the summer of 2017 (field test 2). The supply chain scope covered production until consumption in the home of a consumer. The outside temperature during these periods ranged between 17°C–26°C during the daytime.

The product used in the field study was smoked sliced ham, consumer packed in a modified atmosphere and distributed in plastic return crates stacked together on returnable EU-pallets (see Fig. 3). The printed shelf life was 2016-07-03, i.e., 25 days for field test 1 and 2017-07-26, i.e., 27 days for field test 2. The time range of the field tests started with the production day until up to 36 days after labeled shelf life on the package, i.e., 71–73 days. Perishable food products like smoked sliced ham are primarily spoiled by the growth of microorganisms forming compounds (both volatile and/or particle-bound) that cause off odor and/or taste. Thus, the number of microorganisms is indicative of the quality of the food product. However, this number does not necessarily correlate to the presence of pathogenic microorganisms and the risk of food poisoning.

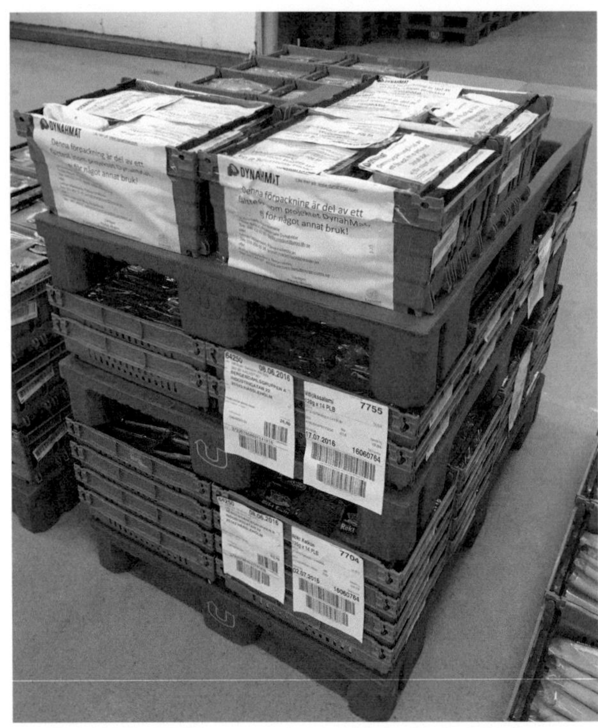

FIG. 3 Image of the primary packages of sliced ham and the return crates (secondary packaging) they were distributed within.

For the field tests, Bluetooth Low Energy sensors (nRF51822, Nordic Semiconductor) were used for measuring the temperature in the FSCs (20 sensors per test). The sensors were placed on the outside of the primary packages in close connection but not on the content. The temperature history being monitored is thus not exactly that in the proximity of the microorganisms, which affects their growth. However, based on other studies (Göransson et al., 2018a) the difference of product surface temperature compared to the surface of the ham is small. The temperature measurements were continuously broadcasted by the sensors and collected by Sony Xperia mobile phones every 10 minutes. The mobile phones transmitted the sensor data, including time stamps and position data (GPS), to a web server, where the data were stored in a relational database and were made available in real time through a Web interface. The data could be retrieved anytime via a PHP-based application and imported to a data analysis tool (e.g., Excel). The calibrations of all sensors were controlled in a thermal incubator and data adjusted based on the calibration results.

The shelf life of the product was predicted from the temperature sensor data using models that first predict microbial growth, which was thereafter secondarily decoded to food quality and remaining shelf life. Thus first the number of microorganisms were predicted from the logged storage temperature and then the days remaining of shelf life were calculated from the number of predicted bacteria and an assumed future storage at +8°C (maximum storage temperature for this product in Sweden).

A set of published candidate prediction models were identified, investigated, and validated concerning functionality for use in the prediction of dynamic shelf life (Kreyenschmidt et al., 2010; Mataragas et al., 2006; Devlieghere et al., 1998, 1999). The selected models all describe how the growth of spoilage bacteria of modified atmosphere packed (MAP) ham, lactic acid bacteria, is affected by storage temperature. Important issues addressed during validation were how well the growth rate of spoilage microorganisms corresponded between the predictive model and the actual product used in the field tests, as well as how initial status in terms of microflora concentration and composition influenced the prediction accuracy.

The validation storage studies were conducted at +8°C. In order to be generally applicable, in this study, assumed best- and worst-case scenarios of the ham product, i.e., before and after a deacidification process during a normal week of production, were compared. From the maximum specific growth rate values for bias (Bf) and accuracy factor (Af) for the selected models were calculated according to Ross (1996) (Table 1).

The model published by Kreyenschmidt et al. (2010) gave a slight underprediction of 13% (B_f), whereas the Mataragas et al. model overestimated bacterial growth at 18%. The model proposed by Devlieghere et al. (1998) gave a strong overprediction. The model by Mataragas et al. (2006) was selected as the most applicable model for the field studies. This decision was based on the fact that it is better, from a consumer point of view, to have an overprediction by the model, since the risk is minimized that the consumer will get a spoiled product even though the dynamic shelf-life labeling says that the ham is of good quality.

TABLE 1 Calculated bias and accuracy factor for modified atmosphere packed (MAP) smoked ham validation data vs. predicted model data at 8°C

	Kreyenschmidt et al. (2010)	Mataragas et al. (2006)	Devlieghere et al. (1998, 1999)
B_f	0.87	1.18	2.54
A_f	1.14	1.18	2.54

6.1 Field study set-up and process

The specific supply chain started at the production site (see Fig. 4) when the ham was sliced (product temperature at 0°C) and packed and placed for storage/dispatch. The packages were then picked up by a transport company and delivered to a wholesaler. After receiving the pallets, the wholesaler stored them (the pallets were split while crates were kept intact). When an order was received from a retail outlet, the number of ordered crates were placed on a new pallet (with other products) and placed for distribution. Another transport company then picked up the new pallet and delivered it to a retail outlet that placed them in storage, followed by placement on retail shelves in the retail outlet (i.e., the primary packages removed from the crates).

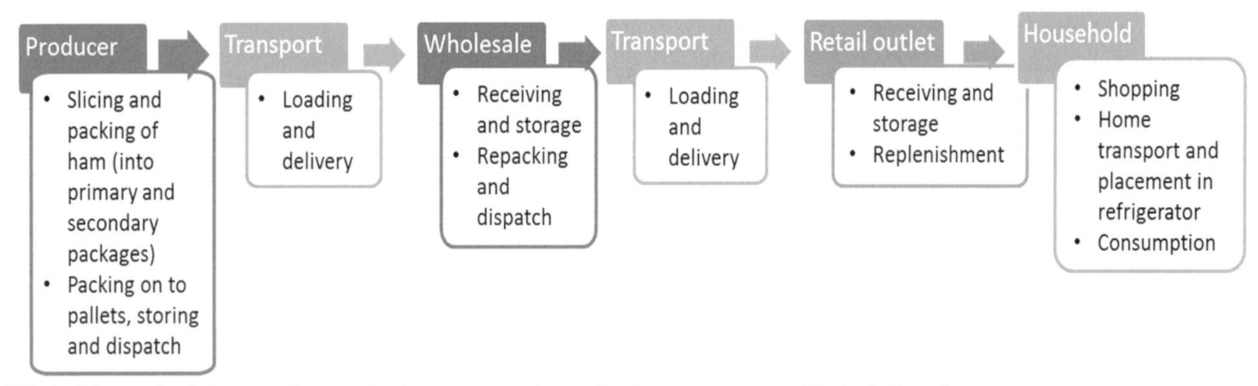

FIG. 4 The supply chain scope, from production to consumption, and main processes covered in the field studies.

At the retail outlet a consumer picked the product off the retail shelf, placed it in a shopping cart, and continued shopping. After paying for the groceries collected, the consumer loaded the products in the trunk of a car and drove home. After arrival at the home, the consumer placed the ham in the refrigerator. In our set-up, half of the packages (20 packages of ham) were then opened and "consumed" while the other 20 were kept in the refrigerator unopened during the field study. The consumption of the ham followed a "typical" breakfast routine, i.e., for time periods between 20 minutes ("normal breakfast") to 2 hours ("weekend breakfast") the ham packages were placed on a kitchen table at normal (20°C–23°C) in-house temperature and then put in the refrigerator again. The "breakfast" procedure was carried out five times during the tests within the first 10 days. See Table 2 for the process steps and the time slots of the field study.

The rationale behind the field study design to open half of the packages and keep half of them sealed was to investigate differences between supply chain and consumer solutions. There are a number of time-temperature solutions on the market for consumer products (e.g., Tempix, 2018) as well as reported research focused on the supply chain part of the cold chain (e.g., Göransson et al., 2018b; Aung and Chang, 2014) that target the reduction of food waste. However, most solutions and studies are only suitable for sealed packages, while most food wastage in the consumer stage relates to packages being opened (Silvenius et al., 2014). Hence, we set out to measure the difference between quality indicators of temperature loading for sealed packaging in the case of the packages being opened vs unopened.

TABLE 2 The process steps and time slots covered in the field study

Supply chain actor	Main processes	Time slots Field test 1	Time slots Field test 2	Process description
Producer	Slicing and packing of ham	June 08 2016 9.00–11.00	June 28 2017 12.00–14.00	45 packages were used in each test (three full secondary packages). Additional six packages were taken for laboratory analysis (initial bacterial load)
	Packing, storing, and dispatch	June 8 11.00–20.00	June 28 14.00–20.00	20 sensors were placed on half of the primary packages and 3 sensors on the outside of the three return crates used. They are then placed on a pallet together with other orders
Transport company	Loading and delivery	June 8 20.00–22.00	June 28 20.00–22.00	The full pallet is loaded as normal on the truck, together with other chilled meat products
Wholesaler	Receiving and storage	June 8 22.00–06.45	June 28 22.00–06.45	Products placed in storage for the night
	Repacking and dispatch	June 9 6.45–8.00	June 29 6.45–8.00	The secondary packages are packed on new pallets together with other chilled products and placed for pick-up
Transport company	Loading and delivery	June 9 8.00–9.30	June 29 8.00–9.30	The transport company distributes the goods
Retail outlet	Receiving and storage	June 9 9.30–11.55	June 29 9.30–14.45	A retail outlet receives the goods and puts them in storage
	Replenishment	June 10 11.55–15.30	June 29 14.45–15.30	In field test 1 the products are kept one day in storage before placed in the store. In field test 2 the products are replenished the same day as arrival
Consumer	Shopping	June 12 15.30–16.00	June 29 15.30–15.45	Shopping is carried out
	Home transport and placement in fridge	June 12 16.00–16.30	June 29 16.30–19.00	In field test 1 the transport took 30 minutes. In field test 2 it took 2 hours (longer distance and stops on the route)
Consumer household	Consumption	June 12–Aug 8	June 29–Aug 16	Half of the packages were opened and half were kept intact and not removed from the refrigerator
	Breakfast 1	June 13 30min	July 1 120min	All opened packages were placed on a kitchen table in room temperature at five occasions with different time periods. For the microbiological analysis of the actual bacteria load on the ham, samples were taken out on several occasions and placed in a freezer
	Breakfast 2	June 14 20min	July 2 30min	
	Breakfast 3	June 15 90min	July 4 60min	
	Breakfast 4	June 16 30min	July 5 20min	
	Breakfast 5	June 17 90min	July 6 30min	

During the field tests, packages of ham were taken at specific time points and frozen in an ordinary consumer freezer (see Table 3). At the end of the field test, all samples were transported to a microbiological laboratory, thawed slowly at refrigerator temperature, and evaluated for the number of lactic acid bacteria. Lactic acid bacteria were grown on MRS (De Man Rogosa Sharpe, Oxoid) for 3 days at 30°C. Freezing inactivates, but does not necessarily decrease, the bacteria present in food. Once thawed, however, the bacteria can again become active and multiply. The effect of freezing on bacteria varies greatly between species, physiological state, food composition, and the rate of freezing and thawing. In this study we are interested in studying the number of lactic acid bacteria and these microorganisms are considered resistant and most probably only slightly decreased when foods are frozen (Lopez et al., 2006).

TABLE 3 Samples taken out for analysis

Samples taken during field test 1	Comments	Samples taken during field test 2	Comments
June 9 2016	Production day	June 28, 2017	Production day
July 3	Date for static shelf life	July 14	Half-time for the static shelf life
July 13	10 days after ssl	July 26	Date for static shelf life
July 25	22 days after ssl	August 6	11 days after ssl
August 8	36 days after ssl	August 16	21 days after ssl

ssl stands for static shelf life and refers to the printed best-before date on the packages.

6.2 Field study findings

Based on follow-up discussions with involved actors (producer, logistics service provider, wholesale and retail outlet), a number of issues were reflected upon and insights gained. A uniform, quality-assured way of measuring temperature along the supply chain from production to retail shelves was found valuable for all parties in the chain as it is now, despite industry recommendations, clearly handled in different ways. In these tests, the inclusion of the consumer fridge was also found interesting but challenging in terms of real applications that could be considered and used, due to several technological and integrity related aspects. However, data provided by a system like the one in this field study were seen to be valuable for further research and development, especially in how the food is really affected throughout the supply chain, and they contribute to FSC research and development.

In both field tests, the overall logistics was efficient. From production to retail outlet via the retail distribution center (a distance of 174 km) took 22 hours in field test 1 and 20 hours in field test 2 (see Fig. 5). Initial surrounding temperature levels, that is, when the products were placed in the dispatch area at the producer, were high and it took almost 6 hours until the temperature of the sensor was below the threshold of +8°C. Transport, storage at wholesale, and distribution to the retail outlet were all at well below the required temperature and most of the time below 4°C.

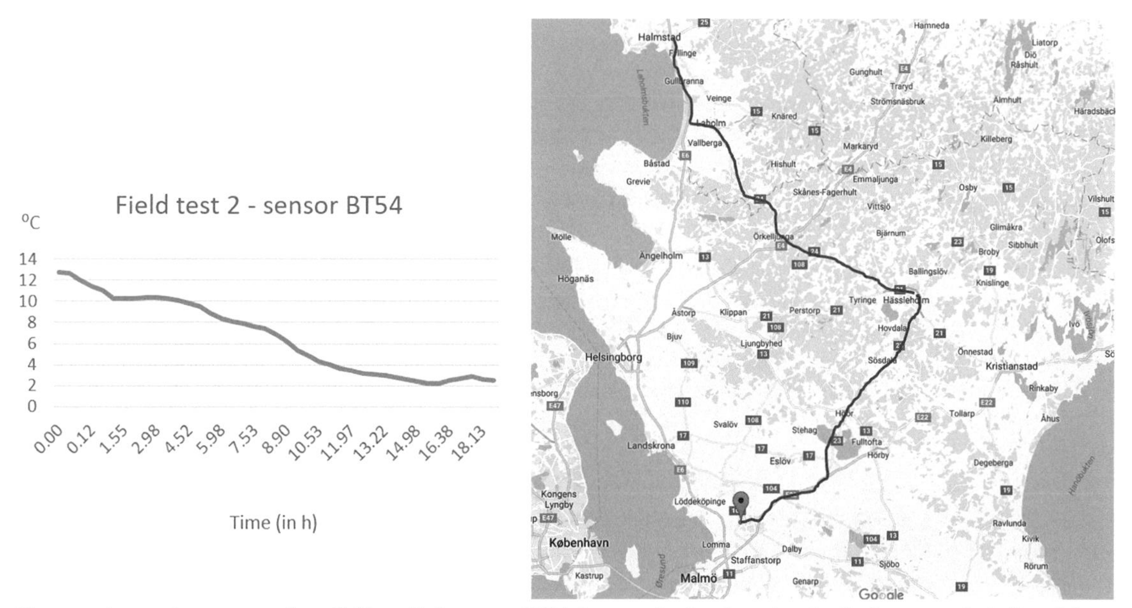

FIG. 5 Distance, time, and temperature from field test 2, for sensor BT54, from production through a distribution center, including night storage, and delivery to a retail outlet.

During the field tests, all the involved supply chain actors had access through a Web interface (see Fig. 6 for a screenshot of the Web interface) to the time and temperature data (dark gray line) for the product, as well as its position. Two dynamic predicted best-before dates were also presented, based on actual temperature impact, and predicted whether kept at the same temperature (in each moment) (light gray line) or if kept at the highest recommended temperature (gray line), i.e., +8°C for the ham.

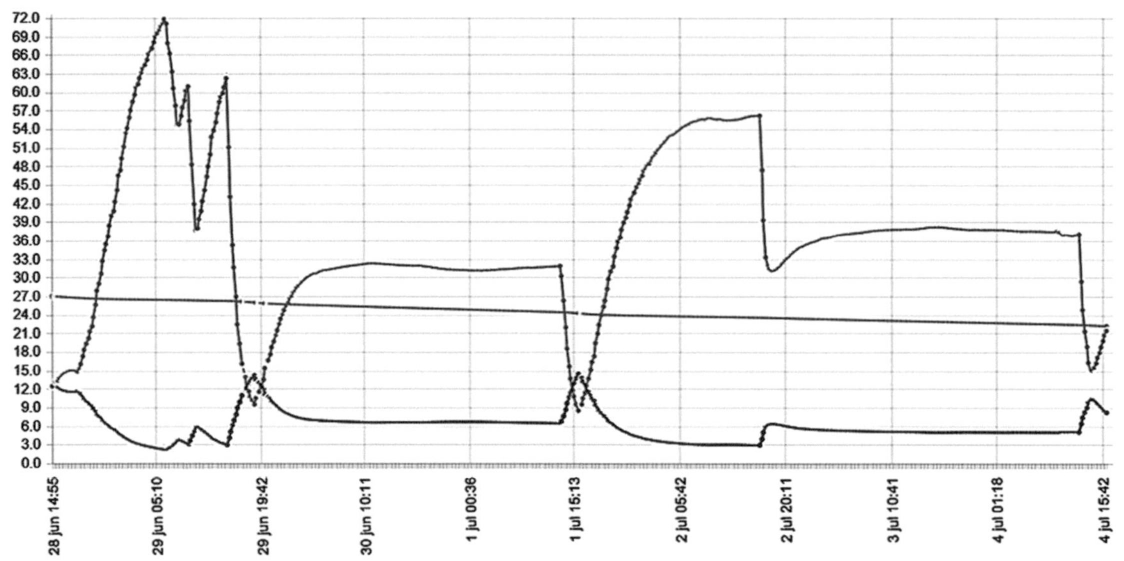

FIG. 6 Screenshot from the Web interface during the field test, where each of the actors in the supply chain could, in real time, follow the temperature and get new predicted best-before dates based on a prediction model. The *dark gray line* shows the actual temperature at each time slot, the *gray line* shows the predicted best-before date if kept in 8C and the *light gray line* shows the predicted best-before date if kept in the actual temperature.

The actors all agreed that monitoring the goods in real time enabled more responsiveness to upcoming deviations of either temperature or position, leading to more effective handling and mitigation of losses, conceivably reducing both direct and indirect costs. Examples of direct costs associated with these kinds of deviations are claims and customer compensation for damaged goods, both economic compensation as well as purchasing and shipping costs for replacement products. Based on more accurate and reliable data, indirect costs such as administrative costs in finding out what, where, and who, i.e., costs for troubleshooting and internal/external follow-up, could be minimized. Also, deductibles and insurance costs were mentioned along with lost sales and dissatisfied end costumers.

The continuous updates of predicted shelf life based on the actual temperature measured, i.e., dynamic shelf life of the product in the tests, were found interesting by the actors. While the applicability of such information for the consumer was reflected upon, a dynamically updated label on each package was found neither feasible nor viable. Instead, based on accurate and reliable information on real temperature exposures of the products in the supply chain, input on how to determine shelf life of the product could be used, and is better aligned to reality. However, such changes would demand a system that could handle abnormalities (e.g., too high or too low temperatures) in order to secure safe products to the consumers.

In Table 4, the dynamically predicted best-before date is presented for one product kept in the refrigerator all the time, and one used for "breakfast," i.e., placed on a kitchen table five times (see Fig. 7 for the time temperature curve of both samples). When comparing the products that have been exposed to a higher temperature impact (e.g., product 1A and 1B in field test 1) both products have longer estimated shelf life but also differences due to the accumulated temperature impact (see Table 4).

Table 4 shows that, due to the initial impact of higher temperature that was measured during field test 2, a shorter shelf life than the static one printed on the package was initially predicted. However, due to low temperature in the rest of the supply chain as well as in the refrigerator, the dynamically predicted shelf life for the ham was prolonged up to 30% longer (11 days) than the static one labeled on the package.

Worth mentioning in relation to the numbers in Table 4 is that, in both field tests, the products used for "breakfast" were also opened, i.e., the microbiological circumstances changed and the protective modified atmosphere was spoiled. Hence, the proclaimed increase in shelf life for product 1B and 2B based on the prediction model is actually shorter than presented in Table 4. The microbiological analysis is presented in the next section.

TABLE 4 Dynamic shelf life in comparison with static (i.e., printed) shelf life for the ham products in the field study based on the prediction models used

		Dsl at wholesale day 2	Dsl during shopping day 3	Dsl in fridge day 7	Dsl in fridge day 16	Dsl in fridge day 23
Field test 1 – ssl 25 days	Product 1a (bt75)—kept in fridge	2017-07-05	2017-07-06	2017-07-10	2017-07-20	2017-07-28
	Change of predicted shelf life	8%	12%	28%	68%	100%
	Product 1b (bt98)—"breakfast"	2017-07-05	2017-07-06	2017-07-10	2017-07-15	2017-07-19
	Change of predicted shelf life	8%	12%	28%	48%	64%
Field test 2 – ssl 27 days	Product 2a (bt54)—kept in fridge	2017-07-25	2017-07-25	2017-07-27	2017-07-31	2017-08-03
	Change of predicted shelf life	−4%	−4%	4%	19%	30%
	Product 2b (bt22)—"breakfast"	2017-07-25	2017-07-25	2017-07-26	2017-07-29	2017-07-31
	Change of predicted shelf life	−4%	−4%	0%	11%	19%

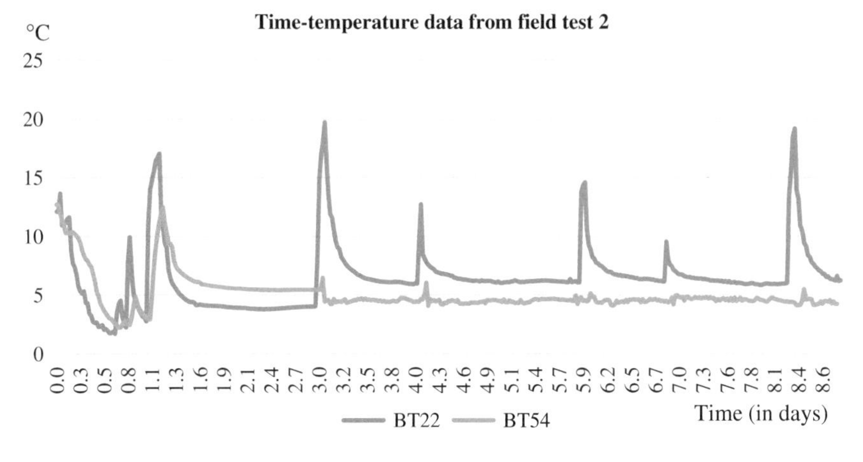

FIG. 7 Time temperature data from two of the sensors, BT 54 for a product being kept in the refrigerator at the consumer level and BT22 being placed at room temperature five times, representing a "breakfast" scenario.

6.3 Results from the microbiological analysis

The results of field test 2, performed in the summer of 2017, are shown in Table 5. The static shelf life (ssl) is set by the producer based on a storage at maximum +8°C (standard storage temperature in Sweden); however, data from the field test illustrates that, due to the low mean temperature along the whole supply chain, a much longer total and/or remaining shelf life than the SSL is predicted based on the prediction models.

As stated previously, the dynamic shelf life is secondarily predicted from the number of spoilage microorganisms which, in the case of MAP sliced ham, is lactic acid bacteria. Thus first the number of lactic acid bacteria is predicted from the logged storage temperature and then the days remaining of shelf life are calculated from the number of predicted

TABLE 5 Field test number 2

Sampling point	Time remaining of static shelf life (days)	Shelf-life prediction (days left until spoilage)	Mean storage temp. (°C)	Prediction no. of lactic acid bacteria (log cfu/g)	Measured no. of lactic acid bacteria[a] (log cfu/g)
Packages taken direct from production					
Production day	28	27	–	1.0	2.0±0.0
Packages keep in refrigerator (unopened)					
Half-time for ssl	12	18	4.2	3.7	5.2±0.8
Date for ssl	0	11	4.5	5.6	7.3±0.5
11 days after ssl	−11	5	4.4	7.2	6.5±0.7
21 days after ssl	−21	1	4.2	8.2	7.0±0.4
Breakfast packages (opened)					
Half-time for ssl	−12	15	6.3	4.5	6.7±0.0
Date for ssl	0	6	6	6.8	7.3±0.2
11 days after ssl	−11	0	6	8.6	7.2±0.8
21 days after ssl	−21	0	6	8.6	7.2±0.1

This table shows the measured mean temperature until sampling point, the remaining days of SSL (a negative value means that shelf life has been passed), predicted shelf life (0 means that the shelf life has been reached), as well as predicted and measured number of spoilage microorganisms, lactic acid bacteria. The data is available for both unopened and opened packages.
[a]*A mean value of five individual samples (packages).*

bacteria and an assumed future storage at +8°C. To evaluate the accuracy of the prediction, the number of lactic acid bacteria was measured during the field test and compared to the prediction. From this comparison two main results were visualized: (1) the number of lactic acid bacteria from the beginning is lower than the dynamic prediction, and (2) the maximum number of bacteria is lower than in the dynamic prediction tool. During the DynahMat project it was realized that fixing the initial number of lactic acid bacteria to a realistic figure is difficult, since the biological variation in-between packages is large. Data from earlier experiments (not shown here) illustrate that the number of lactic acid bacteria in different packages from the same production day and site can vary greatly. It was, e.g., shown that the number of lactic acid bacteria from different packages stored at the same temperature for about 20 days varied from 2 to 7.5 log cfu/g. However, from the same experiment the mean starting number of lactic acid bacteria was 1.1 log cfu/g. It would in the prediction be possible to use, e.g., a Monte Carlo simulation to illustrate this variation; however, being so large the prediction would not be at all useful to the actors of the food chain or the consumer. When bacteria grow, they grow in a logarithmic manner until they reach a stationary phase, the maximum possible number at the set environmental conditions. The stationary phase is limited by nutrients and space, and food spoilage is often reached when the bacteria reach the stationary phase or just after (Adams et al., 2015). The maximum number of lactic acid bacteria in the dynamic prediction was set according to the chosen model (Mataragas et al., 2006) to 8.3–8.9 log cfu/g; however, the field test shows that the actual maximum number of lactic acid bacteria in this ham at these conditions is rather around 7.2–7.3 log cfu/g. Thus this should be investigated further and possibly modified in the dynamic prediction models in order to make more accurate predictions.

6.4 Summary of field study

A starting point of the field study was that the current logic for SSL, i.e., a printed best-before or expiration date labeling, is one of the major reasons for food waste, both in the supply chain and ultimately in households (Lindbom et al., 2014; Rahelu, 2009; WRAP, 2011). The logic guiding the best-before date follows a worst-case scenario in which products are distributed, stored, and handled in the highest allowed temperature, e.g., +4°C, +5°C, or +8°C (depending on national standards and conventions). In this study, it was found that a secured, stable, and cold food supply chain enables longer shelf life for the products investigated. Furthermore, a quality indicator of temperature loading for sealed packaging is not, per se, targeting consumer wastage. With most food waste in households being related to opened packages, other solutions such as smaller-portioned packaging, pricing, and sensors showing how long a package has been opened show more potential to reduce food waste and guide consumers on making better "waste" decisions.

Increased transparency and visibility for supply chain actors regarding actual temperature changes show potential in directly affecting food wastage in the supply chain and potentially also sales in retail stores (e.g., reduced price on products that have passed best-before labeling but still can be sold), and indirectly affecting food wastage by the consumer. This indirect effect is derived from the potential in building trust between food supply chains and consumers (a trust that the many food frauds and scandals have decreased) by being transparent about handling and ensuring cold chains. With increased trust, consumers may consume products over a longer time span. However, as the results from this field study as well as other studies (Göransson et al., 2018b) show, while there are many potentials with new concepts for traceability and transparency in food supply chains, there are a number of organizational and interorganizational challenges that need to be dealt with.

7 Value and challenges with sharing information and providing increased transparency in food supply chains

While technological and conceptual development is a central part of enhanced transparency and visibility in supply chains, the business needs and requirements are the priority for the supply chain actors. Hsiao and Huang (2016, p. 187) state that "Inter-organizational time-temperature sharing could enhance food safety and quality, and further enhance the competitive advantage of food supply chains as a whole." However, Raab et al. (2011) point out that lack of exchange of temperature data between companies is one of the challenges currently remaining in temperature tracking. Furthermore, appropriate alignment among supply chain actors is critical for successful implementation and use of new technology (e.g., sensors) in minimizing waste and increasing safety and quality for consumers. The literature on supply chain alignment, information sharing, and collaboration is vast. While collaboration and information sharing are proven to provide more responsive and agile supply chains that manage changes in demand and meet the customer demand faster (Mentzer et al., 2001), a number of challenges have been reported. Lee (2004) puts forward that lack of alignment is one of the major reasons for failures of supply chain practices, and Lambert and Cooper (2000) argue that risk and gain sharing is critical in supply chain collaboration. Giguere and Householder (2012) discuss supply chain visibility, something that the introduction of new sensors and technologies in one way or another will enable, and conclude that sharing data is primarily a matter of trust, not of technology or data. Furthermore, the alignment of each actor's business strategy with the use and sharing of data in the supply chain is critical for collaboration and in determining on what level the collaboration with different actors should be Giguere and Householder (2012). Hsiao and Huang (2016, p. 186) conclude that interorganizational time-temperature sharing could enhance food safety and quality, and further enhance the competitive advantage of FSCs as a whole.

However, based on a study of information and organization integration in supply chains (Bagchi and Skjoett-Larsen, 2003), it was found that the probable loss of proprietary information and loss of control in sharing business information with suppliers were of major concern for supply chain actors. The main barriers to integration were: (1) the fixed mindset of managers, (2) lack of trust and the fear of sensitive business information falling into competitors' hands; (3) every member of the supply chain not being equally prepared; (4) loss of control; and (5) multiple IT platforms (Bagchi and Skjoett-Larsen, 2003). In their conclusions, (Bagchi and Skjoett-Larsen (2003, p. 104) propose that "the success of a drive to integrate the supply chain depends on the power, influence, motivation and zeal of the prime mover in the supply chain." Kürschner et al. (2008) confirm that collaboration promises mutual benefits for the partners in an SC, but these are rarely realized, due to two main reasons. First, each company has only partial knowledge of other companies' participation, hence incomplete information. Retrieving complete information on the flow of goods requires great effort to locate all actors. Second, the incentives to share sensitive operational data may be lacking. Also, information sharing is usually based on contracts, and to negotiate these contracts with each and every stakeholder, customer, and supplier would be inefficient

(Shah, 2005). Furthermore, Kürschner et al. (2008). (2008) explored possible solutions to work around this problem and find ways to share information between actors in a supply chain. They found that there are five major requirements for a solution for information sharing in an SC: data ownership, security, business relationship independence, organic growth, and quality of service. Hellström et al. (2011, p. 519) state that "even though risk and gain sharing is critical in implementing SCM, there is limited literature on the subject involving more than two supply chain actors."

In line with previous research, it can be concluded that a number of opportunities exist based on more accurate and reliable monitoring of temperature in FSCs. These include systems to handle temperature deviations and alarms quickly and effectively, at the same time as producers base their shelf-life predictions on data obtained from their supply chains. However, for this to happen the protective attitude among FSC actors needs to be reduced and appropriate business models and incentive set-ups developed.

References

Adams, M.R., Moss, M.O., McClure, P.J., 2015. Food Microbiology, fourth ed. Royal Society of Chemistry, Cambridge. ISBN 9781849739603.

Aung, M.M., Chang, Y.S., 2014. Temperature Management for the quality assurance of a perishable food supply chain. Food Control 40, 198–207.

Bagchi, P.K., Skjoett-Larsen, T., 2003. Integration of information technology and organizations in a supply chain. Int. J. Logist. Manag. 14 (1), 89–108.

Banterle, A., Stranieri, S., 2008. The consequences of voluntary traceability system for supply chain relationships. An application of transaction cost economics. J. Food Policy 33, 560–569.

Bendaoud, M., Lecomte, C., Yannou, B., 2007. Traceability systems in the agri-food sector: a functional analysis. Paper presented at the 16th International Conference on Engineering Design, Paris. August.

Bendaoud, M., Lecomte, C., Yannou, B., 2012. A Methodological Framework to Design and Assess Food Traceability Systems. Int. Food Agribusiness Manag. Rev. 15 (1), 103–126.

Bosona, T., Gebresenbet, G., 2013. Food traceability as an integral part of logistics management in food and agricultural supply chain. Food Control 33, 32–48.

Brennan, C., Lunn, W., 2016. Blockchain The Trusted Disrupter. Available at http://www.the-blockchain.com/docs/Credit-Suisse-Blockchain-Trust-Disrupter.pdf. Accessed March 2018.

Bytheway, C.W., 2005. Genesis of FAST. VaLue W O R L D 28 (2), 1–7.

Carter, C.R., Rogers, D.S., 2008. A framework of sustainable supply chain management: moving toward new theory. Int. J. Phys. Distrib. Logist. Manag. 38 (5), 360–387.

Coff, C., Korthals, M., Barling, D., 2008. Ethical traceability and informed food choice. In: Coff, C., Barling, D., Korthals, M., Nielsen, T. (Eds.), Ethical Traceability and Communicating Food. The International Library of Environmental, Agricultural and Food Ethics. In: vol 15. Springer.

Devlieghere, F., Debevere, J., Van Impe, J., 1998. Concentration of carbon dioxide in the water-phase as a parameter to model the effect of a modified atmosphere on microorganisms. Int. J. Food Microbiol. 43 (1-2), 105–113.

Devlieghere, F., Van Belle, B., Debevere, J., 1999. Shelf life of modified atmosphere packed cooked meat products: a predictive model. Int. J. Food Microbiol. 46 (1), 57–70.

Doorey, D.J., 2011. The transparent supply chain: from resistance to implementation at Nike and Levi-Strauss. J. Bus. Ethics 103 (4), 587–603.

van Dorp, K.J., 2002. Tracking and tracing: a structure for development and contemporary Practices. Logist. Inf. Manag. 15 (1), 24–33.

Dubbink, W., Graafland, J., van Liedekerke, L., 2008. CSR, Transparency and the Role of Intermediate Organisations. J. Bus. Ethics 82 (2), 391–406.

EU, 2002. Regulation (EC) No 178/2002 of the European Parliament and of the Council of 28 January 2002 laying down the general principles and requirements of food law, establishing the European Food Safety Authority and laying down procedures in matters of food safety. Off. J. Eur. Communities 31, 1–24.

Fanning, K., Centers, D.P., 2016. Blockchain and Its Coming Impact on Financial Services. J. Corp. Acc. Financ. 27 (5), 53–57.

Fombrun, C.J., 1996. Reputation: Realizing Value From The Corporate Image. Harvard Business School Press, Cop., Boston, MA.

Fraunhofer Press, 2011. Rotten meat doesn't stand a chance. Research News, 4, 5.

Giguere, M., Householder, B., 2012. Supply chain visibility: more trust than technology, Supply chain management review. pp. 20–25.

Gonzalez-Benito, J., Lannelongue, G., Queiruga, D., 2011. Stakeholders and environmental management systems: a synergistic influence on environmental imbalance. J. Clean. Prod. 19 (14), 1622–1630.

Gray, R., 2013. Back to basics: What do we mean by environmental (and social) accounting and what is it for?-A reaction to Thornton. Crit. Perspect. Account. 24 (6), 459–468.

Gustavsson, J., Cederberg, C., Sonesson, U., van Otterdijk, R., Meybeck, A., 2011. Global Food Losses and Food Waste. Food and Agriculture Organization of the United Nations, Rome.

Göransson, M., Jevinger, Å., Nilsson, J., 2018a. Shelf-life variations in pallet unit loads during perishable food supply chain distribution. Food Control 84, 552–560.

Göransson, M., Nilsson, F., Jevinger, Å., 2018b. Temperature performance and food shelf-life accuracy in cold food supply chains – Insights from multiple field studies. Food Control 86, 332–341.

Hancock, M., Vaizey, E., 2016. Distributed ledger technology: beyond block chain. Retrieved from Government Office for Science, https://www.gov.uk/government/news/distributed-ledger-technology-beyond-block-chain.

Hart, S.L., 1995. A natural-resource-based view of the firm. Acad. Manag. Rev. 20 (4), 986–1014.

Hellström, D., Johnsson, C., Norrman, A., 2011. Risk and gain sharing challenges in interorganisational implementation of RFID technology. Int. J. Procure. Manag. 5, 513–534.

Hsiao, H.-I., Huang, K.-L., 2016. Time-temperature transparency in the cold chain. Food Control 64, 181–188.

Hultman, J., Axelsson, B., 2007. Towards a typology of transparency for marketing management research. Ind. Mark. Manag. 36 (5), 627–635.

ISO, 1994. ISO/TC 176/SC 1 8402:1994, Quality management and quality assurance—Vocabulary.

Jensen, S., Båth, K., Lindberg, U., 2013. Vilken effekt skulle sänkt temperatur i kylkedjan få på matsvinnet? Swedish Environmental Protection Agency, Stockholm, Sweden.

Kreyenschmidt, J., Hubner, A., Beierle, E., Chonsch, L., Scherer, A., Petersen, B., 2010. Determination of the shelf life of sliced cooked ham based on the growth of lactic acid bacteria in different steps of the chain. J. Appl. Microbiol. 108 (2), 510–520.

Kürschner, C., Condea, C., Kasten, O., Thiesse, F., 2008. Discovery service design in the EPC Global Network. In: Floerkemeier, C., Langheinrich, M., Fleisch, E., Mattern, F., Sarma, S.E. (Eds.), The Internet of Things. Springer, Berlin, Heidelberg, pp. 19–34.

Lambert, D.M., Cooper, M.C., 2000. Issues in supply chain management. Ind. Mark. Manag. 29 (1), 65–84.

Lee, H.L., 2004. The triple-A supply chain. Harv. Bus. Rev. 10 (11), 102–112.

Lemieux, V.L., 2016. Trusting records: is Blockchain technology the answer? Rec. Manag. J. 26 (2), 110–139.

Lindbom, I., Gustavsson, J., Sundström, B., 2014. Minskat svinn i livsmedelskedjan – ett helhetsgrepp. SR866 Slutrapport, SP, Sweden.

Loessner, M.J., Golden, D.A., Jay, J.M., 2005. Modern Food Microbiology, seventh ed. Springer, California.

Lopez, M.C., Medina, L.M., Jordano, R., 2006. Survival of Lactic Acid Bacteria in Commercial Frozen Yogurt. J. Food Sci. 63 (4), 706–708.

Mataragas, M., Drosinos, E.H., Vaidanis, A., Metaxopoulos, I., 2006. Development of a predictive model for spoilage of cooked cured meat products and its validation under constant and dynamic temperature storage conditions. J. Food Sci. 71 (6), 157–167.

Mena, C., Adenzo-Diaz, B., Yurt, O., 2011. The causes of food waste in the supplier–retailer interface: Evidences from the UK and Spain. Resour. Conserv. Recycl. 55 (6), 648–658.

Mentzer, J.T., DeWitt, W., Keebler, J.S., Min, S., Nix, N.W., Smith, C.D., Zacharia, Z.G., 2001. Defining supply chain management. J. Bus. Logist. 22 (2), 1–25.

Moe, T., 1998. Perspectives on traceability in food manufacture. Trends Food Sci. Technol. 9, 211–214.

Mol, A.P.J., 2015. Transparency and value chain sustainability. J. Clean. Prod. 107, 154–161.

Nakamoto, S., 2008. Bitcoin: A peer-to-peer electronic cash system. Available at https://bitcoin.org/bitcoin.pdf. (Accessed March 2018).

Naturvårdsverket, 2013. Åtgärder för minskat svinn i livsmedelsindustrin - ett industri- och kedjeperspektiv. Swedish Environmental Protection Agency, Stockholm, Sweden.

NVC, 2012. https://www.en.nvc.nl/pasteur-sensor-enabled-rfid/. (Accessed March 2018).

Olsen, P., Borit, M., 2013. How to define traceability. Trends Food Sci. Technol. 29, 142–150.

Olsson, A., Skjöldebrand, C., 2008. Risk Management and Quality Assurance Through the Food Supply Chain – Case Studies in the Swedish Food Industry. Open Food Sci. J. 2, 49–56.

Pasteur Project, 2012. Available at: http://www.catrene.org/web/downloads/results_catrene/PASTEUR_PR-CT204.pdf. (Accessed March 2018).

Pizzuti, T., Mirabelli, G., 2015. The Global Track & Trace System for food: General framework and functioning principles. J. Food Eng. 159, 16–35.

Popper, D.E., 2007. Traceability: Tracking and privacy in the food system. J. Geogr. Rev. 97 (3), 365–389.

Raab, V., Petersen, B., Kreyenschmidt, J., 2011. Temperature monitoring in meat supply chains. Br. Food J. 113 (10), 1267–1289.

Rahelu, K., 2009. Date labeling on food. Nutr. Bull. 34 (4), 388–390.

Regattieri, A., Gamberi, M., Manzini, R., 2007. Traceability of food products: General framework and experimental evidence. J. Food Eng. 81 (2), 347–356.

RipeSense, 2018. Available at: http://www.ripesense.com/index.html. (Accessed March 2018).

Rodriguez-Mozaz, S., Marco, M.P., Lopez de Alda, M.J., Barceló, D., 2004. Biosensors for environmental applications: future development trends. Pure Appl. Chem. 76 (4), 723–752.

Ross, T., 1996. Indices for performance evaluation of predictive models in food microbiology. J. Appl. Bacteriol. 81, 501–508.

SensiTech, 2018. http://www.sensitech.com/en/. (Accessed March 2018).

Shah, N., 2005. Process industry supply chains: advances and challenges. Comput. Chem. Eng. 29 (6), 1225–1235.

Silvenius, E., Grönman, K., Katajajuuri, J.-M., Soukka, R., Koivupuro, H.-K., Virtanen, Y., 2014. The Role of Household Food Waste in Comparing Environmental Impacts of Packaging Alternatives. Packag. Technol. Sci. 27, 277–292.

SmartTrace, 2018. https://smart-trace.com/. (Accessed March 2018).

Svensson, G., 2009. The transparency of SCM ethics: conceptual framework and empirical illustrations. Supply Chain Manag. Int. J. 14 (4), 259–269.

Swan, M., 2016. Blockchain Temporality: Smart Contract Time Specifiability with Blocktime. Rule Technologies: Research, Tools, and Applications. Springer, Cham ISBN 978-3-319-42019-6.

Tempix, 2018. http://tempix.se/. (Accessed March 2018).

Thakur, M., Hurburgh, C.R., 2009. Framework for implementing traceability system in the bulk grain supply chain. J. Food Eng. 95, 617–626.

Tian, F., 2016. An Agri-food Supply Chain Traceability System for China Based on RFID & Blockchain Technology. In: 2016 13th International Conference on Service Systems and Service Management, New York.

Tian, F., 2017. A supply chain traceability system for food safety based on HACCP, blockchain & Internet of things. 14th Int. Conf. on Service Systems and Service Management (ICSSSM), Dalian, China.

Trienekens, J.H., Wognum, P.M., Beulens, A.J.M., van der Vorst, J.G.A.J., 2012. Transparency in complex dynamic food supply chains. Adv. Eng. Inform. 26 (1), 55–65.

Tsai, W.T., Blower, R., Zhu, Y., Yu, L., 2016. A System View of Financial Blockchains. In: Proceedings 2016 IEEE Symposium on Service-Oriented System Engineering, Sose 2016, pp. 450–457.

Yli-Huumo, J., Ko, D., Choi, S., Park, S., Smolander, K., 2016. Where Is Current Research on Blockchain Technology?-A Systematic Review. PLoS ONE 11 (10), 27.

Verdouw, C.N., Wolfert, J., Beulens, A.J.M., Rialland, A., 2016. Virtualization of food supply chains with the internet of things. J. Food Eng. 176, 128–136.

Wognum, P.M., Bremmers, H., Trienekens, J.H., van der Vorst, J., Bloemhof, J.M., 2011. Systems for sustainability and transparency of food supply chains—Current status and challenges. Adv. Eng. Inform. 25 (1), 65–76.

WRAP, 2007. The Food We Waste. London, UK.

WRAP, 2011. Consumer insight: Date labels and storage guidance. London, UK.

Zhou, M., Dong, A., 2011. Bioelectrochemical interface engineering: toward the fabrication of electrochemical biosensors, biofuel cells, and self-powered logic biosensors. Acc. Chem. Res. 44 (11), 1232–1243.

Further reading

European Commission, 2007. Factsheets on Traceability 2007. https://ec.europa.eu/food/sites/food/files/safety/docs/gfl_req_factsheet_traceability_2007_en.pdf.

Storøy, M., Thakur, P.O., 2013. The TraceFood Framework—Principles and guidelines for implementing traceability in food value chains. J. Food Eng. 115 (1), 41–48.

Chapter 16

Forecasting for food demand

Fotios Petropoulos* and Shawn Carver[†]
*School of Management, University of Bath, Bath, United Kingdom, [†]Fiddlehead Technology Inc., Moncton, NB, Canada

Abstract
The sustainability of food supply chains depends on accurately predicting future demand. Forecasting will form the basis for making decisions with regard to replenishment from the distribution centers and ordering from the suppliers. This chapter explores methods and approaches for forecasting for food demand.

1 Introduction

Accurate forecasting of food demand has significant economic and environmental consequences. Unreliable forecasts can result in a multitude of problems that ripple across the food supply chain, ranging from frequent changes to production schedules, expedited shipments, and high inventory carrying costs to poor customer service levels, stock-outs, and significant waste.

This chapter focuses on methods and approaches for forecasting for food demands. In Section 2, we describe methods that could be employed if just univariate time series data are available; such methods are able to capture and model a variety of predictable series patterns, such as trends and seasonality. Then, Section 3 discusses how such methods can be expanded to include causal information, such as promotional activity. The chapter continues (Section 4) with approaches for selecting the best across different forecasting methods for a particular series via validation and cross-validation approaches. Finally, in Sections 5 and 6 we explore how hierarchical structures (both cross-sectional and temporal) can enhance the forecasting process and further improve the forecasting performance. The last section of this chapter concludes and provides a brief discussion on the role of judgment in the forecasting process.

2 Univariate forecasting

Time series data comprise the interaction of several predictable and unpredictable patterns, such as level, trend, seasonality, randomness (noise), and cycle. In some cases, time series may also contain irregular (outlying) or missing values and as such data preprocessing may be required prior to extrapolation. Statistical forecasting targets on the modeling and extrapolation of the predictable series patterns (level, trend, and seasonality) while effectively smoothing out the unpredictable ones. Univariate forecasting methods solely focus on historical observations of the variable of interest (usually the recorded demand patterns for a particular stock keeping unit). In that sense, univariate forecasting can be seen as driving a car just by looking at the rearview mirror. Regardless, numerous empirical studies have demonstrated the effectiveness of univariate forecasting and its importance to demand planning.

Time series data can be widely grouped in four categories with regard to their predictable patterns: level-only (no trend nor seasonality is evident), trend-only, seasonal-only, and trend and seasonal data. More refined categories may be considered for distinguishing between different types of trends (additive or multiplicative; linear or damped) and seasonality (additive or multiplicative). In the next four sections, we explore simple univariate methods to extrapolate time series data from each of the four main categories.

2.1 Forecasting level data

When the past data do not exhibit either trend or seasonality, then level-only methods are considered suitable for extrapolating such signals. Simple level-only methods include:

- Naive method, where the forecast for the next period equals to the last observed actual. Note that this method has no parameters and requires just one past data point. The naive method reacts fast when the level of the data changes; however, it is not robust against outliers as it simply copies forward the noise in the data.

Sustainable Food Supply Chains. https://doi.org/10.1016/B978-0-12-813411-5.00016-8

- Global average, where the forecast for the next period equals to the arithmetic mean of all past observations. As with the naive method, global average has no parameters. Contrary to the naive method, global average is robust against outliers but it is quite slow in reacting to level changes.
- Simple moving average (SMA), where the forecast for the next period equals to the arithmetic mean of the last k observations, where k can take positive integer values in $[1, n]$ and n is the number of available past observations. Note that if $k = 1$ then SMA corresponds to the naive method, whereas if $k = n$ then SMA is equivalent to the global average method. Even if unequal weighting schemes might be considered, in its simpler version SMA assumes equal weights for each of the k last observations.

A more robust method for forecasting for level-only data is provided by the simple exponential smoothing (SES) method (Brown, 1956). SES is the simplest method in the exponential smoothing family of methods. The SES forecast for the next period is calculated as

$$\hat{y}_{t+1} = \alpha y_t + (1-\alpha)\hat{y}_t, \tag{1}$$

where y_t represents the actual value for period t, while $\hat{y}_t$ is the forecast for the same period. Note that Eq. (1) suggests that the forecast for period $t+1$ is a linear combination of the actual at period t and the forecast at period t. In fact, the weights of the combination are controlled by the α smoothing parameter, which reflects to the combination weight of y_t. Note that α takes values in [0, 1]. Also note that when $\alpha = 1$ then SES is equivalent to the naive method.

Let us now rewrite Eq. (1) by replacing t with n, which suggests that we create forecasts for the out-of-sample, and $\hat{y}_n$ with its respective linear combination using again Eq. (1):

$$\hat{y}_{n+1} = \alpha y_n + (1-\alpha)[\alpha y_{n-1} + (1-\alpha)\hat{y}_{n-1}] = \alpha y_n + (1-\alpha)\alpha y_{n-1} + (1-\alpha)^2\hat{y}_{n-1}. \tag{2}$$

By repeating this process and replacing $\hat{y}_{n-1}$, then $\hat{y}_{n-2}$ and so on until we reach the beginning of the data, we get

$$\hat{y}_{n+1} = \alpha y_n + (1-\alpha)\alpha y_{n-1} + (1-\alpha)^2\alpha y_{n-2} + \cdots + (1-\alpha)^{n-2}\alpha y_1 + (1-\alpha)^{n-1}\hat{y}_1. \tag{3}$$

The last equation suggests that:

- the forecast for the next period $n+1$ is a linear combination of all past data (y_t with t in 1, 2, …, n) and the very first forecast ($\hat{y}_1$). Also, given that $0 \leq \alpha \leq 1$, the weights for the actuals exponentially decay. For example, if $\alpha = 0.5$, then the weights for y_n, y_{n-1}, and y_{n-2} are 0.5, 0.25, and 0.125, respectively.
- SES has, apart from the α smoothing parameter, another parameter which is the initial forecast ($\hat{y}_1$); this forecast cannot be calculated through Eq. (1) so it has to be estimated.

The optimal value for the α smoothing parameter is usually automatically selected by the forecasting software by minimizing the one-step-ahead in-sample mean squared error (MSE). Linear or nonlinear optimization techniques may be used for this task. The value of the initial forecast could be either calculated using simple initializations (such as $\hat{y}_1 = y_1$) or optimized along with α.

While in-sample forecasts are usually produced using Eq. (1) for one-step-ahead, multihorizon forecasts can also be calculated as

$$\hat{y}_{t+h} = \alpha y_t + (1-\alpha)\hat{y}_t, \tag{4}$$

which suggests that forecasts for more than one-step ahead are simply equal to the one-step-ahead forecast; or else, the out-of-sample forecasts from SES can be represented as a straight horizontal line.

SES can also be expressed in components form:

$$l_t = \alpha y_t + (1-\alpha)l_{t-1}, \tag{5}$$

$$\hat{y}_{t+h} = l_t. \tag{6}$$

This set of equations suggests that SES estimates just one component (the current level) and the forecast for the next period (s) equals to the estimation of the current level. Note that when expressed in components form, SES initialization parameter is called the initial level (l_0). Eqs. (5), (6) are equivalent to Eq. (1); in the following, more complex exponential smoothing models will be directly expressed in components form.

Fig. 1 presents an example of forecasting a demand series with SES. The values of smoothing parameter and initial level are optimized using the `forecast` package of the R statistical software ($\alpha = 0.565$ and $l_0 = 923.17$).

2.2 Forecasting trended data

If the data exhibit trends (change of the level over time), either upwards or downwards, then SES will produce biased forecasts (forecasts that will systematically be either under or over the real outcomes). In such cases, methods that are able to capture the trend component should be considered as well. One such method is the Holt's exponential smoothing (HES) method (Holt, 1957), which can be expressed as:

$$l_t = \alpha y_t + (1-\alpha)(l_{t-1} + b_{t-1}), \tag{7}$$

$$b_t = \beta(l_t - l_{t-1}) + (1-\beta)b_{t-1}, \tag{8}$$

$$\hat{y}_{t+h} = l_t + hb_t. \tag{9}$$

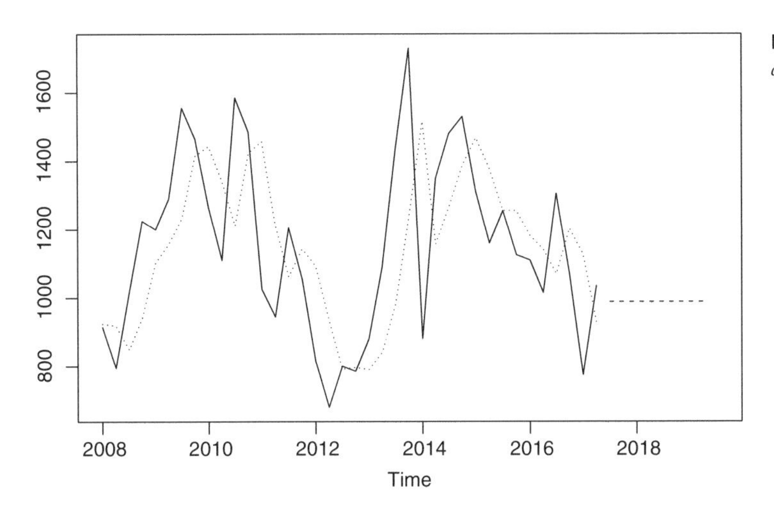

FIG. 1 Forecasting with SES (*solid line*: past actual values; *dotted line*: fitted values; *dashed line*: forecasts).

Compared to Eq. (5), Eq. (7) estimates the current level as a linear combination of the current actual observation, y_t, and the current forecast, $\hat{y}_t = l_{t-1} + b_{t-1}$ (note that the current forecast in SES consisted only of the level, $\hat{y}_t = l_{t-1}$). Eq. (8) estimates the trend component. This works in a similar fashion with the estimation of the level component. The current trend, b_t, is the linear combination of the difference of the latest two-level estimations, $l_t - l_{t-1}$, which can also be seen as the local trend, and the previously estimated trend, b_{t-1}. The weights of the linear combination for trend component are controlled by the β smoothing parameter, which (as was the case with α) takes values in [0, 1]. Finally, the h-step-ahead forecast is the sum of the latest estimated level plus the latest estimated trend multiplied by the respective horizon, h. In other words, HES forecast is a straight line that exhibits a linear trend (either upwards or downwards based on the sign of b_t). HES has in total four parameters: α and β smoothing parameters, initial level, l_0, and initial trend, b_0. We suggest that the initial states are optimized along with the smoothing parameters. Lastly, note that Eq. (8) suggests that the estimation and smoothing of the trend component depends on the smoothing of the level component. At the same time, the estimation of the level component depends on the trend estimation for the previous period. As such, the selection of optimal α and β values should be done concurrently rather than serially.

HES assumes that trend will always be linear and the forecast will increase/decrease with the same rate regardless of the forecasting horizon. However, this is not a reasonable assumption, especially if one considers the typical life-cycle of products. Thus, Gardner and McKenzie (1985) proposed a damped-trend variation of HES, the damped exponential smoothing (DES) method. DES is expressed as follows:

$$l_t = \alpha y_t + (1-\alpha)(l_{t-1} + \phi b_{t-1}), \tag{10}$$

$$b_t = \beta(l_t - l_{t-1}) + (1-\beta)\phi b_{t-1}, \tag{11}$$

$$\hat{y}_{t+h} = l_t + \sum_{i=1}^{h} \phi^i b_t. \tag{12}$$

This set of equations suggests that the estimation of the trend for one-step-ahead forecasts is multiplied by a damping parameter, ϕ, which usually takes values in [0.8, 1]. The trend for multistep-ahead forecasts is further dampened. Assuming $\phi = 0.9$, the trend, b_t, for the one-, two-, and three-step-ahead forecasts of DES is multiplied by $\phi = 0.9$, $\phi + \phi^2 = 1.71$, and

$\phi + \phi^2 + \phi^3 = 2.439$, respectively. By contrast, the trend for the respective forecasts of HES is multiplied by 1, 2, and 3. The DES forecasts do not exhibit a linear trend like HES; the long-term DES forecast is an almost horizontal line. DES falls to HES when $\phi = 1$. Also, note that if we allow $\phi > 1$ then an exponential trend is assumed. The damping parameter, ϕ, is optimized along with the other parameters so that the one-step-ahead MSE is minimized.

Fig. 2 provides a visual example of applying HES and DES on quarterly data that exhibit trend.

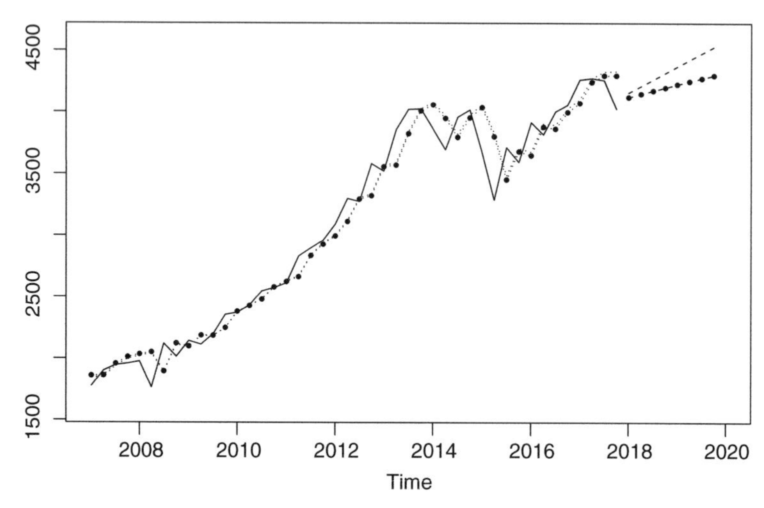

FIG. 2 Forecasting with HES and DES (*solid line*: past actual values; *dotted line: fitted values of HES; dashed line: forecasts of HES; dotted marked line: fitted values of DES; dashed marked line: forecasts of DES*).

2.3 Forecasting seasonal data

Food demand patterns often have single and/or multiple seasonal cycles. For example, monthly demand of ice cream will increase during the summer months. Other products, such as alcoholic beverages, exhibit weekly seasonality when observed on a daily frequency, with the demand being higher on Fridays and Saturdays.

A simple seasonal method arises from an extension of the naive method, where the forecast for the next period (e.g., Saturday if the frequency was daily) is the realized demand of the previous respective period (the demand for last Saturday). However, we suggest the more robust exponential smoothing models, which can be suitably expanded to include such seasonal patterns. For example, a component that smoothes the seasonal component could be added to the SES (Eqs. 5, 6) so that:

$$l_t = \alpha(y_t - s_{t-m}) + (1-\alpha)l_{t-1}, \tag{13}$$

$$s_t = \gamma(y_t - l_t) + (1-\gamma)s_{t-m}, \tag{14}$$

$$\hat{y}_{t+h} = l_t + s_{t+h-m}. \tag{15}$$

Eq. (14) estimates the seasonal component with m representing the number of periods within a full seasonal cycle (12 for monthly data, 4 for quarterly, 7 for daily). A separate smoothing parameter, γ, is used for smoothing the seasonal component; γ, as with α and β, takes values in [0, 1]; however, low values should be avoided unless sufficient cycles of data are available. Eq. (13) suggests that the level is smoothed on the seasonally adjusted data, $y_t - s_{t-m}$ whereas Eq. (14) suggests that the seasonal index is updated based on the value of the local seasonal index, $y_t - l_t$.

There exist m separate seasonal indices (e.g., for monthly data, one for January, one for February, and so on). Each seasonal index is updated every m periods (whereas the level component is updated every single period). Estimation of a seasonal method requires at least two full cycles of data; however, we suggest that three or even four cycles should be used. Note that the method expressed in Eqs. (13)–(15) has in total $m + 3$ parameters, the smoothing parameters (α and γ), initial level (l_0), and initial seasonal indices for each period ($s_{1-m}, s_{2-m}, \ldots, s_0$). The forecast is the sum of the estimation of the level and the seasonal index for the respective period. An example of forecasting using the seasonal SES method described previously is given in Fig. 3.

The seasonality of this method is assumed to be additive in form (i.e., the effect of the seasonal cycles does not interact with the level and the seasonal indices are expressed absolute values); thus, this method is called additive seasonal SES. In any case, a multiplicative seasonal SES can also be expressed as follows:

$$l_t = \alpha(y_t/s_{t-m}) + (1-\alpha)l_{t-1}, \tag{16}$$

$$s_t = \gamma(y_t/l_t) + (1-\gamma)s_{t-m}, \tag{17}$$

$$\hat{y}_{t+h} = l_t s_{t+h-m}. \tag{18}$$

Contrary to the additive seasonal SES, the current seasonally adjusted data and seasonal index are derived by division: y_t/s_{t-m} and y_t/l_t. Similarly, the forecast is the product of the level by the respective seasonal index. Note that in multiplicative seasonal methods, the seasonal indices are expressed in values around unity (representing percentages) and it is assumed that the seasonal effect interacts with the level. Also note that multiplicative seasonal methods are not applicable when the data contain negative or zero values.

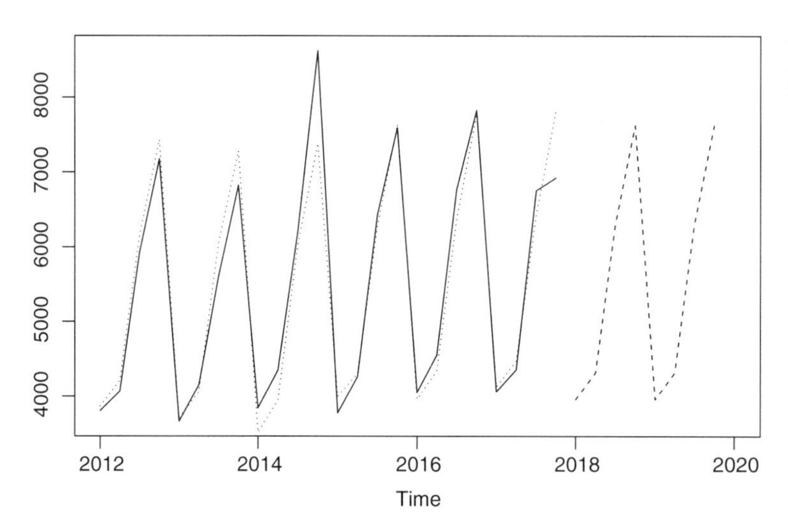

FIG. 3 Forecasting with additive seasonal SES (*solid line*: past actual values; *dotted line*: fitted values; *dashed line*: forecasts).

2.4 Forecasting trended and seasonal data

In the most complex form, exponential smoothing methods have three separate components: level, trend, and seasonality. Such methods are widely known as Holt-Winters (HW) exponential smoothing methods. Similarly to Section 2.3, seasonality may be expressed in either additive or multiplicative forms.

HW method with additive seasonality is expressed as:

$$l_t = \alpha(y_t - s_{t-m}) + (1-\alpha)(l_{t-1} + b_{t-1}), \tag{19}$$

$$b_t = \beta(l_t - l_{t-1}) + (1-\beta)b_{t-1}, \tag{20}$$

$$s_t = \gamma(y_t - l_t) + (1-\gamma)s_{t-m}, \tag{21}$$

$$\hat{y}_{t+h} = l_t + hb_t + s_{t+h-m}. \tag{22}$$

HW method with multiplicative seasonality is expressed as follows.

$$l_t = \alpha(y_t/s_{t-m}) + (1-\alpha)(l_{t-1} + b_{t-1}), \tag{23}$$

$$b_t = \beta(l_t - l_{t-1}) + (1-\beta)b_{t-1}, \tag{24}$$

$$s_t = \gamma(y_t/l_t) + (1-\gamma)s_{t-m}, \tag{25}$$

$$\hat{y}_{t+h} = (l_t + hb_t)s_{t+h-m}. \tag{26}$$

Each component is smoothed separately using a different smoothing parameter. Eqs. (20), (24) are exactly the same as the respective equations in Section 2.2; similarly, Eqs. (21), (25) reflect to the Eqs. (14), (17) of Section 2.3. The estimation of the level is a linear combination of the seasonally adjusted actual, $y_t - s_{t-m}$ or y_t/s_{t-m}, and the seasonally adjusted forecast, $l_{t-1} + b_{t-1}$ (Eqs. 19, 23). Finally, the HW h-step-ahead forecast consists of the sum of the level and trend estimations, $l_t + hb_t$, which is subsequently added to (multiplied by) the respective seasonal index, s_{t+h-m}, when the seasonality is additive (multiplicative). HW method has $m + 5$ parameters to be estimated/optimized, including the three smoothing parameters for the level of the trend and the seasonality. Lastly, note that HW methods could be suitably adjusted to include a damped trend component rather than a linear one. Fig. 4 provides an illustrative example of the HW with additive seasonality.

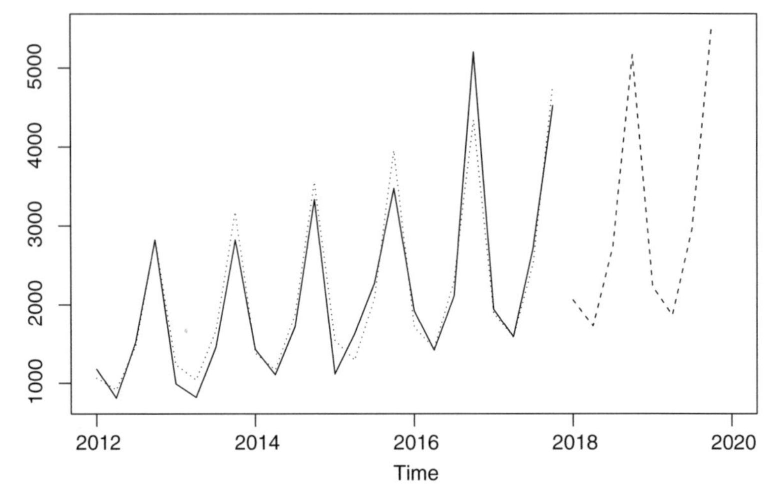

FIG. 4 Forecasting with HW with additive seasonality (*solid line*: past actual values; *dotted line*: fitted values; *dashed line*: forecasts).

3 Including exogenous variables

Quite often, looking just at past demand values to make predictions is not enough. Several drivers may influence the demand, such as weather, own or competitors' promotions, sports events, and other special events or actions. The effect of such drivers should be taken into account when building forecasting models. Traditionally, regression-type models have been used to incorporate the effect of such causal variables. However, we can also expand the formulation of exponential smoothing methods as to include information from external variables (Kourentzes and Petropoulos, 2016). For example, SES can be extended to include additional components for estimating the effect of exogenous variables, X_i with $i = 1, 2, \ldots, N$:

$$l_t = l_{t-1} + \alpha e_t, \tag{27}$$

$$d_{i,t+1} = b_i x_{i,t+1}, \tag{28}$$

$$\hat{y}_{t+1} = l_t + \sum_{i=1}^{N} d_{i,t+1}, \tag{29}$$

where $d_{i,t}$ refers to the effect at time t of a variable X_i with observations $x_{i,t}$ and b_i is the respective coefficient. Similar to regression models, the set of coefficients shows the additive effect of the respective variables. Estimation of the b coefficients is performed together with the estimation of the rest parameters of the exponential smoothing methods (for the example of SES, together with α and initial level) by minimizing the in-sample MSE. The extension shown above for SES can be applied to any of the exponential smoothing methods. Note that the (natural) assumption here is that exogenous information is available for the future periods (as is the case for future planned promotions) or it can be predicted.

In some cases, the correlation between the exogenous variables may be high enough to lead to multicollinearity issues and their negative effects on the estimation of the coefficients. According to Kourentzes and Petropoulos (2016), such issues are especially relevant when temporal aggregation is applied on the data (we will further discuss temporal aggregation in Section 6). As such, Kourentzes and Petropoulos (2016) suggest that the variables are transformed to become orthogonal. Principal components analysis returns a set of principal components orthogonal to each other, that have no redundant information and are ordered in terms of the variance extracted. To maintain the model as simple and at the same time as effective as possible, we suggest the inclusion of the first k principal components, so that the sum of the extracted variance is between 60% and 80%. As the estimated coefficients for the principal components cannot be directly interpreted with regard to the original variables, it is suggested that a back-transformation be applied so that the coefficients of the original variables are derived.

Kourentzes and Petropoulos (2016) showed that exponential smoothing with exogenous variables outperforms significantly univariate methods as well as regression models, both in terms of bias and accuracy.

4 Selecting between methods

Given the plethora of available methods to select from, coupled with the fact that food and beverage-related stock keeping units usually count tens of thousands of items within a retailer, suitable automatic selection strategies seem to be necessitated. In short, there exist three classes of automatic method selection strategies, namely selection based on past performance, selection based on information criteria, and selection based on rules. We suggest that, if enough data are available, selection be performed via the first option, which we also describe in detail in this section. In the last paragraph of this section, we also briefly discuss the other two options.

Selection on past performance can be achieved by suitably splitting the available historical data (in-sample) into two sets, the training set and the test set. Subsequently, using the training set as input, forecasts that correspond to the periods of the test set are produced from all available methods. Finally, the forecasts of each method are compared against the withheld actuals on the training set observations and the methods are ranked in terms of performance given a cost function, such as MSE, mean absolute error (MAE), and mean absolute percentage error. More formally, assume that the available historical data are presented in a vector y with values $y_1, y_2, \ldots, y_n$. Also assume that the in-sample is divided so that the training set consists of the $n - p$ first observations of y whereas the test set consists of the last p observations. The value of p can be decided so as to match the required forecast horizon, h; however, other options are available (such as $p = m$). In any case, the training set should include an adequate number of observations that would allow fitting of the applied forecasting methods. The p-steps-ahead forecasts of method j, $\hat{y}^j_{n-p+1}, \hat{y}^j_{n-p+2}, \ldots, \hat{y}^j_n$, are produced using the first $n - p$ observations and withholding the last p. These are then compared against the actuals; for instance, if MAE is used as a cost function, then $MAE_j = p^{-1}\sum_{i=n-p+1}^{n}|y_i - \hat{y}^j_i|$. This process is repeated for every one of the J forecasting methods available, or $j = 1, 2, \ldots, J$. Finally, the method, j, with the minimum value of MAE_j is selected. Once a method is selected, forecasts beyond the end of the in-sample data (n) are produced using all available observations, $y_1, y_2, \ldots, y_n$. The approach described here is known as fixed origin evaluation and sometimes is also referred to as validation for time series data. Forecast origin is the period from which the forecasts originate; in the case of validation, there is only one forecast origin referring to the selection process, which is the last period of the training set, $n - p$. This process is illustrated in Fig. 5.

However, a fixed origin evaluation has the disadvantage of overly focusing on a single validation window, on which irregularities might occur. A better way to select between methods based on their past performance would be through a multiple validation approach, also called cross-validation or rolling origin evaluation. Forecasts for each method are not produced just once from one origin (reflecting to a single validation period) but multiple times (rolling origin). Most usually, overlapping validation windows are considered and the rolling through the origins takes place by one period at a time. However, nonoverlapping windows can also be defined. The case of overlapping cross-validation can be formally described as follows. The in-sample data may be divided in the initial training set, consisting of $n - p$ observations, and the cross-validation set, consisting of p observations. Forecasts for the next h' periods (where $h' < p$) are produced by each method using the initial training set. Then, the training set increases by one period, so that it now contains $n - p + 1$ observations; h'-periods-ahead forecasts are again produced from the new origin (y_{n-p+1}). This process is repeated until the origin y_{n-p+q} where $p - q = h'$ so that all h'-step-ahead forecasts can be evaluated. Consequently, $q + 1$ sets of h' forecasts each are produced for each of the J available methods. These forecasts are compared against the respective actuals of the cross-validation period that were withheld. Finally, the method with the best cross-validated performance is selected. The process of selecting via cross-validation is graphically depicted in Fig. 6.

For the curious readers who would like to see more details on fixed versus rolling origin evaluation schemes, we suggest the paper by Tashman (2000).

5 Refocusing the scope: Cross-sectional aggregation

Data within an organization are often organized in cross-sectional hierarchical structures. Each level of the hierarchy refers to a different degree of data aggregation. For example, possible hierarchical levels include sales by stock keeping units (SKUs), category, location, channels, clients, or combinations of these. A simple hierarchy is depicted in Fig. 7. This hierarchy consists of three levels: company, department, and SKU. The first department consists of three SKUs, while the second department consists of two SKUs. Data management processes should generally focus on the systematic recording and storing of data on the most granular level of the hierarchy within a company. In the case of Fig. 7, that would be the SKU-level; however, in other cases the most granular level could be a combination of SKU per location sales.

Historical data within a hierarchical structure are consistent. This means that for every period t the sum of the values of all the bottom level nodes is equal to the value of the top-level node. Data summation also holds for every intermediate aggregation level. However, this is not true for forecasts. Following the structure of Fig. 7, the sum of the forecasts for SKUs 4 and 5 (S_4 and S_5) is not generally equal to the forecast of the respective department, D_2. This leads to two important issues:

- Given that different functions of the organization (operations, sales, marketing, finance, etc.) focus on different levels of aggregations, separate sets of forecasts are independently produced within the company. Such forecasts are the basis for decisions within the different functions. However, forecast inconsistency will inevitably lead to decision inconsistency. In other words, there is a need for approaches that lead to forecast consistency.
- While naturally forecasts needed at the SKU-level would be produced at the same level, this process does not guarantee maximization of the forecasting performance. Time series at higher aggregation levels will generally exhibit less noise and lower levels of intermittence; at the same time, individual series characteristics are better captured and modeled at lower aggregation levels. So, a process to identify optimal levels of aggregation is required.

Four hierarchical approaches can lead to forecast consistency:

- The *bottom-up* approach is possibly the most widely applied hierarchical approach. Forecasts are produced on the most granular hierarchical level (SKU level for the hierarchy depicted in Fig. 7). Consequently, forecasts for all other levels are calculated as the respective sums of the lowest level forecasts following the hierarchical structure.
- In the *top-down* approach, forecasts are produced only for a single hierarchical node that represents the highest hierarchical level (company level for the hierarchy depicted in Fig. 7). Forecasts for lower aggregation levels are calculated as proportions of the forecasts of the highest level. Historical or forecasted proportions may be considered (Athanasopoulos et al., 2009). In the former case, the top-down approach requires model fitting, parameterization, and forecasting of a single series (time series data on the very top level), rendering it the least expensive hierarchical approach.

- The *middle-out* approach is conceptually a combination of bottom-up and top-down approaches. Initially, forecasts are produced at a middle hierarchical level (department level for the hierarchy depicted in Fig. 7). Then, forecasts for higher aggregation levels are calculated using structurally imposed summations (similarly to the bottom-up approach). On the other hand, forecasts for lower levels are derived by appropriate forecast decomposition (similarly to the top-down approach). Fiddlehead's fieldwork suggests that this is the most commonly used hierarchical approach among the food and beverage manufacturers.
- The most recently proposed hierarchical approach is called *optimal combination* (Hyndman et al., 2011). In this approach, forecasts at all levels are generated. These forecasts are subsequently reconciled in a statistically optimal way. Optimal combination is the single hierarchical approach that directly takes into account forecasts produced at all levels. However, this also renders it the most computationally intensive approach.

Regardless of the fact that all four aforementioned approaches lead to forecast consistency that is essential to consistency in terms of decisions, still a selection between these four approaches is required so that forecasting performance is maximized. Such a selection can follow the guidelines for forecasting method selection presented in Section 4. The performance of the various hierarchical approaches may be measured through a validation or cross-validation fashion. The approach with the lowest error through a validation set is taken forward. Hierarchical approach selection can be completed by calculating the performance across the hierarchy or by focusing at specific hierarchical levels or even by just focusing on a single hierarchical level. If the latter is the case, it should be noted that the optimal aggregation level might not necessarily match the level where the cost-function is to be minimized. This suggests that forecasting through cross-sectional hierarchical structures allows for refocusing the scope. Decision makers are invited not to blindly extrapolate the historical data that directly refer to the variable of interest but to also investigate the predictability of alternative cross-sectional aggregation levels. This process is usually referred to as hierarchical level optimization. However, it is worth noting that the role of portfolio segmentation, the trade-off between accuracy and the number of forecasting units to be maintained at each level, as well as the level at which judgmental adjustments are most frequently applied are key elements to take into account when selecting the most appropriate hierarchical level.

6 Extracting information from the data: Multiple temporal aggregation

On top of the cross-sectional aggregation of the time series data, as explored in the previous section, temporal aggregation can also be considered. Temporal aggregation refers to within series data transformation that focuses on changing the observed frequency. For instance, a monthly time series can be transformed to a quarterly one via nonoverlapping temporal aggregation that considers time buckets of size 3 (the aggregation level equal in this case is 3). It suggested that instead of focusing on a single frequency of the observed data (which usually matches the frequency at which the data are collected), to transform the series in many different frequencies (Kourentzes et al., 2014). For instance, a monthly time series may be transformed to bimonthly, quarterly, four-monthly, semesterly, etc. up to the yearly or even biannual frequency.

Temporal aggregation of the originally observed series will help us view the same data through different lenses. Specific series characteristics are amplified or attenuated in different frequencies. When dealing with fast-moving series, within-year periodicity and seasonality is more apparent and better modeled in high-frequency time series whereas the long-term trend is easier to be identified in the lower-frequency (temporally aggregated) data (Petropoulos and Kourentzes, 2014). If the data in hand are intermittent, then temporal aggregation will reduce the degree intermittence, allowing the application of the fast-moving methods described in Section 2 (Nikolopoulos et al., 2011). Moreover, the variability of the data will decrease as we move toward higher temporal aggregation levels (Spithourakis et al., 2014; Petropoulos et al., 2016b). Essentially, temporal aggregation is a cheap and efficient method that allows us to extract more information from the data in hand.

Selection of the optimal aggregation level might be difficult in practice. However, it has been empirically demonstrated that producing a forecast via combining the series component estimates from multiple aggregation levels can lead to increased forecasting performance, especially for the longer horizons (Kourentzes et al., 2014). Improvements in performance through multiple temporal aggregation have also been observed in the cases where exogenous information is available (Kourentzes and Petropoulos, 2016). Such improvements are linked with tackling the model and parameter uncertainty, which refers to the holy grail of forecasting: optimally selecting model and set of parameters for the data in hand. However, an even more important benefit that arises from the application of the multiple temporal aggregation approach (on top of the improved accuracy) is the alignment of decisions on different levels: operational (usually a few weeks to a few months ahead), tactical (usually one to two quarters ahead), and strategic (usually 1–2 years ahead).

In a recent study, Athanasopoulos et al. (2017) proposed that multiple temporal aggregation could be expressed as hierarchical structured, the so-called temporal hierarchies. An example of a temporal hierarchy is depicted in Fig. 8, where the data are observed in the quarterly frequency (lowest aggregation level) and are temporally aggregated to semesters and years. Structuring multiple temporal levels as hierarchies allows the application of the hierarchical approaches discussed in Section 5. More importantly, one can also consider super-hierarchies that incorporate both cross-sectional and temporal information, or cross-temporal hierarchies. Such hierarchies will allow the calibration of decisions with regard to different functions of the company (sales, operations, logistics, manufacturing, marketing, finance, upper-level management), different divisions and business operating units as well as different horizons (short- vs. long-term decisions). In other words, cross-temporal hierarchies will allow for the "one number forecast" that will lead to harmonized decisions within the company.

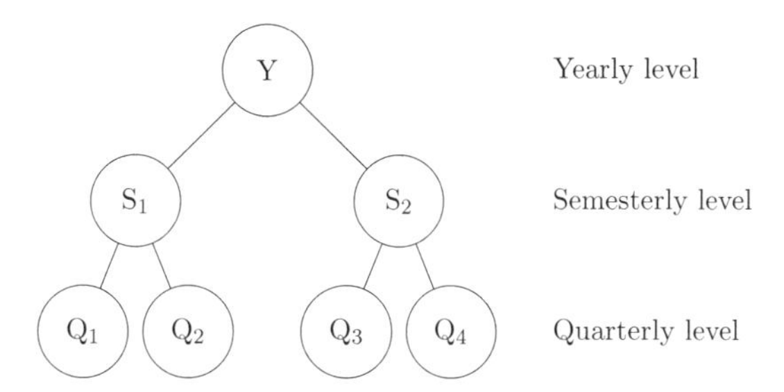

FIG. 8 A temporal hierarchy.

Regardless of the advantages of multiple temporal aggregation approaches, empirical evidence from Fiddlehead's work has identified a few cases where the original series should be used instead. These included sudden level changes, extreme seasonality, and localized trends. In such cases, preprocessing of the original data might be considered, such as seasonal adjustments. At the same time, the performance of multiple temporal aggregation increases together with the length of the series, while depromotionalizing the data, allows for better capturing of the underlying characteristics.

7 Concluding remarks and the role of judgment

In this chapter, we explored methods and approaches to statistically produce forecasts for food demand. We presented univariate forecasting techniques suitable to model a variety of predictable series patterns but we also discussed how exogenous information may be incorporated into such methods. We also described approaches of automatically selecting the most suitable method for each time series individually, through validation and cross-validation processes. Finally, we explored the role of hierarchical structures, how cross-sectional and temporal aggregation can lead to increased performance, and discussed issues with forecast reconciliation.

We would also like to very briefly touch on the role of managerial judgment in the forecasting process. It is usually observed that the final forecast within any company is not solely based on statistical models. In fact, managerial judgment may appear in different stages of the forecasting function. Sometimes, a purely judgmental forecast is produced and taken forward for decision making. We advise against such practice, as numerous studies have demonstrated that such judgmental forecasts are inferior to statistical ones. Other times, the statistical output is adjusted/modified so that the impact of anticipated special circumstances and conditions (such as forthcoming promotions or other special events) are taken into account. Several recent studies have empirically examined the conditions under which such adjustments might be beneficial in practice (see, e.g., Fildes et al., 2009; Petropoulos et al., 2016a). In any case, we suggest that a forecast-value-added analysis (Gilliland, 2013) be applied so that the benefit of performing such judgmental adjustments can be analyzed and monitored. Finally, a recent work has showed that judgment can complement the method selection phase of the forecasting process (Petropoulos et al., 2017). In fact, wisdom of crowds (weighted combinations from the judgmental method selections of multiple experts) or a 50%–50% combination of statistics + expert can bring significant benefits in terms of forecast accuracy.

References

Athanasopoulos, G., Ahmed, R.A., Hyndman, R.J., 2009. Hierarchical forecasts for Australian domestic tourism. Int. J. Forecast. 0169-2070. 25 (1), 146–166. https://doi.org/10.1016/j.ijforecast.2008.07.004.

Athanasopoulos, G., Hyndman, R.J., Kourentzes, N., Petropoulos, F., 2017. Forecasting with temporal hierarchies. Eur. J. Oper. Res. 03772217. 262 (1), 60–74. https://doi.org/10.1016/j.ejor.2017.02.046.

Brown, R.G., 1956. Exponential Smoothing for Predicting Demand. Arthur D. Little Inc., Cambridge, MA.

Fildes, R., Goodwin, P., Lawrence, M., Nikolopoulos, K., 2009. Effective forecasting and judgmental adjustments: an empirical evaluation and strategies for improvement in supply-chain planning. Int. J. Forecast. 0169-2070. 25 (1), 3–23. https://doi.org/10.1016/j.ijforecast.2008.11.010.

Gardner, E.S., McKenzie, E., 1985. Forecasting trends in time series. Manag. Sci. 0025-1909. 31 (10), 1237–1246.

Gilliland, M., 2013. FVA: a reality check on forecasting practices. Foresight Int. J. Appl. Forecast. 29, 14–18.

Holt, C.C., 1957. Forecasting seasonals and trends by exponentially weighted moving averages. O.N.R. Memorandum 52/1957.

Hyndman, R.J., Ahmed, R.A., Athanasopoulos, G., Shang, H.L., 2011. Optimal combination forecasts for hierarchical time series. Comput. Stat. Data Anal. 55 (9), 2579–2589. https://doi.org/10.1016/j.csda.2011.03.006.

Kourentzes, N., Petropoulos, F., 2016. Forecasting with multivariate temporal aggregation: the case of promotional modelling. Int. J. Prod. Econ. 0925-5273. 181A, 145–153. https://doi.org/10.1016/j.ijpe.2015.09.011.

Kourentzes, N., Petropoulos, F., Arenas, J.R.T., 2014. Improving forecasting by estimating time series structural components across multiple frequencies. Int. J. Forecast. 30 (2), 291–302.

Nikolopoulos, K., Syntetos, A.A., Boylan, J.E., Petropoulos, F., Assimakopoulos, V., 2011. An aggregate–disaggregate intermittent demand approach (ADIDA) to forecasting: an empirical proposition and analysis. J. Oper. Res. Soc. 62 (3), 544–554.

Petropoulos, F., Kourentzes, N., 2014. Improving forecasting via multiple temporal aggregation. Foresight 34, 12–17.

Petropoulos, F., Fildes, R., Goodwin, P., 2016. Do "big losses" in judgmental adjustments to statistical forecasts affect experts' behaviour? Eur. J. Oper. Res. 0377-2217. 249 (3), 842–852. https://doi.org/10.1016/j.ejor.2015.06.002.

Petropoulos, F., Kourentzes, N., Nikolopoulos, K., 2016. Another look at estimators for intermittent demand. Int. J. Prod. Econ. 0925-5273. 181A, 154–161. https://doi.org/10.1016/j.ijpe.2016.04.017.

Petropoulos, F., Kourentzes, N., Nikolopoulos, K., Siemsen, E., 2017. Judgmental selection of forecasting models. Working Paper.

Spithourakis, G., Petropoulos, F., Nikolopoulos, K., Assimakopoulos, V., 2014. A systemic view of ADIDA framework. IMA Manag. Math. 25, 125–137.

Tashman, L.J., 2000. Out-of-sample tests of forecasting accuracy: an analysis and review. Int. J. Forecast. 0169-2070. 16 (4), 437–450. https://doi.org/10.1016/S0169-2070(00)00065-0.

Chapter 17

Food systems sustainability: The complex challenge of food loss and waste

Matteo Vittuari, Fabio De Menna, Laura García-Herrero, Marco Pagani, Laura Brenes-Peralta and Andrea Segrè
Department of Agricultural and Food Sciences, Alma Mater Studiorum—University of Bologna, Bologna, Italy

Abstract

Food loss and waste (FLW) represents a major challenge for food systems sustainability and a growing concern in the political agenda of national governments and international organizations. This interest is primarily driven by the interrelated implications that FLW has with food security, human health, economic development, and ecosystems. The staggering amount of FLW currently generated along supply chains exacerbates global food system challenges, such as world population growth, change of dietary habits, energy intensity of agro-food systems, and food supply vulnerability due to climate change. In a life-cycle perspective, beyond representing a missed opportunity to feed the growing world population, FLW also puts a huge pressure on natural capital, in terms of natural resources consumption (e.g., fossil energy, water, fish stocks, agricultural land), environmental pollution (water, air, soil), GHG emissions, ecosystem resilience, and biodiversity loss.

This chapter aims at analyzing food systems sustainability through the lens of the interrelated implications and impacts of FLW on production and consumption. First, it provides an overview of the debate around FLW definition and quantification. Then, it focuses on the causes and impacts of FLW and the related methodological challenges. Additionally, recent policy initiatives are discussed in a comparative perspective. Finally, challenges and future research needs are highlighted.

1 Sustainability in food systems: A resource nexus challenge

A food system encompasses all stages of the food supply chain (FSC) needed to feed the population: growing, harvesting, packing, processing, distributing, consuming, and disposing (Aragrande et al., 2001). Its sustainability entails the provision of adequate, safe, affordable, and healthy food that meets nutritional needs in culturally and socially acceptable ways, while safeguarding natural resources and limiting environmental damage (MacDonald and Reitmeier, 2017). Food systems sustainability represents the result of the interactions and mutual influences between the agents in charge of the different stages: farms, firms, suppliers of goods and services, distributors, large corporations, catering services, consumers, public institutions (Brunori and Bartolini, 2013; Nestle, 2013; Paarlberg and Paarlberg, 2013).

The structural changes experienced by agricultural, food, and rural systems in many regions during the past decade largely influenced the sustainability of food systems (Paarlberg and Paarlberg, 2013; Knickel et al., 2017). For instance, in large parts of Europe and other areas, agricultural production has become highly specialized and capital-intensive, while food processing and retailing have grown exponentially in scale (Knickel et al., 2017). These changes have contributed to an abundant supply of relatively inexpensive food for consumers in some regions and have supported the development of nonagricultural sectors by releasing labor resources and creating a steady demand for machinery, production inputs, and diversified services. However, these trends have caused an increasing decoupling of contemporary food production from natural processes and a consequential dependence on industrially produced inputs and fossil fuels. The negative social and environmental outcomes, such as rural out-migration, conversion of natural habitats to agricultural land, and the associated impact on biodiversity, soil degradation from intense exploitation, air and water pollution, and climatic changes effects, have been widely analyzed and documented (Hazell and Wood, 2008; WHO, 2012; Gomiero, 2016). The energy intensity of modern food systems represents a crucial challenge in a scenario of decreasing fossil resources, control of greenhouse gases (GHG) emissions, growing population, increasingly industrialized agriculture, and uncertain biofuels policies (Pimentel and Pimentel, 2007; Cuéllar and Webber, 2010; Vittuari et al., 2016).

The impact of climate change on agriculture is increasing, negatively affecting both crop and livestock systems worldwide. Many regions are experiencing natural and man-made disasters of higher frequency and intensity. Strong vulnerability for developing countries is predicted in the next decades, since productive areas will be exposed to extreme

Sustainable Food Supply Chains. https://doi.org/10.1016/B978-0-12-813411-5.00017-X

events such as tropical cyclones, floods, excessive precipitations, or drought (UNFCCC, 2007), with limited mechanisms to mitigate, adapt, or overcome the after-event damages. At the same time, human practices in agricultural systems and forestry can exacerbate the intensity of negative impacts (IPCC, 2014). The agricultural sector is a major emitter of GHG emission: 17% directly through agricultural activities and an additional 7% to 14% through land use changes (OECD, 2015).

Improved technologies for food production and processing, more efficient distribution chains, more effective knowledge transfer, as well as (foreign) capital investments contributed to the growth and consolidation of agro-industrial processes. This created opportunities for producing and exporting higher value crops. At the same time, several questions have emerged relating to the role of small farmers in industrialization processes and the effects on nutrition and environment (Reardon and Barrett, 2000; Paarlberg and Paarlberg, 2013). In addition, international agro-food trade significantly expanded in the past decades with structural changes in favor of developing countries, which in some cases became important players. Global food commodities trade counts now for more than USD520 billion per year and would be enough to feed approximately two billion people (Ghosh and Guven, 2006; MacDonald et al., 2015). Nevertheless, food markets are increasingly unstable and characterized by high price volatility, as demonstrated by the international price shock of 2008 and the associated effects on food security (Kalkuhl et al., 2016; OECD-FAO, 2016).

Therefore the forthcoming increase of global population and related demand for food, which is expected to grow 70% by 2050 (FAO, 2012), will represent a huge challenge for food security and sustainability. Food consumption is changing as well. Obesity is becoming a concern, just like hunger, in some medium-low income countries, like Mexico, Brazil, Chile, or South Africa (OECD, 2017). The growing demand of meat products in newly industrialized countries, with China counting for more than 30%, is posing harmful effects on the environment, public health, and economic systems (OECD-FAO, 2016). Thus, many countries are calling for lower meat consumption and implement dedicated awareness campaigns and policy measures (Paarlberg and Paarlberg, 2013; Lang and Heasman, 2015). Currently, about 793 million people are undernourished globally as the UN reported in its Sustainable Development Goals, and about 10 million die of hunger or hunger-related diseases (Nellemann, 2009), while around one-third of food produced is wasted. It is therefore urgent to address the food waste issue.

2 Defining and measuring food loss and waste

The debate around food waste definitions has been influenced by the perspective by which definitions are developed, targeting among others food security, resource efficiency, and/or nutritional quality aspects of food production and consumption. There are also changes over time and geographic differences, reflecting the development of the discussion as well as the different cultural approaches. FAO (1981) provided an early definition including all agro-food products allocated to human consumption that are instead discarded, lost, degraded, or consumed by pests, at any stage of the FSC. In 2011, FAO (2011a) revised this definition by emphasizing the distinction between food loss—the decrease in edible food mass available for human consumption throughout the different stages up to distribution—and food waste—the edible food discarded at the retail and consumer level. Food loss is mainly caused by logistical and infrastructural limitations, while food waste is mainly related to behavioral factors (FAO, 2011a).

In its attempt to identify a comprehensive definition, FAO released a *Definitional Framework of Food Loss* and analyzed "Food losses and waste in the context of sustainable food systems" in a report developed by the High Level Panel of Experts on Food Security and Nutrition within the Committee on World Food Security (CFS). According to the *Definitional Framework of Food Loss* (FAO, 2014), FLW includes:

- all resource flows including by-products or secondary products that are meant for human consumption, but that in specific supply chains cannot be transformed;
- food that is fit to enter the FSC, but intentionally discarded or redirected to nonfood use in the preharvest phase;
- food that is harvest-mature and unintentionally getting spoilt in the preharvest phase;
- food that is fit to proceed in the FSC, but redirected to nonfood use or discarded in the postharvest phase of sorting and grading (fruits, fish discards, etc.) without getting spoilt or spilled;
- food that is redirected to animal feed or compost; food fit for consumption that has been left to spoil or expire by choice or as a result of negligence by an actor—predominantly, but not exclusively the final consumer at household level.

The report *Food losses and waste in the context of sustainable food systems* moves behind food losses and adopts a food security and nutrition lens defining FLW as "a decrease, at all stages of the food chain from harvest to consumption, in mass, of food that was originally intended for human consumption, regardless of the cause" (HLPE, 2014, p. 11). As with previous FAO works, this report also distinguishes between losses occurring in the early stages of the food chain and waste occurring at the final stages. The major advancement is the introduction of the concept of food quality loss or waste

(FQLW), which refers to the decrease of a quality attribute of food (nutrition, aspect, etc.), linked to the degradation of the product, at all stages of the food chain from harvest to consumption (FAO, 2014).

Other institutions, operating in different settings and focusing mostly on food waste, emphasize the difference between avoidable and unavoidable food waste (WRAP, 2009) or suggest including overnutrition, i.e., the difference between nutritional needs and actual food intake (Serafini and Toti, 2016).

At the European level a first definition was crafted within the FP7 Project FUSIONS (Food Use for Social Innovation by Optimizing Waste Prevention Strategies), which defined food waste as: any food, and inedible parts of food, removed from the food supply chain to be recovered or disposed, including composted, crops ploughed in/not harvested, anaerobic digestion, bio-energy production, co-generation, incineration, disposal to sewer, landfill or discarded to sea (Östergen et al., 2014). Table 1 compares the different definitions of food waste provided by FUSIONS and FAO.

TABLE 1 Comparative table: FUSIONS vs FAO definitions

Definitions	HLPE report	FAO D. framework	FUSIONS D. framework
Edible food fractions fit to enter the FSC, but intentionally discarded or redirected to nonfood use in the preharvest phase	√	√	
Edible food *not* being valorized[a]	√	√	√
Inedible parts of food[b] *not* being valorized[c]			√
Edible parts to be valorized including feed	√	√	
Inedible part to be valorized including feed			

[a]*Including food that is harvest-mature and unintentionally getting spoiled in the preharvest phase.*
[b]*Including inedible parts of food that is harvest-mature and unintentionally getting spoiled in the preharvest phase.*
[c]*Practically special rules are given on how to apply this rule in the FUSIONS manual to make the cut in a reasonable and practical way.*
Source: Authors' elaboration from HLPE report, FAO Framework and Fusions Framework.

The diversity of resources engaged in the agro-food system and the complex routes characterizing and connecting the different stages of the FSC are summarized in Fig. 1. This complexity explains the difficulties encountered by national governments and international organizations in setting clear boundaries and identifying a common definition.

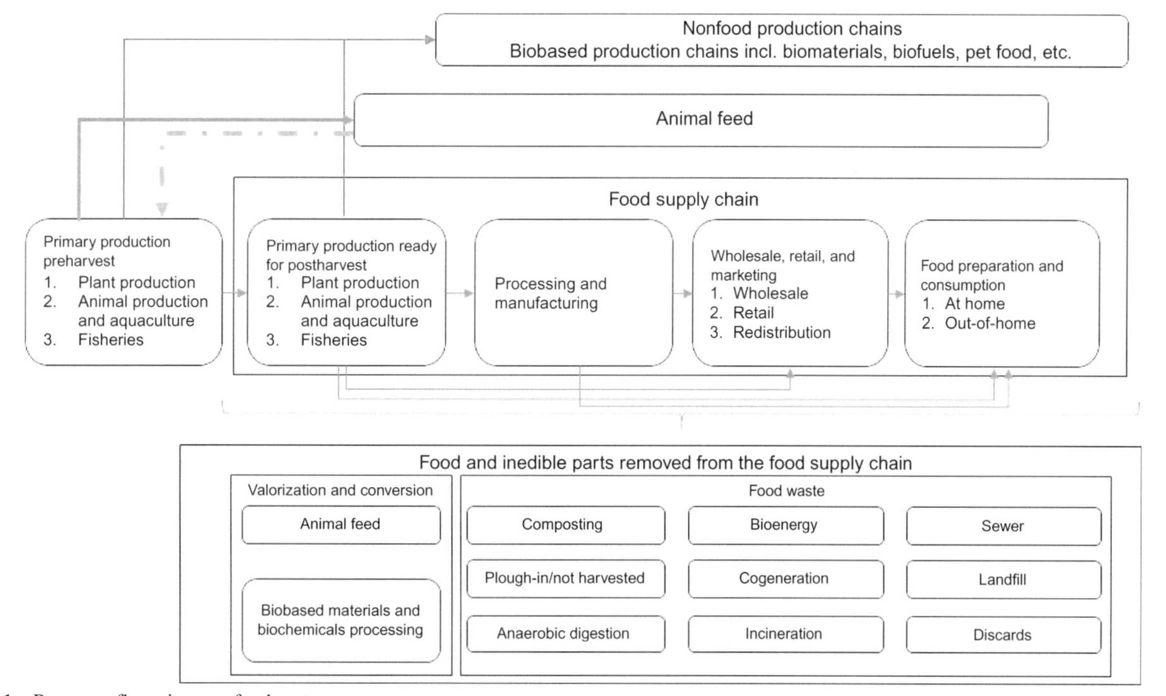

FIG. 1 Resource flows in agro-food systems.

Defining FLW has direct consequences on the design, effectiveness, and comparability of quantification methodologies and on their related results. What finally emerges from the estimates proposed by national governments and international organizations is an array of different data that are contributing to understanding the magnitude of the phenomena, but that are not comparable one with another.

At the global level, the widely adopted FAO estimate suggests that roughly one-third of the edible parts of food produced for human consumption, about 1.3 Gt, is lost or wasted globally (FAO, 2011b). This figure shows a major polarization between low-income countries, where the agricultural and the postharvest handling of products generate most of the losses, and medium-high income countries, where food is wasted predominantly in the downstream segments of the chain. In fact, in low-income countries, FLW amounts to 120–170 kg per capita per year, with about 40% occurring during post-harvest and processing. In industrialized countries, FLW reaches 280–300 kg per capita per year, with about 40% happening during distribution and consumption (FAO, 2011b). However, these numbers do not always reflect the complexity of each FSC, while FLW depends also on product-specific characteristics and management conditions and practices. In general, the amount of FLW tends to be higher for products where quality standards for commercialization cause a strict grading, as in the case of vegetables, fruit, and tubers (FAO, 2011b).

In the EU in 2015, according to estimates based on the FUSIONS definition, FLW amounted to 146–200 kg per person with between 46.7% and 63.5% of the total, or between 83 and 101 kg, generated from households (Stenmarck et al., 2016).

In the United States, the ReFED study (2016) proposes a significant revision of the previous estimates of food waste that ranged from 35 million to 103 million tons per year (Bellemare et al., 2017). ReFED collected one of the broadest datasets to date to establish a map of where food is wasted. The baseline amount of US FLW for 2015 is approximately 62.5 million tons annually: 52.4 million tons disposed of annually in landfills and incinerators, and 10.1 million tons from unharvested crops and packhouses.

Table 2 summarizes the global FLW estimates proposed by FAO (2011a,b), the European estimates developed by FUSIONS (Stenmarck et al., 2016), and the US estimate from ReFED (2016).

TABLE 2 Comparison of major studies on United States, EU, and LAC food waste baseline

	Food waste estimates		
	Annual (million tons/year)	Per capita (kg/persons/year)	Source
United States	63	196	ReFED (2016)
	103	321	FAO (2011a,b, 2017)
	67	208	USDA (2014)
EU	89	120	FAO (2011a,b, 2017)
	88	119	FUSIONS (2016)
LAC	127	220	FAO (2011a,b, 2017)

3 Understanding the causes

Understanding the causes of food losses and waste is particularly crucial for the identification of targeted interventions. Most of the studies investigating the causes of FLW are focusing on the final segments of the food chain, where most of the food waste is generated in industrialized countries, i.e., retail, food services, and especially consumers/households (Aschemann-Witzel et al., 2015; Neff et al., 2015; Parizeau et al., 2015; Stancu et al., 2015; Van Geffen et al., 2016). The analyses aiming at identifying the main FW determinants for the entire FSC generally represent part of wider studies including other relevant objectives such as FLW definition, quantification, environmental and economic impacts, legislative barriers, and policy recommendations (FAO, 2011a,b, 2014; Bio Intelligence Service, 2010; Parfitt et al., 2010; Thyberg and Tonjes, 2016; Adam, 2015; Waarts et al., 2011).

As suggested in Fig. 2 the causes of FLW in the earlier part of the food chain (production and harvest, first processing, industrial processing) are mostly related to technical and technological limitations, poor storage infrastructures, climate and weather conditions. Governmental regulations such as rigid food protocols and standards play an important role. In general, public policies and subsidies can have unintended consequences, as in the case of direct subsidies causing overproduction or, in the case of fruits and vegetables. market withdrawals ordered to control prices under the Common Market Organization of the Common Agricultural Policy of the European Union. Specific market conditions such as missing markets, price transmission, and imperfect information, and standards set by private stakeholders as retailers (i.e., cosmetic standards for horticultural products) represent a relevant driver for food waste generation (Parfitt et al., 2010; Mena et al., 2011; Lebersorger and Schneider, 2014; Canali et al., 2017).

At the distribution and retail level FLW can be related to inefficiencies of the supply chain logistic, errors in order forecasting, deterioration of products and packaging, aggressive marketing strategies (i.e., special offers) often increasing the probability of generating food waste in the household, and governmental regulations such as date marking and food hygiene and safety. Emerging issues include unfair practices occurring, in particular in those situations characterized by unbalanced market power relations (Parfitt et al., 2010; Mena et al., 2011; Lebersorger and Schneider, 2014; Canali et al., 2017).

Household food waste is mostly related to behavioral factors with a rather significant influence exerted by demographics, geographical and social elements, and knowledge encompassing all food routines: planning, purchasing, storing, preparing, serving, disposing (Koivupuro et al., 2012; Parizeau et al., 2015; Stancu et al., 2016; Setti et al., 2016; Van Geffen et al., 2016; Gaiani et al., 2018). External conditions influencing consumer behavior as potential opportunities (i.e., a more efficient appliance in the kitchen) and unpredictable situations might also represent important factors (Van Geffen et al., 2016).

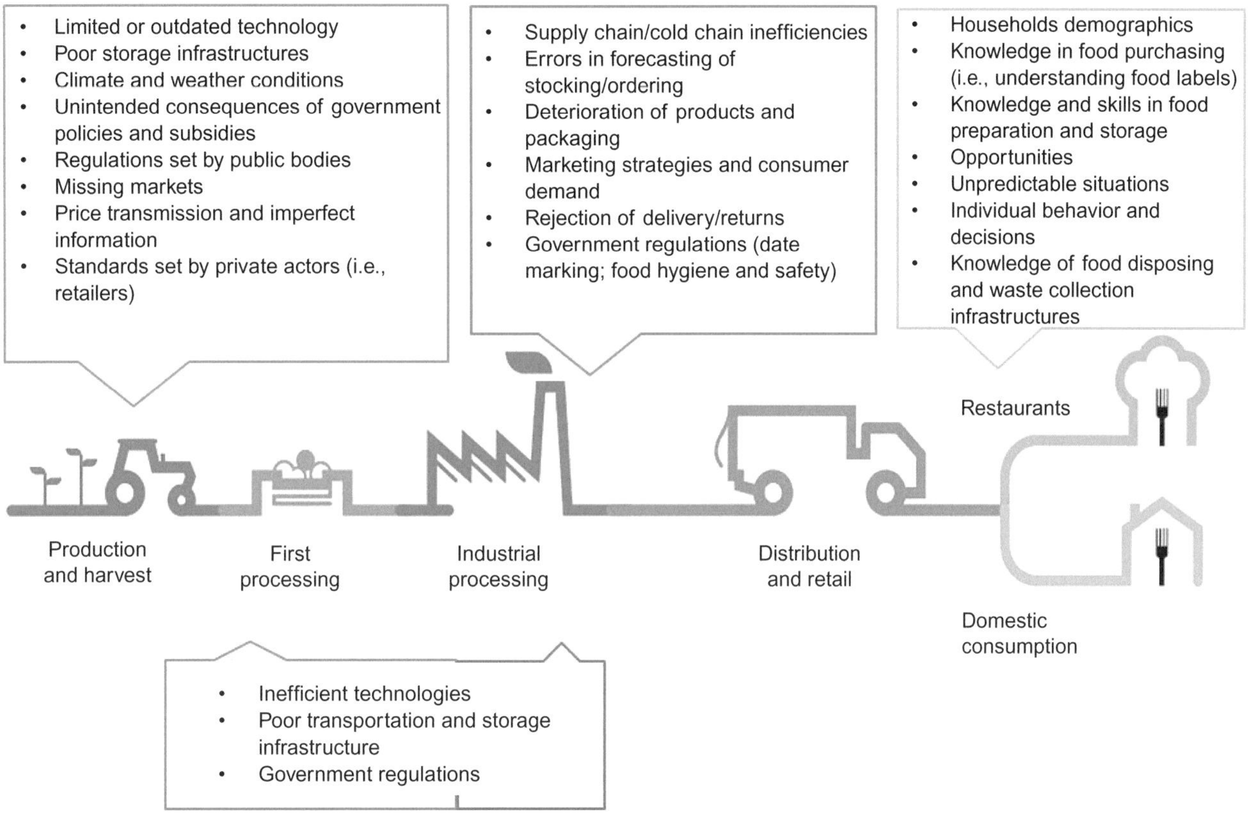

FIG. 2 Major causes of FLW. *(Authors' elaboration from Buchner et al., 2012; Canali et al., 2017, and Van Geffen et al., 2016.)*

4 Analyzing the impacts: A life cycle thinking perspective

The relevance of FLW is intrinsically connected with the complex challenges of food systems sustainability outlined at the beginning of the chapter. The food waste scenario has two dimensions (Papargyropoulou et al., 2014). On one hand, the global population is rising, food waste generation is not diminishing, and food security is becoming an increasingly urgent issue (FAO, 2011b). On the other hand, agriculture and the food sector greatly contribute to climate change, due to the energy intensity, water use, and land degradation. It is estimated that 22% of the GHG emissions in the EU are attributable to the food sector (Papargyropoulou et al., 2014). Food waste aggravates all these challenges and presents several direct and indirect environmental, economic, and social impacts.

FLW management in all segments of the supply chain usually requires proper handling and disposal. Some end-of-life treatments, such as landfilling or incineration, have substantial impacts on the environment. Likewise, disposal and the loss of value of wasted food products are examples of direct economic impacts. In addition, FLW has important direct consequences on food security and poverty. However, as underlined by FAO (2014), this is only a relatively small share of the real footprint of FLW, since it should also include the hidden indirect impacts caused by the production and distribution of uneaten food. For example, the carbon footprint of food wastage is estimated in about 4.4 Gt CO_2 equivalent per year, ranking as the third top emitter after the United States and China, while the blue water footprint is about 250 km^3 (FAO, 2014).

When adopting a full-cost accounting perspective, the global cost of food waste, including economic, environmental, and social costs, is about USD2.6 trillion per year, roughly equivalent to the GDP of France, or approximately twice the total annual food expenditure in the United States (FAO, 2014). While almost USD1 trillion per year are due to the economic cost of FLW, environmental costs reach around USD700 billion and social costs around USD900 billion. Therefore, appropriate strategies of FLW prevention and valorization could avoid or mitigate such embodied impacts.

As argued by Östergren et al. (2017), urged by the relevance of resource efficiency and circular economy agenda, policymakers and different stakeholders are seeking alternatives for current surplus food or side flows within the food supply chain. Any new valorization or intervention aimed to prevent food waste will, however, be associated with impacts (monetary and environmental). To allow informed decision making at all levels—from individual stakeholder to policy level—robust, consistent, and science-based approaches are required.

Therefore to understand the impact and interrelation between FLW created at the production and consumption level, a methodology that provides the environmental and economic impact of the waste in order to reconsider the different interventions that producers/consumers and policy makers are suggesting, and in this sense, life cycle thinking (LCT) techniques could respond to this need. Life cycle thinking allows to go beyond traditional linear focuses on production and disposal, by including environmental, social, and economic impacts of a product over its entire life cycle (from cradle to grave), and consequently allowing a more holistic assessment, highlighting, comprehension and possible prevention/reduction of FLW. In this sense, the concepts of life cycle management and sustainable production and consumption modify the way in which "waste" is framed, underlining the gap between waste prevention and waste management (Papargyropoulou et al., 2014). LCT can:

- Prevent/avoid food waste shifting.
- Identify trade-offs or win-win scenarios as well as cost distribution of food waste prevention or valorization.
- Be useful in comparing two or more options for product choice/route or deciding what to do with a product at the end of its life.

Life cycle assessment (LCA) is a compilation and evaluation of the inputs, outputs, and the potential environmental impacts of a product system throughout its life cycle. The International Standards Organisation (ISO) (2006) describes the principles and framework for LCA, including definition of the goal and scope, inventory analysis, impact assessment, and interpretation. As far as costs are regarded, life cycle costing (LCC), as De Menna et al. (2016) discusses, is a consolidated methodology aimed at calculating the overall cost of a product or a service over its lifespan or life cycle. Despite being used for a long time by both decision makers and businesses, LCC has been standardized only with reference to specific product categories. Several approaches can be found in the literature, mainly differing in terms of perspective, costs included, and potential application.

The combination of both methodologies (E-LCC and LCC) would allow the consumer and the producer to understand how every impact is produced and where and help policymakers to act in specific segments of the chain and be more effective in the fight against food waste.

5 Designing policies to address FLW

National governments and other public bodies increasingly recognize FLW as a major challenge and have responded in a number of ways, designing and implementing policies on biowaste management, resource efficiency, consumer awareness, and food donation. Most of these interventions have been designed according to the food waste hierarchy framework, which is prioritizing food waste prevention, followed by food recovery, and finally recycling solutions.

In 2015, a major step was taken by the international community adopting a set of new goals to reinforce the fight against poverty and hunger, protect the planet and ensure peace and prosperity for all. In this frame, Sustainable Development Goal (SDG) number 12 aims at "Ensuring sustainable consumption and production patterns" and suggests an additional focus

(SDG 12.3) on food waste, stating: "By 2030, halve per capita global food waste at the retail and consumer levels and reduce food losses along production and supply chains, including post-harvest losses." The introduction of this goal represents a formal recognition of food waste as a global priority to tackle with a combination of top-down and bottom-up approaches. This initiative of the United Nations stimulated a further interest in FLW that has emerged as a priority on the global, regional, and national sustainability agenda.

The following section focuses on the policy framework of the European Union, an early starter in the fight against FLW; on the United States, which started with some delays led by the United States Department of Agriculture and the US Environmental Protection Agency; and on Latin America, where the work on FLW is supported by FAO.

5.1 The policy framework in the European Union

In 2015 *Closing the loop—An EU Action Plan for the Circular Economy* was published by the European Commission. The plan presents the European Commission's ambitions on stimulating Europe's transition towards a circular economy with a view to boost global competitiveness, foster sustainable growth, and generate new jobs. Food waste prevention is included as an integral part of achieving these ambitions.

Within the so-called Circular Economy package, EU and Member States are committed to meeting the United Nations' Sustainable Development Goals (SDGs), including target 12.3. To support achievement of the SDG targets for food waste reduction in the EU, the Commission foresaw:

- the elaboration of a common EU methodology to consistently measure food waste in cooperation with Member States and stakeholders;
- the creation of a new platform involving both Member States and actors in the food chain in order to help define measures needed to achieve the food waste SDG, facilitate intersector cooperation, and share best practices and results achieved;
- measures to clarify EU legislation related to waste, food, and feed, and facilitate food donation and the use of former foodstuffs and by-products from the food chain for feed production, without compromising food and feed safety, and examine ways to improve the use of date marking by actors in the food chain and its understanding by consumers, in particular "best-before" labeling.

The importance of food waste reduction within the context of EU strategies on resource efficiency has been highlighted and addressed by both the European Commission and the EU Parliament through some key nonlegislative acts. The EU's *Roadmap to a Resource Efficient Europe* (European Commission, 2011) identified food as a key sector where resource efficiency should be improved, and set an aspirational goal to halve the disposal of edible food waste in the EU by 2020. In 2012, the European Parliament adopted a nonlegislative resolution on how to avoid food wastage (European Commission, 2011) that called for action to halve food waste by 2025 and improve access to food by the needy. The Resolution highlighted the importance of food waste prevention policies within the context of EU strategies on resource efficiency. It explicitly asked the Commission to take concrete actions to reduce food wastage within the context of the resource-efficient Europe flagship initiative. Moreover, the need for a common EU strategy against food waste has been further highlighted within the 7th Environmental Action program (European Commission, 2011) where the EU Commission is required to "present a comprehensive strategy to combat unnecessary food waste and work with Member States in the fight against excessive food waste generation."

The measures addressed for food waste reduction, foreseen within the Circular Economy Package, should be intended as a starting point toward the definition and implementation of a more common EU strategy to address food waste.

The different EU states are developing their own strategies to avoid food waste, from facilitating food donations to raising public awareness. Since most of the food waste in developed countries originates at the retail and consumer level, most of the actions focus on these stages (Table 3).

TABLE 3 Interventions to avoid food waste in Italy and France

Level	Italy	France
National	**Name of practice:** National Food Waste Prevention Plan (PINPAS) **Launched**: in 2014. Implementation ongoing **Aim**: to get food waste prevention at the center of the political agenda from the local to the European level, increasing	**Name of practice**: National Pact Against Food Waste **Launched**: in 2013. Implementation ongoing **Aim**: reducing food waste by half by 2025, developing actions with all stakeholders: farmers, wholesale markets, processing industries, retailers, catering, and local/

Continued

TABLE 3 Interventions to avoid food waste in Italy and France—cont'd

Level	Italy	France
	knowledge about the environmental, social and economic impacts of food waste and raising awareness among consumers **Priorities:** – Prevention – Human consumption (donation or transformation)	regional authorities raising their awareness **Priorities:** – Prevention – Human consumption (donation or transformation) – Feed valorization – Use for composting or energy recovery
Donation	**Law 166/2016:** aims at reducing food waste for each of the phases of production, transformation, distribution, and delivery of food, pharmaceuticals and other products through the achievement of priority goals **Goal:** to cut at least one-fifth of the estimated five million metric tons of food wasted by making food donation simpler. They will also receive tax credits based on how much food they donate **Key elements:** – Provisions concerning the donation and distribution of food and pharmaceutical products – Businesses will be able to record their donations on a monthly basis and will not be subject to penalties for giving away food past its sell-by date	**Law 2016-138:** aims at reducing food waste established that large supermarkets (with more than 400 m^2 of commercial space) are no longer allowed to throw away good quality food approaching its "best-before" date **Goal:** reduce food waste by half by 2025 through the second priority, human consumption (food donation) **Key elements:** – It includes additional actions such as information and education on combating food wastage in schools – Allows the integration in combating food waste of companies' social and environmental reporting

Source: Author's elaboration.

5.2 The policy framework in the United States

Similarly to the EU, in the United States policies related to food waste exist at the federal, state, and municipal levels. These interventions have different characteristics in terms of objectives and design, including the promotion of best practices, the penalization of certain behavior, the simplification of date labeling regulations, donation liability protections and tax incentives, animal feed, and waste regulations (ReFED, 2016).

In alignment with SDG 12.3, the United States Department of Agriculture (USDA) and the US Environmental Protection Agency (EPA) announced the first-ever domestic goal to halve FLW by the year 2030. By taking action on the US 2030 Food Loss and Waste Reduction goal (2030 FLW reduction goal), the United States can help feed the hungry, save money for families and businesses, and protect the environment. Led by USDA and EPA, the federal government is seeking to work with communities, organizations, and businesses along with other partners in state, tribal, and local governments to reduce food loss and waste to reach the goal. Table 4 shows the different interventions to followed in two US States to avoid food waste.

TABLE 4 Interventions to avoid food waste in Alabama and Oregon

Level	Alabama	Oregon
Prevention	**Data labeling:** the aim of the regulations is to provide useful and clear information to consumers to avoid food waste. Data label is *not* required for any specific product	**Data labeling:** the aim of the regulations is to provide useful and clear information to consumers to avoid food waste. Data label is required for perishables
Recovery policy	– Ala. Code 1975 § 20-1-6 protects the donor, as a good faith, and sets the Standards for donated food as the following: "apparently fit for human consumption includes food not readily marketable due to appearance, freshness, grade, or surplus." **Tax incentives** not applied additional to Federal law tax incentives/deductions in state of Alabama	**Liability protection:** regarding the Bill Emerson Good Samaritan Food Donation Act, which provides a strong federal baseline of protection for food donors. It covers individuals, businesses, nonprofit organizations, the officers of businesses and nonprofit organizations, and gleaners

TABLE 4 Interventions to avoid food waste in Alabama and Oregon—cont'd

Level	Alabama	Oregon
		– The regulation Or. Rev. Stat. § 30.890 set the level of protection, stakeholders and the Standards for Donated Food **Tax incentives:** – Or. Rev. Stat. § 315.154 and 315.156: The benefit is established as 10% of wholesale market price, the eligible donors are taxpayers or corporations that grow crops or livestock (eligible food crops and livestock) and the recipients Food Bank or other charitable organization in OR that distributes food without charge
Recycling	**Animal feed:** particular regulations on the safety of food provided to swine, Ala. Code § 2-15-211 (2015), with the enforcement of "whoever feeds garbage to swine will be guilty of a misdemeanor and fined up to USD500 and imprisoned for up to 6 month. 2-15-211 (2015)"	**Animal feed:** particular regulations on the safety of food provided to swine, by Or. Rev. Stat. 600.010–.120 (2015). Including the following enforcement: "The state department of agriculture may quarantine any swine that are being fed garbage or offal. 600.105 (2015). The state also may restrain any actual or threatened violation. 600.120 (2015)"

Source: Authors' own elaboration based on ReFED database.

5.3 The policy framework in the Latin American and Caribbean region

There is an agreed-upon concern about the urgency to address FLW in this region. Some assessments besides FAO estimations have been done in Costa Rica, the Dominican Republic, Colombia, Guyana, Trinidad and Tobago, Saint Lucia, and Argentina, in collaboration with food banks, planning agencies, research institutes and universities, among others (Feliz, 2014; Brenes-Peralta et al., 2015; DNP, 2016; FAO-RLC, 2016). Although comparisons are not possible among these due to different methodologies and targets, the information begins to point to the urgency to tackle FLW. FAO is currently working on the Global Food Loss Index (GFLI) indicator, which is expected to set the baseline and expected progress according to target 12.3, covering losses occurring on farm, during transport, in storage, and during processing. At this moment, losses in retail and household are not covered; however, decision makers and stakeholders have begun to consider elements that would help determine a baseline and report parameters for this index in the nearby future, together with possible interventions in different stages of the food systems.

Institutions, researchers, and technicians had been addressing postharvest losses, but the view of FLW in a wider dimension had not been commonly considered in the region. It was not until the agreements made in the 33rd FAO Regional Conference for Latin America and the Caribbean that a facilitating environment was set for further analysis. Consequently, the region held its First Regional Consultation regarding this matter in November 2014 in Chile (FAO-RLC, 2015), resulting in the creation of a network of experts that would later advise committees, government agencies, and local initiatives. Facts like the limited knowledge on the magnitude of FLW, isolated efforts, and the absence of a quantification methodology are being taken into consideration.

Two main guidelines are currently steering the Regional FLW actions. One is the commitment of the CELAC (Comunidad de Estados Latinoamericanos y del Caribe—Community of Latin American and Caribbean States) to execute the Plan for Food Security, Nutrition and Hunger Eradication 2025, which emerged as an effort of its 33 members to eradicate hunger and poverty from the region by 2025. The plan consists of four pillars and ten main lines of action, including FLW (SEPSA, 2016), and stresses the need to work together with small-scale family farmers, as vehicles of rural development, to tackle poverty and food insecurity as well. In this sense, the CELAC Plan expects to guide the region in a harmonized manner with SDG12.3 towards the reduction of FLW. On the other hand, as a response to the prior agreement and from a more technical and hands-on approach, the Region celebrated a series of Regional Dialogues for FLW called by the Regional Office of FAO, where the network of experts and representatives from the private sector as well as the legislative and the executive branches of these countries, have come together to discuss the matter and validate the proposal of an

International Code of Conduct to prevent and reduce FLW. The technical background argument was presented at the COAG by Mexico and Costa Rica during 2018 (FAO, 2018; FAO-RLC, 2017).

Other initiatives at the national level regarding environmental scope and competitiveness approaches are also enabling further discussion and action to reduce FLW. As a result, national committees have begun operation in order to act through interinstitutional and intersectorial liaisons in terms of awareness, research, alliances, and governance, expecting that the region will commit to addressing FLW.

6 The way ahead

Despite attention from the academic world, civil society, and policymakers, the debate on FLW is still challenged by contrasting definitions, scope boundaries, drivers (Canali et al., 2017), and the lack of common quantification and reporting methods along the FSC (Tostivint et al., 2016). Moreover, as policies and policy proposals are emerging, there is a greater need to establish criteria to be used for the evaluation of their impact and effectiveness (Burgos et al., 2016). After over a decade of national and international efforts to address FLW, some challenges remain open.

Food losses and waste have a complex nature. Their causes are different across sectors, segments, geographical boundaries, and levels of governance. This diversity calls for targeted approaches and for dedicated analytical tools to identify the hotspots where FLW are generated along the food supply chain and implement dedicated interventions. Food waste policies are not yet fully integrated with other sectorial policies. Their integration into urban food policies, dietary guidelines, and waste management plans represent an opportunity for all the different levels of governance.

Companies do not always prioritize food waste prevention and reduction among their core commitments, through the establishment of clear targets and collaboration with their commercial networks. Several categories of consumers are not fully aware of the magnitude and impacts of FLW. There is wide room to engage them and nudge them towards changing behavior.

There is a crucial need for analytical tools focusing both on specific issues (target value chains, consumer behavior) and on more systemic approaches considering the entire food supply chain and all the potential valorization routes. The road to reduce food waste might have been paved with good intentions, but it also requires significant results that represent the knowledge base to address one of the major challenges for the sustainability of our food system.

References

Adam, A., 2015. Drivers of food waste and policy responses to the issue: the role of retailers in food supply chains. Working Paper, Institute for International Political Economy Berlin, No. 59/2015, Inst. for International Political Economy, Berlin.

Aragrande, M., Argenti, O., Lewis, B., 2001. Studying Food Supply and Distribution Systems to Cities in Developing Countries and Countries in Transition. Database. Available from: http://www.fao.org/docrep/003/X6996E/X6996E00.HTM.

Aschemann-Witzel, J., et al., 2015. Consumer-related food waste: causes and potential for action. Sustainability 7 (6), 6457–6477. https://doi.org/10.3390/su7066457.

Bellemare, M.F., et al., 2017. On the measurement of food waste. Am. J. Agric. Econ. 99 (5), 1148–1158. https://doi.org/10.1093/ajae/aax034.

Brenes-Peralta, L., Jiménez-Morales, M., Gamboa-Murillo, M., 2015. Diagnóstico de Pérdidas y Desperdicio alimenticio en dos canales de comercialización de la agrocadena de tomate costarricense para su posterior disminución. Available from: http://www.fao.org/3/a-i4655s.pdf.

Bio Intelligence Service, 2010. Preparatory study on food waste across EU 27. https://doi.org/10.2779/85947.

Brunori, G., Bartolini, F., 2013. Local agri-food systems in a global world: market, social and environmental challenges. Eur. Rev. Agric. Econ. 40 (2), 408–411. Available at: http://erae.oxfordjournals.org/content/40/2/408.short.

Buchner, B., Fischler, C., Gustafson, E., Reilly, J., Riccardi, G., Ricordi, C., Veronesi, U., 2012. Food waste: causes, impacts and proposals. Barilla Cent. Food Nutr. 53–61. https://doi.org/45854585.

Burgos, S., et al., 2016. Policy Evaluation Framework. FUSIONS project. ISBN: 978-94-6257-719-0.

Canali, M., et al., 2017. Food waste drivers in Europe, from identification to possible interventions. Sustainability. 9(1). https://doi.org/10.3390/su9010037.

Cuéllar, A.D., Webber, M.E., 2010. Wasted food, wasted energy: the embedded energy in food waste in the United States. Environ. Sci. Technol. 44 (16), 6464–6469. ACS Publications.

De Menna, F., et al., 2016. Methodology for Evaluating LCC of Natural Resources and Life Sciences Acknowledgments & Disclaimer. Available from: http://eu-refresh.org/sites/default/files/REFRESH_D5_2_Meth_for_ev_LCC_Final_formatted_0.pdf.

DNP, 2016. Pérdida y Desperdicio de Alimentos en Colombia, Estudio de la Dirección de Seguimiento y Evaluación de Políticas Públicas. Departamento Nacional de Planeación, Bogotá, Colombia.

European Commission, 2011. Communication from the Commission to the European Parliament, the Council, the European Economic and Social Committee and the Committee of the Regions Roadmap to a Resource Efficient Europe COM (2011) 571 final.

FAO, 2011a. Global food losses and food waste: extent, causes and prevention. International Congress: Save Food!, p. 38. https://doi.org/10.1098/rstb.2010.0126.

FAO, 2011b. Global food losses and food waste - Extent, causes and prevention. Save Food: An initiative on Food Loss and Waste Reduction. https://doi.org/10.1098/rstb.2010.0126.

FAO, 2012. Global Trends and Future Challenges for the Work of the Organization—Web Annex. In: FAO Regional Conference for Europe (ERC). Available from: http://www.fao.org/docrep/meeting/025/GT_WebAnnex_RC2012.pdf.

FAO, 2014. Definitional Framework of Food Loss, pp. 1–18. Available from: http://www.fao.org/fileadmin/user_upload/save-food/PDF/FLW_Definition_and_Scope_2014.pdf.

FAO-RLC, 2015. 2° Boletín Pérdidas y Desperdicio de Alimentos en América Latina y el Caribe.

FAO, 2018. Report of the 26th Session of the Committee on Agriculture. Rome, October 1–5, 2018, pp. 1–18.

FAO-RLC, 2016. 3° Boletín: Pérdidas y Desperdicio de Alimentos en América Latina y el Caribe. Available from: http://www.fao.org/3/a-i5504s.pdf.

FAO-RLC, 2017. 4° Boletín: Pérdidas y Desperdicio de Alimentos en América Latina y el Caribe.

FAO-UNEP, 1981. Food Loss Prevention in Perishable Crops. (ISBN 92-5-101028-5). Available from: http://www.fao.org/docrep/s8620e/S8620E00.htm.

Feliz, L., 2014. Sondeo Pérdida de Alimentos en la República Dominicana para la ampliación del Proyecto Banco de Alimentos.

Ghosh, B.N., Guven, H.M., 2006. Globalization and the Third World. Practice. https://doi.org/10.1057/9780230502567.

FUSIONS, 2016. Estimates of European food waste levels. FUSIONS Project. FUSIONS Project. ISBN 9789188319012.

Gaiani, S., et al., 2018. Food wasters: profiling consumers' attitude to waste food in Italy. Waste Manag 72, 17–24. https://doi.org/10.1016/j.wasman.2017.11.012.

Gomiero, T., 2016. Soil degradation, land scarcity and food security: reviewing a complex challenge. Sustainability 8 (3), 1–41. https://doi.org/10.3390/su8030281.

Hazell, P., Wood, S., 2008. Drivers of change in global agriculture. Philos. Trans. R. Soc. Lon. B Biol. Sci. 363 (1491), 495–515. https://doi.org/10.1098/rstb.2007.2166.

HLPE, 2014. Food Losses and Waste in the Context of Sustainable Food Systems. A Report by the High Level Panel of Experts on Food Security and Nutrition of the Committee on World Food Security, pp. 1–6 (June), https://doi.org/65842315.

IPCC (2014) Climate Change 2014: Impacts, Adaptation, and Vulnerability. Summaries, Frequently Asked Questions, and Cross-Chapter Boxes. A Contribution of Working Group II to the Fifth Assessment Report of the Intergovernmental Panel on Climate Change [Field, C.B., V, Climate Change 2014: Impacts, Adaptation, and Vulnerability. Summaries, Frequently Asked Questions, and Cross-Chapter Boxes. A Contribution of Working Group II to the Fifth Assessment Report of the Intergovernmental Panel on Climate Change] doi:10.1016/j.renene.2009.11.012.

Kalkuhl, M., Von Braun, J., Torero, M., 2016. Food Price Volatility and Its Implications for Food Security and Policy. (34), pp. 1–2. https://doi.org/10.1007/978-3-319-28201-5.

Knickel, K., et al., 2017. Between aspirations and reality: making farming, food systems and rural areas more resilient, sustainable and equitable. J. Rural. Stud. https://doi.org/10.1016/j.jrurstud.2017.04.012.

Lang, T., Heasman, M., 2015. Food Wars: The Global Battle for Mouths, Minds and Markets. Routledge.

Koivupuro, H.K., et al., 2012. Influence of socio-demographical, behavioural and attitudinal factors on the amount of avoidable food waste generated in Finnish households. Int. J. Consum. Stud. 36 (2), 183–191.

Lebersorger, S., Schneider, F., 2014. Food loss rates at the food retail, influencing factors and reasons as a basis for waste prevention measures. Waste Manag. 34, 1911–1919.

MacDonald, R., Reitmeier, C., 2017. Understanding Food Systems, Agriculture, Food Science and Nutrition in the United States, pp. 287–338.

MacDonald, G.K., et al., 2015. Rethinking agricultural trade relationships in an era of globalization. Bioscience 65 (3), 275–289. https://doi.org/10.1093/biosci/biu225.

Mena, C., Adenso-Diaz, B., Yurt, O., 2011. The causes of food waste in the supplier-retailer interface: evidences from the UK and Spain. Resour. Conserv. Recycl. 55, 648–658.

Neff, R.A., Spiker, M.L., Truant, P.L., 2015. Wasted food: U.S. consumers reported awareness, attitudes, and behaviors. PLoS One 10 (6), 1–16. https://doi.org/10.1371/journal.pone.0127881.

Nellemann, C., 2009. The environmental food crisis. Pediatr. Diabetes. https://doi.org/10.1111/j.1399-5448.2011.00774.x.

Nestle, M., 2013. Food Politics: How the Food Industry Influences Nutrition and Health. Univ of California Press.

OECD, 2015. The Economic Consequences of Climate Change. https://doi.org/10.1787/9789264235410-en.

OECD, 2017. Obesity update 2017. Diabetologe. https://doi.org/10.1007/s11428-017-0241-7.

OECD-FAO, 2016. 2015–2024. In: OECD-FAO.http://www.oecdilibrary.org/agriculture-and-food/oecd-fao-agricultural-outlook_1999114.

Östergen, K., Gustavsson, J., et al., 2014. FUSIONS Definitional Framework for Food Waste.

Östergren, K., et al., 2017. Food Supply Chain Side Flows Management Through Life Cycle Assessment and Life Cycle Costing: a Practitioner's Perspective, pp. 300–303.

Paarlberg, R., Paarlberg, R.L., 2013. Food Politics: What Everyone Needs to Know. Oxford University Press.

Papargyropoulou, E., et al., 2014. The food waste hierarchy as a framework for the management of food surplus and food waste. J. Clean. Prod. 76, 106–115. https://doi.org/10.1016/j.jclepro.2014.04.020.

Parfitt, J., Barthel, M., Macnaughton, S., 2010. Food waste within food supply chains: quantification and potential for change to 2050. Philos. Trans. R. Soc. Lon. B Biol. Sci. 365 (1554), 3065–3081. https://doi.org/10.1098/rstb.2010.0126.

Parizeau, K., von Massow, M., Martin, R., 2015. Household-level dynamics of food waste production and related beliefs, attitudes, and behaviours in Guelph, Ontario. Waste Manag. 35, 207–217. https://doi.org/10.1016/j.wasman.2014.09.019.

Pimentel, D., Pimentel, M.H., 2007. Food, Energy, and Society. CRC press.

Reardon, T., Barrett, C.B., 2000. Agroindustrialization, globalization, and international development an overview of issues, patterns, and determinants. Agric. Econ. 23 (3), 195–205. https://doi.org/10.1016/S0169-5150(00)00092-X.

ReFED, 2016. A Roadmap to Reduce U.S. Food Waste by 20 Percent.

SEPSA, 2016. Plan Nacional para la Seguridad Alimentaria, Nutrición y Erradicación del Hambre 2025: Plan SAN-CELAC Costa Rica, I Quinquenio. San José, Costa Rica.

Serafini, M., Toti, E., 2016. Unsustainability of obesity: metabolic food waste. Front. Nutr. 3, 1–5. https://doi.org/10.3389/fnut.2016.00040 (October).

Stancu, V., Haugaard, P., Lähteenmäki, L., 2015. Determinants of consumer food waste behaviour: two routes to food waste. Appetite 96, 7–17.

Setti, M., et al., 2016. Italian consumers' income and food waste behaviour. Br. Food J. 118 (7), 1731–1746.

Stenmarck, Å., Jensen, C., Quested, T., Moates, G., 2016. Estimates of European Food Waste Levels.

Stancu, V., et al., 2016. Determinants of consumer food waste behaviour: two routes to food waste. Appetite 96, 7–17. https://doi.org/10.1016/j.appet.2015.08.025.

The International Standards Organisation, 2006. International Standard Assessment—Requirements and Guidelines. Iso 14044, pp. 1–48. https://doi.org/10.1007/s11367-011-0297-3.

Thyberg, K.L., Tonjes, D.J., 2016. Drivers of food waste and their implications for sustainable policy development. Resour. Conserv. Recycl. 106, 110–123. https://doi.org/10.1016/j.resconrec.2015.11.016.

Tostivint, C., et al., 2016. Food Waste Quantification Manual to Monitor Food Waste Amounts and Progression. BIO by Deloitte.

Van Geffen, L.E.J., van Herpen, E., van Trijp, J.C.M., 2016. Causes & Determinants of Consumers Food Waste. Project Report. In: EU Horizon 2020 REFRESH.

UNFCCC, 2007. Climate Change: Impacts, Vulnerabilities and Adaptation in Developing Countries. United Nations Framework Convention on Climate Change. https://doi.org/10.1029/2005JD006289.

USDA, 2014. The Estimated Amount, Value, and Calories of Postharvest Food Losses at the Retail and Consumer Levels in the United States. EIB-121. U.S. Department of Agriculture, Economic Research Service. https://doi.org/10.2139/ssrn.2501659.

Vittuari, M., De Menna, F., Pagani, M., 2016. The hidden burden of food waste: the double energy waste in Italy. Energies. 9(8). https://doi.org/10.3390/en9080660.

Waarts, Y.R., Eppink, M., Oosterkamp, E.B., Hiller, S.R.C.H., van der Sluis, A.A., Timmermans, T., 2011. Reducing Food Waste: Obstacles Experienced in Legislation and Regulations.

WHO, 2012. Our Planet, Our Health, Our Future. Public Health & Environment Department (PHE), p. 64. Available from: www.who.int/globalchange/.../reports/health_rioconventions.pdf.

WRAP, 2009. Household Food and Drink Waste in the UK. Report prepared by WRAP, Banbury.

Further reading

Canali, M., et al., 2014. Drivers of Current Food Waste Generation, Threats of Future Increase and Opportunities for reduction. ISBN 978-94-6257-354-3. FUSIONS Project.

European Commission and Report, T, 2010. Preparatory Study on Food Waste Across Eu 27. October. https://doi.org/10.2779/85947.

Vittuari, M., et al., 2015. Review of EU Legislation and policies with implications on food waste. FUSIONS Report. Available from: http://www.eu-fusions.org/index.php/publications. (Accessed 11 November 2015).

Chapter 18

Handling food waste and losses: Criticalities and methodologies

Rodolfo García-Flores*, Pablo Juliano[†] and Karolina Petkovic[‡]

*CSIRO Data61, Melbourne, VIC, Australia [†]CSIRO Agriculture and Food, Melbourne, VIC, Australia [‡]CSIRO Manufacturing, Clayton South, VIC, Australia

Abstract

There is no doubt that the world will need more food in the future, as demand is driven by a growing population and rising incomes. However, this increase in food demand will be constrained by scarcer resources, more attention to food security, and changing dietary habits. Today, the most effective way to increase the amount of food produced is through efficiency gains, as clearing new land for agricultural activities stopped being sustainable a long time ago. The obvious and most effective way to gain efficiencies is to reduce the amount of food that is produced but not consumed. Food losses may happen before or after the retail point, and these are distinguished by calling the former *food loss* and the latter *food waste*. Although the problems related to food waste are just as severe as those related to food loss and there have been considerable efforts to reduce food waste, the success of these initiatives relies to a great extent on public awareness and policies aimed at changing consumer behavior. By contrast, avoiding food loss is more amenable to the use of quantitative techniques and digital technology, and for this reason, the discussion in this chapter will be mostly focused on food loss, although our review provides pointers to causes and current progress in food waste prevention.

In this chapter, we review the potential impact of some new digital technologies, including quantitative logistic optimization models, to avoid food loss: we explain the importance of having good quality data, discuss methodologies to obtain and validate the data, and briefly review new technologies like the Internet of Things to obtain data updates in real time. After the review, we present a case study of a vegetable supply chain to illustrate the usefulness of the approach, and end with a summary of new technologies to reduce food loss.

1 Introduction

It is expected that food consumption will continue to grow for at least the next 40 years (Godfray et al., 2010), driven by a growing population and rising incomes. However, this increase in food demand will be constrained by scarcer resources, more attention to food security, and changing dietary habits. Today, the most effective way to increase the amount of food produced is through efficiency gains, as clearing new land for agricultural activities stopped being sustainable a long time ago.

Although there are no universally agreed definitions of food loss and food waste, we present the following as introduced in Thyberg and Tonjes (2016). *Food loss* is the decrease in edible food mass throughout the part of the supply chain that specifically leads to edible food for human consumption, whereas *food waste* is food that was originally produced for human consumption but then was discarded or not consumed by humans. Waste includes food that spoiled prior to disposal and food that was still edible when thrown away.

Reducing food loss, as explained, requires reducing handling inefficiencies at any stage of the supply chain from farm to retail point, including production, transportation, processing, storage, preservation, and packaging, and is affected by climate factors, infrastructure limitations, and quality, esthetic, and safety standards. Gaining efficiency while performing these tasks is becoming more important by the day, and as a consequence, food logistics needs to evolve from applying unidirectional, general supply chain management methodologies to addressing specific problems of the food industry such as sustainability, safety, and perishability. Logistics and planning must find new opportunities to add value to food losses from farm to retail, by considering the impact of food supply systems on the triple bottom line of economic, environmental, and social indicators.

In this chapter we argue that, in order to cover all these requirements, the industry must leverage new food processing and digital data capturing technologies such as sensors to apply advanced quantitative decision-making techniques in real time. In the following sections, we review the potential impact of some new digital technologies in quantitative logistic optimization models to avoid food loss: we explain the importance of having good quality data, discuss methodologies to obtain (e.g., sensor networks) and validate (e.g., through blockchain) the data in real time, and briefly review new

Sustainable Food Supply Chains. https://doi.org/10.1016/B978-0-12-813411-5.00018-1

technologies like Digital Agriculture (DA) and the Internet of Things (IoT) to obtain updates in real time. The review's central theme is the effect these technologies are having on increasing the accuracy and effectiveness of optimization models. The emphasis of this chapter is mostly on methodologies to avoid food loss, although some of the review refers to food waste. The interested reader may consult Thyberg and Tonjes (2016) for an extensive review on the causes and current progress in prevention of food waste. After the review, we present a case study of a vegetable supply chain to illustrate the usefulness of the approach, and end with a summary of the review and future trends in food logistics.

2 The challenges

It is estimated that one-third of food produced for human consumption is lost or wasted globally, which amounted in 2011 to 1.3 billion tons per year (Gustavsson et al., 2011). The impact of these losses goes beyond the deprivation of nourishment to a growing world population, as the water, energy, labor, and other resources needed to produce this food go to waste as well. Furthermore, the production processes of unconsumed produce contribute to increased greenhouse gases, and unfortunately the agricultural industry is a major producer not only of carbon dioxide, but also of methane and nitrous oxide, which have a more potent greenhouse effect than CO_2.

The causes and occurrence patterns of food loss and waste are many and varied. The data from the Food and Agriculture Organization of the United Nations (FAO), shown in Fig. 1, demonstrate that food loss tends to occur in developing countries, whereas in the developed world food waste tends to be a more serious problem. The FAO estimates that the per-capita waste by consumers in Europe and North America is 95–115 kg per year, while this figure in sub-Saharan Africa and South/Southeast Asia is only 6–11 kg per year. The higher proportion of food loss in the developing world is normally attributed to the poor state or lack of infrastructure, limited know-how, and unreliable economic and market environments (see, e.g., Gogo et al., 2017). In contrast, developed countries can count on well-developed infrastructure, abundant expertise, and a well-defined regulatory framework, but because food is cheap and readily accessible, households and hospitality services do not have an incentive to avoid waste beyond the retail point (Godfray et al., 2010).

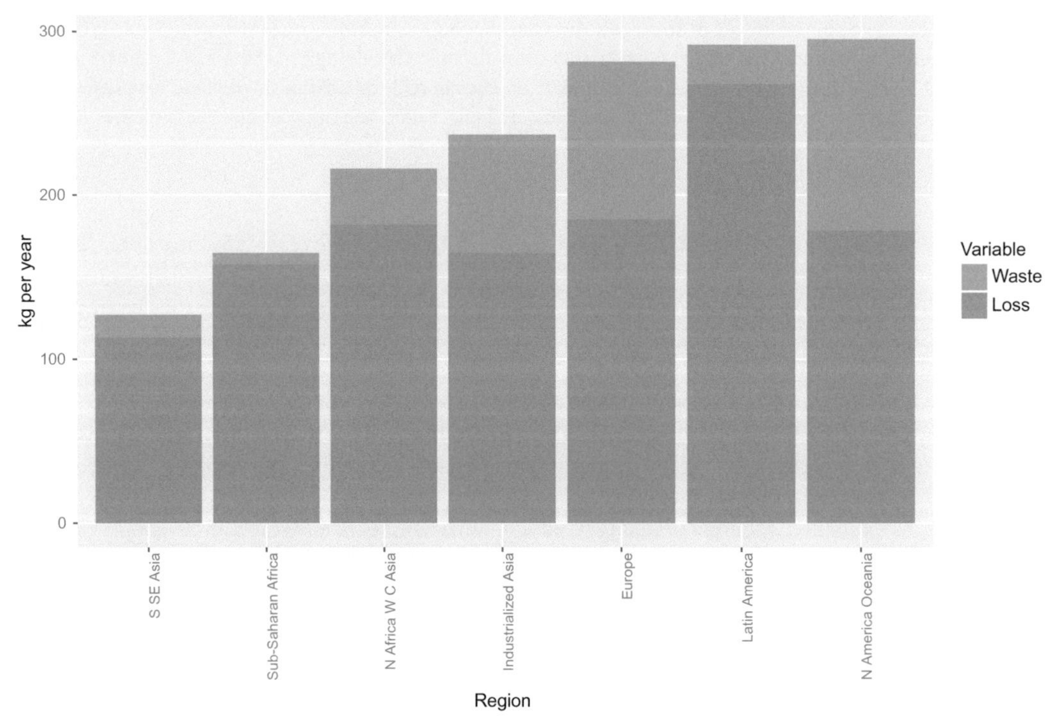

FIG. 1 Food losses and waste per capita in different regions of the world. *(Data from Gustavsson, J., Cederberg, C., Sonesson, U., Otterdijk, R., Meybeck, A., 2011. Global Food Losses and Food Waste. Food and Agriculture Organization of the United Nations, Rome.)*

Closely related to food loss is the concept of *yield gap*. Yield gap refers to the difference between realized productivity and the best that can be achieved using currently available technologies and management (Godfray et al., 2010). The "best" productivity that can be achieved depends of course on a number of factors, including climate, water availability, soil quality, plant varieties and animal breeds, machinery and tools, pest control, biodiversity, and knowledge. Estimating yield gaps in order to close them is not easy, as many of these factors are subjective, dynamic, and site-specific, but most of them can be ameliorated with better governance. For example, soils degrade over time unless taken care of adequately, and pests develop resistance to control methods, but innovation can offset these problems and better policies can improve existing infrastructure and design markets and social interactions with adequate incentives to reduce the yield gap. Although analyzing yield gap falls outside of the scope of this chapter, we are certain that all the technologies that we present in the next sections will eventually be used to increase food productivity and reduce yield gaps. A recent review of yield gap explaining factors from the point of view of data availability and collection can be found in Beza et al. (2017).

3 The Food Loss Bank and Digital Agriculture

Following Petkovic et al. (2017), we will use the Food Loss Bank (FLB) in the context of DA as the unifying concepts to guide the development of this chapter (Fig. 2). The FLB is a networked supply chain concept that is meant to take the edible food loss biomass suitable for human consumption and classify it, determine its origin, check its chemical and biological safety, stabilize it from further degradation, and facilitate its storage and diversion back into the food supply chain. The FLB is made up of physical and virtual stage gates whereby intelligent decision making is applied at each stage, as shown in Fig. 2A. To work as intended, the FLB combines data of different formats and sources relating to biomass type and safety, as well as stability and supply chain logistics, in order to determine how to best divert biomass at each point. The data related to physical

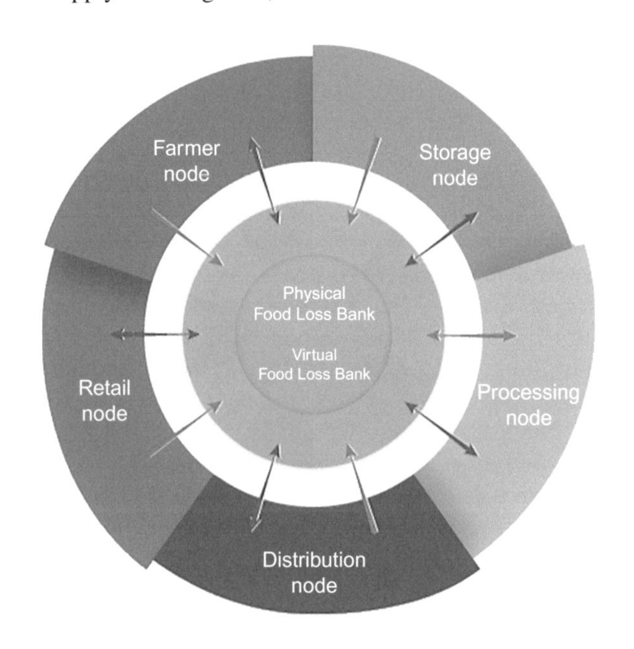

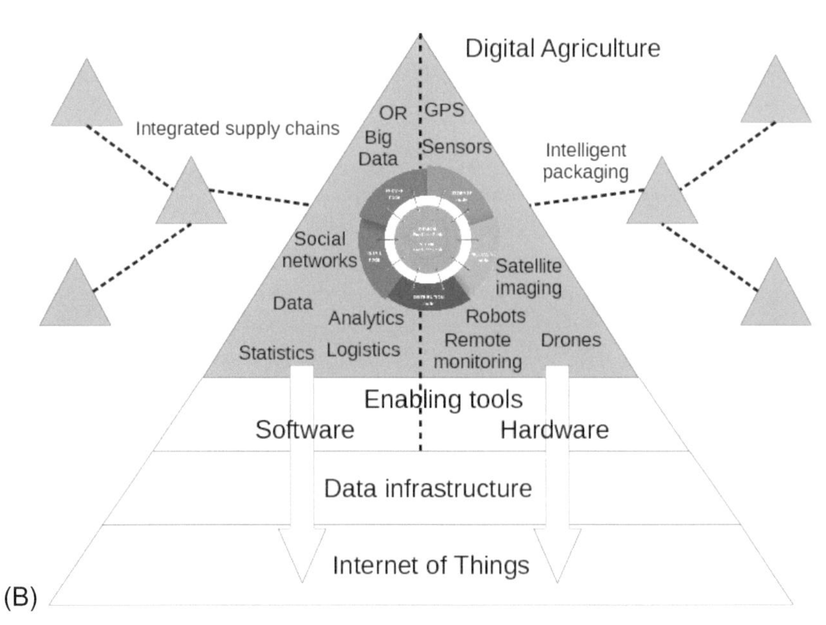

FIG. 2 The digital FLB is used as the unifying concept to introduce technologies to reduce food loss. FLB uses the enabling technologies of DA to deliver on its promise. Data refers to market, agricultural, climate, and social science data (e.g., opportunities to leverage collaboration clusters). (A) Schematic of the Food Loss Bank. (B) DA with the FLB as its computational engine. *(Part (A) from Petkovic, K., Fox, E., García-Flores, R., Chandry, S., Sangwan, P., Sanguansri, P., Augustin, M., 2017. A glimpse into the future: a Food Loss Bank for transforming the food supply chain? Food Aust. (submitted).)*

biomass can be assessed at each stage of the food supply chain by lab-on-a-drone or other mobile, highly integrated laboratories that test for microbial, chemical, and physical properties and safety. The data collected can then be sent to the virtual food loss-streaming platform for fit-for-purpose or use matching.

Petkovic et al. (2017) suggested that the new agriculture and food value chain should be a dynamic network with the FLB as a central node (Fig. 2A). The network consists of both physical and virtual components: the physical component is the store of edible food loss biomass, which can be diverted back into the food supply chain, together with the sensors, IoT components, mobile laboratories, and other hardware; the virtual component is a repository of knowledge designed to support decisions made by all participants in the food supply chain. The virtual FLB optimizes the new, dynamic food chain by using algorithms, simulations, and a trade-off analysis, which compares quality to higher operation costs. It consolidates a variety of other data streams, including information about perishability and traceability.

The FLB is intended to serve as the decision-making engine to avoid food loss (Fig. 2B), as the FLB can leverage the technological enablers of DA and precision agriculture to accomplish its purpose. The digital infrastructure needed to fulfill the FLB is supported by the IoT, which is defined as the interconnection of sensing, actuating, and computing devices that enable everyday objects (the *things*) to collect and exchange data. The presence of these devices provides the management and integration of data from multiple sources on cloud storage platforms. The selection, cleaning up and preparation, quality checking, security, and accessibility of food production can then be performed. For transportation of perishable goods among companies, digital integration of the supply chain and smart packaging will also contribute to reduce food loss. Next, we summarize the ways in which the main components of the framework depicted in Fig. 2B can help reduce food loss.

3.1 The agricultural IoT

Agriculture is a rapidly expanding domain within the IoT (Esenam, 2017), with many research groups experimenting and developing practical applications. For example, Popović et al. (2017) implemented a private IoT cloud platform to be used for research, with emphasis on sustainable agriculture, crop and ecosystem monitoring, pollution reduction, and analysis and standardization of food products. Examples of agricultural research projects prototyped in Popović et al.'s (2017) platform include wine production and water measurement. Noletto et al. (2016) investigated the extent of the use and awareness of IoT in Brazil with emphasis on intelligent packaging and IoT for reduction of food loss, and found that the lack of knowledge about these technologies was an important barrier for adoption. Ram et al. (2015) developed an intelligent water monitoring platform framework and system structure for a facility agriculture ecosystem based on IoT, and Nukala et al. (2016) presented a brief review of applications of the IoT "from farm to fork," stressing that much of the work in this area is still experimental.

Regarding the sensing technologies to feed the IoT, miniaturization and the widespread availability of interconnected computer power are beginning to make a difference in food production operations. Hanyu et al. (2011) reported on the use of a sensor network especially designed for food safety management, and tested it on a confectionery manufacturing plant, and Zou et al. (2017) reviewed the technologies related to RFID-enabled wireless networks for food packaging and logistics, emphasizing applications in fresh food tracking and monitoring. Rateni et al. (2017) recently presented a review of analytical devices able to enhance the ordinary smartphone with diagnostic capabilities, which is a natural trend given that smartphones are ubiquitous and are equipped with numerous components that can be employed for measurement and detection. The reviewed devices cover a variety of techniques including spectroscopy and biosensors, and the authors note that, despite all the progress in device miniaturization and portability, reliability and reproducibility are still a challenge for much of this technology. As the preceding discussion shows, the potential of the IoT to avoid food loss is only beginning to be understood.

3.2 Data infrastructure and consolidation

In addition to the new and portable sensing technology, other technological improvements such as increased digitization of transport operations, advances in satellite imagery, increased data storage capacities, and increased access to publicly available data resources have produced a wealth of information not previously available. This information has helped many research and commercial projects produce improved and more accurate solutions but, at the same time, requires novel ways to design data infrastructure and encourage data consolidation. The necessary data infrastructure to produce this wealth of information should be composed of elements possessing the properties of *interoperability*, *portability*, *substitutability*, and *use of open software standards* (Biggins et al., 2013).

Some examples of the important role that data are playing in avoiding food loss include Bahadur et al. (2016), who use newly collected data to study food loss in developing countries, where food loss is more of an issue when compared with developed countries. Food loss in developing countries happens mostly due to inefficient storage and processing facilities, although it also occurs due to poor supply chain infrastructure. For example, some regions of the world have poor crop management that creates crop losses on farms, or poor sanitary conditions that create produce of poor quality, short shelf life and, potentially, safety issues, or roads in bad condition, etc. Bahadur et al.'s (2016) study presented a multivariate analysis where the volume of food loss in 93 countries over 20 years was used as the dependent variable and a range of socioeconomic factors were used as independent variables. Variables, such as wealth, agricultural machinery, transportation, and telecommunications, were significant in explaining the amount of lost food, and the insights were used to propose more efficient food handling policies. Bahadur et al.'s (2016) data came from the UN's Food and Agricultural Organization (FAO), which provides information on a range of crop, livestock, and fishery loss in many countries, and the World Bank. These authors also note that access to mobile phones and other wireless devices in developing countries is helping control agricultural diseases and providing access to information and education. Another example is Higgins et al. (2013), who assessed the effect of infrastructure investments in the cattle and beef supply chain in the northern states of Australia. This project used newly collected data sets capturing cattle movements along the cattle supply chain that operates in the north of the country.

Buzby and Hyman's (2012) study complemented Bahadur et al. (2016) in that it focuses on food inefficiencies in a developed country. In the United States, after recycling materials, such as paper and paperboard, food waste was the single largest component of municipal solid waste in 2009, more than plastics, metals, wood, or yard waste. Food waste is also a major contributor to greenhouse gas emissions, especially methane. Buzby and Hyman (2012) use, as data sources, the Loss-Adjusted Food Availability (LAFA) data from the US Department of Agriculture's Economic Research Service (ERS) (2011), although they note that the structure and limitations of these databases somewhat limit their aim of estimating avoidable food loss. Ridout et al. (2010) assessed the carbon and water footprints of mango losses in Australia. Heller and Keolian (2014) also studied food losses, mostly in the United States, but focus on the environmental costs, particularly in the greenhouse gas emissions produced. The data sources these authors used are based on existing life-cycle analyses, many of which are also based on the LAFA database. Heller and Keolian (2014) pointed out additional limitations of this database, such as adopting a top-down approach to estimating food loss and food consumption that only provides a gross average of the US diet and overestimates per capita caloric intake.

Joutsela and Korhonen (2015) opted for online research via crowdsourcing in order to collect information, user experiences, and ideas from a community of packaging industrialists in order to build a knowledge database that helps to understand the relation between packaging and food loss. The online community of 137 members provided a large amount of quantitative and qualitative data that could be used for research and packaging design purposes. This work demonstrates that crowdsourcing may be a fast and inexpensive way to collect niche knowledge domain data.

An alternative, more traditional, data collection process is presented in Willersinn et al. (2015), who undertook a comprehensive study of losses in the whole Swiss potato supply chain. These authors used quantitative data collected from field trials, from structured interviews with wholesalers, processors, and retailers, and from consumer surveys in combination with a 30-day diary study. As in many developed countries, much of the loss in Switzerland is due to quality standards. Lebersorger and Schneider (2014) collected data about potential influencing factors and reasons for food losses, in order to provide a basis for the development of waste prevention measures. These authors pointed out that, although food waste from retail has been investigated in a number of studies, there is still a lack of reliable data. They presented a study based on a survey sent to 612 food retailers in Austria, 10.7% of the total, noting that records of loss of bulk fruits and vegetable are quite inaccurate. An important conclusion of this study is that food loss rates decrease with increased numbers of purchases per year and increased sales at retail outlets, suggesting a reduction akin to economies of scale. However, correlations were low and these variables explain not more than 33% of the variation of food loss rates, which indicated that food loss rates are rather influenced by human factors.

Jood and Cai (2016) advocate for prevention through increased awareness to reduce food loss, and point out that industrialized Asia is the major region contributing to food loss, with 27.9%, followed by South and Southeast Asia with 21.3%. Jood and Cai (2016) also observe that, in general, food loss is larger in developing regions mainly due to the losses occurring during handling and storage, whereas food waste is significantly higher in developed countries. To reduce food loss, they articulate the need to fill the data gaps, together with quantifying the impact of the food loss reductions. To achieve the former objective, they stress that it is crucial to harmonize definitions and methodologies used in different contexts, so this is probably an opportunity for the development of ontologies and dictionaries, some of which have been implemented in the ASPIRE system in the context of industrial ecology (Stock et al., 2015). This paper also cites concrete examples of digital disruptions to reduce food loss in developing countries, as for example by providing mobile tools allowing thousands of farmers in rural regions to access market price information, weather alerts, and advice on crop management, which are all related to postharvest food loss prevention. Encouraging information exchange between suppliers and retailers in order to achieve more accurate and timely automated demand forecasting is considered a potential strategy to reduce food waste at the retail stage.

The papers reviewed in this section indicate a wide variety of methodologies used to gather data related to food loss. We note that almost none discusses a centralized data and decision-making repository for food loss similar to the FLB, and that many of the data collection approaches suggest methodologies that could be automated to construct the FLB repository.

3.3 Shared ledger technologies: The blockchain

Although still in its early stages of adoption, the shared ledger paradigm is finding its way in a number of applications that require high reliability and shared, trusted, privacy-preserving, nonrepudiable data repositories (Hull et al., 2016). In a few words, shared ledgers are databases where control over the data's evolution is shared between entities. One well-known type of shared ledger is the blockchain, for which an excellent introduction can be found here.[1]

At the time of writing, shared ledgers are being used mostly in the banking and finance sector (see Cocco et al. (2017) for an overview of applications in the banking system), but many are realizing its potential to address common problems in the food industry. For example, it is well known that food safety is a major concern in China, with consumers having serious misgivings about food provenance and quality after incidents related to poisoning due to counterfeit and adulterated vegetables, drinks, and baby formula. To address these problems, Tian (2016) proposed a system that combines RFID tags with blockchain technology to develop a fresh agri-food supply chain traceability system, with the ultimate purpose of guaranteeing food safety. Global food businesses are also leveraging the technology, with IBM reported to implement shared ledgers to help them reduce contamination of their products.[2] Even though these initiatives are aimed at increasing food safety, the increased confidence these projects give to the information related to food origin, condition, and movement will definitely be useful to reduce food losses along the supply chain.

3.4 Smart packaging and flexible expiry dates

Better packaging represents a very cost-effective way to prevent food loss. If well designed, it can help maintain product quality and extend shelf life, as well as being resource-efficient and energy-smart. Packaging serves a number of purposes. In the context of fresh vegetables, these include protecting against mechanical damage, promoting rapid precooling, facilitating uniform cooling (which implies low resistance to airflow), and allowing the exchange of metabolic gases such as oxygen and carbon dioxide (Patare and Opara, 2014, cited by Defraeye et al., 2015). Defraeye et al. (2015) provides an overview of (possibly conflicting) package design objectives.

Beyond the physical properties that food packaging enhances and preserves, digital technology in packaging can also assist in supply chain management tasks, especially when leveraging effective rules and regulations in this role. For example, assume that there was a flexible regulatory framework in place that allowed for dynamic expiry dates, or different classes of food quality (e.g., "premium" and "regular") without compromising on the quality and safety to the final consumer, instead of the monolithic, fixed due dates that are common today. If this were the case, dynamic expiry dates could be handled with sensing technology built into packaging, so as to give the consumers a better idea of the quality of the produce in real time. Some of this technology is reviewed in Mohebi and Marquez (2014).

A complete economic and logistic analysis of diminishing returns with flexible expiry dates does not exist at present, although some aspects of this problem have been analyzed. Diminishing quality was incorporated into the harvesting model presented by Ferrer et al. (2008), who focused on wine production. Wines were placed in categories according to quality and harvesting grapes away from the optimal time can result in the produced wine being of a lower grade. Ferrer et al. (2008) presented a wine grape harvest scheduling optimization problem where a cost was imposed for harvesting at nonoptimal times. The associated reduction in wine quality was expressed in a quality loss function, which relates relative cost to the deviation from the optimal harvest day. Similarly, Li (2013) investigated dynamic pricing strategies in relation to customer satisfaction levels and willingness to trade perishable food quality for a reduced price. This study concluded that a strategy of offering smaller discounts but with more shelf-life remaining would be more satisfactory to consumers than the current strategy of marking down the price of a product only when the expiry date is imminent. Li (2013) suggests that more dynamic pricing can both increase customer satisfaction and also stimulate the purchase of products with a shorter remaining shelf-life, which in turn would reduce the likelihood of food needing to be discarded due to expiration.

Recent research into smart packaging, and food tracing technology in general, is dominated by RFID, which is considered to be the successor of bar codes (Bibi et al., 2017). RFID readers are around 10 times more expensive than bar code readers,

1. See http://anders.com/blockchain/. Accessed 22 September 2017.
2. See https://www.cnbc.com/2017/08/22/ibm-nestle-unilever-walmart-blockchain-food-contamination.html. Accessed 22 September 2017.

but they can also store much more information (Ezeike and Hung, 2009). Abad et al. (2009) developed a smart tag using RFID capable of monitoring temperature and relative humidity for use in cold supply chains. Lorite et al. (2017) developed a prototype critical temperature indicator smart sensor combining a microfluidic temperature sensor with an RFID. This device can monitor whether a critical temperature has been reached with a response time of approximately 6 minutes.

The ubiquitous use of RFID and wireless sensor networks will be accompanied by heavy data traffic, making it necessary to implement the enabling technologies of the IoT already reviewed in Section 3.1. Gnimpieba et al. (2015) present a platform based on IoT and cloud computing to accomplish continuous traceability and real-time tracking, which are difficult to achieve due to the heterogeneous nature of the different platforms and technologies used at different points in the food supply chain. Grunow and Piramuthu (2013) investigated the use of RFID on a per-item basis, considering the utility of such implementation for retailers and customers. Grunow and Piramuthu (2013) also discussed the future possibility of replacing the traditional printed expiry date with remaining shelf-life information, enabled through RFID technology and information sharing, to minimize wastage.

3.5 Reverse logistics models

Reverse logistics refers to the operations of reusing products and materials after delivery to the consumer. The study presented by French and LaForge (2006) provides general insight into reuse and waste disposal in process industries (as opposed to discrete industries such as manufacturing), which also include food processing. The study identified sources of materials that may be reused, as well as the processing options available to manufacturers. Basing their discussion on a framework introduced by Fleischmann (1997), French and LaForge (2006) discuss the five categories of reverse logistics flows in the context of the process industries:

1. *End-of-use returns.* These are returns where there is economic value, due to environmental regulation, or asset protection goals. In general, this does not apply to food products, although it may apply to some particular instances (e.g., recovery of avocado stones or mango seeds to produce cosmetics).
2. *Commercial returns.* Undoing a preceding business transaction.
3. *Warranty returns.* Refers to failed products returned to the original sender due to failure, damage, or security hazard. Repair or disposal are alternative actions. In the context of the food industry, this mostly applies to product recalls due to food security risks or failure to comply with promised specifications.
4. *Production scrap and by-products.* Excess material reintroduced to the process, off-spec material is reworked to meet requirements. By-products may join other supply chains. French and LaForge (2006) note that internal returns are more significant in process industries than in manufacturing industries, although this is not always the case in the food industry. It may happen when food has exceeded its shelf life. An example in food production would be apple pomace, which is generally disposed of, that is, considered a scrap product, even though it does have a lot of nutritional value.
5. *Packaging.* This is the oldest and most common subject in reverse logistics studies. An example is González-Torre et al. (2004), who compare environmental and reverse logistics practices in Spain and Belgium.

Hasani et al. (2012) discussed the problem of how to actually extract value from food products beyond their prime in the context of reverse logistics. Although the motivation of the paper is to deal with uncertainty in lifetime and price using case studies in electronics and food products, some discussion is given on the possible uses of food products that are not of premium quality but can still be safely consumed. For example, returned cans of tomato can be used to produce stew. Hasani et al. (2012) also distinguished between reverse logistics, which considers reverse flow of materials from customers, and closed-loop supply chains, which simultaneously consider forward and backward flows of materials along the supply chain.

Kim et al. (2014) analyzed uncertainty on the so-called returnable transport items, or RTIs. These comprise containers, pallets, or crates. This makes the problem somewhat similar to the reverse logistics study of packaging. The aim is to analyze how delays in the return of RTIs affect the expected total cost of the supply chain and to develop guidelines to assist decision makers in minimizing the expected total cost. Kovačić et al. (2015) used a methodology called Extended Material Requirements Planning to evaluate the effect of perturbations in lead time and temperature on the net present value of perishable agricultural goods using as a case study the production and distribution of baby formula.

Food product recalls may also be considered an exercise in reverse logistics. Kumar (2014) proposed a system to manage food product recalls which, in addition to serving as a tool to manage food safety, also limits liabilities for corporate negligence and avoids damage caused by negative publicity. The value of Kumar's paper for the current discussion is that it proposes a knowledge-based system that organizes cumulative knowledge and expertise related to supply chain integration, standardization, and procedurization that may be useful in the context of recovering food loss, rather than in food product recalls. The idea of an FLB as proposed in this test-bed is very similar to Kumar's knowledge-based system, and for such a system to exist, Kumar notes that it is important that there is transparency and traceability of information at all levels. RFID and tracing technology is invaluable to achieve these.

3.6 Other technologies

An extensive review of existing technologies to eliminate food losses and waste is beyond the scope of this paper. For a review on other emerging, nondigital food technologies, we refer the reader to other chapters of the present book and to Galanakis (2015). For other digital technologies, Tian et al. (2016) may be consulted.

4 Case study: A vegetables supply chain

We now present a case study of *Brassica* production in Australia. Broccoli and broccolini are grown in most Australian states, of which Victoria is the largest producer, with 48% share of total production. Fig. 3 shows the amount and value of broccoli in the supply chain: for the year ending in June 2015, 68,571 t were produced, representing $188.7 million in value (all figures in Australian dollars), with wholesale value of the fresh supply of $210.7 M. Exports represent $10.5 M, 69% of which go to Singapore (Horticultural Innovation Australia Limited, 2015).

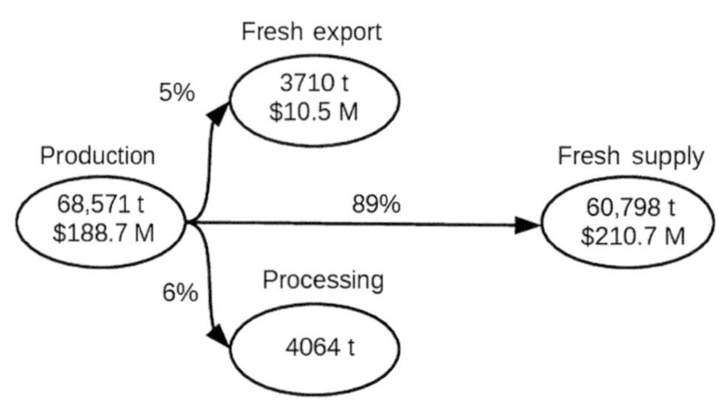

FIG. 3 Overview of *Brassica* production in Australia. For the year ending in June 2015, 68,571 t were produced, representing $188.7 million dollars in value, with wholesale value of the fresh supply of $210.7 M. Exports represent $10.5 M, 69% of which go to Singapore (Horticultural Innovation Australia Limited, 2015).

The mathematical model introduced in this section extends the model proposed by Atallah et al. (2014) as a Fixed Charge Location Problem (FCLP) (Current et al., 2004). Atallah et al.'s (2014) original optimization model aimed to minimize the total production and transportation costs of fresh broccoli, and analyzed the effect of seasonality and of changing the set of production regions, located in the eastern United States, over these costs. By contrast, our model emphasizes the need to process the product that is not of the right quality for retail by *determining the optimal location and number of processing facilities*, using relevant production data for the Australian market and supply chain conditions (Horticultural Innovation Australia Limited, 2015). The aim of these processing facilities is to reduce food loss by producing broccoli-based edible products (e.g., snacks or food supplement powder) with extended shelf life and high nutritional value. Fig. 4 is a map of the region under study showing the location and demand and supply of consumers and producers of broccoli, respectively. Broccoli producers are marked in yellow and consumer markets are marked in red (for interpretation of the references to color in this figure legend, the reader is referred to the electronic version of this chapter). The producers that can be selected as processing sites are marked in blue, and the size of the circle is proportional to the volume in tonnes of broccoli produced or consumed.

The full formulation of the problem can be found in Appendix A.2. The problem can be stated as follows: determine the flows of vegetable from producers to consumers and to sites that can process the food losses along the supply chain, as well as the locations of these processing sites, while minimizing the operation and transportation costs, subject to demand, production, site setup, and capacity constraints. Objective function (A.1) is the expression of the total costs, which include production and transportation costs, as well as site setup costs. A summary of the constraints follows.

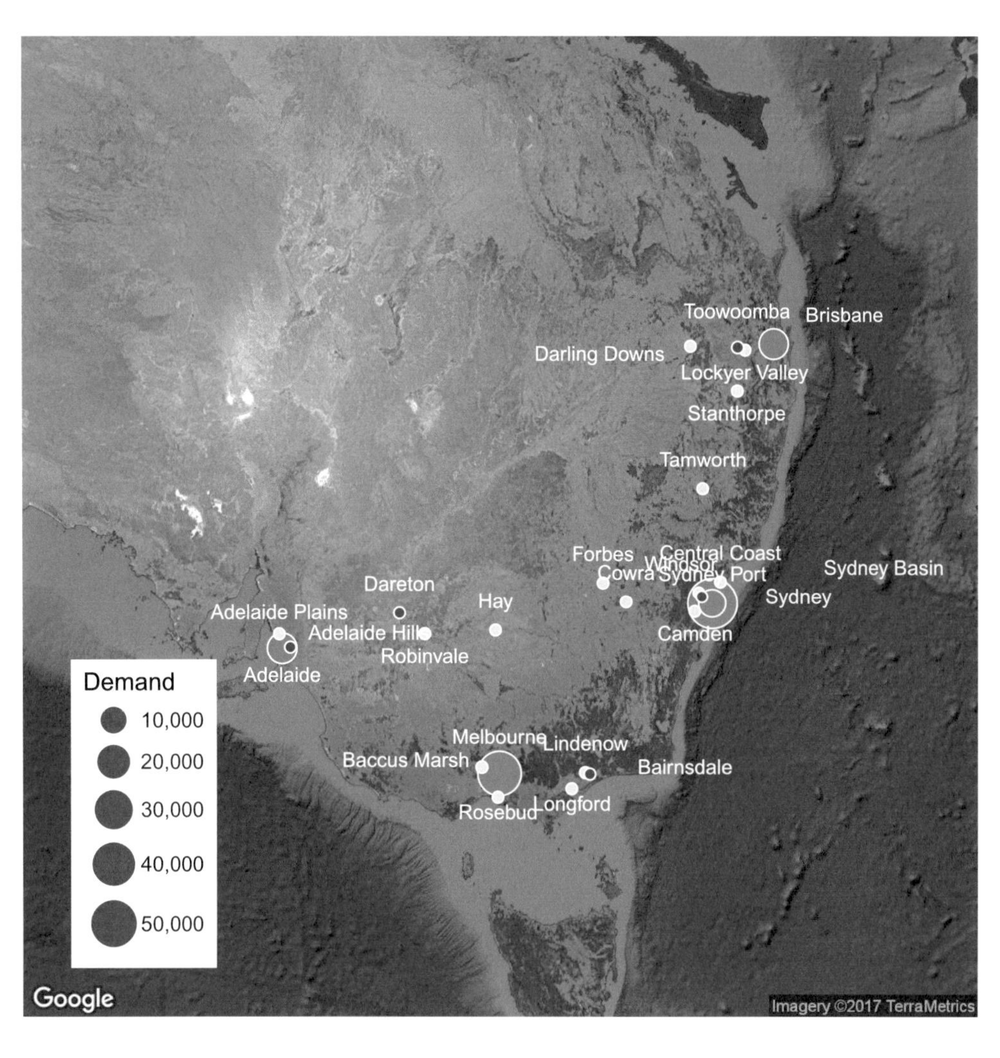

FIG. 4 Locations of broccoli producers (*yellow*) and consumer markets (*red*) in continental Southeast Australia (for interpretation of the references to color in this figure, the reader is referred to the electronic version of this chapter). Candidate processing sites are also producers and are marked in *blue*. The size of the circle is proportional to the volume in tonnes of broccoli produced or consumed.

1. The *seasonal supply* constraints (A.2) ensure that the amount of product transported from a producer to a consumer equals the amount produced minus the production losses.
2. The *yield* constraints (A.3) state the maximum amount of product that can be produced by unit of land area.
3. The *demand* constraints (A.4) state that the transported product must satisfy the demand of the consumers.
4. Constraints (A.5) ensure that every producer sends all its lost amount of product to only *one candidate site*.
5. Constraints (A.6) state that food losses must end up in *one selected candidate processing site*.
6. The selected candidate sites are constrained in their *capacity* to process food loss, as expressed in constraints (A.7).
7. The model assumes that losses on the field are irrecoverable and stay there, so the only losses that can be processed come from transportation. Constraints (A.8) *connect* the network flow and FCLP constraints, and state that, if there is no loss in transportation from producer i, then the indicator variable v_{ij} must also be zero.
8. The number of allowed processing sites may be limited, constraint (A.9).

The model assumes that:

1. There may be one or more selected processing site from among the candidates. There is no limit to the number of candidate processing sites that can be selected.
2. The losses from a producer end up in only one selected processing site.
3. The candidate sites have a limited capacity to process the losses.
4. Losses can happen in transportation and in production. Both can be sent to processing sites.
5. The weight on selecting candidate processing sites is on their "demand," that is, on the ability to process lost product from nearby producers.

Additionally, and in this state of development, we are assuming a time horizon of 3 years in quarterly periods, all with equal (but representative) production and equal demand, and that the set of candidates includes Adelaide Hills (in the state of South Australia), Bairnsdale (Victoria), Dareton, Sydney Basin (New South Wales), and Towoomba (Queensland). The base case problem has 1622 variables, of which 1512 are real and 110 are binary, and 5004 constraints. The model was implemented in version 1.8 of the Clojure[3] language. We obtained all the following results using version 5.5 of the lp_solve[4] optimizer in a 64-bit Intel Xeon CPU with two processors of four cores (2.27 GHz) each and 8 GB of RAM. All experiments were carried out in Ubuntu 16.04.

Fig. 5 shows the results when the number of candidate sites to be selected is constrained. If constrained to two, the selected sites are Bairnsdale and Toowoomba; to three, the sites are Adelaide Hills, Bairnsdale, and Toowoomba; to four, Adelaide Hills, Bairnsdale, Dareton, and Toowoomba. Fig. 5A shows the layout of the supply chain when all five candidates are selected. When the number of sites is constrained to one, Bairnsdale is the selected site (not shown in the figure).

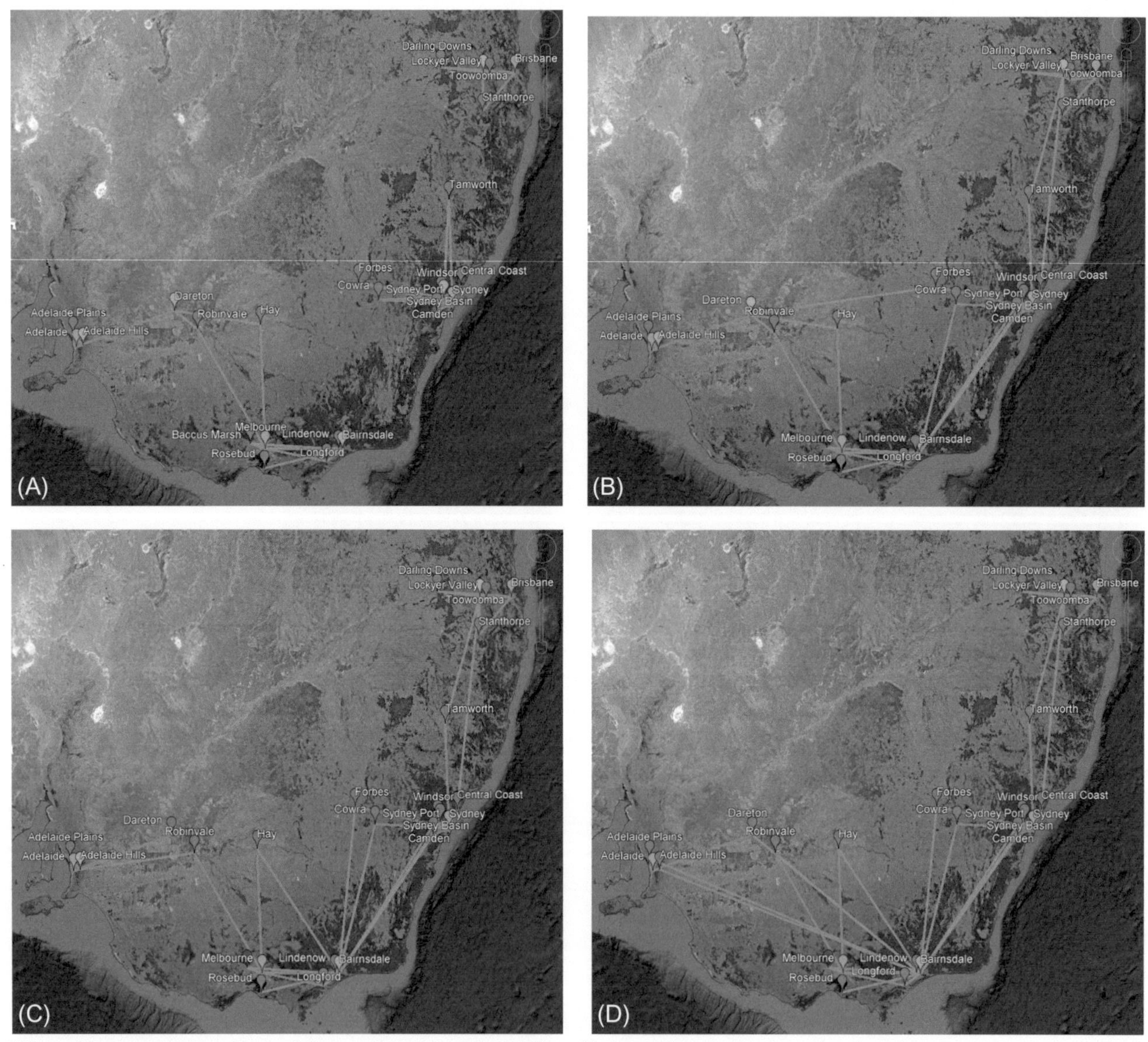

FIG. 5 The transportation of broccoli from producers to markets, indicated by *green lines*, and to processing sites, shown as *orange paths* (for interpretation of the references to color inthis figure legend, the reader is referred to the electronic version of this chapter). Selected candidate sites are shown in *light green*. When the maximum number of selected sites is one, the selected candidate is Bairnsdale; the corresponding map is not shown. (A) All sites selected. (B) Four sites selected. (C) Three sites selected. (D) Two sites selected.

3. See http://clojure.org/. Accessed 7 June 2017.
4. See http://lpsolve.sourceforge.net/5.5/. Accessed 24 October 2017.

Following Atallah et al. (2014), we also use Weighted Average Source Distance (WASD) as a measure of the nonlocal origin of particular consumption patterns. WASD is an indicator originally introduced by Carlsson-Kanyama (1997) to produce a single distance figure that combines information on distances from the producers to the consumers and the amount of products consumed. Fig. 6 shows the value of this measure applied to the broccoli losses when the number of selected candidate sites increases from one to five. It is clear that the gains in this measure decrease very quickly as more candidate sites are selected, with a reduction of 68% of WASD when passing from building one to two sites, to 26% for two to three sites, to 9.3% for three to four sites, to 5% for five sites. However, the value of the objective function shows a similar decrease (not shown) as the number of selected candidate sites increases while keeping all other parameters equal, suggesting that the gains from reducing the transportation costs of lost vegetables are greater than the capital costs of building new processing sites. This implies that it pays off to have as many processing sites close to the source as needed, at least with the current site setup cost estimate of $50,000. A better estimate will require more detailed information about the products to be obtained from the broccoli losses. The expression used to calculate WASD is listed in Appendix B.

To end this section, we illustrate the application of the technologies reviewed in this chapter to the *Brassica* supply chain. The stakeholders could, for example:

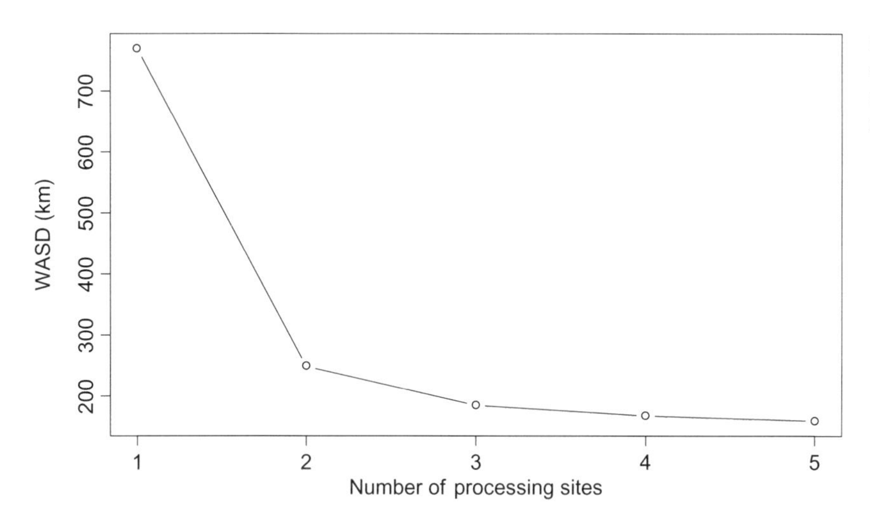

FIG. 6 The WASD measure for the food losses as a function of the number of selected processing sites. WASD may be interpreted as the degree of nonlocal origin of particular consumption patterns.

1. Host a fruit and vegetable FLB in an area in northern Victoria, close to the majority of the *Brassica* producing regions. The physical components of the FLB could be built on the same sites identified as optimal for processing the food loss; these would receive the food loss biomass suitable for human consumption, store it, process it, and commercialize it back into the food supply chain.
2. The virtual component of the FLB could operate in a decentralized fashion and incorporate a variety of optimization and simulation models for decision making, including supply chain, forecasting, socioeconomical, and environmental models.
3. Incorporate real-time tracking of the fresh produce and losses to be further processed using GPS and RFID technology. This tracking information could be used to continuously update production and transportation scheduling in the virtual component of the FLB (i.e., reactive scheduling).
4. Build support data infrastructure and collect timely and relevant information to understand better the social, economic, environmental, and productive processes that affect the fruit and vegetables supply chain. This infrastructure would be useful to inform decisions at all levels, including long-term political and investment decisions.
5. Implement smart packaging technologies that not only extend the food's expiry date, but also inform the final customer on provenance and nutritional value of the fresh produce as well as of the processed vegetables that would have otherwise become losses.

5 Concluding remarks

The necessary increase in food production is constrained as much by scarce resources as it is from increased attention from governments and the general public on food security, provenance, and changing dietary habits. The natural first step to address these challenges is by reducing, and eventually eliminating, food losses and waste. In this chapter, we have provided an overview of methodologies that could be applied to reduce food loss, and to a lesser extent, food waste, using the FLB and DA as unifying concepts, and then described the usefulness of the related enabling digital technologies. We also introduced a case study of the Australian *Brassica* supply chain and used it to exemplify some of the enabling technologies. The research question we posed for this case study is: What is the optimal number and location of processing sites where the broccoli that would have otherwise been considered as losses could be processed and converted into high value-added, nutritious products such as snacks or supplement powder?

These enabling technologies reviewed in this chapter are only now starting to come together, and will require strong mathematical and analytical foundations to fulfill their promise. For example, the variability and uncertainty inherent in most food supply chains are important barriers to technologies such as precision agriculture. So far, the success of precision agriculture has been limited, partly because of agricultural and food production systems' inherent complexity caused by their biological nature. This complexity is reflected in high variability of food quality and costs seen at various points in the supply chain, the difficulty in incorporating sustainability metrics, and the critical importance of timing to process and market (García-Flores et al., 2015). Incorporating this uncertainty is an important extension to the broccoli case study presented. Finally, we should note that management of food supply chains shares many of these features with general management of natural resources, including forestry and mining (Bjorndal et al., 2012). As such, much of the discussion in this chapter also applies to the latter industries.

Acknowledgments

The first author wishes to acknowledge Mr Peter Taylor from Data61 and Ms Sabrina Cohan for their valuable contributions.

Acronyms

DA Digital Agriculture
FAO Food and Agriculture Organization of the United Nations
FCLP Fixed Charge Location Problem
FLB Food Loss Bank
IoT Internet of Things
LAFA Loss-Adjusted Food Availability
WASD Weighted Average Source Distance

Appendix A Mathematical model

A.1 Nomenclature

The notation used is as follows:

Sets

$\mathcal{C}$ The set of candidate sites, $\mathcal{C} \subset \mathcal{S}$
$\mathcal{K}$ The set of seasons
$\mathcal{L}$ The set of valid arcs between $i \in \mathcal{S}$ and $j \in \mathcal{T}$
$\mathcal{N}$ The set of all sites
$\mathcal{S}$ The set of sources
$\mathcal{T}$ The set of sinks

Decision variables

v_{ij} Indicator variable that takes the value 1 if the food loss from site $i \in \mathcal{S}$ is transported to candidate site $j \in \mathcal{C}$, 0 otherwise.
x_{ik} Production in kilograms at site $i \in \mathcal{S}$ in season $k \in \mathcal{K}$.
y_{ijk} Amount of transported product in kilograms from $i \in \mathcal{S}$ to j in season $k \in \mathcal{K}$.
z_i Indicator variable that takes the value 1 if site $i \in \mathcal{S}$ is selected, 0 otherwise.

Parameters

$\boldsymbol{\eta_i}$	yield in kilograms produced per hectare produced at site $i \in \mathcal{S}$
$\hat{\boldsymbol{H}}_{\boldsymbol{i}}$	$=\sum_{k\in\mathcal{K}} H_{ik}$, food loss in kilograms to be processed from site $i \in \mathcal{S}$ along the whole time horizon
$\boldsymbol{A_{ik}}$	area in hectares at site $i \in \mathcal{S}$ in season $k \in \mathcal{K}$
$\boldsymbol{C_i}$	capacity in kilograms to process losses of candidate site $i \in \mathcal{C}$
$\boldsymbol{D_{ij}}$	distance in kilometers between sites $i \in \mathcal{S}$ and $j \in \mathcal{T} \cup \mathcal{C}$
$\boldsymbol{F_i}$	cost of setting up a food loss processing facility at site $i \in \mathcal{C}$
$\boldsymbol{H_{ik}}$	$= (TL_{ik} + PL_{ik} - TL_{ik}PL_{ik})\eta_i A_{ik}$, food loss in kilograms to be processed from site $i \in \mathcal{S}$ and season $k \in \mathcal{K}$
$\boldsymbol{L_{ik}}$	fraction lost at site $i \in \mathcal{S}$ and season $k \in \mathcal{K}$
$\boldsymbol{MAX}$	maximum number of processing sites allowed
$\boldsymbol{PC_{ik}}$	production cost per kilogram at site $i \in \mathcal{S}$ in season $k \in \mathcal{K}$
$\boldsymbol{PL_{ik}}$	loss as percentage in production at site $i \in \mathcal{S}$ and season $k \in \mathcal{K}$
$\boldsymbol{TC}$	transportation cost per kilogram per kilometer
$\boldsymbol{TL_{ik}}$	loss as percentage in transportation at site $i \in \mathcal{S}$ and season $k \in \mathcal{K}$
$\boldsymbol{W_{ik}}$	demand for product at site $i \in \mathcal{T}$ in period $k \in \mathcal{K}$

Subindexes

$\boldsymbol{i, j, l}$	Site
$\boldsymbol{k}$	Season

A.2 Mathematical model

The problem requires the following decision variables. Let x_{ik} be the production at site i, y_{ijk} the amount of transported product from site i to j at season k, z_i an indicator variable taking the value 1 if a processing center is selected at a candidate site i and 0 otherwise, and v_{ij} an indicator variable that takes the value 1 if the food loss is transported from i to j and 0 otherwise. The objective function is

$$\begin{aligned}\text{Minimize} \sum_{i\in\mathcal{S}}\sum_{k\in\mathcal{K}} PC_{ik}x_{ik} + TC \sum_{(i,j)\in\mathcal{L}}\sum_{k\in\mathcal{K}} D_{ij}y_{ijk} \\ + TC\sum_{i\in\mathcal{S}}\hat{H}_i\sum_{j\in\mathcal{C}} D_{ij}v_{ij} + \sum_{i\in\mathcal{C}} F_i z_i.\end{aligned} \tag{A.1}$$

Subject to:

$$\sum_{(i,j)\in\mathcal{L}} y_{ijk} \le x_{ik} - PL_i\eta_i A_{ik} \quad \forall i \in \mathcal{S}, \forall k \in \mathcal{K}, \tag{A.2}$$

$$x_{ik} \le \eta_i A_{ik} \quad \forall i \in \mathcal{S}, \forall k \in \mathcal{K}, \tag{A.3}$$

where η_i is the yield in kilograms per unit of area and A_{ik} is the land area available for cultivation.

$$\sum_{(i,j)\in\mathcal{L}} y_{ijk} \ge W_{jk} \quad \forall j \in \mathcal{J}, \forall k \in \mathcal{K}, \tag{A.4}$$

$$\sum_{j\in\mathcal{C}} v_{ij} = 1 \quad \forall i \in \mathcal{S}, \tag{A.5}$$

$$v_{ij} - z_j \le 0 \quad \forall i \in \mathcal{S}, j \in \mathcal{C}, \tag{A.6}$$

$$H_{ik}v_{ij} - C_j z_j \le 0 \quad \forall i \in \mathcal{S}, \forall j \in \mathcal{C}, \forall k \in \mathcal{K}, \tag{A.7}$$

where C_j is the capacity of candidate processing site j, and parameter

$$\begin{aligned} H_{ik} &= TL_{ik}(1-PL_{ik})\eta_i A_{ik} + PL_{ik}\eta_i A_{ik} \\ &= (TL_{ik}+PL_{ik}-TL_{ik}PL_{ik})\eta_i A_{ik} \quad \forall i \in \mathcal{S}, \forall j \in \mathcal{C}, \forall k \in \mathcal{K} \end{aligned}$$

is the sum of food losses at the production and transport stages from producer i at season k. Because v_{ij} is binary,

$$v_{ij} \le TL_i \sum_{l \in \mathcal{T}} y_{ilk} \quad \forall i \in \mathcal{S}, \forall j \in \mathcal{C}, \forall k \in \mathcal{K}, \tag{A.8}$$

$$\sum_{i \in \mathcal{C}} z_i \le MAX, \tag{A.9}$$

where MAX is the maximum number of processing sites allowed.

Appendix B Weighted Average Source Distance

Using the notation in Appendix A.1, we calculate WASD of broccoli losses as

$$W\,ASD^{\text{loss}} = \frac{\sum_{i \in \mathcal{S}} \hat{H}_i \sum_{j \in \mathcal{C}} D_{ij} v_{ij}}{\sum_{i \in \mathcal{S}} \hat{H}_i \sum_{j \in \mathcal{C}} v_{ij}}. \tag{B.1}$$

References

Abad, E., Palacio, F., Nuin, M., de Zárate, A.G., Juarros, A., Gómez, J.M., Marco, S., 2009. RFID smart tag for traceability and cold chain monitoring of foods: demonstration in an intercontinental fresh fish logistic chain. J. Food Eng. 93, 394–399.

Atallah, S., Gómez, M., Björkman, T., 2014. Localization effects for a fresh vegetable product supply chain: Broccoli in the eastern united states. Food Policy 49, 151–159.

Bahadur, K., Haque, I., Legwegoh, A., Fraser, E., 2016. Strategies to reduce food loss in the global south. Sustainability. 8 (7). Article No. 595.

Beza, E., Silva, J., Kooistra, L., Reidsma, P., 2017. Review of yield gap explaining factors and opportunities for alternative data collection approaches. Eur. J. Agron. 82, 206–222.

Bibi, F., Guillaume, C., Gontard, N., Sorli, B., 2017. A review: RFID technology having sensing aptitudes for food industry and their contribution to tracking and monitoring of food products. Trends Food Sci. Technol. 62, 91–103.

Biggins, D., Chi, C., Compton, M., Georgakopoulos, D., Jayaraman, P., Palmer, D., Smith, G., Salehi, A., Taylor, P., Terhorst, A., Zaslavsky, A., 2013. Sensing middleware architecture: WP2 state of the art, methodology, components, interfaces. CSIRO.

Bjorndal, T., Herrero, I., Newman, A., Romero, C., Weintraub, A., 2012. Operations research in the natural resource industry. Int. Trans. Oper. Res. 19, 39–62. https://doi.org/10.1111/j.1475-3995.2010.00800.x.

Buzby, J., Hyman, J., 2012. Total and per capita value of food loss in the United States. Food Policy 37, 561–570.

Carlsson-Kanyama, A., 1997. Weighted average source points and distances for consumption origin-tools for environmental impact analysis? Ecol. Econ. 23 (1), 15–23.

Cocco, A., Pinna, A., Marchesi, M., 2017. Banking on blockchain: costs savings thanks to the blockchain technology. Future Internet. 9 (3). Article 25.

Current, J., Daskin, M., Schilling, D., 2004. Facility Location: Applications and Theory. In: second ed. Springer, New York, pp. 81–118.

Defraeye, T., Cronjé, P., Berry, T., Opara, U., East, A., Hertog, M., Verboven, P., Nicolai, B., 2015. Towards integrated performance evaluation of future packaging for fresh produce in the cold chain. Trends Food Sci. Technol. 44 (2), 201–225.

ERS, 2011. Loss-Adjusted Food Availability Data. Economic Research Service (ERS), US Department of Agriculture, Washington, DC.

Esenam, A., 2017. Overview of Digital Agriculture: making growers' lives more productive. Int. Sugar J. 119 (1422), 466–470.

Ezeike, G., Hung, Y., 2009. Refrigeration of Fresh Produce From Field to Home: Refrigeration Systems and Logistics. In: Elsevier, London, pp. 513–537.

Ferrer, J., Cawley, A., Maturana, S., Toloza, S., Vera, J., 2008. An optimization approach for scheduling wine grape harvest operations. Int. J. Prod. Econ. 112, 985–999.

Fleischmann, M., 1997. Quantitative models for reverse logistics: a review. Eur. J. Oper. Res. 103 (1), 1–17. https://doi.org/10.1016/S0377-2217(97)00230-0.

French, M., LaForge, R., 2006. Closed-loop supply chains in process industries: an empirical study of producer re-use issues. J. Oper. Manag. 24, 271–286. https://doi.org/10.1016/j.jom.2004.07.012.

Galanakis, C., 2015. Food Waste Recovery—Processing Technologies and Industrial Techniques. Elsevier, London.

García-Flores, R., de Souza Filho, O., Martins, R., Martins, C., Juliano, P., 2015. Using Logistic Models to Optimize the Food Supply Chain. In: Woodhead Publishing, Cambridge, UK, pp. 307–330.

Gnimpieba, D., Nait-Sidi-Moh, A., Durand, D., Fortin, J., 2015. Using Internet of Things technologies for the collaborative supply chain: application to tracking of pallets and containers. Procedia Comput. Sci. 56, 550–557.

Godfray, H., Beddington, J., Crute, I., Haddad, L., Lawrence, D., Muir, J., Pretty, J., Robinson, S., Thomas, S., Toulmin, C., 2010. Food security: the challenge of feeding 9 billion people. Science 327 (5967), 812–818.

Gogo, E., Opiyo, A., Ulrichs, C., Keil, S., 2017. Nutritional and economic postharvest loss analysis of African indigenous leafy vegetables along the supply chain in Kenya. Postharvest Biol. Technol. 130, 39–47.

González-Torre, P., Adenzo-Díaz, B., Artiba, H., 2004. Environmental and reverse logistics policies in European bottling and packaging firms. Int. J. Prod. Econ. 88, 95–104.

Grunow, M., Piramuthu, S., 2013. RFID in highly perishable food supply chains—remaining shelf life to supplant expiry date? Int. J. Prod. Econ. 146, 717–727.

Gustavsson, J., Cederberg, C., Sonesson, U., Otterdijk, R., Meybeck, A., 2011. Global Food Losses and Food Waste. Food and Agriculture Organization of the United Nations, Rome.

Hanyu, H., Shimura, T., Fukui, T., 2011. Sensor network for HACCP food safety management. In: Zhang, Y., Wang, W., Wang, W. (Eds.), In: Proceedings of 2011 International Conference on Communication Technology and Application ICCTA2011, IET, pp. 666–670.

Hasani, A., Zegordi, S., Nikhbash, E., 2012. Robust closed-loop supply chain network design for perishable goods in agile manufacturing under uncertainty. Int. J. Prod. Res. 50 (16), 4649–4669.

Heller, M., Keolian, G., 2014. Greenhouse gas emission estimates of us dietary choices and food loss. J. Ind. Ecol. 19 (3), 391–401.

Higgins, A., Watson, I., Chilcott, C., Zhou, M., García-Flores, R., Eady, S., McFallan, S., Prestwidge, D., Laredo, L., 2013. A framework for optimising capital investment and operations in livestock logistics. Rangel. J. 35 (2), 181–191.

Horticultural Innovation Australia Limited, 2015. Australian Horticulture Statistics Handbook 2014/2015. Horticultural Innovation Australia Limited, Sydney, NSW.

Hull, R., Batra, V.S., Chen, Y.-M., Deutsch, A., Heath, F., Vianu, V., 2016. Towards a shared ledger business collaboration language based on data-aware processes. In: Zheng, Q., Stroulia, E., Tata, S., Bhiri, S. (Eds.), Lecture Notes in Computer Science. Service-Oriented Computing—Proceedings of the 14th International Conference ICSOC 2016, 9936. Springer, Banff, Canada, pp. 18–36.

Jood, S.M., Cai, X., 2016. Reducing food loss and waste to enhance food security and environmental sustainability. Environ. Sci. Technol. 50, 8432–8443. https://doi.org/10.1021/acs.est.6b01993.

Joutsela, M., Korhonen, V., 2015. Capturing the user mindset—using the online research community method in packaging research. Packag. Technol. Sci. 28, 325–340.

Kim, T., Glock, C., Kwon, Y., 2014. A closed-loop supply chain for deteriorating products under stochastic container return times. Omega 43, 30–40.

Kovačić, D., Hontoria, E., Ross-McDonell, L., Bogataj, M., 2015. Location and lead-time perturbations in multi-level assembly systems of perishable systems in Spanish baby food logistics. Cent. Eur. J. Oper. Res. 23, 607–623.

Kumar, S., 2014. A knowledge based reliability engineering approach to manage product safety and recalls. Expert Syst. Appl. 41, 5329–5339.

Lebersorger, A., Schneider, F., 2014. Food loss rates at the food retail, influencing factors and reasons as a basis for waste prevention measures. Waste Manag. 34, 1911–1919.

Li, J., 2013. The prospective impact of a multi-period pricing strategy on consumer perceptions for perishable foods. Br. Food J. 115, 377–393.

Lorite, G., Selkälä, T., Sipola, T., Pelanzuela, J., Jebete, E., Vinuales, A., Cabanero, G., Grande, H., Tuominen, J., Uusitalo, S., Hakalahti, L., Kordas, K., Toth, G., 2017. Novel, smart and RFID assisted critical temperature indicator for supply chain monitoring. J. Food Eng. 193, 20–28.

Mohebi, E., Marquez, L., 2014. Intelligent packaging in meat industry: an overview of existing solutions. J. Food Sci. Technol. 52 (7), 3947–3964.

Noletto, A., Loureiro, S., Castro, R., Lima, O., 2016. Intelligent packing and the Internet of Things in Brazilian food supply chains: the current state and challenges. In: Lecture Notes in Logistics. Dynamics in Logistics, LDIC 2016, Springer, Bremen, Germany, pp. 173–183.

Nukala, R., Panduru, K., Shields, A., Riordan, D., Doody, P., Walsh, J., 2016. Internet of Things: a review from "farm to fork". In: Proceedings of the 27th Irish Signals and Systems Conference (ISSC), IEEE. Ulster University, Derry, Northern Ireland.

Patare, P., Opara, U., 2014. Structural design of corrugated boxes for horticultural produce: a review. Biosyst. Eng. 125, 128–140.

Petkovic, K., Fox, E., García-Flores, R., Chandry, S., Sangwan, P., Sanguansri, P., Augustin, M., 2017. A glimpse into the future: a food loss bank for transforming the food supply chain? Food Aust. 42–44 (Jan/Feb 2017).

Popović, T., Latinović, N., Pešić, A., Zečević, Z., Krstajić, B., Djukanović, S., 2017. Architecting and IoT-enabled platform for precision agriculture and ecological monitoring: a case study. Comput. Electron. Agric. 140, 255–265.

Ram, V., Visal, H., Dhanalakshmi, S., Vidya, P., 2015. Regulation of water in agriculture field using Internet of Things. In: Proceedings 2015 IEEE International Conference on Technological Innovations in ICT for Agriculture and Rural Development TIAR 2015, IEEE, Chennai, India, pp. 112–115.

Rateni, G., Paolo, D., Cavallo, F., 2017. Smartphone-based food diagnostics technologies: a review. Sensors. 17 (6). Article No. 1453.

Ridout, B., Juliano, P., Sanguansri, P., Sellahewa, J., 2010. The water footprint of food waste: case study of fresh mango in Australia. J. Clean. Prod. 18, 1714–1721.

Stock, F., Dunstall, S., Ayre, M., Ernst, A., Nazari, A., Thiruvady, D., King, S., 2015. Using optimisation to suggest alternative supply chains in the context of industrial symbiosis. In: Weber, T., McPhee, M., Anderssen, R. (Eds.), In: MODSIM2015, 21st International Congress on Modelling and Simulation. Modelling and Simulation Society of Australia and New Zealand, pp. 1766–1772. Broadbeach, QLD.

Thyberg, K., Tonjes, D., 2016. Drivers of food waste and their implications for sustainable policy development. Resour. Conserv. Recycl. 106, 110–123.

Tian, F., 2016. An agri-food supply chain traceability system for China based on RFID and blockchain technology. In: Proceedings of the 13th International Conference on Service Systems and Service Management (ICSSSM), IEEE, pp. 1–6.

Tian, J., Bryksa, B., Yada, R., 2016. Feeding the world into the future—food and nutrition security: the role of food science and technology. Front. Life Sci. 9 (3), 155–166.

Willersinn, C., Mack, G., Mouron, P., Keiser, A., Siegrist, M., 2015. Quantity and quality of food losses along the Swiss potato supply chain: stepwise investigation and the influence of quality standards on losses. Waste Manag. 46, 120–132. https://doi.org/10.1016/j.wasman.2015.08.033.

Zou, Z., Chen, Q., Uysal, I., Zheng, L., 2017. Radio-frequency identification enabled wireless sensing for intelligent food logistics. Philos. Trans. R. Soc. A 372, 20130313.

Chapter 19

Sustainable urban food planning: Optimizing land-use allocation and transportation in urban-rural ecosystems

Stefano Penazzi*,† and Riccardo Accorsi*

*Department of Industrial Engineering, Alma Mater Studiorum—University of Bologna, Bologna, Italy †The Logistics Institute, University of Hull, Hull, United Kingdom

Abstract

Over recent years, faster urbanization and natural resources depletion have compelled an increasing attention to environmental conservation and the careful management of soil and land use. Sustainable Land-Use Planning (LUP) is the activity of allocating resources and uses to specific sites and areas, with different purposes that depend on the decision maker. Examples of these include maximizing crop yields or the profits of landowners and minimizing environmental impacts such as water consumption or greenhouse gases (GHGs) emissions. Several environmental drivers and stressors can be considered in LUP, but one of the goals of environmental conservation is undoubtedly the reduction of carbon emissions. Considering the rapid growth of cities, which centralize resources, population, and consumers and thereby contribute to climate change, sustainable LUP planning of integrated urban and rural ecosystem can be the key to addressing such environmental and social issues.

This chapter builds on the well-known LUP problem by investigating the relationship between transportation of people, i.e., mobility, and of food resources within a closed urban-rural ecosystem, and by determining how this affects the overall carbon emissions. The proposed LUP approach is based on a model that integrates agricultural properties such as soil features and crop yields, with the renewable energy potential of the area and the urban needs of accessibility and density. The discussed mix integer linear programming (MILP) model is implemented into a planning-support tool aimed at studying the optimal allocation of land uses in an urban-rural ecosystem with the goal of minimizing the carbon emissions. The model takes into account a closed ecosystem where agriculture, energy production, and residential areas satisfy different internal demands and needs. Some results obtained by the application of this tool are illustrated and discussed, showing the potential of the proposed approach in supporting the planning of low-carbon urban-rural ecosystems.

1 Introduction

The world population will grow by about 3 billion people within 2050, and 66% of humanity will live in urban areas, as compared to 54% today (UN, 2014). This urbanization process is fueled by both population growth and by rural-urban migrating flows (Deng et al., 2014). As a consequence, there will be a rising demand for new residential areas and for production of food commodities and energy (Madlener and Sunak, 2011). The rationale of this chapter is the fact that, while cities cover 2% of the earth's surface, they are responsible for about 75% of the overall resources consumption (Pacione, 2009). Furthermore, cities and urban areas are also responsible for about 80% of the global energy consumption (IEA, 2008).

In order to fulfill such demand, more and more lands are being converted and reclassified from rural to urban areas, and further lands are being cleared for exploitation through deforestation. If not properly controlled and regulated, this rapid development may result in a set of negative environmental impacts that is not sustainable over the long term. Land use and cover change are responsible for one-third of the world's cities' carbon dioxide emissions (CO_2) (Denman et al., 2007). Since CO_2 sink is closely related to the presence of forests and natural landscapes, the potential of CO_2 sequestration of an area depends on how much natural land is converted to human use. Other factors affecting the natural carbon cycle are reduction of biodiversity, soil acidification, and soil erosion (Brussard et al., 2010; Lu et al., 2015). Pauleit and Duhme (2000) studied the interaction between carbon emissions and land use in urban areas. Xu and Lin (2014) discussed the impacts of industrialization and urbanization on CO_2 emissions in China. Haas et al., 2015 investigated urbanization

Sustainable Food Supply Chains. https://doi.org/10.1016/B978-0-12-813411-5.00019-3

and its potential environmental consequences in the Shanghai and Stockholm metropolitan areas over two decades. These studies all highlight the need to manage the urbanization process and to minimize the resulting environmental impacts. Meanwhile, given the rapid growth of the world population, the literature sustains the importance of reducing the environmental impacts of food production and distribution processes from farm to table (Accorsi et al., 2017a,b, 2018; Desjardins et al., 2007; Gallo et al., 2017).

In the last decades, several concepts, models, and tools have been developed to describe and support sustainable urbanization, helping planners with this difficult task. Wolman (1965) introduced the concept of urban metabolism. He observed the city as an organism with the equivalent of metabolic processes, studying how freight, energy, food, and other inputs flow into the organism and which outgoing flows are generated. Recently, Zhang (2013) defined urban metabolism as a process of resources consumption, transformation, and waste generation to account for the generation, collection, and disposal of resources and wastes by the city.

The urban metabolism paradigm provides an effective way to track the social, economic, and natural phenomena that occur in cities and to quantify the greenhouse gas (GHG) emissions due to transformation, handling, transportation, and consumption activities. Tracking these activities at a quantitative layer means properly defining the spatial scale of analysis. Urban ecosystems exchange various sorts of flows. These are interurban flows, i.e., between cities, and intraurban flows, between the resources and actors within the city itself. The analysis of the interflows between cities (i.e., nodes) indeed becomes a question of where to establish cities and which city (or node) is responsible to supply which product, resulting in fact in a very renowned formulation of the location-allocation problem (see, for instance, in the field of supply chain management Accorsi et al., 2015). When the focus shifts to urban zones, the flows between utilizers as well as any other transformation activity (i.e., processing) must be mapped to a particular area. This class of problems, called land-use allocation problems, aims to allocate different activities or uses to specific units of area within a geospatial environment, balancing several, often conflicting, factors with each other (Aerts et al., 2003; Caparros-Midwood et al., 2015, 2017). Moreover, they have to consider the long-term nature of these factors. Once a decision is made to associate an activity with a zone (e.g., building, industry, agriculture, leisure), it is very difficult to make changes without increasing the cost and the impacts (Casiroli, 2007).

An example of an effort to merge these two classes of problems to plan the agro-food supply chains as zero-carbon closed ecosystems is illustrated in Accorsi et al. (2016). Still, modeling of the intraurban and interurban flows represents an open challenge, particularly in the field of urban carbon metabolism. Indeed, to address the growing concerns and worries of climate change, researchers are focusing on urban carbon metabolism (Sovacool and Brown, 2009).

This chapter embeds the land-use allocation problem and the spatial distribution of the land uses (i.e., intraurban flows) into the urban carbon metabolism framework. Land uses, their relationships, connectivity, and energy, food, and people flows generated within the urban system are modeled as a closed ecosystem. The main objective of this chapter is to optimize the spatial distribution of land uses and the flows between the established lands by minimizing the overall CO_2 emissions generated by the urban-rural ecosystem. As can be seen from Table 1 and Fig. 1, CO_2 is by far the greatest contributor to GHGs and hence our focus is on CO_2.

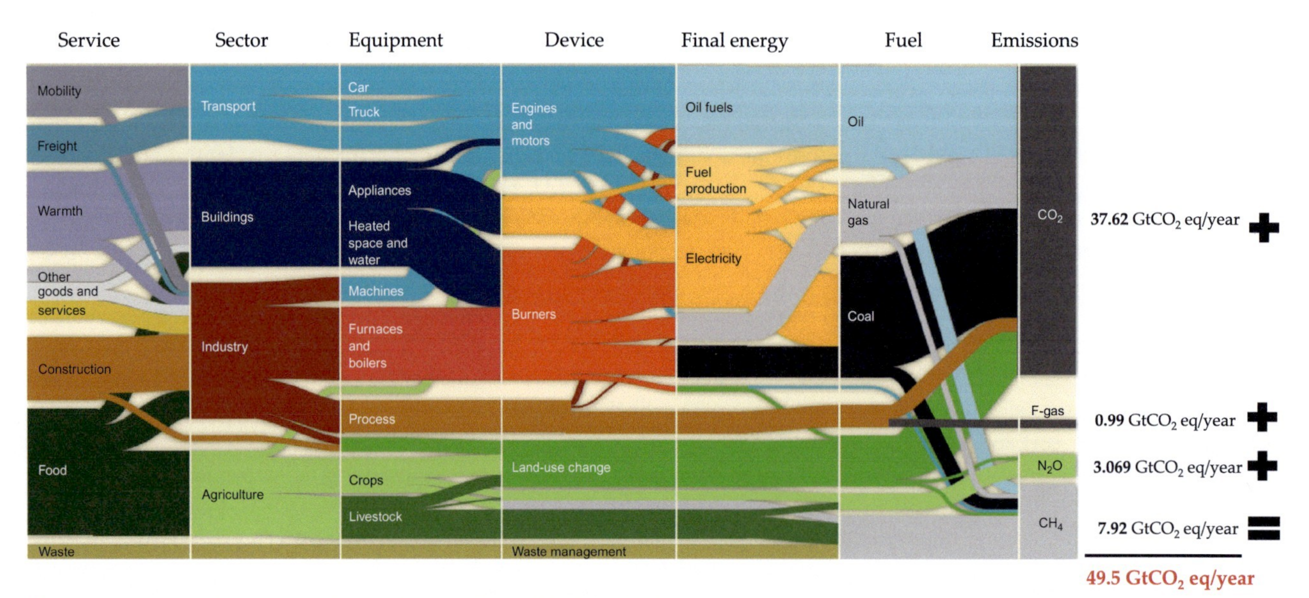

FIG. 1 Carbon emissions sources: Sankey diagram (IEA, 2008; IPCC 2014).

TABLE 1 IPCC 2014, main impact of GHG pollutants

Impact category	Percentage (%)
HFC, PFC, SF_6	2
N_2O	6.2
CH_4	16
FOLU	11
CO_2 from fossil fuel and industrial processes	65

While some sectors, like agriculture or power generation, contribute massively to land use, others, like transportation, while requiring little land use, still generate major CO_2 emissions as a result of the flows between actors, nodes, and geographic areas. Since urban areas are responsible for 80% of the overall GHG emissions, consideration of these flows is necessary to implement a sustainable urban planning based on the actual spatial distribution pattern.

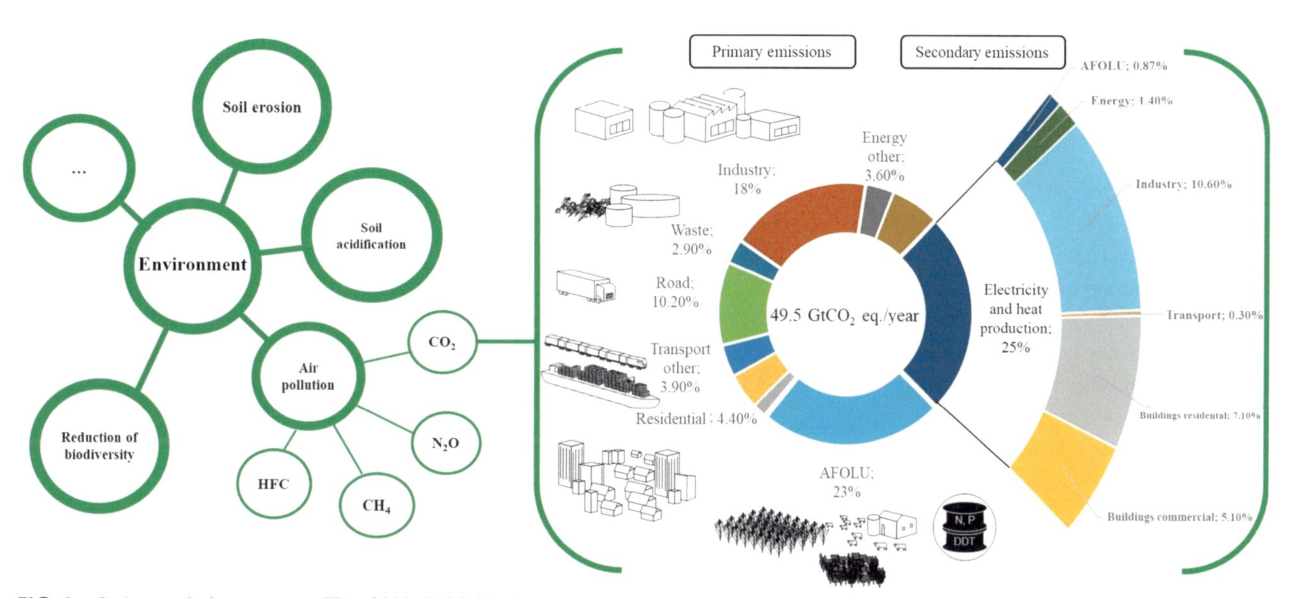

FIG. 2 Carbon emissions sources (IEA, 2008; IPCC 2014).

According to the sources illustrated in Fig. 2, in this chapter the following four sectors are considered as main contributors to CO_2 emissions: (i) electricity and power supply, (ii) residential buildings, (iii) agriculture, forestry and other land use, and (iv) transportation. Therefore these four sectors are the entities/processes involved in our urban-rural closed ecosystem.

This chapter discusses and implements a method to support sustainable urban-rural planning by using an optimization model built within the carbon urban metabolism paradigm and integrating the land-use allocation problem with the aforementioned spatial distribution pattern. In addition to energy supply, the model also considers freight transportation and urban mobility together with the associated flows. This is a departure from the various urban planning models presented in the literature, where freight flows are not taken into account. With regard to freight, the model only considers the movements associated with agro-food goods with the purpose of fulfilling human primary needs (as opposed to items such as clothing, etc.).

The remainder of the chapter is organized as follows: Section 2 briefly overviews the literature on the design and planning of urban areas under the urban metabolism paradigm; Section 3 describes the main characteristics of the model; Section 4 illustrates a methodology for the application of the model to support real-world planning and presents some exemplifying results; lastly, conclusions are presented in Section 5.

2 Short literature review

The aim of the proposed planning-support model is to minimize the environmental impact derived from the main CO_2 polluters of an urban-rural ecosystem identified as food transportation from rural to urban areas, energy supply to industry and buildings, and urban mobility. These are considered as pillars of the urbanization process. In comparison with other Land-Use Planning (LUP) models (Yewlett, 2001), our model involves urban flows of freight and people and energy supplies to evaluate the optimal spatial distribution for a set of elements/resources, within the same problem.

Urbanization requires more and more lands to be converted and reclassified from rural to urban areas. This leads to an increase in distances between manufacturing facilities and markets; wider and more interconnected supply networks; and an increment in vehicle ownership and mobility demand (Rodrigue et al., 2006).

This indicates that the urbanization and its related features, such as urban forms and urban density, affect travel demand and transport energy for both freight and passenger movements (Poumanyvong et al., 2012). The review of the literature shows lack of integrated planning, since transportation is not considered in LUP. Most of the attempts involve passenger mobility but not freight.

Urban mobility, i.e., moving people from one place to another, is also a significant source of GHGs. Despite the huge impact of urban mobility on the GHG emissions, its incorporation in LUP is still rare in the literature. Optimization is poorly utilized to model and plan urban mobility flows since the origin-destination routes, the associated flows, and the number of decision makers increase exponentially with the complexity of the problem. In the literature, the management of urban mobility is approached via analytical techniques such as the study of stochastic patterns or as multiagent simulation systems. In our case, urban mobility is embedded within the optimization model via a density function that is highly correlated with almost all metrics of urban sprawl (Ewing and Cervero, 2001). Several studies have been done to examine the relationship between population density and carbon emissions caused by car usage, public transport, domestic heating, and so on (Brownstone and Golob, 2009; Glaeser and Kahn, 2010). According to Pickrell and Schimek (1999), the usage of private vehicles (i.e., cars) increases with an increase in urban sprawl and decreases as density increases.

Pauleit and Duhme (2000) studied the interaction between carbon emissions and land use in urban areas, concluding that LUP is a tool to affect the carbon imbalance. The need to balance the environmental impacts and the consumers and flows of an urban ecosystem is widely studied under the paradigm of urban metabolism.

Early, Wolman (1965) introduced the concept of urban metabolism. He considers the city as an organism with the equivalent of metabolic processes, studying how freight, energy, and other inputs flow into the organism and which outgoing flows this process generates. Recently, Zhang (2013) defined urban metabolism as a process of resources consumption, transformation, and waste generation to account for the generation and collection/disposal of resources and wastes by the city.

It is well known that urban ecosystems exchange various sorts of flows. These are classified as interurban flows and intraurban flows. The spatial distribution of the intraurban flows affects the flow nature and its impact on the urban metabolism. Among carbon metabolic fluxes, even transportation flows deserve to be incorporated in urban planning.

The transport sector is one of the main causes of the carbon emissions in urban areas. From 1970 to 2004, the transport sector was responsible for 23% of all energy-related CO_2 emissions (Denman et al., 2007). Yan and Crookes (2010) pointed out that the transport sector gets the fastest growth rate in CO_2 emissions of any other sector. It is projected that transport CO_2 emissions will grow by more than 80% by 2050 (IEA, 2008). Therefore, land-use and transport development have a mutual influence in terms of environmental impact.

The IEA (2008) describes the methods and approaches to understand the link between transportation investment and land development; moreover, it indicates that transportation agencies are recognizing induced land development as an impact of transportation capacity projects. Over the past 50 years, several studies have focused on the development of models to understand and predict in which ways and how much the land use affects daily travel and transport mode.

Lowry and Balling (2009) presented a new approach to facilitate the cooperation between city and regional planners. The approach was accomplished in two sequential stages: a regional planning stage and a city planning stage. In both stages, land-use and transportation planning are cast as multiobjective design problems and solved using a genetic algorithm. Waddell et al. (2007) presented a case study on the application of a detailed land-use simulation model system and its integration with a regional travel demand model. Su et al. (2014) developed a land-use allocation methodology called the three stages-two-feedback method for both land-use allocation and transportation policy options. Their objective is to minimize the vehicle miles traveled (VMT) given the constraints such as land-use allocation and the multimodal transport assignment. Schwarz et al. (2016) used an agent-based simulation model to investigate the effects that urban development scenarios have on building infrastructure quality and the population dynamics.

The literature in this field agrees that the carbon cycle represents one of the most important actors in an urban ecosystem. Carbon footprint of an urban-rural ecosystem depends on the imbalance between carbon emissions and sequestration. An assessment and planning of these two flows is crucial to get a sustainable urban plan.

To address the growing concerns and worries about climate change, this chapter develops a methodology to integrate different urban metabolic fluxes into an LUP model intended to manage the urbanization process and minimize the resulting environmental impacts.

3 Modeling low-carbon urban ecosystems

To cope with the carbon imbalance of an urban-rural ecosystem, the proposed model takes into account both positive and negative CO_2 emission flows: positive for emissions and negative for CO_2 sequestration that comes from green areas. The plants over such areas sink the carbon in their leaves, branches, bark, and roots; approximately half the dry weight of a tree biomass is made up of carbon. A good balance between human-induced CO_2 flows and CO_2 sink flows of plants is necessary to pursue a sustainable urban carbon metabolism.

For this reason, the proposed model fosters a process of urbanization by controlling the environmental impact in term of carbon emissions. Hence, the developed multisite land-use allocation model allows urban planners to develop an integrated urban-rural closed ecosystem involving energy supply, building, food provision, and transportation, while minimizing the overall CO_2 emissions.

The model seeks the best allocation of several entities or services (agricultural, residential, green areas, and also areas reserved for renewable energy production such as solar, wind, biofuels) over a closed ecosystem. A scalable grid splits the area into several cells, i.e., spatial units, called lands. Each land is a homogeneous spatial unit with the same shape and the same surface area. The feasible combinations of lands and entities establish the set of potential ecosystem configurations. The model finds the spatial configuration that minimizes the CO_2 emissions while fulfilling the demands of population in terms of housing, food, and energy requirements. The objective function balances carbon emissions due to freight and urban transportation and carbon sequestration.

The basic assumptions of the model are that we have a closed ecosystem in a stationary situation, where all the parameters are preliminarily known and constant during the planning period with no external influences. The model is ideal for a situation where a new urban area needs to be created for a known population count. Two basic types of demand are considered: (i) primary demand consisting of food and energy to support daily living; (ii) secondary demand in terms of energy to produce the food required. Fig. 3 shows the link between resource flows, network entities, and types of demand, illustrating the change of paradigm required from the typical planning approaches and the proposed one.

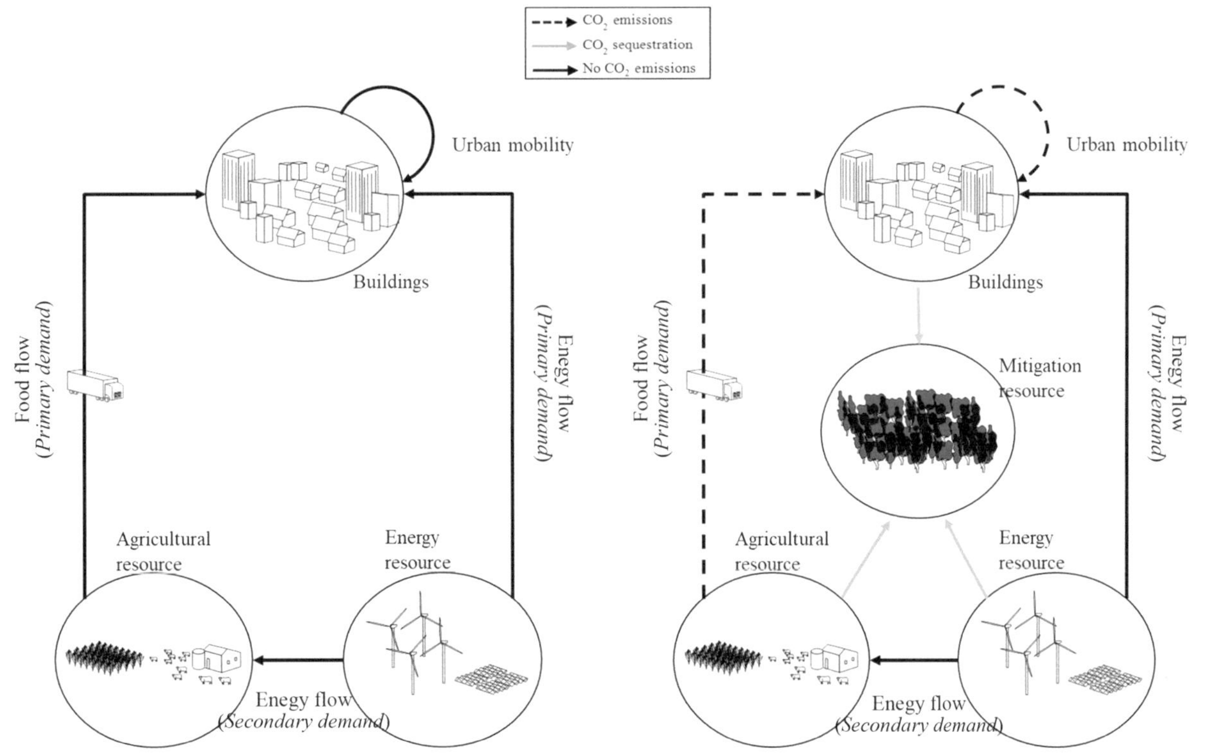

FIG. 3 Physical flows of food and dwellers to satisfy primary and secondary demand. On the left typical flows/entities are considered in the planning problem without accounting for carbon emissions. On the right, carbon emissions from transportation and land-use change are considered as well as mitigation services provided by carbon plantings.

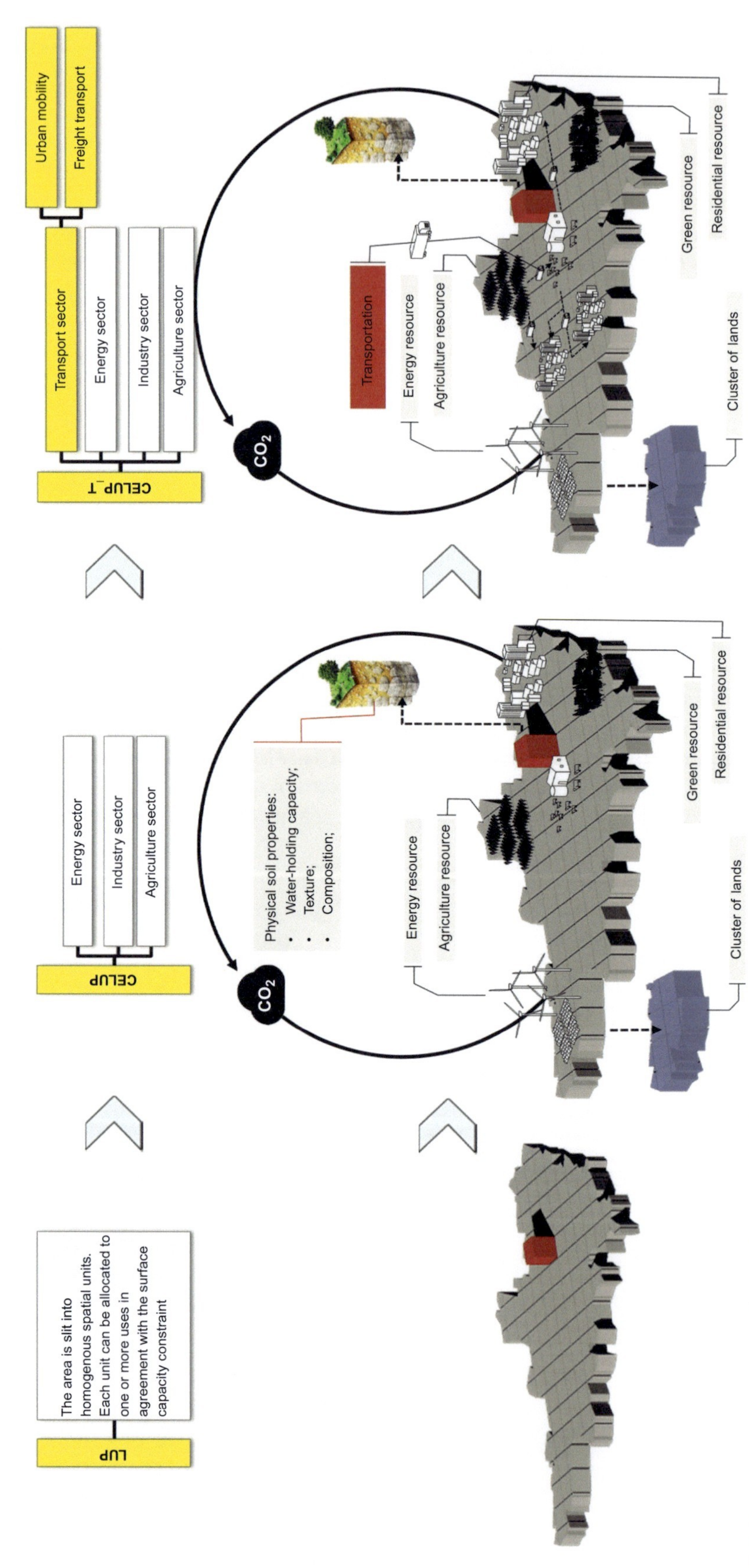

FIG. 4 Urban-rural metabolic flows and land-use planning problems.

Food resources are provided by agricultural productions, i.e., farms and crops, and supplied using road transportation to the urban areas, i.e., residential and commercial. The energy resources are generated by two forms of land-intensive sources: clean renewables, i.e., solar and wind, and CO_2 emitting renewables, i.e., biofuels, given by the cultivation of specific crops. Despite their origin, energy resources fuel a primary demand from buildings (i.e., for energy and heat consumption) and a secondary demand from rural areas (i.e., energy for plants, equipment, and machineries). In the current planning scenario (on the left of Fig. 3), the carbon emissions generated by food distribution and dweller mobility (i.e., overall transportation) are not considered, as conversely happens in the proposed pattern (on the right), which also uses mitigation services, like green areas, to sink CO_2.

By illustrating the evolution of the problem from the current to the proposed paradigm, Fig. 4 represents how different land uses contribute to fulfill the urban-rural metabolic flows. The allocation of these land uses is typically managed through LUP models. In this chapter new shapes of LUP are implemented with the aim of minimizing the carbon footprint of the land-use planning, i.e., carbon emission land-use planning (CELUP), also involving transportation impacts (CELUP_T).

The decisions made by the model are where to locate producing lands vs. demanding lands, taking distances into account. For the lands, the model also considers geographical factors such as soil properties, weather conditions, and accessibility (see Fig. 4).

To address urban mobility, the model uses the relationship between population density and the vehicle use to estimate the CO_2 emission connected with people moving. This relationship is typically a concave nondecreasing function in terms of the population, the slope of which represents the CO_2 emissions per capita, which decreases as the population density increases.

The model considers the allocation of green spaces in lands to act as CO_2 sinks. The so-called green resource is the fourth entity involved in the model (see the mitigation resource in Fig. 3). These generate negative CO_2 flows through the process of photosynthesis. While the complete formulation of the model is beyond the scope of this chapter, the objective function and an overview of the main constraints are shown here only for the sake of completeness.

Sets and indices:

L	Set of total lands
RF	Set of food resources
BF	Set of biofuel energy resources
Q	Set of clusters Gq
W	Set of population density values

Parameters:

CS	Carbon dioxide sequestration per green area (kg CO_2/ha)
CS_{bf}	Carbon dioxide sequestration to produce biofuel from plants (kg CO_2/ha)
$CECT$	Carbon dioxide produced by transport of food resource r (kg CO_2/ha)
$CEUT_{qw}$	Carbon dioxide produced by urban mobility in the cluster q with density w (kg CO_2/ha)
$D_{ll'}$	Distance between land l and land l' (km)

Decision variables:

y_{lr}	1 if the resource r is produced in the land l. 0 otherwise
y_{lcs}	1 if the resource plants are allocated in the land l. 0 otherwise
h_{qw}	1 if in the subset G_l there is at least one land with the urban resource
$x_{ll'r}$	flow from land l to land l' to satisfy the demand of the resource r in land l

$$\min \sum_{l\in L}\sum_{l'\in L}\sum_{r\in RF} x_{ll'r}\cdot D_{ll'}\cdot CECT + \sum_{q\in Q}\sum_{w\in W} h_{qw}\cdot CEUT_{qw} - \left(\sum_{l\in L} y_{lcs}\cdot CS + \sum_{l\in L}\sum_{r\in BF} y_{lr}\cdot CS_{bf}\right)$$

s.t. Linear constraints
Decision variables domain

The objective function minimizes the sum of CO_2 emissions resulting from transport of food and energy supply and urban mobility, considering the negative contribution of the green spaces acting as carbon sinks.

The model is completed by a set of linear constraints divided into demand fulfilment, capacity, and green closeness. First, constraints are required to satisfy the primary and secondary demand. Primary demand is considered per capita and it is composed of food commodities, considered nonsubstitutable for each other, and household energy. Indeed, secondary demand is considered per product type and production unit (i.e., energy per crop hectare).

The constraints on the maximum productivity capacity of food commodities and energy affect every cell in the area. Maximum productivity of food commodities is connected with several environmental factors such as soil properties and weather conditions. The same applies to energy productivity in the case of renewable sources (i.e., solar irradiation, wind speed).

In order to maximize the carbon sequestration, the green areas have to be established beside the CO_2 sources. Nowak and Crane (2001) compare urban forest with nonurban forest, finding that nonurban forest typically has about twice the tree density as urban greens and about half the average carbon density per unit of tree cover (urban green cover: 9.25 kg CO_2/ha vs forest: 5.3 kg CO_2/ha). For this reason, the model is constrained to allocate a minimum number of green areas inside of a subset of lands, which we call cluster. In the observed geographical area, each single land is contained in one or more clusters. The number of green areas allocated inside each cluster is such that it offsets a set percentage of the total emission of that cluster.

Fig. 5 summarizes the rationale of the model and exemplifies some carbon metabolic flows and their direction as accounted for by the objective function.

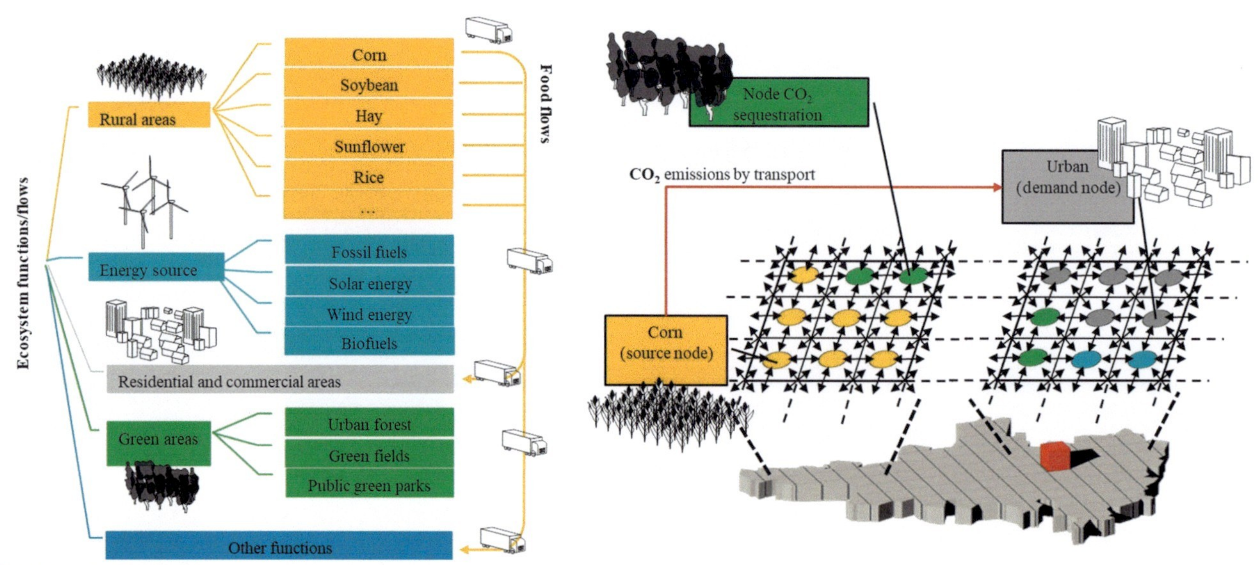

FIG. 5 Model rationale.

4 Methodology and planning-support tools

The introduced model seeks the best allocation of several entities or resources (i.e., agricultural, residential, energy sources, plantings) to minimize the carbon emissions of the urban-rural ecosystem and address the primary population needs (i.e., housing, energy, food, mobility). Given the complexity of the issues involved, the implementation of the model is incorporated into a wider methodology, illustrated in Figs. 6 and 7, which leads the decision maker through the analysis.

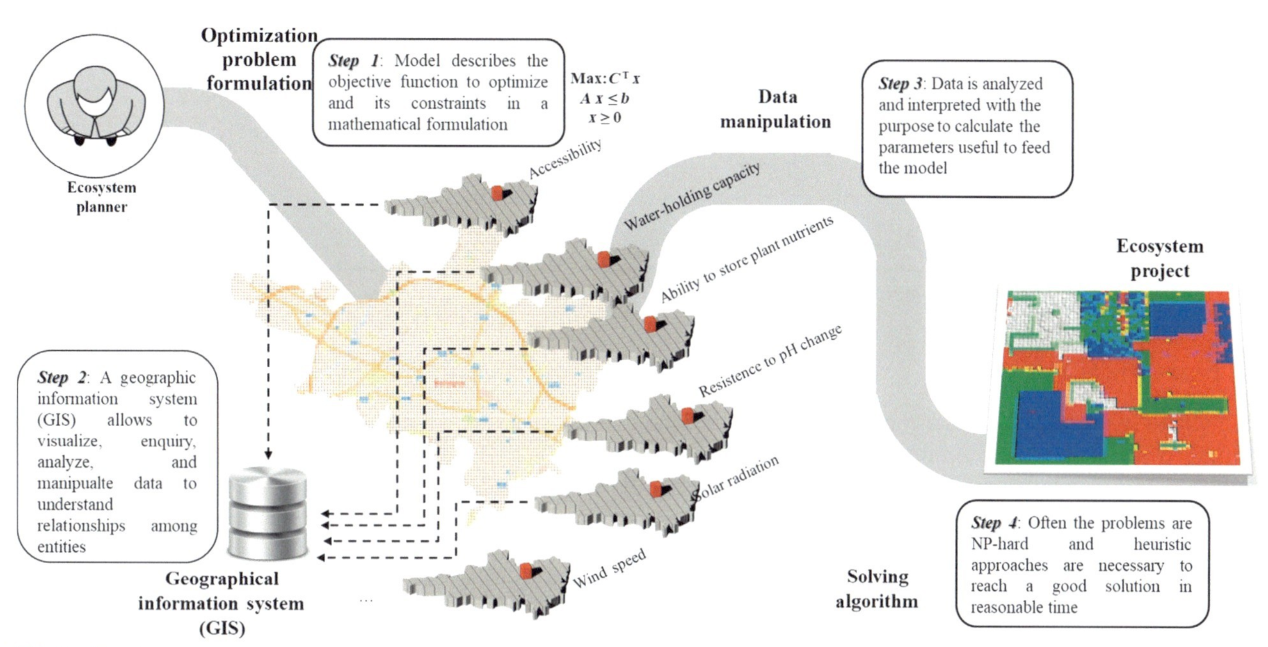

FIG. 6 Planning steps.

The first step is to define the set of linear constraints that limits the objective function. In the previous section, a brief definition of the minimum set of constraints needed is given. Others can be inserted according to the specific instance or changed in agreement with a new objective function. For example, new formulation of the model may include other objective functions (e.g., cost-based functions) rather than involve a multiobjective approach. Some dedicated (e.g., AMPL, GAMS) or general-purpose (e.g., C, Phyton) languages can be used to write the model and make it solvable by a solver.

To fuel the model, a multidisciplinary dataset is then collected and manipulated through a relational database and SQL scripts (e.g., MySQL) to quantify the parameters. Given the nature of the data, geographic information systems (GISs) can be used either to gather information on soil (e.g., GrassGIS), energy potentials (e.g., PVGIS, WindGIS), and accessibility of an area, or to store and show such records on a map (see "graphic output" in Fig. 7). At this step the GISs can also be used to identify the most promising areas as to where to trigger an urbanization process from a rural environment.

Once the parameters of the model have been set, this can be solved by using a commercial mixed-integer linear solver (e.g., Gurobi, Cplex). These typically offer some effective well-known solving algorithms (e.g., branch-and-bound, cutting-plane method, or branch-and-cut) which find an optimal solution for the MILP problem with a medium instance in reasonable time.

By solving the CELUP_T model, the optimal land-use allocation of the observed urban-rural environment is provided to the planner, who is then responsible for interpreting it and deriving from it a feasible and practicable configuration of the real-world area. The methodology concludes by returning the solution as a layer of the GIS, allowing future review of the planning.

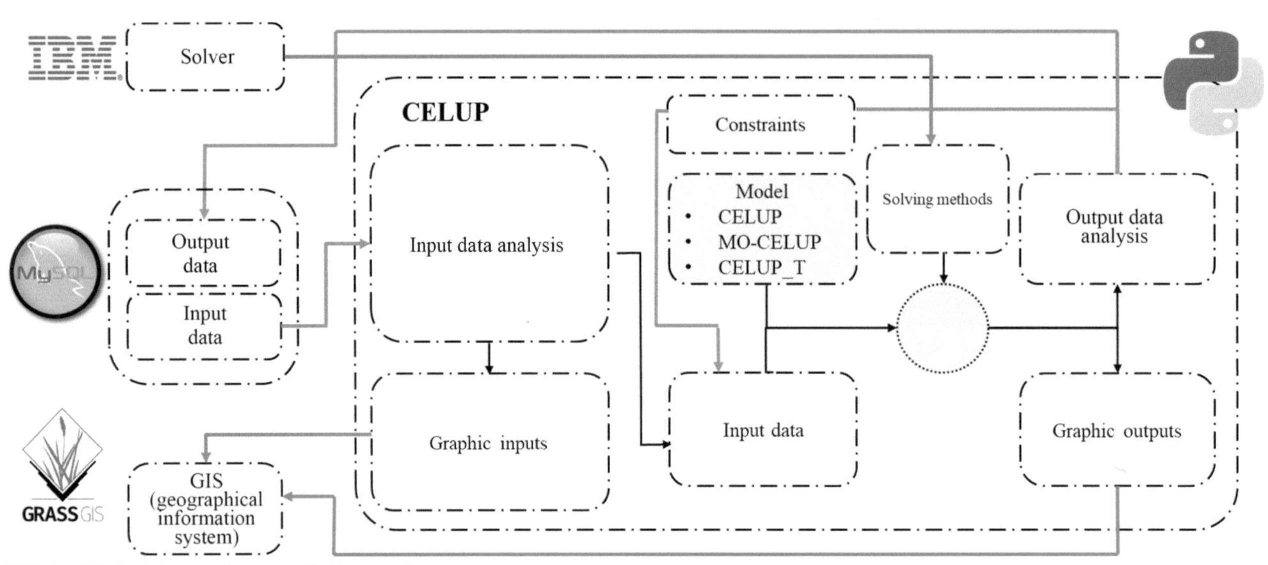

FIG. 7 Methodology and support-decision tool.

5 Numerical example

The following example illustrates how an urbanization process can be aided by the proposed model and how it can be influenced by transportation and resources distribution. For purposes of presentation, consider a cleared area in which each entity/resource can be allocated lands, where each land is 0.01 skm. Table 2 summarizes the food demand per capita per year for each commodity. Both primary and secondary consumptions are included. For instance, Italy produces 1 million tons of corn per year. Of this, 85% is used for animal fodder, 10% is used for pharmaceutical companies, and only the remaining 5% is used directly in the food industry to fulfil the primary demand. Energy for production is required by agriculture, mainly for irrigation and equipment powering, and can be satisfied by gas combustion or electricity. Table 2 also shows the mean and standard deviation of the energy required for each crop, per kilogram per year, for each land.

TABLE 2 Food demand per capita and energy demand to produce food resources

		Food demand	Energy requirements by crop	
	Product	Weight (kg/year)	Mean (kWh/kg year)	SD (kWh/kg year)
1	Corn	150	0.78	0.195
2	Soybean	61	1.5	0.375
3	Hay	20	0.6	0.15
4	Sunflower	11	2.26	0.565
5	Durum wheat	25	1.486	0.371
6	Soft wheat	30	0.7527	0.189
7	Sorghum	18	1.23	0.307
8	Rice	9	1.8	0.45
9	Potatoes	35	0.45	1.112
10	Tomatoes	58	2.27	0.56
11	Oranges	18.5	1.54	0.385
12	Apples	18.5	1.67	0.417

Food resources productivity depends on the soil features, like water-holding capacity, aeration, susceptibility to water erosion, suitability for tillage after rain, as well as on weather and climate characteristics (input data as for Accorsi et al., 2016 and Accorsi, 2019). To generate the productivity data for each land, we consider a normal distribution, while the metric homogeneity measures the spatial distribution of the parameter. Low homogeneity most affects the efforts to find an optimal solution due to an increase of possible optimal scenarios. Table 3 shows mean values, standard deviation, and homogeneity used to generate food productivity per resource per land and year.

TABLE 3 Data productivity per food resource

Product	Mean (kg/year ha)	SD (kg/year ha)	Homogeneity rate
Corn	15,000	5250	0.12
Soybean	2500	875	0.12
Hay	3600	1260	0.12
Sunflower	3000	1050	0.12
Soft wheat	2500	875	0.12
Durum wheat	2750	962.5	0.12
Sorghum	1800	630	0.12
Rice	4400	1540	0.12
Potatoes	17,400	6090	0.12
Tomatoes	11,500	4025	0.12
Oranges	6000	2100	0.12
Apples	4800	1680	0.12

Lands can be dedicated to energy production. Although it's possible to consider different energy resources, this case study takes into account solar and biogas. Table 4 shows mean values, standard deviation, and the homogeneity rate used to generate energy productivity per resource per land and year.

TABLE 4 Energy resources

Type of energy production	Mean energy product (kWh/year ha)	SD (kWh/year ha)	Homogeneity rate
Solar	1,320,000	594,000	0.25
Bioethanol/biogas (from rapeseed oil)	37,000	16,650	0.12

Households require energy for different purposes: building heating, water heating, air conditioning, electric and electronic devices. Building and water heating are satisfied through gas combustion while air conditioning and electronic devices are satisfied through electricity. Household energy demand is strongly correlated with seasonal temperatures and climate. For this example, we consider the average climate parameters at a latitude of 44°. The household energy demands per capita are shown in Table 5.

The other parameters related to the general characteristic of the observed geographical area are summarized in Table 5, while Table 6 reports the breakpoints of the concave curve for the relationship between population density and average CO_2 emission due to urban mobility. Finally, Fig. 8 illustrates the flows resulting within the urban-rural ecosystem and the resulting carbon urban metabolism process.

The model is solved on a computer running a dual core 64-bit Intel Xenon processor at 2.7 GHz with 64 GB RAM using an IBM CPLEX 12.5 solver. Fig. 8 shows the optimal solution of the model. To simplify the visualization of the output, products belonging to the same resource category are grouped under the same color. Fig. 8 also shows a detailed view of a residential area, where the best population density has been selected for every single land.

TABLE 5 General model's input data

Parameter	Value
Number of lands	6400
Single land dimension	0.01 skm
Max number of people per land	117
Total number of people to insert	35,000
Number of lands per cluster	64
Number of clusters	100
Minimum rate of carbon sequestration per cluster	0.08
Electric energy demand per capita	5000 kWh/year
Biogas energy demand per capita	800 kWh/year
Carbon sequestration per green land	5000 kg CO_2/ha year
Carbon sequestration from agriculture products	2300 kg CO_2/ha year
Transport carbon emission	0.12 kg CO_2/ton km
Number of food resources	12
Number of energy resources	2

TABLE 6 Density-emission curve

Density (People/skm)	0	1671	3342	5013	6684	8355	10,026	11,697
t CO_2/year skm	0	4317	8402	12,398	16,337	20,234	24,096	27,930

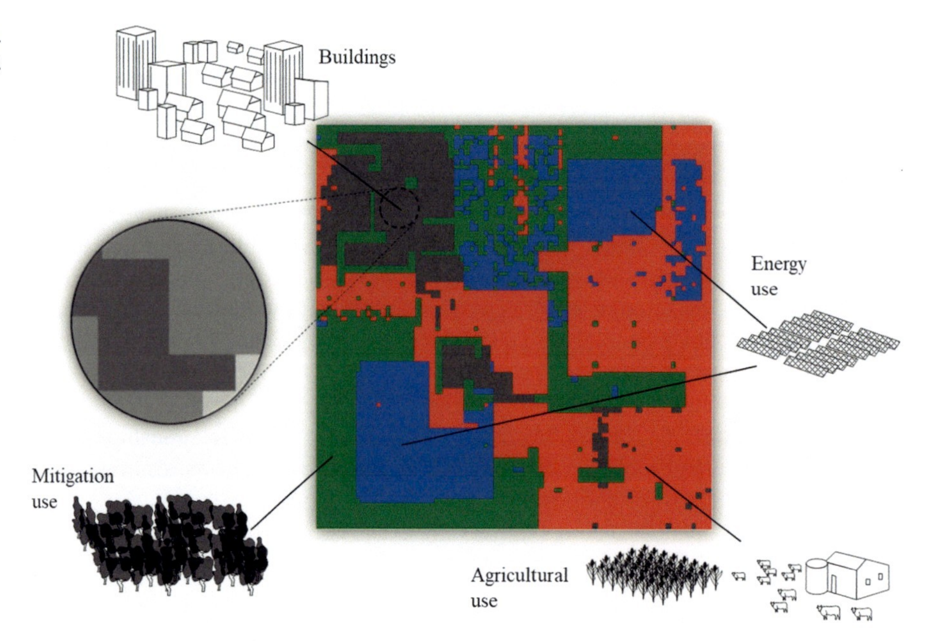

FIG. 8 Land-uses allocated by planning. *Gray*: residential/buildings; *blue*: energy use; *red*: agricultural use; *Green*: green cover.

While the solution of such a planning problem provides the optimal configuration of land uses over the observed geography, it also highlights the relationship between carbon emissions and transportation. The carbon emissions associated with the transport of food from agricultural sites to residential/commercial areas and those from the mobility of citizens are both considered and accounted for by the objective function. In particular, it is assumed that the emissions from the urban mobility increase slower-than-linear with the dweller density according to the piecewise function tracked in Table 6 (Kim and Brownstone, 2013).

In order to minimize the overall carbon emissions, two conflicting behaviors/strategies result, as follows. The reduction of emissions from food (freight) transport is obtained by reducing the distance from the rural areas and the residential commercial areas. This strategy enhances the accessibility between such uses but reduces the density of buildings (residential/commercial areas). Conversely, to minimize the emissions from dweller mobility the city density should be increased as much as possible, by reducing the level of accessibility of buildings with the other land uses. The model is also intended to find the optimal compromise between these two strategies. The intensity of food flows in comparison with the urban mobility, as well as the carbon emissions per traveled kilometer of the vehicle used to distribute food resources may affect the configuration of the urban-rural ecosystem, as exemplified in Fig. 9.

In Fig. 9, high emissions per kilometer from freight transportation (solutions previously shown) push to reduce the distance between supply areas and consuming areas, thus enforcing the sprawl of land uses. Conversely, when the emissions per kilometer for mobility increase, the urban areas are more dense and clustered.

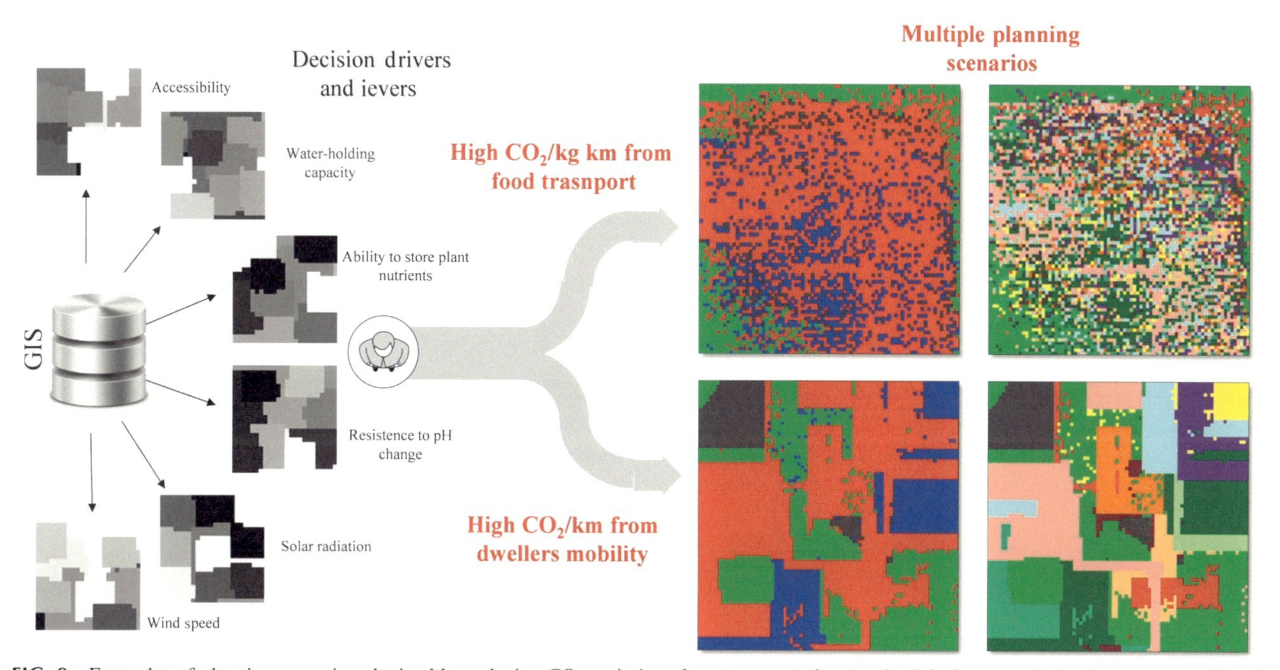

FIG. 9 Examples of planning scenarios obtained by reducing CO_2 emissions from transportation (on the left, four generic land uses are considered: buildings, mitigation, energy, and agricultural uses; on the right up to 13 uses are involved, where different crops are differently colored).

6 Conclusion

This chapter illustrates a methodology based on a land-use planning model that integrates transport activities and flows within the urbanization process to achieve a low-carbon urban-rural environment. We consider the transport and distribution process, both with regard to food resources from the agricultural areas to the residential/commercial areas as well as urban mobility. Furthermore, the illustrated model highlights the importance of handling connectivity between land uses and entities within an urban-rural ecosystem.

Since generalization is impossible, at least in the early model introduction, only food transport is considered at the moment, but further extensions are expected by adding other resources involving other industrial and economic sectors, toward the mutual planning of low-carbon urban areas and supply chains.

This model is by no means comprehensive. By introducing such a model, the authors hope that it will stimulate research across various disciplines to provide tools and techniques for the sustainable design of food ecosystems (Accorsi et al., 2017c) and for the land-use planning supporting the urbanization process.

References

Accorsi, R., 2019. Planning sustainable food supply chains to meet growing demands. In: Encyclopedia of Food Security and Sustainability.vol. 3, pp. 45–53. Reference Module in *Food Sciences*, Elsevier, https://doi.org/10.1016/B978-0-08-100596-5.22028-3.

Accorsi, R., Cholette, S., Manzini, R., Tufano, A., 2018. A hierarchical data architecture for sustainable food supply chain management and planning. J. Clean. Prod. 203, 1039–1054.

Accorsi, R., Cholette, S., Manzini, R., Pini, C., Penazzi, S., 2016. The land-network problem: ecosystem carbon balance in planning sustainable agro-food supply chains. J. Clean. Prod. 112 (1), 158–171.

Accorsi, R., Manzini, R., Pini, C., Penazzi, S., 2015. On the design of closed-loop networks for product life cycle management: economic, environmental and geography considerations. J. Transp. Geogr. 48, 121–134.

Accorsi, R., Manzini, R., Pini, C., 2017a. How logistics decisions affect the environmental sustainability of modern food supply chains. A case study from an Italian large scale-retailer. In: Bhat, R. (Ed.), Sustainability Challenges in the Agrofood Sector, first ed, pp. 179–200. © 2017 John Wiley & Sons Ltd.

Accorsi, R., Gallo, A., Manzini, R., 2017b. A climate-driven decision-support model for the distribution of perishable products. J. Clean. Prod. 165, 917–929.

Accorsi, R., Bortolini, M., Baruffaldi, G., Pilati, F., Ferrari, E., 2017c. Internet-of-things paradigm in food supply chains control and management. Procedia Manuf. 11, 889–895.

Aerts, J., Eisinger, E., Heuvelink, G., Stewart, T.J., 2003. Using linear integer programming for multi-site land-use allocation. Geogr. Anal. 35 (2), 148–169.

Brownstone, D., Golob, T.F., 2009. The impact of residential density on vehicle usage and energy consumption. J. Urban Econ. 65, 91–98.

Brussard, L., Caron, P., Campbell, B., Lipper, L., Mainka, S., Rabbinge, R., Babin, D., Pulleman, M., 2010. Reconciling biodiversity conservation and food security: scientific challenges for a new agriculture. Curr. Opin. Environ. Sustain. 2, 34–42.

Caparros-Midwood, D., Barr, S., Dawson, R., 2015. Optimised spatial planning to meet long term urban sustainability objectives. Comput. Environ. Urban. Syst. 54, 154–164.

Caparros-Midwood, D., Barr, S., Dawson, R., 2017. Spatial optimization of future urban development with regards to climate risk and sustainability objectives. Risk Anal. 37 (11), 2164–2181.

Casiroli, F., 2007. The changeable shape of the city. Public Transp. Int. 6 (5), 6–9.

Deng, X., Huang, J., Rozelle, S., Zhang, J., Li, Z., 2014. Impact of urbanization on cultivated land changes in China. Land Use Policy 45, 1–7.

Denman, K.L., Brasseur, G., Chidthaisong, A., Ciais, P., Cox, P.M., Dickinson, R.E., Hauglustaine, D., Heinze, C., Holland, E., Jacob, D., Lohmann, U., Ramachandran, S., da Silva Dias, P.L., Wofsy, S.C., Zhang, X., 2007. Couplings between changes in the climate system and biogeochemistry. In: - Solomon, S., Qin, D., Manning, M., Chen, Z., Marquis, M., Averyt, K.B., Tignor, M., Miller, H.L. (Eds.), Climate change 2007: The Physical Science Basis. Contribution of Working Group I to the Fourth Assessment Report of the Intergovernmental Panel on Climate Change (IPCC). Cambridge University Press, Cambridge/New York, NY, pp. 499–587.

Desjardins, R.L., Sivakumar, M.V.K., De Kimpe, C., 2007. The contribution of agriculture on the state of climate: workshop summary and recommendations. Agric. For. Meteorol. 142, 314–324.

Ewing, R., Cervero, R., 2001. Travel and the built environment: a synthesis. Transp. Res. Rec. 1780, 87–113.

Gallo, A., Accorsi, R., Baruffaldi, G., Manzini, R., 2017. Designing sustainable cold chains for long-range food distribution: energy-effective corridors on the Silk Road belt. Sustainability 9 (11), 2044.

Glaeser, E.L., Kahn, E.M., 2010. The greenness of cities: Carbon dioxide emissions and urban development. J. Urban Econ. 67 (3), 404–418.

Haas, J., Furberg, D., Ban, Y., 2015. Satellite monitoring of urbanization and environmental impacts—a comparison of Stockholm and Shanghai. Int. J. Appl. Earth Obs. Geoinf. 38, 138–149.

International Energy Agency (IEA), 2008. World Energy Outlook.

IPCC, 2014. In: Core Writing Team, Pachauri, R.K., Meyer, L.A. (Eds.), Climate Change 2014: Synthesis Report. Contribution of Working Groups I, II and III to the Fifth Assessment Report of the Intergovernmental Panel on Climate Change. IPCC, Geneva. 151 pp.

Kim, J., Brownstone, D., 2013. The impact of residential density on vehicle usage and fuel consumption: evidence from national samples. Energy Econ. 40, 196–206.

Lowry, M., Balling, R.J., 2009. An approach to land-use and transportation planning that facilitates city and region cooperation. Environ. Plann. B. Plann. Des. 36 (3), 487–504.

Lu, Q., Gao, Z., Ning, J., Bi, X., Wang, Q., 2015. Impact of progressive urbanization and changing cropping systems on soil erosion and net primary production. Ecol. Eng. 75, 187–194.

Madlener, R., Sunak, Y., 2011. Impacts of urbanization on urban structures and energy demand: what can we learn for urban energy planning and urbanization management? Sustain. Cities Soc. 1 (1), 45–53.

Nowak, D., Crane, D., 2001. Carbon Storage and Sequestration by Urban Trees in the USA. USDA Forest Service, Northeastern Research Station, 5 Moon Library, SUNY-ESF, Syracuse, NY 13210, USA.

Pacione, M., 2009. Urban Geography: A Global Perspective. Routledge Ed, London/New York.

Pauleit, S., Duhme, F., 2000. Assessing the environmental performance of land cover types for urban planning. Landsc. Urban Plan. 52 (1), 1–20.

Pickrell, D., Schimek, P., 1999. Growth in motor vehicle ownership and use: evidence from the national personal transportation survey. J. Transp. Stat. 2, 1–17.

Poumanyvong, P., Kaneko, S., Dhakal, S., 2012. Impacts of urbanization on national transport and road energy use: evidence from low, middle and high income countries. Energy Policy 46, 268–277.

Rodrigue, J.P., Comtois, C., Slack, B., 2006. The Geography of Transport Systems. Routledge, Abingdon, Oxon, USA and Canada.

Schwarz, N., Flacke, J., Sliuzas, R., 2016. Modelling the impacts of urban upgrading on population dynamics. Environ. Model. Softw. 78, 150–162.

Sovacool, B.K., Brown, M.A., 2009. Twelve metropolitan carbon footprints: a preliminary comparative global assessment. Energy Policy 38 (9), 4856–4869.

Su, H., Wu, J.H., Tan, Y., Bao, Y., Song, B., He, X., 2014. A land use and transportation integration method for land use allocation and transportation strategies in China. Transp. Res. A Policy Pract. 69, 329–353.

United Nations, Department of Economic and Social Affairs, Population Division, 2014. World Urbanization Prospects: The 2014 Revision. Highlights (ST/ESA/SER.A/352).

Waddell, P., Ulfarsson, G., Franklin, J., Lobb, J., 2007. Incorporating land use in metropolitan transportation planning. Transp. Res. A Policy Pract. 41, 382–410.

Wolman, A., 1965. The metabolism of cities. Sci. Am. 213 (3), 179–190.

Xu, B., Lin, B., 2014. How industrialization and urbanization process impacts on CO_2 emissions in China: evidence from nonparametric additive regression models. Energy Econ. 48, 188–202.

Yan, X., Crookes, R.J., 2010. Energy demand and emissions from road transportation vehicles in China. Prog. Energy Combust. Sci. 36 (6), 651–676.

Yewlett, C., 2001. OR in strategic land-use planning. J. Oper. Res. Soc. 52, 4–13.

Zhang, Y., 2013. Urban metabolism: a review of research methodologies. Environ. Pollut. 178, 463–473.

Further reading

Kennedy, C., et al., 2010. Methodology for inventorying greenhouse gas emissions from global cities. Energy Policy 38 (9), 4828–4837.

Chapter 20

Sustainable operations in reusable food packaging networks

Giulia Baruffaldi*, Riccardo Accorsi*, Luca Volpe*, Riccardo Manzini* and Fredrik Nilsson[†]

*Department of Industrial Engineering, Alma Mater Studiorum—University of Bologna, Bologna, Italy [†]Packaging Logistics, Department of Design Sciences, Lund University, Lund, Sweden

Abstract

This chapter explores the processes of a typical logistics network for the management of reusable packaging for food products. While the impacts associated with packaging waste in the food sector are well known, the adoption of reusable crates or handling systems for food items entails many logistics processes such as storage, transportation, and cleaning, whose impact needs to be quantified and assessed with cost-benefit analysis. This chapter thus presents a methodology and a decision-support tool used to quantify the logistic and environmental impacts associated with packaging distribution in the closed-loop network between growers, retailers, and the pooler.

This methodology allows quantifying of the performance of the as-is scenario and predicting the savings of choosing intermodal transport solutions (i.e., railways, seaways) for the delivery and collection of reusable plastic crates (RPCs) for fruits and vegetables. The methodology is applied to a multiscenario what-if analysis of a case study provided by an Italian pooler operating in the retail food supply chain. The results are generated through a decision-support tool, which embeds a geographic information system (GIS) and realizes data-driven assessment of storage and distribution operations experienced by RPCs.

We quantified some categories of impacts among the set of greenhouse gases emissions (GHGs) resulting from transportation and associated costs. The results showcase a total transportation cost reduction of 11.7% in the to-be scenario, while the number of kilograms of CO_2eq decreases by 9.2%.

The contribution of this chapter lies in the investigation of the environmental sustainability of a packaging closed-loop network (CLN) for food products. Moreover, we decided to limit the boundaries of the analysis to the transport process, which is often neglected and underrated in typical life cycle assessment (LCA). Findings from this chapter represent practical suggestions and strategic guidelines for managers and practitioners of reusable package systems toward more sustainable operations.

1 Introduction

As a fundamental part of the food industry, packaging plays an important role in the European economy (FOODDRINK Europe, 2016). The collection and processing of virgin materials like plastic and carton board used for new packages contribute greatly to energy demand and volume of waste. To address such impacts, in the last two decades reusable packaging networks for food products have captured the attention of both scholars and practitioners (Tonn et al., 2014). Nowadays, the market share associated with the usage of reusable packaging for the transportation of fruits and vegetables is estimated to be around 40% in Europe (Stiftung Initiative Mehrweg, 2009). The implementation of a reusable packaging system requires the creation of a closed-loop network to manage the package along the entire life cycle (Daniel et al., 2009). However, the reverse flows of packaging from consumers to suppliers increase the complexity of the logistics network when compared to the traditional one-way system (Wu and Dunn, 1995). Moreover, the package travels along a wider network, resulting in higher transportation costs and greater environmental impacts (Accorsi et al., 2014).

In a closed loop network (CLN), the proper management and planning of transport operations is necessary to reduce costs from both forward and reverse logistics. Albrecht et al. (2013) provide evidence that the cost of reverse flows increases with network sprawl. They demonstrate that one-way packaging is preferable in fruit and vegetable distribution characterized by long distances, while reusable packages are convenient with shorter distances, although the geography of the network is not the only factor to consider when choosing the packaging system to adopt. For example, different shipping contracts as well as the optimization of routes or transport mode may contribute to decrease the total impact of CLNs, aiding managers in evaluating the convenience of reusable packaging accurately.

This chapter focuses on the analysis of the logistics and environmental impacts of a reusable package CLN for fruits and vegetables. The observed system is shown in Fig. 1, which represents the supply of closed crates from the pooler to the food

Sustainable Food Supply Chains. https://doi.org/10.1016/B978-0-12-813411-5.00020-X

vendors, the distribution of packed food to the retailers, and the collection of empty (and closed) crates from these by the pooler. The following study is primarily motivated by a lack in the literature of integrated studies that tie together the sustainability goal and the management of the operations in a packaging CLN. Moreover, the following investigation is based on the question as to whether the reduction of packaging waste and use of virgin material offset the environmental impact of the transport, storage, and washing processes required.

Managers can adopt the levers of network design and process optimization to reduce the impact of packaging CLN. In both cases, this entails an in-depth understanding of the structure of the network, with the purpose of identifying room for improvement and the adequate lever to try. In comparison with life cycle assessment (LCA) analyses (ISO 14040, 2006), this chapter limits the boundaries of the system to the transport process only, the real impact of which is often neglected due to the level of uncertainty of transport records (Heijungs and Huijbregts, 2004).

The analysis is implemented through a decision-support tool (DST) developed for the analysis of the performance and impacts of logistics networks (Accorsi et al., 2018). This tool supports what-if multiscenario analysis that compares different network scenarios, quantifying some environmental categories of impacts (e.g., GHGs) as well as the traveled distance. A case study from an Italian pooler operating in the retail supply chain for fruits and vegetables is illustrated in these pages.

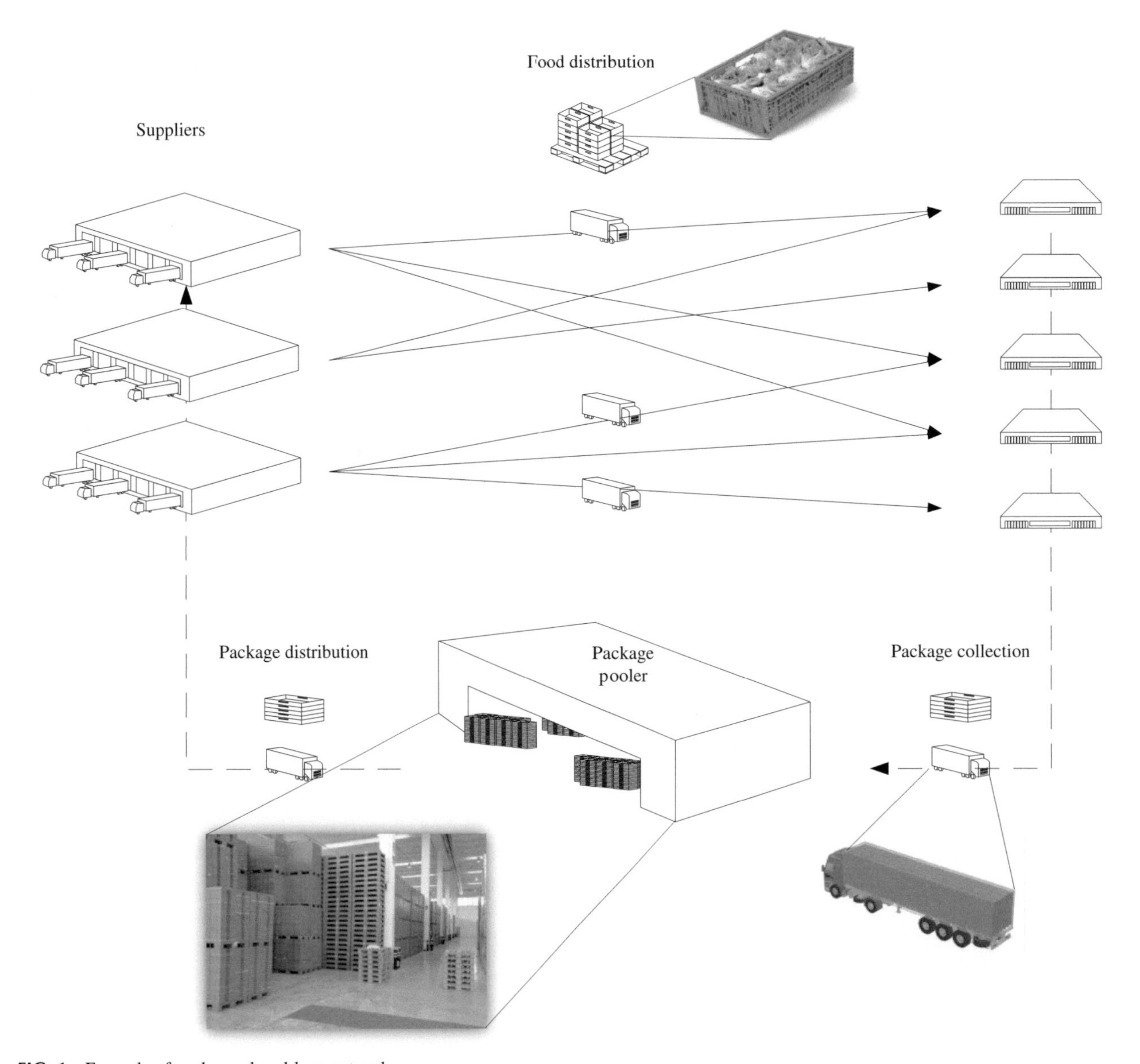

FIG. 1 Example of package closed-loop network.

The remainder of this chapter is organized as follows: Section 2 provides a literature review, while Section 3 illustrates the methodology; Section 4 reports the results from the case study, which are further discussed in Section 5; finally, after the description of another the nation-wide reusable packaging network from the retail Swedish industry (Section 6), Section 7 concludes the chapter, summarizing the main contents.

2 Literature review

The literature on the operations of reusable package systems is limited. Govindan et al. (2015) present a survey on reverse logistics and CLN, which underlines a gap in the literature on the impact that such networks have on the environment. Some studies compare the performance of reusable and one-way packages in order to find the most environmentally friendly solution (Bortolini et al., 2018; Pålsson et al., 2013). Levi et al. (2011) reveal that the volume of virgin materials processed is not the only lever involved, and the management of the supply chain operations of the CLN plays a key role in addressing sustainability goals. The transport process, which often discourages the adoption of reusable packaging (Pearce, 1997), is an important environmental stressor, and its proper management and planning is necessary before considering the adoption of reusable packaging (Levi et al., 2011). According to Koskela et al. (2014), the most important factors that contribute to environmental impacts of reusable packaging systems are:

- *traveling distance* of the shipments;
- the adopted *transport mode* (i.e., railways, seaways);
- the *freight load of the package*.

Addressing the question of how to select packaging systems from a sustainability perspective, Pålsson et al. (2013) concluded that geographical distance and fill rates were the most influential factors in the selection process.

From a supply chain perspective, a number of factors are important for the suitability of reusable packaging systems. According to Twede and Clark (2004) the structure and relationships within a supply chain or network are highly influential; these include strong channel leaders with cost-saving incentives; short shipping distances; efficient sorting, cleaning, and tracking systems in place; and industry consortia responsible for standardization. Furthermore, reusable packaging systems can be designed based on different purposes and scopes, which also affect the usage and environmental impacts. A system for single product use can be efficient if volumes are high, while a multiproduct design enables wider spread and adoption. In a similar way, the design choices of single-, multi-, and open-loop set-ups have direct implications on the system-wide effects to be gained. While a single-loop system works well in a specific (often short-distance) relationship, a multi- and open-loop set-up enables more actors to use and share the same resources within an industry. The challenge lies then in the coordination and governance structures in place for maintenance and development.

Furthermore, even the network sprawl may discourage the use of reusable packaging (Albrecht et al., 2013; Accorsi et al., 2014). The coordination between forward and reverse flows affects the performance of the CLN (i.e., in terms of inventory and logistics costs optimization, as well as service level enhancements) as demonstrated by Elia and Gnoni (2015). More coordination between the network's actors (i.e., suppliers, pooler, and retailers) can drive the design of an effective and sustainable reusable packaging system (Accorsi et al., 2015). The choice of control strategies is also found critical for successful use of reusable packaging. Hellström and Johansson (2010) conclude, based on their study of returnable transport items, that having management systems or switch pool set-ups to control reusable packages significantly lowers the shrinkage levels, which has both financial and environmental implications. Ross and Evans (2003) underline that environmental impacts can be reduced by changing the geographical location of some CLN nodes and the related allocation of the processes. Moreover, coordination may support the consolidation of shipments among the actors and facilitate optimized transport planning, reducing the overall GHG emissions (Pan et al., 2013). Lastly, intermodal transport is well known both among practitioners and scholars as a lever to reduce GHG emissions (Bontekoning et al., 2008). Chapman (2007) provides a survey in which the contributions of the different modes of transport on climate change are discussed.

As a consequence, a beneficial effect on environmental impacts is expected by also using intermodality in reusable packaging CLNs. However, as this entails more complex processes and infrastructural requirements, such benefits should be assessed and quantified *ex ante* before implementing an intermodal transportation system in practice. This chapter addresses to this aim.

3 Methodology

This chapter evaluates the impact on the logistics and environmental performance of different packaging network configuration and transportation scenarios. We consider alternative transport modality choices to address the reduction of the GHG emissions.

The proposed analysis is implemented through a multistep methodology and applied to a case study provided by an Italian pooler operating in the retail supply chain for fruits and vegetables. The pooler supplies food growers/vendors with reusable plastic crates (RPCs) of different size and capacity, and then collects empty packages from the retailers and organizes storage and cleaning processes, building up a physical connection among the supply chain actors.

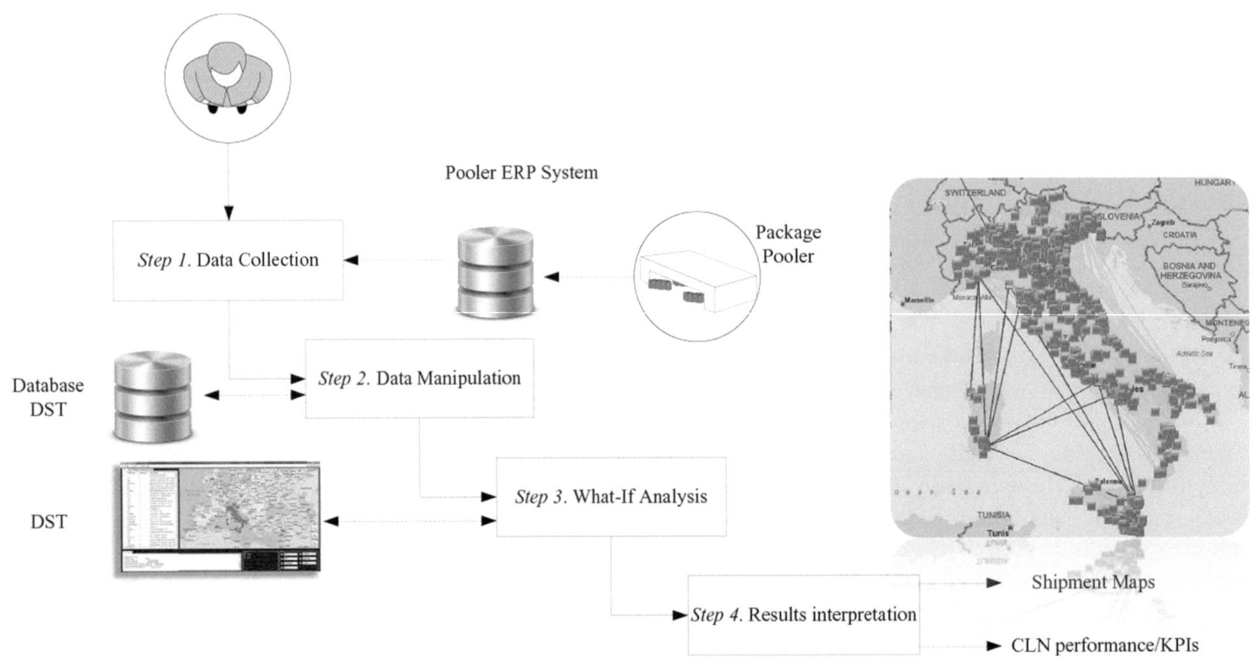

FIG. 2 Methodology.

Fig. 2 illustrates the methodology implemented for the analysis of the packaging CLN. An initial phase of data collection (*Step 1.*) allows information to be gathered on the distribution of the nodes of the network, the shipments of food and packages among the nodes, and the flows of raw materials necessary for the production of the crates along a time horizon (i.e., 1 year). Data is mainly extracted from the ERP systems of the companies involved in the network (i.e., the pooler and the suppliers). Examples of the tables and data fields collected are proposed in Fig. 3 and further discussed in Baruffaldi et al. (2019), who also suggest how to perform manipulation activities (see *Step 2.*) in terms of SQL queries.

In the third step (*Step 3.*), a what-if analysis is performed in order to compare the environmental impact generated by the current CLN (as-is) and that generated by a new network scenario (to-be) involving transport intermodality. In such a scenario, the shipments are rearranged in order to enhance, when possible, the use of intermodality.

The results are obtained by virtualizing the transportation flows onto a decision-support tool written in C#. Net, Network Analyzer 3.0 (see Accorsi et al., 2018), which imports shipment records from a structured database and, through a geographic information system (GIS), quantifies the traveling and the environmental impacts associated with each route. For this reason, the collected data should be previously organized according to the database entity-relationship diagram (*Step 2.*). As reported in Fig. 3, the main tables include information on products (e.g. products code, products description, sizes), shipments $s \in S$ (e.g., order code, order quantity), nodes (e.g., node name, address) and modes of transport (e.g., railway, seaway) with their load capacity cap_m and GHG emissions GHG_m^t (of type $t \in T$).

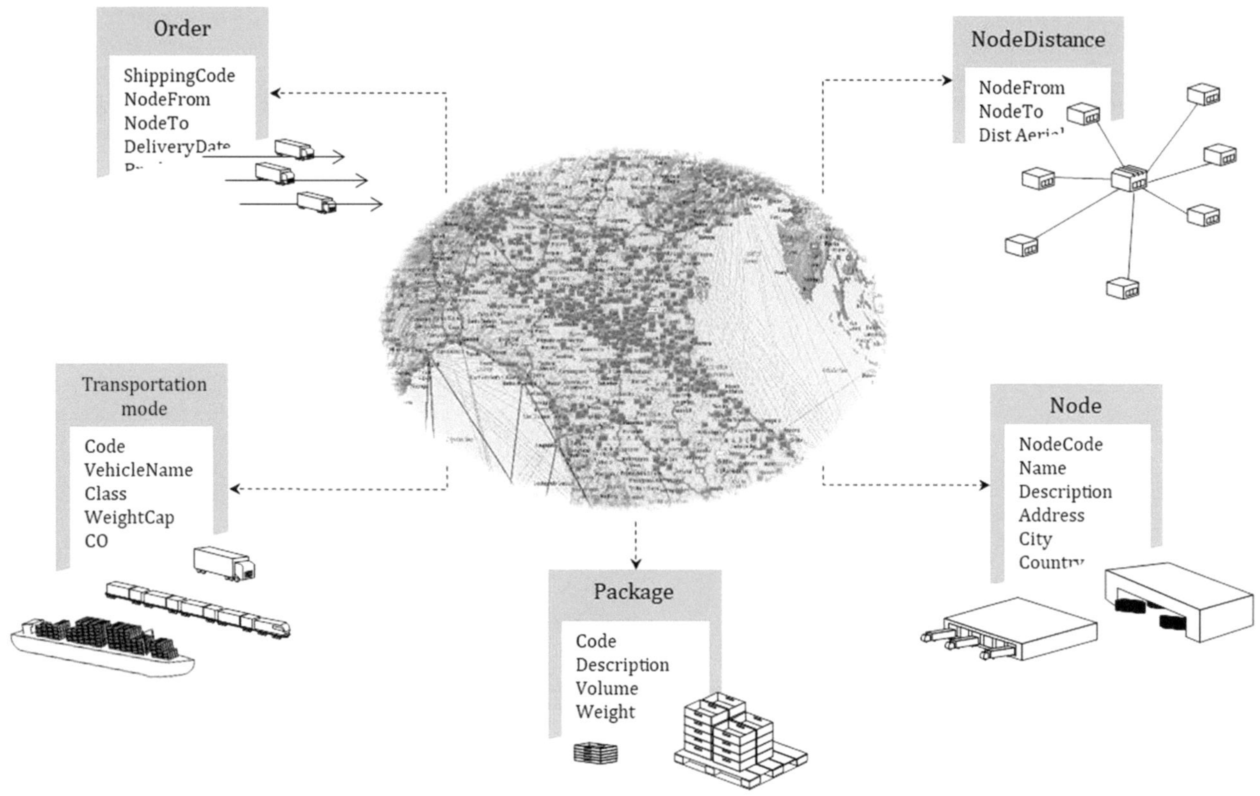

FIG. 3 Data collection: Data fields and tables.

The tool, whose main graphic user interface is reported in Fig. 4, imports the records from the database as well as the geographical coordinates of the nodes. Then, it calculates the distances d_{ijm} traveled by mode m between each pair of nodes $i, j \in N$, building a from-to chart. The shipment records are thus virtualized and quantified over the simulated time horizon in terms of traveled kilometers and shipped loads $x^{pkg}{}_{ijm}$. A set of user-friendly graphical user interfaces (GUIs) allow visualization of the flows on the map, as shown in Fig. 4. For the sake of readability, flows are mapped by Euclidean distances.

The tool quantifies the main GHG emissions, enabling the user to compare the effects of different logistics and network scenarios along *Step 4*. Eq. (1) describes how the environmental impacts are calculated for each shipment $s \in S$ and category of impact $t \in T$:

$$Impact(s,t) = \sum_{m \in M} x^{pkg}{}_{sijm} \cdot d_{ijm} \cdot \frac{\left(GHG^{Full}{}_{m}{}^{t} - GHG^{Empty}{}_{m}{}^{t}\right)}{cap_m} + d_{ijm} \cdot GHG^{Empty}{}_{m}{}^{t} \quad \forall s \in S, t \in T \quad Impact(w,t) = \sum_{s \in S} Impact(s,t) \quad \forall t \in T \tag{1}$$

For example, the tool manages to calculate the GHG emissions generated by scenarios $w \in W$ (Eq. 1) characterized by different transport modes and different levels of cargo utilization (i.e., whether full or partially loaded). Different transport-emission databases can be used for the analysis (for example, Lipasto Database (VTT Technical Research Centre of Finland, 2015)).

In the same way, the tool quantifies the transport costs associated with each shipment and cumulates this per scenario. The unit cost for this analysis has been deduced from the report of Confcommercio(2015), which monitored the average transportation cost (tons·Euro·km) in Italy over a time horizon of 6 years (from 2007 to 2013) for the different transport modes.

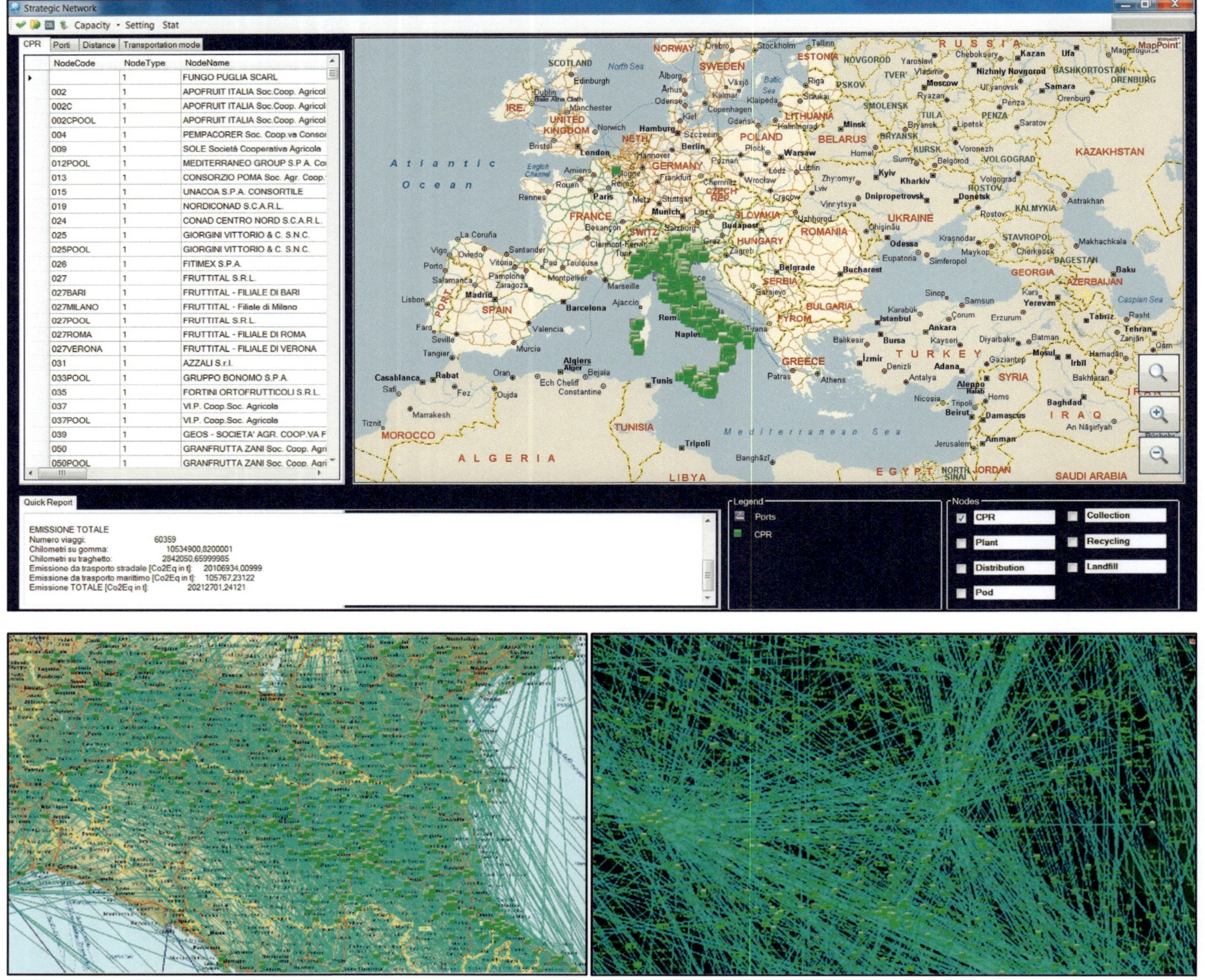

FIG. 4 GUI of Network Analyzer and resulting maps.

4 Case study

The analyzed pooling network involves more than 1500 nodes spread nationwide in Italy. Fig. 5 summarizes the different types of actors and nodes within the CLN and describes the relationships among them in terms of material flow exchange.

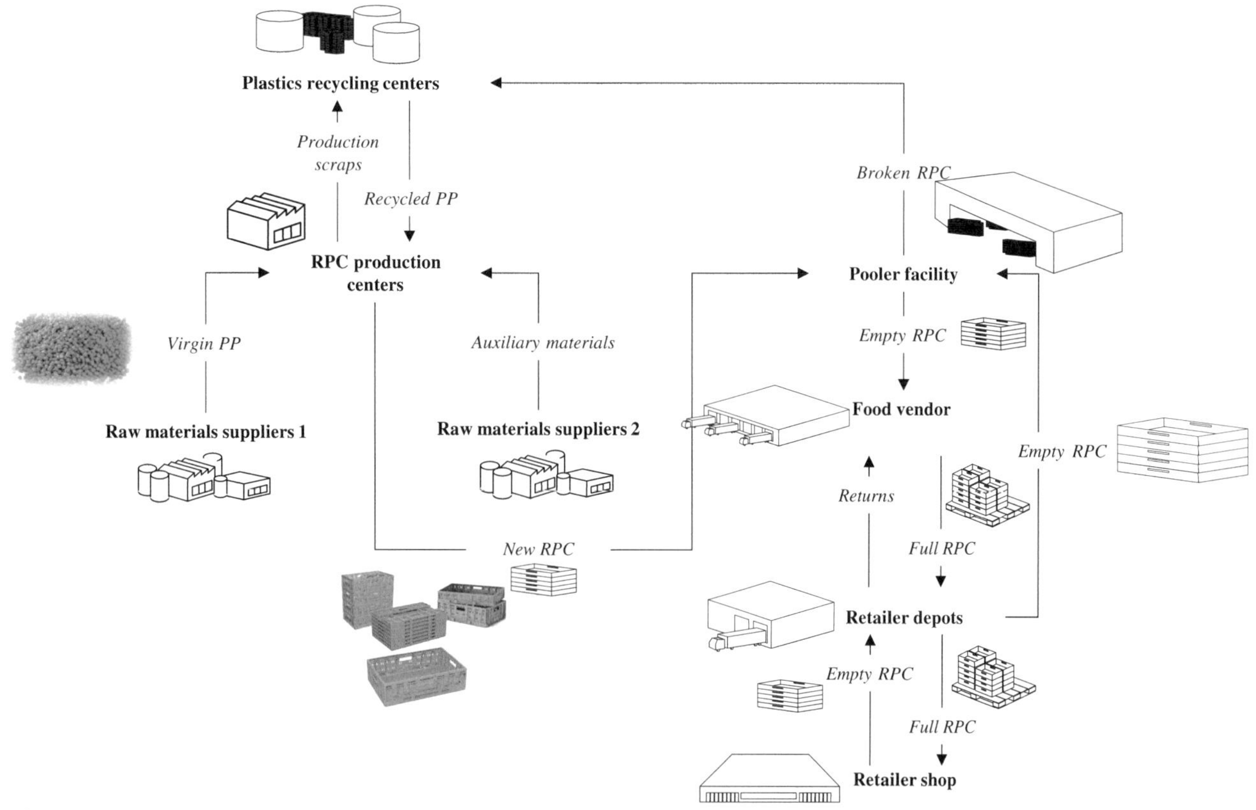

FIG. 5 Pooling network overview.

The pooler facilities supply 1068 vendors of vegetables and fruits with empty crates, which are then filled with fresh food and shipped to the retailer depots. These depots serve the retailers' stores with full RPCs and collect the dirty packages, which are again available when the food is sold. It is worth noting that closed crates reduce their volume by about 70%. Therefore, the number of shipped crates per truck increases in the reverse flow. The retailer depots return the dirty crates to the pooler facilities for storage and washing processes. Due to logistics issues, the pooler currently washes 50% of the dirty crates over a year, but the company would bring this percentage to 100% to improve the service for its clients.

Crates are designed for a lifespan of about 10 years. However, at the end of their life, they are not disposed of but recovered to produce recycled polypropylene (PP). A similar process is carried out with broken crates. New RPCs are produced by the RPC production centers, which combine the recycled PP with virgin PP and auxiliary materials, thus enhancing the material flows required throughout the CLN (see Fig. 5).

In the current scenario (as-is), transport activities are devoted to third-party logistics (3PL) providers that face the issue of balancing the crates' flows between production areas, which concentrate the majority of food vendors, and the consumption areas, which conversely concentrate retailer depots and stores. Specifically, vendors are concentrated in the south of the country and retailers in the north. Therefore, the geographic distribution of the pooler's facilities with respect to the other food supply chain nodes has a crucial impact on the logistics and environmental performance of the packaging system.

Moreover, 3PL providers seek to ship RPCs with other goods in order to increase load utilization of trucks and to exploit economies of scale. Although the trucks may be more saturated, the RPCs may travel on longer routes to meet the distribution planning of the 3PL provider. The boundaries of the analysis include the supply chain actors involved in the pooling network, while the goods carried for other 3PL customers are neglected as they do not contribute to the performance of the packaging system.

Table 1 reports the main parameters of the as-is CLN scenario referred to the analyzed time horizon between 1/1/2015 and 12/31/2015.

TABLE 1 Network parameters.

Scenario Parameters			
Time horizon	01/01/2015–12/31/2015	*Modes of transports*	*Capacity (tons)*
		Tipper truck	25
		Tanker	25
Nodes	*n.*		
Plastics recycling centers	1		
RPC production centers	2	*Shipments*	*n.*
Raw materials suppliers	10	Number of shipments	671,173
Pooler facilities	34	Shipped freight (tons)	965,308
Vegetables and fruits producers	1068		
Retailer depots and stores	487		

Two CLN scenarios are virtualized and compared according to the scheme of Fig. 6. Both scenarios account for the collection and distribution of 965,308 tons of goods (i.e., empty and full crates, raw materials, broken crates) across the network. We consider a tipper truck with a maximum capacity of 25 tons. Similarly, goods carried by seaway are stowed on tankers of 25 tons of capacity. These assumptions are valid for both scenarios, so the results are comparable.

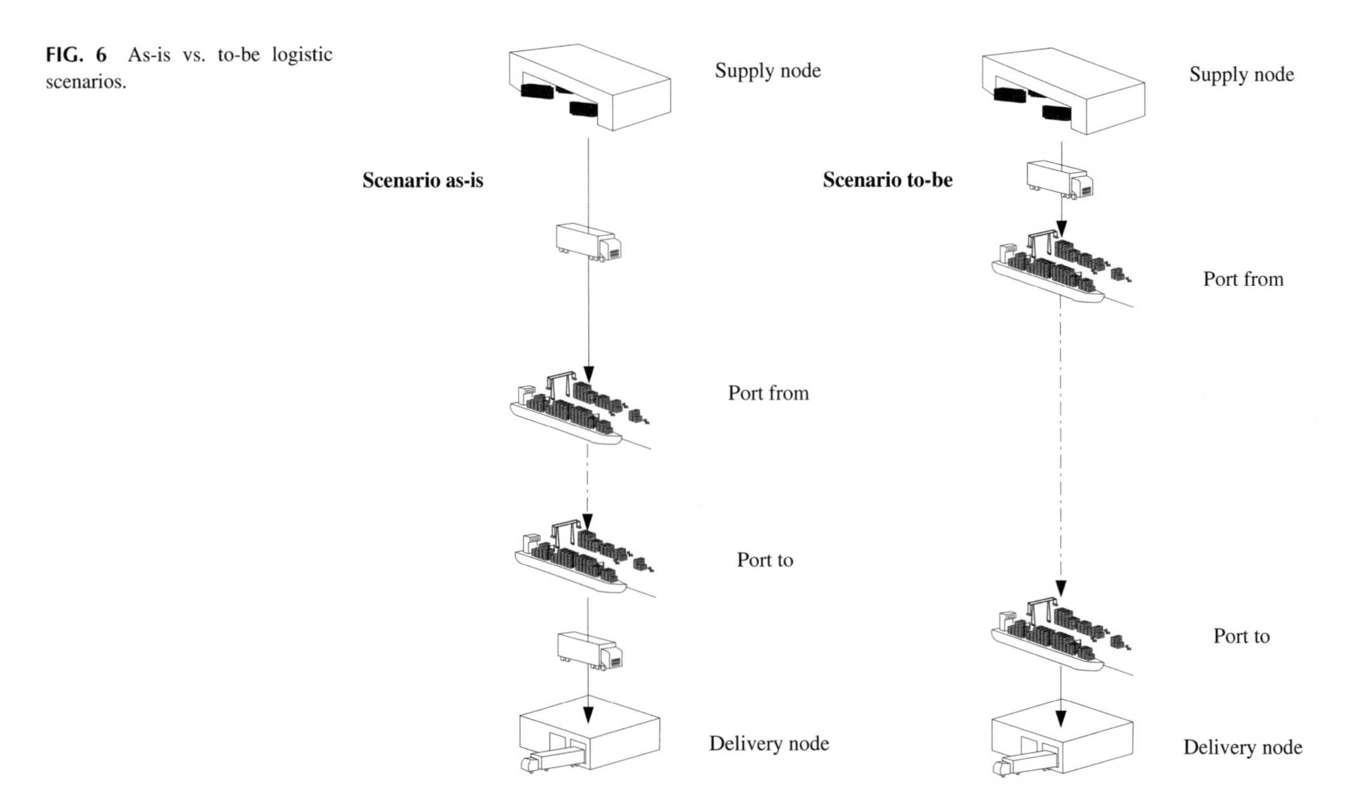

FIG. 6 As-is vs. to-be logistic scenarios.

In the as-is scenario, most of the shipments to/from Sicily, which together account for 8.7%, pass through the port of Naples even though the departure/destination nodes are located in the north of Italy. In some cases, crates travel on trucks up to the Strait of Messina.

Conversely, the proposed to-be scenario suggests reducing the road traveling by favoring the seaway. While maritime distribution is mainly used for overseas freight, while road transportation represents a more viable option in a small country like Italy, use of the seaway is indeed considered to be a more environmentally friendly transport solution (Schipper and Fulton, 2003). As shown in the network maps of Fig. 7, three new seaways connect Sicily with the port of Genoa, the port of Leghorn, and the port of Rome. Compared to the as-is scenario, the to-be scenario generates a reduction of total traveled distance of around 811,000km, while the number of kilometers traveled by the vessels increases 90.7%. According to Confcommercio (2015), vessel transportation presents the lowest unit cost per shipped ton compared to the other transport modes. Thus, the to-be scenario generates a total transport cost reduction of 11.7%, as reported in Fig. 8.

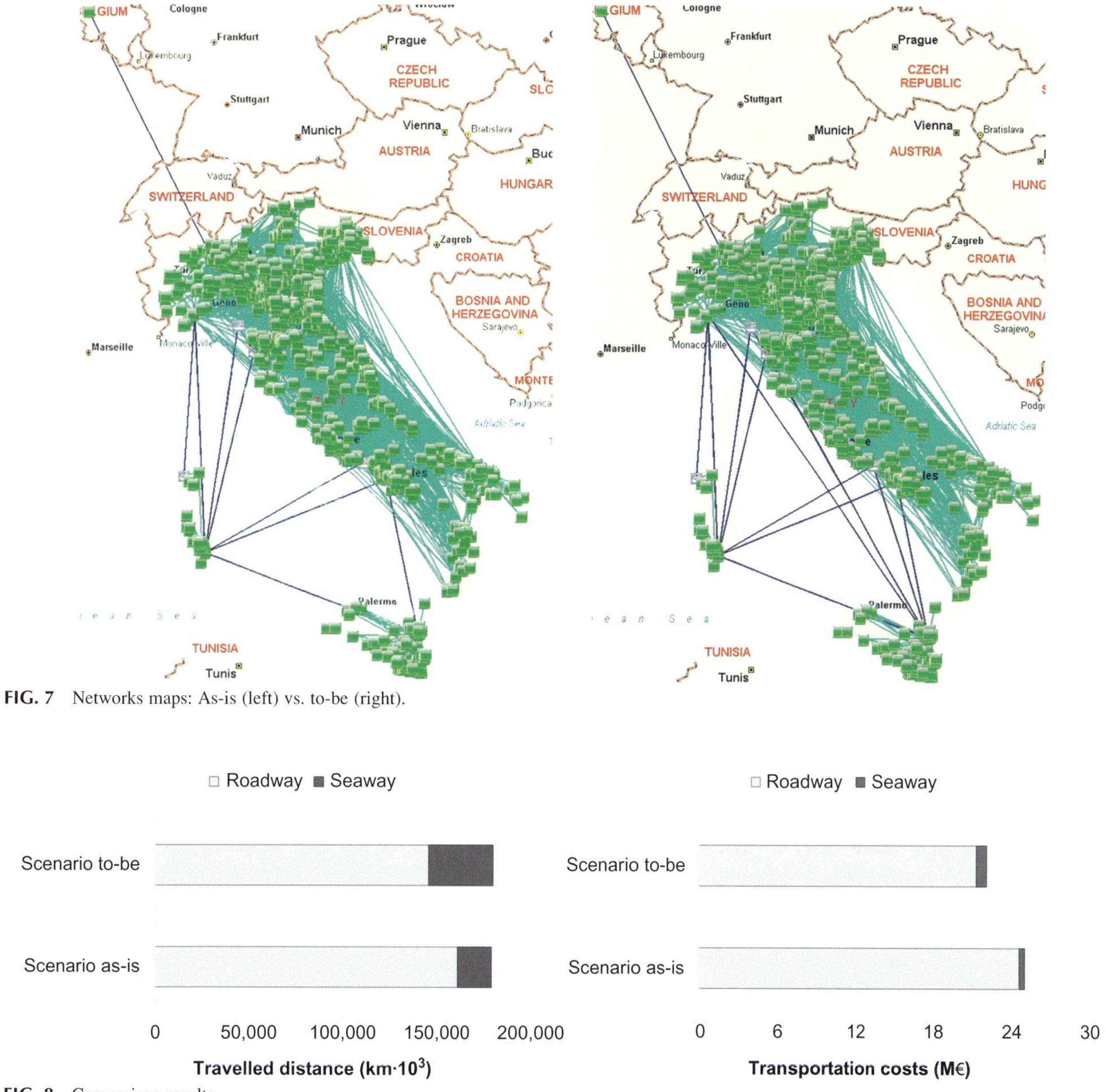

FIG. 7 Networks maps: As-is (left) vs. to-be (right).

FIG. 8 Comparison results.

The obtaining environmental GHG emissions are reported in Table 2. As a consequence of the reduced traveling, the value of all categories of GHG emission decreases in the to-be scenario. As expected, the only exception is represented by sulfur particles, which are largely emitted by vessels (Chapman, 2007). Overall, the to-be scenario generates a significant reduction of the kilograms of CO_2eq, i.e., total global warming potential, of 9.2%, highlighting the importance of exploiting intermodality to address crucial environmental issues like climate change.

TABLE 2 GHG emissions: As-is vs. to-be scenario.

GHG emission *t*	Scenario as-is (kg)	Scenario to-be (kg)	Δ (%)
CO	26,324.6	24,243.4	−7.9
HC	14,926.5	13,573.4	−9.1
NO_x	952,738.8	873,325.8	−8.3
PM	8408.7	7891.3	−6.2
CH_4	1171.0	1103.6	−5.8
N_2O	4564.4	4139.9	−9.3
NH_3	806.1	728.8	−9.6
SO_2	5042.1	9035.7	79.2
CO_2	123,537,369.7	112,180,200.4	−9.2
CO_2eq	124,995,568.7	113,498,245.0	−9.20

5 Discussion

Despite the need for further analysis to validate these findings, some considerations rise from this study. First, the results showcase how the to-be scenario achieves the twofold objective of reducing the total transportation cost while decreasing the environmental impact. Further analysis may involve the evaluation of the railway opportunities, which are currently under investigation by the pooler's management.

The data-gathering process represents the most time-intensive activity. The proposed approach requires a high level of accuracy of input data in order to provide reliable results. Therefore in order to replicate this study with other networks or supply chains, a large companies' involvement is required.

This chapter illustrates the first step of a more ambitious project, attempting to provide managers of the packaging company with reliable, tailored, user-friendly, decision-support tools for CLN management. Furthermore, simulation tools may support fruit and vegetable vendors in deciding whether or not to adopt reusable packaging, or the type of package to adopt, or the most convenient pooling network to deal with.

6 Experience from other reusable packaging networks

In this section, the origin and impact of an industry-wide multiparty packaging system are presented, including the challenges as well as results achieved from usage of the system in the food and retail industries since its market introduction.

The idea of an industry-wide reusable packaging system was born in 1992 out of insights gained from the specific emphasis in the Swedish food industry on resource inefficiencies related to high packaging handling costs, quality issues, and product/packaging waste. After a period of studies, trials, taskforces, and development projects, the idea materialized into the forming of an industry-owned company, Svenska Retursystem (SRS), in 1997 (Gustafsson et al., 2006). Since its market introduction, the business has grown and in 2017 more than 15 million return crates are used by 1500 customers (half of the Swedish fresh produce deliveries), with a turnover of approximately 60 million euros. It has been estimated that the use of the reusable crates instead of disposable alternatives reduced carbon dioxide emissions by more than 30,000 tonnes during 2018. The mission of SRS is to "develop and operate an efficient reusable system that simplifies and improves our customers' logistics and distribution of goods" (www.retursystem.se, October 18, 2018). The set-up of the reusable packaging system is similar to the Italian pooling network: that is, food producers/suppliers order reusable crates filled with various products, and these are then delivered to different wholesalers for further distribution to retail outlets. For the return flow, empty reusable crates are sent back to the wholesalers and from there transported to SRS facilities for washing and redistribution. In this case, the same system (six different sizes of crates designed to fit in a modularized way on pallets of EU pallet size) is used for fruits and vegetables, meat, fish, ready-made food, etc., with the purpose of enabling processes as efficiently as possible at each stage of the supply chain. One of the major challenges, especially related to the environmental impact, is the shrinkage, or the loss of crates (often the crates are used for other purposes in retail outlets and other sites), which produces a need for new crates. The estimated lifetime of the crates is more than 100 uses. Hence, the replacement of crates, either with virgin or recycled material, has a major impact on the environmental performance of the system.

One of the key success factors of SRS is its governance structure. Being equally owned by the Trade Association for Grocery of Sweden (SvHD) and the Swedish Food and Drinks Retailers Association (DLF) has enabled standardization as well as adoption of the reusable packaging system. Hence, in the development of sustainable packaging systems, one of the key enablers from both investment and implementation perspectives is a proper governance structure. With growing needs and interest in concepts such as circular economy and sustainable development, the role of reusable packaging systems could benefit whole industries.

7 Conclusions

This chapter illustrates a methodology and a support-decision tool that allows the analysis and quantification of the costs, environmental impacts, and benefits resulting from different logistics scenarios adopted in reusable packaging networks for food products. Through a year-long case study, the proposed analysis assesses the environmental impact reduction generated by a greater use of intermodality in reusable packaging networks. The case study is provided by an Italian RPC pooler, operating in a network of more than 1600 nodes around Italy. Results showcase a total transportation costs reduction of 11.7% in the to-be scenario, while the kilograms of CO_2eq decrease by 9.2%. Experiences from other nationwide (Sweden) reusable food packaging systems are then illustrated to highlight the impact of similar networks in different countries. Despite the need for further analysis, the authors believe that this chapter represents a valuable proof of concept of the role that logistics plays in the sustainability of food packaging systems.

Acknowledgment

The authors would like to sincerely thank the company CPR System involved in the study and particularly Dr. Enrico Frigo, Dr. Monica Artosi, and Dr. Sabrina Pagnoni for their valuable support in this research project.

References

Accorsi, R., Cascini, A., Cholette, S., Manzini, R., Mora, C., 2014. Economic and environmental assessment of reusable plastic containers: a food catering supply chain case study. Int. J. Prod. Econ. 152, 88–101.

Accorsi, R., Manzini, R., Pini, C., Penazzi, S., 2015. On the design of closed-loop networks for product life cycle management: economic, environmental and geography considerations. J. Transp. Geogr. 48 (2015), 121–134.

Accorsi, R., Cholette, S., Manzini, R., Tufano, A., 2018. A hierarchical data architecture for sustainable food supply chain management and planning. J. Clean. Prod. 203, 1039–1054.

Albrecht, S., Brandstetter, P., Beck, T., Fullana-i-Palmer, P., Grönman, K., Baitz, M., Deimling, S., Sandilands, J., Fischer, M., 2013. An extended life cycle analysis of packaging systems for fruit and vegetable transport in Europe. Int. J. Life Cycle Assess. 18, 159–1567.

Baruffaldi, G., Accorsi, R., Volpe, L., Manzini, R., 2019. A data architecture to aid life cycle assessment in closed-loop reusable plastic container networks. Procedia Manuf. (in press).

Bontekoning, Y.M., Macharis, C., Trip, J.J., 2008. Is a new applied transportation research field emerging?—a review of intermodal rail–truck freight transport literature. Transp. Res. A 38, 1–34.

Bortolini, M., Galizia, F.G., Mora, C., Botti, L., Rosano, M., 2018. Bi-objective design of fresh food supply chain networks with reusable and disposable packaging containers. J. Clean. Prod. 184 (2018), 375–388.

Chapman, L., 2007. Transport and climate change: a review. J. Transp. Geogr. 15, 354–367.

Confcommercio, 2015. Analisi e previsioni per il trasporto merci in Italia.

Daniel, V., Guide, R., Van Wassenhove, L.N., 2009. OR FORUM—the evolution of closed-loop supply chain research. Oper. Res. 57 (1), 10–18.

Elia, V., Gnoni, M.G., 2015. Designing an effective closed loop system for pallet management. Int. J. Prod. Econ. 170, 730–740.

FOODDRINK Europe, 2016. Data & Trends: EU Food and Drink Industry. Brussels, Belgium.

Heijungs, R., Huijbregts, M.A.J., 2004. A review of approaches to treat uncertainty in LCA. In: Proceedings of the Second Biennial Meeting of IEMSs, Complexity and Integrated Resources Management, Osnabrück, Germany, pp. 332–339.

Govindan, K., Soleimani, H., Kannan, D., 2015. Reverse logistics and closed-loop supply chain: a comprehensive review to explore the future. Eur. J. Oper. Res. 240 (3), 603–626.

Gustafsson, K., Jönson, G., Smith, D., Sparks, L., 2006. Retailing Logistics and Fresh Food Packaging— Managing Change in the Supply Chain. Kogan Page, London.

ISO 14040, 2006. Environmental Management—Life Cycle Assessment—Principles and Framework. International Organization for Standardization.

Hellström, D., Johansson, O., 2010. The impact of control strategies on the management of returnable transport items. Transp. Res. Part E 46, 1128–1139.

Koskela, S., et al., 2014. Reusable plastic crate or recyclable cardboard box? A comparison of two delivery systems. J. Clean. Prod. 69, 83–90.
Levi, M., et al., 2011. A comparative life cycle assessment of disposable and reusable packaging for the distribution of Italian fruit and vegetables. Packag. Technol. Sci. 24 (7), 387–400.
Pan, S., et al., 2013. The reduction of greenhouse gas emissions from freight transport by pooling supply chains. Int. J. Prod. Econ. 143, 86–94.
Pålsson, H., Finnsgård, C., Wänström, C., 2013. Selection of packaging systems in supply chains from a sustainability perspective: the case of Volvo. Packag. Technol. Sci. 26, 289–310.
Pearce, F., 1997. Burn me. New Scientist 156 (2109), 30–34.
Ross, S., Evans, D., 2003. The environmental effect of reusing and recycling a plastic-based packaging system. J. Clean. Prod. 11, 561–571.
Schipper, L.J., Fulton, L., 2003. Carbon dioxide emissions from transportation: trends, driving factors, and forces for change. In: Hensher, D.A., Button, K.J. (Eds.), Handbooks in Transport. In: Handbook of Transport and the Environment, vol. 4. Elsevier, pp. 203–225.
Stiftung Initiative Mehrweg, 2009. The sustainability of packaging system for fruit and vegetable transport in Europe based on life-cycle analysis. Available at http://www.stiftung-mehrweg.de/en/oekobilanzstudie.php.
VTT Technical Research Centre of Finland, 2015. LIPASTO—A Calculation System for Traffic Exhaust Emissions and Energy Use in Finland.
Tonn, B., Frymier, P.D., Stiefel, D., Skinner, L.S., Suraweera, N., Tuck, R., 2014. Toward an infinitely reusable, recyclable, and renewable industrial ecosystem. J. Clean. Prod. 66, 392–406.
Twede, D., Clarke, R., 2004. Supply chain issues in reusable packaging. J. Market. Channel 12 (1), 7–26.
Wu, H., Dunn, S.C., 1995. Environmentally responsible logistics systems. Int. J. Phys. Distrib. Logist. Manag. 25 (5), 20–38.

Chapter 21

A model to enhance the penetration of the renewables to power multistage food supply chains

Marco Bortolini, Riccardo Accorsi, Mauro Gamberi and Francesco Pilati

Department of Industrial Engineering, Alma Mater Studiorum—University of Bologna, Bologna, Italy

Abstract

Food supply chains (FSCs) allow the effective and safe delivery of food products from farmed crops to consumer forks. The challenging properties of many varieties of food, that is, perishability, quality decay, and short shelf life, require shipping and storage conditions able to guarantee high standards of safety and quality for the final consumers. Behind these conditions lies the demand for large amounts of energy to power refrigerated storage and shipping modules, to speed handling and transportation, etc. A switch from fossil fuels to renewables is mandatory to increase the sustainability of modern FSCs.

Traditionally, no integration has taken place between the renewable power system design and the supply chain infrastructure location and management. The attention placed on FSC designs for green energy is rising. Thus this chapter provides a high-level analytic model for efficient design of FSCs in the direction of integrating renewable plants, for example, solar, photovoltaics, wind, biomass, etc., as a key input for green operations, so that the node location and flow allocation is driven by both network efficiency and the green energy supply possibilities.

1 Introduction and background

Interest in the optimal design and management of food logistic networks crosses agriculture and industry, connecting producers to consumers, with efficiency, safety, and sustainability targets (Amiri, 2006). The topic of food supply chains (FSCs) is well studied, being addressed by scientists and researchers from a wide range of scientific sectors (Ahumada and Villalobos, 2009). Looking at food engineering and management science, recent research efforts have been in the direction of optimizing FSCs against a major set of metrics expressed through qualitative and quantitative key performance indicators (KPIs) (Manzini and Accorsi, 2013; Ferreira and Arantes, 2015; Kirwan et al., 2017). Such KPIs refer, primarily, to food demand satisfaction at top quality levels of safety and security, with no deficiencies or health damage to the final consumers. This is especially critical within developing and rural locales. The "problem tree" proposed by Caniato et al. (2017) extensively presents the causes and effects of bad practices in food supply, their impacts on the population, and the factors and barriers easing or complicating the FSC role. Among them, infrastructures and energy availability are of primary importance and are known to be potential technological barriers. In addition, the so-called governance barriers set loss loops in the creation of favorable conditions for the safe distribution and supply of food.

Multiple studies in the recent literature stress the intrinsic complexity of FSCs because of the existence of strong interactions among the involved actors and the environment, that is, externalities. The integrated frameworks proposed by Vlajic et al. (2012), Miranda-Ackerman et al. (2017), Zilberman et al. (2017), Flores and Villalobos (2018), and Govindan (2018), from different perspectives, highlight the dimensions of FSCs, suggesting strategies for their robust and sustainable long-term design and indicating the factors of success for the FSC decision makers.

Sustainability is at the center of attention because of the increasing environmental and social concerns from the final market, companies, and institutional players (Allaoui et al., 2018). Multiple studies have targeted FSC design against sustainability metrics, focusing attention on specific resources to be preserved. As an example, Bergendahl et al. (2018) propose an integrated multilevel analysis scheme on the transdisciplinarity of food, energy, and water for ecological and modern supply chains. Similarly, Karan et al. (2018) present a stochastic approach to optimize sustainable food-energy-water systems. Recent models that include sustainability within FSC multiobjective design and planning are found in Accorsi et al. (2016), Bortolini et al. (2016a,b), Catalá et al. (2016), Jonkman et al. (2017), Mogale et al. (2017), and Utomo et al. (2017), while Djekic et al.

Sustainable Food Supply Chains. https://doi.org/10.1016/B978-0-12-813411-5.00021-1

(2018) present a comprehensive review of environmental models in the food chain. These research efforts aim at overcoming the existing integration gap between sustainability and supply chain function best design, clearly expressed by Ala-Harja and Helo (2014): "Environmental concerns have been examined and treated separately in supply chain functions and there is as yet no integrative approach or mechanism that measures, controls, and improves the environmental aspects of an entire supply chain; a limitation that does not facilitate optimizing the green performance of a supply chain."

According to Zanoni and Zavanella (2012), FSCs require large amounts of energy to treat, process, and preserve food. The energy intensity of FSCs is a hot topic to consider, while the energy availability and sources are elements to include within the FSC design problem. When considering the energy and sustainability issues, the potential role of renewables comes immediately to mind. The renewables, for example, solar, photovoltaics, wind, biomass, etc., are known to be sustainable, distributed, and smart sources with many environmental strengths, bringing the possibility of closely connecting the points of production and use and the opportunity to reduce use of the grid, which is still based on fossil fuels since national energy mixes are typically made up of coal, oil, and gas.

Increasing attention is being placed on renewable sources, distributed through smart energy systems at a multiscale level. Basic definitions are found in Lund et al. (2017), while interesting overviews on the role of smart grids and renewable energies within the supply chain are found in Abdullah et al. (2014), Hossain et al. (2016), Fichera et al. (2017), Marquant et al. (2017), Saavedra et al. (2018) and Zame et al. (2018). In addition, Chen et al. (2017) and Malik et al. (2018) present and apply optimal allocation strategies of energy and energy storage systems within smart distribution networks, given the available renewable energy potential and facilities. The application of these strategies generates optimized distributed renewable energy availability within their reference design area. Such an element needs to be integrated into FSC design so that an increasing amount of green energy powers the food networks and, particularly, the energy intensive storage, handling, and processing plants.

1.1 Aim and scope of this research

Starting from the crucial role of energy to power FSCs and the parallel attention on sustainability and green energy practices, this research aims at merging these two topics with a final goal of including the energy supply dimension in the FSC network design. Particularly, a "design for green energy" concept is followed so that the traditional FSC design models based on efficiency and economic aspects are integrated into a focus on the energy supply issues, easing the use of the available distributed and renewable sources to reduce grid dependency. Fig. 1 outlines the aim and scope of this research, while Fig. 2 illustrates the main entities involved and the rationale of the proposed framework.

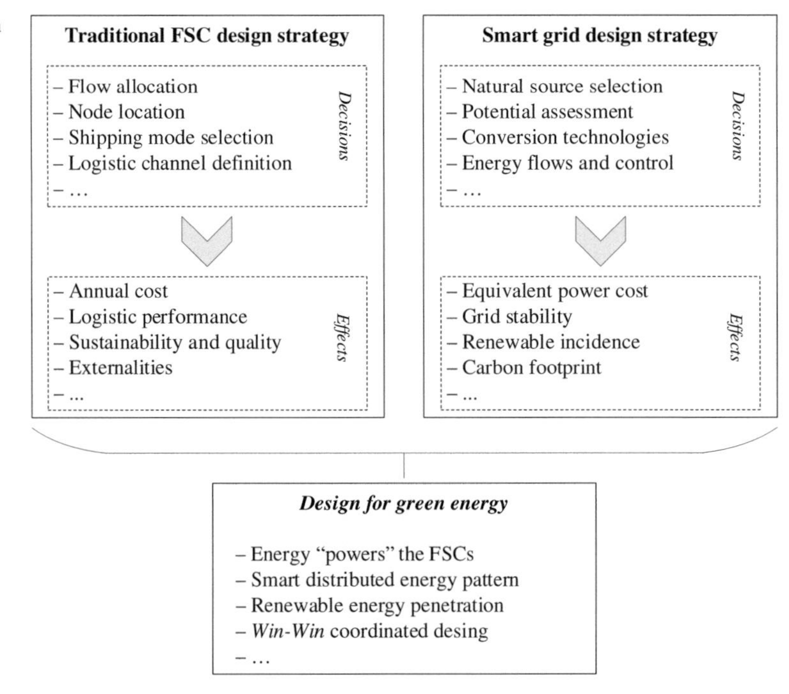

FIG. 1 Integrated approach to the "design for green energy" of FSCs.

Food supply chain network design	Renewable power-grid design
- Supply chain nodes optimally located to minimize transporation costs. - Operations management to minimize production, storage, Handling and distribution costs.	- Renewable energy plants optimally located on the bases of the climate conditions of the observed geography. - Renewable power plants and grid located and managed to reduce costs of the supplied energy and the power losses experienced across the network.

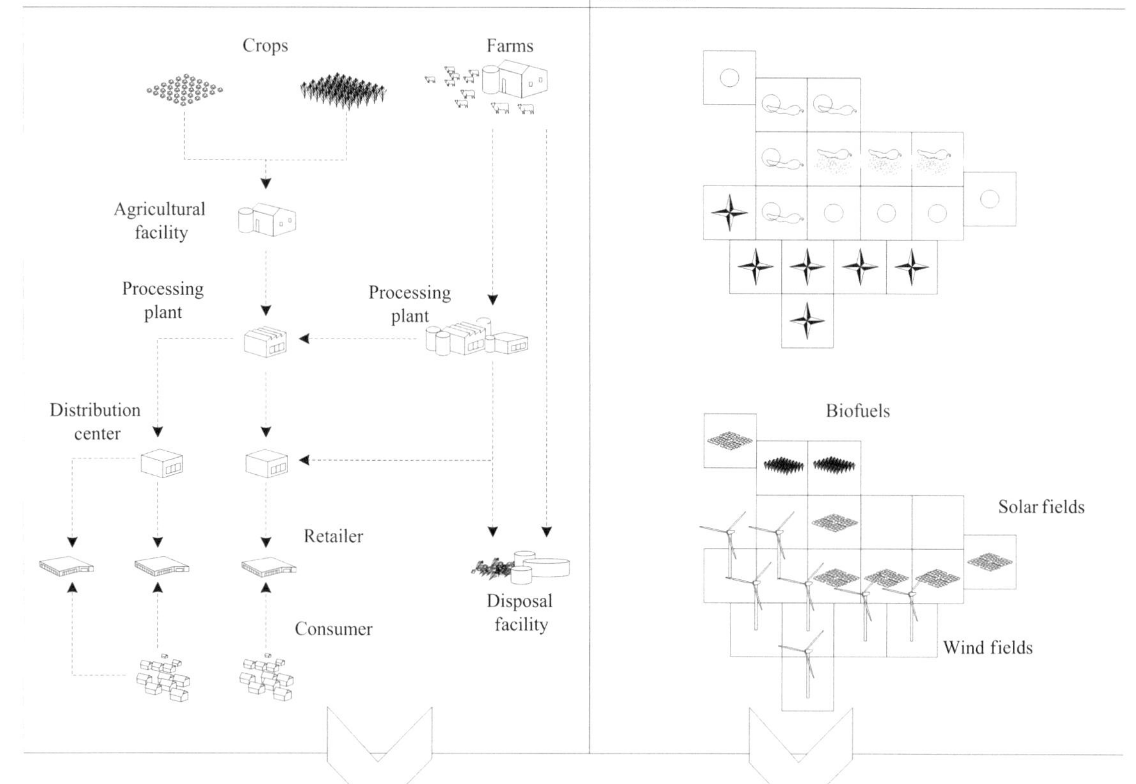

Integrated FSC and renewable power-grid network design

– Supply chain nodes and renewable power plants optimally colocated to minimize operational costs and together improve the sustainability of FSC operations.

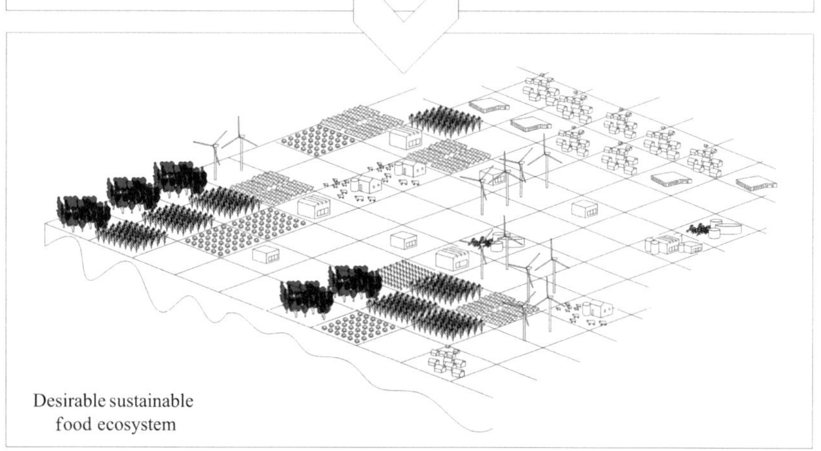

FIG. 2 Entities involved in the planning framework.

1.2 Methodology

The methodology for modeling the FSC best design is the so-called location allocation problem (LAP), meaning the location of the supply chain facilities and the allocation of the flows of products and energy. LAP is widely adopted in logistics and solved through optimal and heuristic algorithms leading to the supply chain structure definition. Recent reference examples of use of this methodology are found in Manzini et al. (2013), Mota et al. (2017), and Sampat et al. (2017), while Lima-Junior and Carpinetti (2017) revise quantitative models for the supply chain design and performance evaluation.

2 Problem formulation and entities

The proposed model aims at georeferencing an FSC to supply the market. The focused area is big enough to include producers, plants, and a major set of consumers. The final goal of the model is to locate, optimally, the storage, handling, and processing plants to minimize the internal and external costs for shipments and energy to feed the processes. The following list of statements provides the conceptual formulation behind the proposed model:

1. The considered area is divided into a set of nonoverlapped geozones representing the problem geographical resolution. Each geozone is uniform in demand and energy potential, while connections are between geozones;
2. Food production is carried out at farmers/crops located in given geozones;
3. The final destination of food is groceries/final markets located in given geozones;
4. The FSC is from farmers/crops to groceries/final markets via storage, handling, and processing plants. Three levels and a two-stage network is considered;
5. The number, best location, and product flow for the storage, handling, and processing plants is to be optimized;
6. The FSC manages a uniform product, measured in quantity;
7. Each geozone has some available smart and distributed renewable energy sources and facilities to convert it into usable electric power;
8. The available smart and distributed renewable energy can be used within the geozone of production or in other geozones by paying a specific connection cost;
9. The national grid is the energy backup;
10. The specific benefits and costs of energy include externalities, equivalent social and environmental cost, computed from a life cycle cost (LCC) perspective;
11. The shipped food product has its own quality loss function, which is a function of the travel time. The cost of loss of quality is among the cost drivers.

Starting from this introduced problem formulation, a major set of notations is necessary to express, analytically, the proposed model, covered in the following sections.

2.1 Indices, sets, parameters, and variables

The indices refer to the model entities. For simplicity, they are numbered progressively. Different coding is possible with small changes in the model formulation.

Sets and indices

$f=1, \ldots, F$	Index for farmers/crops, origin of flows
$g=1, \ldots, G$	Index for groceries/final markets, destination of flows
$m=1, \ldots, M$	Index for shipping modes
$s=1, \ldots, S$	Index for renewable smart energy sources
$w=1, \ldots, W$	Index for storage, handling, and processing plants
$z=1, \ldots, Z$	Index for geozones

Among the entities, w is of primary importance because it refers to the entities to be located in the considered area.

The model parameters are the information available and they contribute to defining both the objective function and the feasibility constraints.

Parameters

a_{fz}	1 if farmer/crop f is in geozone z, 0 otherwise
a_{gz}	1 if grocery/market g is in geozone z, 0 otherwise
a_{mz}	1 if shipping mode m is available in geozone z, 0 otherwise
a_{wz}	1 if plant w can be located in geozone z, 0 otherwise
b_s	Benefit of energy source s with respect to the grid, including externalities (€/kWh)
ce	Energy grid cost, including externalities (€/kWh)
cq	Cost of quality loss (€/q.ty)
c_m	Specific shipping cost for mode m (€/km)
c_{zz_1}	Equivalent cost for energy connections between geozone z and z_1 (€/kWh)
d_g	Aggregate product demand of grocery/market g (q.ty/period)
e_w	Specific energy need of plant w (kWh/q.ty)
rq_{sz}	Maximum renewable energy potential for source s in geozone z (kWh/period)
t_{mzz_1}	Travel distance between geozone z and z_1 for shipping mode m (km)
v_{mzz_1}	Commercial speed of mode m on route $z \rightarrow z_1$ (km/h)
$\varphi(\tau)$	Quality loss density function, where τ is the travel time (%)

The following decision variables are introduced.

Decision variables

k_{szw}	Renewable energy supply, source s, from geozone z to plant w (kWh/period)
x_{fwn}	Product flow from farmer/crop f to plant w via mode m (q.ty/period)
x_{wgm}	Product flow from plant w to grocery/market g via mode m (q.ty/period)
y_{wz}	1 if plant w is located in geozone z, 0 otherwise

The decision variables are of three types: k_{szw} concerns the distributed renewable energy flows, x_{fwm} and x_{wgm} define the product flow along the FSC, y_{wz} locates the storage, handling, and processing plants. Such variables immediately represent the target decisions that the model addresses.

2.2 Data collection

When conceiving analytic models to support any decision-making process, specific attention must be paid to the data-collection phase. This means that the researcher must propose models with available parameters (see Section 2.1) or ones that can feasibly be acquired at reasonable cost in terms of time and data preprocessing.

Within this context, specific attention is placed on this aspect and, by adopting a practical viewpoint, the input parameters are simplified, even if this means an increase in model complexity. The boundary conditions of the considered area and geozones allow the binary parameters a_{fz}, a_{gz}, a_{mz} and a_{wz} to be obtained immediately. The energy cost and benefit come from the current grid energy tariffs and the levelized cost of electricity (LCOE) for renewable smart systems when installed in a specific geozone. Generally, the cost of the quality loss, cq, is expressed in terms of revenue decrease because low-quality products have lower market margins. The cost of collecting and wasting unsold product is added. The shipping cost, c_m, is among the most frequently monitored drivers by companies. When such cost depends on a specific route, the parameter c_m becomes c_{mzz_1} with little change in the model. The cost for energy connection between two different geozones includes the differential for the grid cost to sell energy to the grid with discontinuity and at low quality level, for example, reactive power. It also includes the equivalent cost to connect the distributed renewable system to the grid. Low difficulty occurs in the collection of the aggregate demand data and the plant energy intensity, because such data are an input of a major set of decisions. Concerning the energy potential of the geozones for each source, rp_{sz}, multiple studies and models are available, for both energy resource availability and energy conversion system efficiency. Finally, studies on the quality loss curves for (non-)perishable products are available in the literature.

Globally, despite the amount of the input data is relevant, the data availability is consistent and the collection process is feasible for wide regions around the globe.

3 Model presentation

The following equations express the analytic formulation of the model through the presentation of the cost objective function and the related feasibility constraints.

3.1 Objective function

$$\min \Psi = \sum_{z=1}^{Z}\sum_{z_1=1}^{Z}\sum_{m=1}^{M}\sum_{f=1}^{F}\sum_{w=1}^{W} c_m \cdot a_{fz} \cdot y_{wz_1} \cdot t_{mzz_1} \cdot x_{fwm} \tag{1}$$

$$+\sum_{z=1}^{Z}\sum_{z_1=1}^{Z}\sum_{m=1}^{M}\sum_{w=1}^{W}\sum_{g=1}^{G} c_m \cdot y_{wz} \cdot a_{gz} \cdot t_{mzz_1} \cdot x_{wgm} \tag{2}$$

$$+\sum_{s=1}^{S}\sum_{z=1}^{Z}\sum_{z_1=1}^{Z}\sum_{w=1}^{W} c_{zz_1} \cdot k_{szw} \cdot y_{wz_1} \tag{3}$$

$$-\sum_{s=1}^{S}\sum_{z=1}^{Z}\sum_{w=1}^{W} b_s \cdot k_{szw} \tag{4}$$

$$+\sum_{w=1}^{W} ce \cdot \left(e_w \cdot \sum_{f=1}^{F}\sum_{m=1}^{M} x_{fwm} - \sum_{s=1}^{S}\sum_{z=1}^{Z} k_{szw} \right) \tag{5}$$

$$+\sum_{z=1}^{Z}\sum_{z_1=1}^{Z}\sum_{m=1}^{M}\sum_{F=1}^{F}\sum_{w=1}^{W} cq \cdot x_{fwm} \cdot a_{fz} \cdot y_{wz_1} \cdot \int_0^{\frac{t_{mzz_1}}{v_{mzz_1}}} \varphi(\tau)d\tau \tag{6}$$

$$+\sum_{z=1}^{Z}\sum_{z_1=1}^{Z}\sum_{m=1}^{M}\sum_{w=1}^{W}\sum_{g=1}^{G} cq \cdot x_{wgm} \cdot y_{wz} \cdot a_{gz_1} \cdot \int_0^{\frac{t_{mzz_1}}{v_{mzz_1}}} \varphi(\tau)d\tau \tag{7}$$

The objective function minimizes a major set of cost drivers presenting divergent trends, seeking the best balance among them. Here, (1) is the shipment cost from farmers/crops to the storage, handling, and processing plants, while (2) is the same cost for the second stage of the FSC, that is, from the intermediate plants to the groceries/final markets; (3) includes the cost to use the renewable smart energy in a geozone different from the energy production area; (4) accounts for the benefit of using renewable energy instead of the energy from the grid and must include the externality cost due to lower emissions, while (5) is the grid energy cost, along with externality cost, to fully supply the plant energy intensity; and, finally, (6) computes the cost of quality loss at the first stage of the FSC, and (7) is the same cost for the second stage of the network.

3.2 Feasibility constraints

$$\sum_{w=1}^{W}\sum_{m=1}^{M} x_{wgm} \geq d_g \quad \forall g \tag{8}$$

$$\sum_{f=1}^{F}\sum_{w=1}^{W} a_{fz} \cdot x_{fwm} \leq a_{mz} \cdot \mathcal{M} \quad \forall m,z \tag{9}$$

$$\sum_{f=1}^{F}\sum_{w=1}^{W} y_{wz}\cdot x_{fwm} \le a_{mz}\cdot \mathcal{M} \quad \forall m,z \tag{10}$$

$$\sum_{w=1}^{W}\sum_{g=1}^{G} y_{wz}\cdot x_{wgm} \le a_{mz}\cdot \mathcal{M} \quad \forall m,z \tag{11}$$

$$\sum_{w=1}^{W}\sum_{g=1}^{G} a_{gz}\cdot x_{wgm} \le a_{mz}\cdot \mathcal{M} \quad \forall m,z \tag{12}$$

$$\sum_{f=1}^{F}\sum_{m=1}^{M} x_{fwm} \ge \sum_{g=1}^{G}\sum_{m=1}^{M} x_{wgm} \quad \forall w \tag{13}$$

$$\sum_{w=1}^{W} k_{szw} \le rp_{sz} \quad \forall s,z \tag{14}$$

$$y_{wz} \le a_{wz} \quad \forall w,z \tag{15}$$

$$x_{fwm}, x_{wgm} \ge 0 \quad \forall f,g,m,w \tag{16}$$

$$y_{wz} \text{ binary } \quad \forall w,z \tag{17}$$

The constraint in (8) guarantees the demand of each grocery/market is met, while (9) matches the farmers/crops, their geozones, and the available shipping modes so that product flows at the first stage have their origin at the farmer/crop geozone and travel by an available shipping mode; (10) and (11) focus on the storage, handling, and processing plants enabling the entry and exiting flows in correspondence with their geo-zone; (12) is similar to (9) with reference to the groceries/final markets; the constraint in (13) balances the product flows that enter and exit the storage, handling, and processing plants, while (14) limits the renewable smart energy for each source to the potential of the corresponding geo-zone; (15) prohibits locating a plant in an unavailable geozone; and, finally, (16) and (17) provide consistence to the decisional variables.

3.3 Complexity

The proposed methodology for the FSC design consists of a quadratic programming optimization model with $2\cdot(F\cdot G\cdot M\cdot W)$ continuous nonnegative variables and $W\cdot Z$ binary variables. The feasibility constraints are $Z\cdot(4\cdot M+S)+G+W$ excluding the constraints giving consistence to the decisional variables.

Quadratic programming for the supply chain design is not as popular as linear programming, but multiple contributions exist, together with a major set of solving methods. Recent examples are in Ashtab et al. (2014) and Ba et al. (2016), while Zhao and Liu (2017) discuss possible solution methods.

Within the present model, a feasible path to solve the quadratic complexity is through model linearization, by cutting off the binary variables y_{wz}, matching the storage, handling, and processing plants to the geozones, and including the geozone as an attribute of each plant w changing the parameter a_{wz} as in the following:

a_{wz} 1 if plant w is available in geozone z, 0 otherwise.

In this way, the index w refers to the specific storage, handling, and processing plant of geozone z and not to a generic plant to be located in any feasible geozone. Following this linearization method, the objective function becomes the following:

$$\min \Psi = \sum_{z=1}^{Z}\sum_{z_1=1}^{Z}\sum_{m=1}^{M}\sum_{f=1}^{F}\sum_{\substack{w=1:\\ a_{wz_1}=1}}^{W} c_m\cdot a_{fz}\cdot t_{mzz_1}\cdot x_{fwm} \tag{1'}$$

$$+\sum_{z=1}^{Z}\sum_{z_1=1}^{Z}\sum_{m=1}^{M}\sum_{\substack{w=1:\\ a_{wz}=1}}^{W}\sum_{g=1}^{G} c_m\cdot a_{gz}\cdot t_{mzz_1}\cdot x_{wgm} \tag{2'}$$

$$+\sum_{s=1}^{S}\sum_{z=1}^{Z}\sum_{z_1=1}^{Z}\sum_{\substack{w=1:\\ a_{wz_1}=1}}^{W} c_{zz_1}\cdot k_{szw} \tag{3'}$$

$$-\sum_{s=1}^{S}\sum_{z=1}^{Z}\sum_{w=1}^{W} b_s\cdot k_{szw} \tag{4'}$$

$$+\sum_{w=1}^{W} ce\cdot\left(e_w\cdot\sum_{f=1}^{F}\sum_{m=1}^{M}x_{fwm}-\sum_{s=1}^{S}\sum_{z=1}^{Z}k_{szw}\right) \tag{5'}$$

$$+\sum_{z=1}^{Z}\sum_{z_1=1}^{Z}\sum_{m=1}^{M}\sum_{F=1}^{F}\sum_{\substack{w=1:\\ a_{wz_1}=1}}^{W} cq\cdot x_{fwm}\cdot a_{fz}\cdot\int_0^{\frac{t_{mzz_1}}{v_{mzz_1}}}\varphi(\tau)d\tau \tag{6'}$$

$$+\sum_{z=1}^{Z}\sum_{z_1=1}^{Z}\sum_{m=1}^{M}\sum_{\substack{w=1:\\ a_{wz}=1}}^{W}\sum_{g=1}^{G} cq\cdot x_{wgm}\cdot a_{gz_1}\cdot\int_0^{\frac{t_{mzz_1}}{v_{mzz_1}}}\varphi(\tau)d\tau \tag{7'}$$

Concerning the feasibility constraints, following the introduced changes, previous constraints (10) and (11) become the following, while constraints (15) and (17) are cut off:

$$\sum_{f=1}^{F}\sum_{\substack{w=1:\\ a_{wz}=1}}^{W} x_{fwm}\le a_{mz}\cdot\mathcal{M}\ \ \forall m,z \tag{8'}$$

$$\sum_{\substack{w=1:\\ a_{wz}=1}}^{W}\sum_{g=1}^{G} x_{wgm}\le a_{mz}\cdot\mathcal{M}\ \ \forall m,z \tag{9'}$$

Finally, to make the model linear, the quality loss density function, $\varphi(\tau)$, has to be linear too. The reader can refer to Gallo et al. (2017) for a feasible way to tackle this issue.

4 Discussion

The issue of integrating the energy supply within the FSC design requires some serious attention and effort in order to enhance the state of the art and, mainly in this case, to include renewables, available within distributed smart energy systems.

Evidence exists behind this proposed model, supported by studies and data, for the high energy intensity of the food sector, with an incidence of about 15%–20% of the global energy consumption both in the EU and US areas (Canning et al., 2010; Monforti-Ferrario et al., 2015). Consequently, the cost of energy is of major importance when studying FSCs. This means that local energy availability at low cost is a plus and the FSC design can be driven by energy together with the other logistics and cost drivers.

The model in Section 3 goes in this direction, finding a global cost optimum between the traditional logistics cost of shipments and the energy cost to supply the processes. It encourages both shortening the chain, so as not to generate costly routes and have to pay for product quality loss, and, at the same time, to take advantage of the renewable energy sources available in the same (or near) consumption geozone(s). Exploiting renewables has the positive effect of saving grid energy and cost, with the additional benefit expressed as saved cost of externality. Such a cost is calculated beginning with the impact of the electric energy grid chain, a function of the national energy mix. Major differences exist among countries, as demonstrated in recent studies by Heiskanen and Matschoss (2017) and Kounetas (2018) for the EU area, by Shi (2016) for

Southeast Asia, by Ponce-Jara et al. (2017) for American developed and developing regions, and by Lima et al. (2017) for the EU versus American areas. Consequently, the massive adoption of renewables has a positive effect that is different among countries. The general trends show that developing countries could receive greater advantages because of their high fossil fuel–based energy mix. Nevertheless, in all regions a benefit does exist.

Finally, to clarify the holistic focus of the model, the system boundaries, defining the perimeter of the analysis and, consequently, the drivers of cost to be calculated, are of primary importance for obtaining consistent results. In this sense, specific attention must be paid to externality costs due to environmental emissions and the social impact of FSCs and, particularly, the energy sources feeding them. It is important not to include only partial costs of this type, which would generate partial results. In fact, despite recent studies demonstrating the economic sustainability of renewable energy (Ueckerdt et al., 2013; Bortolini et al., 2014), differences among contexts exist, and neglecting the externality cost may switch the benefit of renewables use into a cost increase, creating a vicious circle with negative consequences.

5 Conclusion and next steps

Including the energy dimension of powering multistage FSCs is of interest for increasing system performance and enhancing the penetration of renewable sources, to reduce the environmental impact of food logistics.

This chapter focuses on such an aspect, presenting a high-level model for FSC design, that is, the logistics node location, flow allocation, energy source selection, and energy flow dispatching, to minimize a cost objective function that includes the energy cost together with the traditional cost of outbound logistics. A quadratic programming model optimizes the FSC structure, locating the storage, handling and processing plants and allocating the product and energy flows, from the farmers/crops to the groceries/final markets. Furthermore, model linearization has been presented to ease the solution phase. Finally, a detailed discussion of the data collection phase and the model impact on the operational contexts is presented to close the proposed model to the practitioners.

The next feasible research steps go in complementary directions. On one side, the model application to one or more specific geographical areas and food products is expected, to draw comparisons with the as-is scenarios that are fully based on the grid energy supply. In parallel, model refinement and extension to different FSC configurations and product mixes will allow an expansion of the range of the covered operational environments. Future investigations may also involve the implementation of such integrated planning models into comprehensive decision-support tools (Accorsi et al., 2017, 2018) intended for the simulation and optimization of complex and sustainable FSC ecosystems.

References

Abdullah, M.A., Agalgaonkar, A.P., Muttaqi, K.M., 2014. Assessment of energy supply and continuity of service in distribution network with renewable distributed generation. Appl. Energy 113, 1015–1026.

Accorsi, R., Cholette, S., Manzini, R., Pini, C., Penazzi, S., 2016. The land-network problem: ecosystem carbon balance in planning sustainable agro-food supply chains. J. Clean. Prod. 112, 158–171.

Accorsi, R., Bortolini, M., Baruffaldi, G., Pilati, F., Ferrari, E., 2017. Internet-of-things paradigm in food supply chains control and management. Proc. Manufact. 11, 889–895.

Accorsi, R., Cholette, S., Manzini, R., Tufano, A., 2018. A hierarchical data architecture for sustainable food supply chain management and planning. J. Clean. Prod. 203, 1039–1054.

Ahumada, O., Villalobos, J.R., 2009. Application of planning models in the agri-food supply chain: a review. Eur. J. Oper. Res. 195, 1–20.

Ala-Harja, H., Helo, P., 2014. Green supply chain decisions—case-based performance analysis from the food industry. Transp. Res. E 69, 97–107.

Allaoui, H., Guo, Y., Choudhary, A., Bloemhof, J., 2018. Sustainable agro-food supply chain design using two-stage hybrid multi-objective decision-making approach. Comput. Oper. Res. 89, 369–384.

Amiri, A., 2006. Designing a distribution network in a supply chain system: formulation and efficient solution procedure. Eur. J. Oper. Res. 171 (2), 567–576.

Ashtab, S., Caron, R.J., Selvarajah, E., 2014. A binary quadratic optimization model for three level supply chain design. Proceedings of the 47th CIRP Conference on Manufacturing Systems, Procedia CIRP, 28–30 April 2014, Windsor, Ontario, Canada. Vol. 17, 635–638.

Ba, B.H., Prins, C., Prodhon, C., 2016. Models for optimization and performance evaluation of biomass supply chains: an operational research perspective. Renew. Energy 87, 977–989.

Bergendahl, J.A., Sarkis, J., Timko, M.T., 2018. Transdisciplinarity and the food energy and water nexus: ecological modernization and supply chain sustainability perspectives. Resour. Conserv. Recycl. 133, 309–319.

Bortolini, M., Gamberi, M., Graziani, A., 2014. Technical and economic design of photovoltaic and battery energy storage system. Energ. Conver. Manage. 86, 81–92.

Bortolini, M., Faccio, M., Ferrari, E., Gamberi, M., Pilati, F., 2016a. Fresh food sustainable distribution: cost, delivery time and carbon footprint three-objective optimization. J. Food Eng. 174, 56–67.

Bortolini, M., Faccio, M., Gamberi, M., Pilati, F., 2016b. Multi-objective design of multi-modal fresh food distribution networks. Int. J. Logist. Syst. Manag. 24 (2), 155–177.

Caniato, M., Carliez, D., Thulstrup, A., 2017. Challenges and opportunities of new energy schemes for food security in humanitarian contexts: a selective review. Sustain. Energy Technol. Assess. 22, 208–219.

Canning, P., Charles, A., Huang, S., Polenske, K.R., Waters, A., 2010. Energy use in the U.S. food system. United States Department of Agriculture, Economic Research Report Number 94.

Catalá, L.P., Moreno, M.S., Blanco, A.M., Bandoni, J.A., 2016. A bi-objective optimization model for tactical planning in the pome fruit industry supply chain. Comput. Electron. Agric. 130, 128–141.

Chen, M., Zou, G., Jin, X., Yao, Z., Liu, Y., Yin, H., 2017. Optimal allocation method on distributed energy storage system in active distribution network. Proceedings of the 4th International Conference on Power and Energy Systems Engineering (CPESE 2017), Energy Procedia, 25–29 September 2017, Berlin, Germany. Vol. 141, 525–531.

Djekic, I., Sanjuán, N., Clemente, G., Jambrak, A.R., Djukić-Vuković, A., Brondnjak, U.V., Pop, E., Thomopoulos, R., Tonda, A., 2018. Review on environmental models in the food chain—current status and future perspectives. J. Clean. Prod. 176, 1012–1025.

Ferreira, L.M., Arantes, A.J., 2015. Proposal of a framework to assess the supply chain performance in the agri-food sector. Proceedings of the 4th International Conference on Operations Research and Enterprise Systems (ICORES 2015), 10–12 January 2015, Lisbon, Portugal.

Fichera, A., Frasca, M., Volpe, R., 2017. Complex networks for the integration of distributed energy systems in urban areas. Appl. Energy 193, 336–345.

Flores, H., Villalobos, J.R., 2018. A modeling framework for the strategic design of local fresh-food systems. Agr. Syst. 161, 1–15.

Gallo, A., Accorsi, R., Baruffaldi, G., Manzini, R., 2017. Designing sustainable cold chains for long-range food distribution: energy-effective corridors on the silk road belt. Sustainability 9 (11), 2044.

Govindan, K., 2018. Sustainable consumption and production in the food supply chain: a conceptual framework. Int. J. Prod. Econ. 195, 419–431.

Heiskanen, E., Matschoss, K., 2017. Understanding the uneven diffusion of building-scale renewable energy systems: a review of household, local and country level factors in diverse European countries. Renew. Sustain. Energy Rev. 75, 580–591.

Hossain, M.S., Madlool, N.A., Rahim, N.A., Selvaraj, J., Pandey, A.K., Khan, A.F., 2016. Role of smart grid in renewable energy: an overview. Renew. Sustain. Energy Rev. 60, 1168–1184.

Jonkman, J., Bloemhof, J.M., van der Vorst, J.G.A.J., van der Padt, A., 2017. Selecting food process designs from a supply chain perspective. J. Food Eng. 195, 52–60.

Karan, E., Asadi, S., Mohtar, R., Baawain, M., 2018. Towards the optimization of sustainable food-energy-water systems: a stochastic approach. J. Clean. Prod. 171, 662–674.

Kirwan, J., Maye, D., Brunori, G., 2017. Acknowledging complexity in food supply chains when assessing their performance and sustainability. J. Rural Stud. 52, 21–32.

Kounetas, K.E., 2018. Energy consumption and CO_2 emissions convergence in European Union member states. A tonneau des Danaides? Energy Econ. 69, 111–127.

Lima, F., Lopes Nunes, M., Cunha, J., Lucena, A.F.P., 2017. Driving forces for aggregate energy consumption: a cross-country approach. Renew. Sustain. Energy Rev. 68, 1033–1050.

Lima-Junior, F.R., Carpinetti, L.C.R., 2017. Quantitative models for supply chain performance evaluation: a literature review. Comput. Ind. Eng. 113, 333–346.

Lund, H., Østergaard, P.A., Connolly, D., Mathiesen, B.V., 2017. Smart energy and smart energy systems. Energy 137, 556–565.

Malik, A.S., Albadi, M., Al-Jabri, M., Bani-Araba, A., Al-Ameri, A., Al Shehhi, A., 2018. Smart grid scenarios and their impact on strategic plan—a case study of Omani power sector. Sustain. Cities Soc. 37, 213–221.

Manzini, R., Accorsi, R., 2013. The new conceptual framework for food supply chain assessment. J. Food Eng. 115, 251–263.

Manzini, R., Accorsi, R., Bortolini, M., 2013. Operational planning models for distribution networks. Int. J. Prod. Res. 52 (1), 89–116.

Marquant, J.F., Evins, R., Bollinger, L.A., Carmelier, J., 2017. A holarchic approach for multi-scale distributed energy system optimisation. Appl. Energy 208, 935–953.

Miranda-Ackerman, M.A., Azzaro-Pantel, C., Aguilar-Lasserre, A.A., 2017. A green supply chain network design framework for the processed food industry: application to the orange juice agrofood cluster. Comput. Ind. Eng. 109, 369–389.

Mogale, D.G., Dolgui, A., Kandhway, R., Kumar, S.K., Tiwari, M.K., 2017. A multi-period inventory transportation model for tactical planning of food grain supply chain. Comput. Ind. Eng. 110, 379–394.

Monforti-Ferrario, F., Dallemand, J.-F., Pinedo Pascua, I., Motola, V., Banja, M., Scarlat, N., Mederac, H., Castellazzi, L., Labanca, N., Bertoldi, P., Pennington, D., Goralczyk, M., Schau, E.M., Saouter, E., Sala, S., Notarnicola, B., Tassielli, G., Renzulli, P., 2015. Energy use in the EU food sector: state of play and opportunities for improvement. Science and Policy Report, European Commission Joint Research Centre, Luxembourg.

Mota, B., Gomes, M.I., Carvalho, A., Barbosa-Povoa, A.P., 2017. Sustainable supply chains: an integrated modeling approach under uncertainty. Omega. https://doi.org/10.1016/j.omega.2017.05.006.

Ponce-Jara, M.A., Ruiz, E., Gil, R., Sancristóbal, E., Pérez-Molina, C., Castro, M., 2017. Smart grid: assessment of the past and present in developed and developing countries. Energ. Strat. Rev. 18, 38–52.

Saavedra, M.M.R., Fontes, C.H.O., Freires, F.G.M., 2018. Sustainable and renewable energy supply chain: a system dynamics overview. Renew. Sustain. Energy Rev. 82, 24–259.

Sampat, A.M., Martin, E., Martin, M., Zavala, V.M., 2017. Optimization formulations for multi-product supply chain networks. Comput. Chem. Eng. 104, 296–310.

Shi, X., 2016. The future of ASEAN energy mix: a SWOT analysis. Renew. Sustain. Energy Rev. 53, 672–680.

Ueckerdt, F., Hirth, L., Luderer, G., Edenhofer, O., 2013. System LCOE: what are the costs of variable renewables? Energy 63, 61–75.

Utomo, D.S., Onggo, B.S., Eldridge, S., 2017. Applications of agent-based modelling and simulation in the agri-food supply chains. Eur. J. Oper. Res. https://doi.org/10.1016/j.ejor.2017.10.041.

Vlajic, J.V., van der Vorst, J.G.A.J., Haijema, R., 2012. A framework for designing robust food supply chains. Int. J. Prod. Econ. 137, 176–189.

Zame, K.K., Brehm, C.A., Nitica, A.T., Richard, C.L., Schweitzer III, G.D., 2018. Smart grid and energy storage: policy recommendations. Renew. Sustain. Energy Rev. 82, 1646–1654.

Zanoni, S., Zavanella, L., 2012. Chilled or frozen? Decision strategies for sustainable food supply chains. Int. J. Prod. Econ. 140, 731–736.

Zhao, Y., Liu, S., 2017. Global optimization algorithm for mixed integer quadratically constrained quadratic problem. J. Comput. Appl. Math. 319, 159–169.

Zilberman, D., Lu, L., Reardon, T., 2017. Innovation-induced food supply chain design. Food Policy. https://doi.org/10.1016/j.foodpol.2017.03.010.

Chapter 22

Decision support models for fresh fruits and vegetables supply chain management

Omar Ahumada* and J. Rene Villalobos†

*Autonomous University of Occident, Culiacan, Mexico †International Logistics and Productivity Improvement Laboratory, Arizona State University, Tempe, AZ, United States

Abstract

In this chapter we present planning tools for the planning and coordination of the supply chain of fresh fruits and vegetables. In particular, we introduce an integrated planning environment that seeks to maximize the profits for a farmer-shipper with the capability of advancing vertically in the supply chain structure to position his production at different levels of the supply chain, including direct shipment to terminal markets. The models presented address different levels of planning scope, from operational, through tactical, to strategic. However, the focus is on tactical and operational planning. Both levels of planning incorporate the perishability of the underlying products into the models. While each fruit and vegetable presents its own peculiarities, we use the case of tomato as a proxy for the rest of the fresh fruits and vegetables.

1 Introduction

One of the greatest challenges of the 21st century is meeting the needs of a growing world population, projected to increase 35% by 2050. Such an increase will require a 70%–100% increase in current food production, given projected trends in diets, consumption, and income (van Wart et al., 2013), which will be difficult to achieve considering observable changing climate patterns. Increased food consumption includes a higher demand for staple products, such as rice, wheat, and beans, as well as an increase in demand for high-value products such as fresh fruits and vegetables (FFVs) fueled by better economic conditions and a more health-conscious consumer.

Among agricultural supply chains (ASCs), those for FFVs are some of the most dynamic sectors of the industry, yet also one of the hardest to manage (Huang, 2004). For FFVs, the complexity of managing the supply chain is compounded by the highly perishable nature of the product, which can result in significant product waste and other inefficiencies, such as excess of energy consumption. For instance, in the United States alone, over 50% of fresh vegetables are wasted, with 30% of total production lost before reaching the consumer (Gustavsson et al., 2011). Much of this loss is caused by supply chain inefficiencies, and the impact of such high levels of waste is reflected in several ways, including financial impacts at the farm level, on the environment, less product availability for consumption, and ultimately higher food prices for the consumer (IMechE, 2013).

Much of the losses throughout the supply chain can be attributed to a lack of production coordination that can efficiently match supply and demand at the grower (IMechE, 2013). The inability of growers to coordinate is partly due to a lack of planning capabilities at the farm level, such as applications of decision support tools that can guide growers through corrective measures for eliminating losses. In this chapter, we present some decision support tools that help farmers plan their production in a way that is more attuned to the market needs and the existing supply chain infrastructure. While the objective of this chapter is to present available planning tools for fresh fruits and vegetables, the focus is on tactical and operational planning tools for fresh vegetables and annual fruits. The main difference between planning tools is that, in the case of most fruits produced on trees and bushes, planting decisions are usually not made every year because of their perennial nature. Another difference between the planting models presented in this chapter and those for fruit trees is that there is less flexibility in affecting harvesting windows through planning the planting process. However, the underlying mathematical tools used to build the planning models are very similar to those used in the planning models for the supply chains of fresh fruits. For examples of planning models for the supply chain of fresh fruits, the reader is referred to Soto-Silva et al. (2016).

Sustainable Food Supply Chains. https://doi.org/10.1016/B978-0-12-813411-5.00022-3

2 General planning models for fruits and vegetables

2.1 Planning process

The fresh produce industry has become very competitive and complex, with increasing variety and requirements for producers, based on consumer and retailer demands. As an example of this complexity, we can consider the planning environment for a tomato crop destined for the fresh market. In tomato production, growers go through several planning phases before they ship their product to the market. For instance, there is a preseason in which growers need to make the initial decisions, such as which production technology to use, crop planning, and land requirements. A second phase comprises those decisions made during the growing season, from planting to cultivation activities to get the crop ready for harvest. The third phase takes place during the harvesting season, during which the producer makes production and packing decisions, and finally there is a postharvest phase, which includes all marketing, storage, distribution, and transportation decisions (Fig. 1).

Another factor that needs to be considered is when to plant the product, so that harvest can take place at a convenient time. This aspect is important because it may determine the crop price when harvested. For instance, Table 1 presents the case of the tomato crop in the state of Sinaloa, in Mexico, where different planting dates from August to February result in expected harvesting dates from November to June. This table presents historical information that provides not only the total expected production for each planting cycle, but also the percentage of the total production expected for each week of harvest. With this information, it is possible to estimate the amount of each crop to plant and their planting dates to cover the expected market's demand for the season.

The competitiveness of growers is dependent on using the best production technology according to the customer's requirements, as well as the farmer's geographical location, technological capacity, and financial resources (Fig. 1).

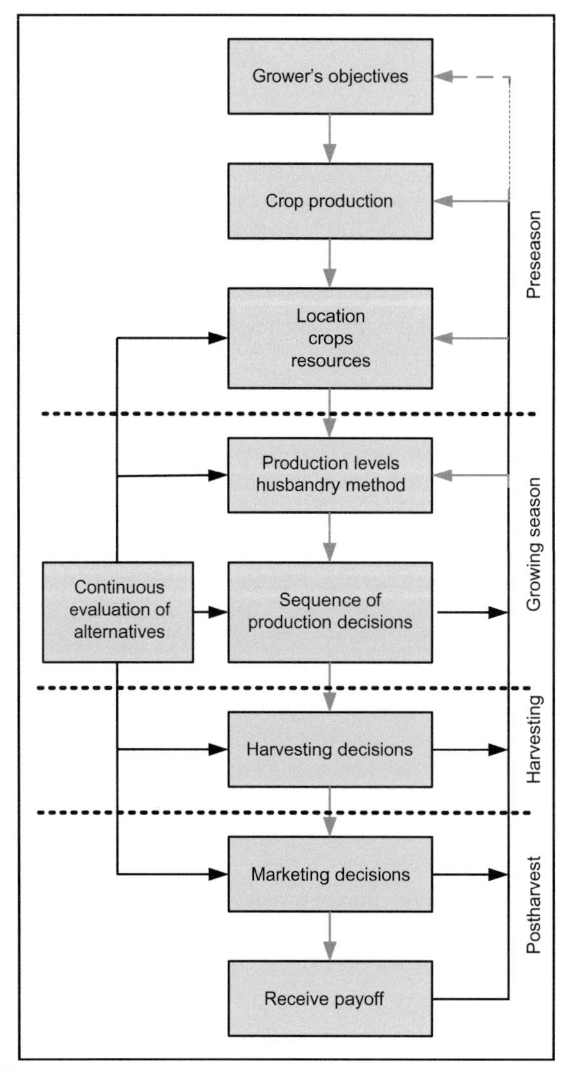

FIG. 1 Agricultural planning (Rae, 1977).

As is the case with the production system selection, some planting and growing decisions are also made at the preseason phase (before the season starts). Some of the planting decisions involve the selection of the best tomato variety, the selection of the total area to plant of each variety, and the planting period. These decisions are, for the most part, made based on forecasts of the demand and prices expected to be effective in the season ahead. Unfortunately, prices and demand for fresh vegetables are not fixed and have shown historically high levels of variability. For instance, Fig. 2 presents the monthly statistics for tomato prices taken over a period of 25 years. The information reported in the table shows the maximum, minimum, and the 75th and 25th percentiles of the prices, showing the variability of the market prices. Fig. 3 presents the weekly prices at the retail level in the United States for tomatoes for the years 2014–18. Thus the initial planning of the production season is marked by uncertainty, complicating the farmer's planning problem.

TABLE 1 Production estimates of different crop cycles

		Harvest by week																
		November				December				January				February				
Date of plant	Production	1	2	3	4	1	2	3	4	1	2	3	4	1	2	3	4	%
15-Aug	1662			5	5	10	10	10	10	9	9	8	8	8	8			100
30-Aug	1828					5	5	10	10	10	10	9	9	8	8	8	8	100
14-Sep	2373					5	5	6	10	10	10	10	10	9	9	8	8	100

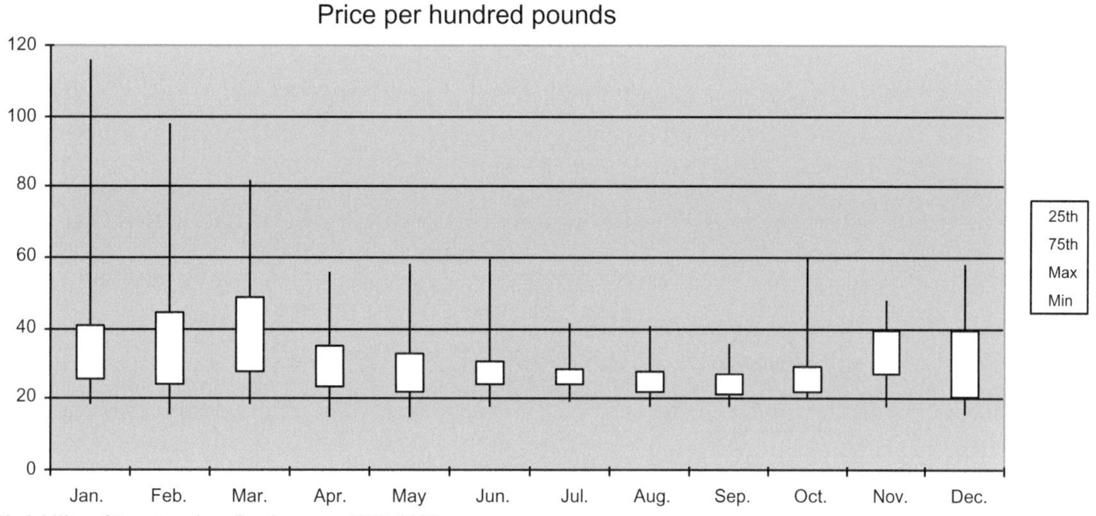

FIG. 2 Variability of tomato prices for the years 1979–2003.

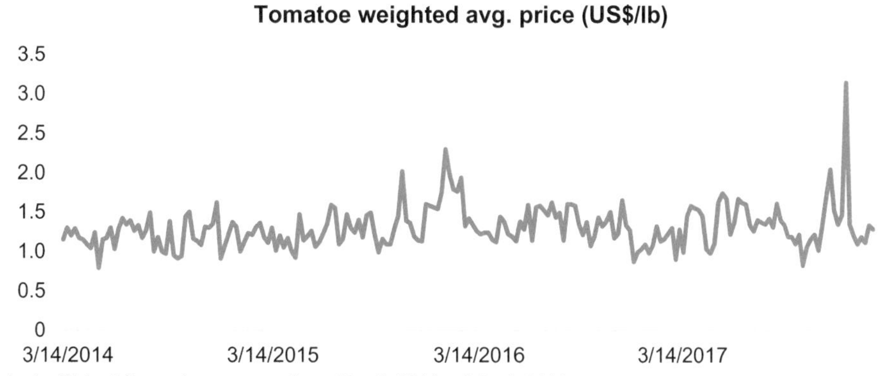

FIG. 3 Retail prices in the United States for tomatoes from March 2014 to March 2018.

During the growing season, the farmer first needs to plant the crops. Later, there are other activities that need to be scheduled as demanded by the crops, such as fertilization and irrigation. The planning models need to consider labor, equipment, and other resources available to perform all these operations. One feature of these decisions is that they are made once the season advances and the weather conditions and the crop development materializes. For instance, during the growing season, the farmer can rely on better weather forecasts and the feedback obtained from the crop itself to estimate the expected yields from the crop. The farmer can also obtain better information about the prevailing market prices. In this progression to the harvesting season, the farmer will prepare a desired harvest schedule, and estimate the amount of transportation equipment required to deliver the crops to the selected markets. Finally, once the harvesting phase arrives, the farmer schedules the amount to harvest per day of each crop. Then the farmer decides, after getting market information, when and what to ship (sell) to each customer. A variable market (prices change almost every week) and a perishable product affect decisions in the harvesting phase, so the farmer has a limited flexibility to determine selling and shipping decisions.

The previous description of the planning environment is based on fresh produce products such as tomatoes; however, the models developed for this type of problem can be easily extended to other crops, since they have similar planning processes. The aim of this chapter is to introduce a general tool that can be used as a starting point for a more detailed planning application that can be adapted to the needs of different growers. The next section provides the general definition of such a tool and the requirements that it must meet to satisfy the needs of producers.

2.2 Envisioned planning environment

The grower/shipper term applies to those individuals or companies who produce, commercialize, and sometimes distribute their own products. The supply chain environment of growers/shippers (Fig. 4) is formed by external and internal factors that affect the way the operations are conducted. The external factors include governmental policies and the distribution channels (retailers and processors) available. Those over which the farmer has control provide the internal factors, which include distribution channels and market demand for the products, among others.

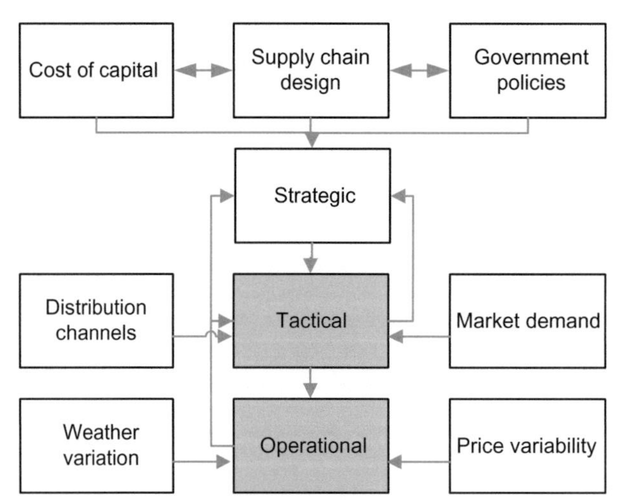

FIG. 4 Growers' supply chain environment.

In the supply chain of agricultural products, strategic decisions are those decisions whose outcome influences the long-term performance of growers, such as the development of production infrastructure, site selection, and market selection. In contrast, tactical decisions deal with medium-term activities, which include what, when, and how much to produce. Operational decisions deal with the planning of all the activities needed for the growing and harvesting season. Operational decisions include when to irrigate and fertilize, when to harvest, how much to harvest, how much to ship, etc. Operational decisions influence only short periods of time, such as weekly or even daily decisions.

2.3 Hierarchical planning approach

One way to handle the decision-making process is to use a hierarchical approach, which separates the decisions into strategic, tactical, and operational plans. The tactical plan deals with decisions performed before the growing season starts, such as planting decisions. On the other hand, the operational plan deals with those decisions that are made after the planting of the crops.

The reason for taking a hierarchical approach is because the decisions made at the strategic level impact the decisions made at the tactical level, and those at the operational level are impacted by both strategic and tactical decisions. Another feature is that, when dealing with decisions that involve a shorter time (like operational planning), the decisions are made with less uncertainty, since there is additional information available; if compared to the information available at the time the strategic and tactical decisions were made, this new information should be taken into consideration in the development of the new strategic tactical and operational plans (there should be feedback from the results obtained). Thus the adequate planning and coordination of plans (e.g., tactical and operational) are required for improving the survival and profitability of growers/shippers.

Fig. 5 shows a schematic of the two-phase approach, where typical inputs for Phase I (tactical) include historical price and production yield information, while the decision outputs include how much and when to plant of each potential crop. These two outputs serve as input for Phase II (operational). Typical outputs for the second phase include irrigation programs, fertilization plans, harvesting, and shipment schedules, with the respective destination of the shipments.

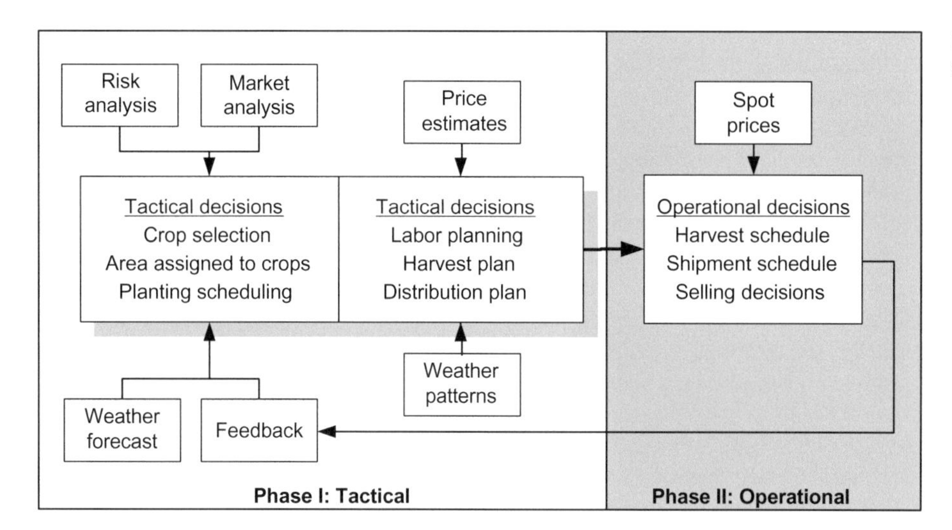

FIG. 5 Proposed agricultural planning environment.

The planning environment to be presented includes two analytical models, one designed for developing the tactical plans and another for the operational plan. The tactical model takes the relevant information and renders a plan for growing (when, what, and how much to produce) in the tactical phase, considering the limitations in terms of resources such as land, investment, expected yield, and expected market prices. Then, in the second model, the decisions involve the harvest and distribution of the products considering the prices at different markets, the logistical costs, and the life expectancy of the product. The models do not address intermediate decisions related to the growing of the crops such as irrigation and fertilization of the crops.

One potential risk of using a hierarchical approach is trying to solve the tactical and operational models independently, which might result in plans that restrict the overall performance of the supply chain. To avoid this suboptimization, it is proposed to link the tactical and operational models by including at the tactical level the most relevant information to be used in the operational model. An example of the information to be included in the tactical model is the weekly expected demand, the estimated transportation cost, and the expected market prices. With detailed information about demand, price, and transportation, it is possible to determine a tactical plan, which includes the amount of land to plant for each crop and the quantity to harvest and ship every week. Using this information, the tactical model has the necessary information to develop a detailed plan for the planting phase, but with lower resolution for the production and distribution plans, since a more detailed plan (or higher resolution) will be developed with the operational model. As mentioned before, the development of a detailed plan for the operational and tactical decisions at this stage would be impractical given the time gap from planting to harvesting. This gap might render a detailed plan for harvesting and distributing the crops impractical, given the time to its implementation and the variability presented by climatic conditions as well as the normal changes in the markets. The next section provides the planning objectives and the main contributions of the research presented in this chapter.

3 Tactical decision making under uncertainty

3.1 Planning environment

As mentioned before, the main tactical decisions involve planting decisions throughout the growing season. In the literature, there are several planning models for tactical decisions that assume there is absolute certainty in the production and distribution of fresh agricultural products. However, the assumptions that the prices and the yield of the crops harvested can be estimated with their expected values are rarely adequate for fresh produce, since experience indicates that these factors are highly dependent on the weather and market conditions (Lowe and Preckel, 2004). Furthermore, price and yield uncertainty are particularly relevant for those growers that sell their crops on the open market. For these growers, the prices of their products depend on the effects of supply and demand along the season, and, in the case of fresh produce, the observed variability tends to be higher than that of other crops, because of the limited opportunities for storage throughout the supply chain. For this reason, a model that captures these uncertainties can result in making better planning decisions regarding the production and distribution of these products.

Ahumada et al. (2012) developed a model that deals with the uncertainty encountered in the fresh produce industry. The model includes uncertain parameters to account for some of the variability experienced by the producers of tomatoes. The main motivation for building the model was to develop planting and distribution plans that are robust to the uncertain effects of the market and weather. From the grower's perspective, the model should help the producers achieve their goals in the fresh produce supply chain, whether these goals include maximizing the expected income of growers, also known as a risk neutral approach, or reducing the probability of losing the investment made in growing these products, which is also known as the risk-averse approach. The modeling approach they selected for modeling this problem was a two-stage stochastic program (Birge and Louveaux, 1997) in which the decisions in the first stage are set to meet the uncertain outcomes in the second stage. The use of a two-stage planning model comes naturally to agricultural planning, given the time elapsed from planting to the time of harvest (10–15 weeks). Most of the time, the long lead time between the time of planting and the time of harvest requires that all planting decisions be made before there is a single crop harvested. These physical constraints prevent the tactical plan from being reevaluated, thus reducing the planning decisions to a single stage. An envisioned application of the two-stage approach is shown in Fig. 6. At the beginning of the planting season, the farmer should decide on how much of each product will be planted without having certain information of weather and market conditions.

In accordance with the two-stage approach, the information available to the farmer at the time of making the tactical decisions is divided into two sets. The first-stage set incorporates the planting constraints and the costs associated with the planting decisions. In the second stage the information available is the historical distribution of crop prices and crop yields. Also, in the second stage, the model handles transportation, harvesting, and distribution decisions. Other relevant features in the second stage include demand requirements that must be met, such as preexisting contracts, market demand, and transportation available during the harvesting period. The solution for the two-stage problem is then dependent on the first-stage decisions (planting), the random realizations (crop yield and prices), and the second stage decisions (harvesting and distribution).

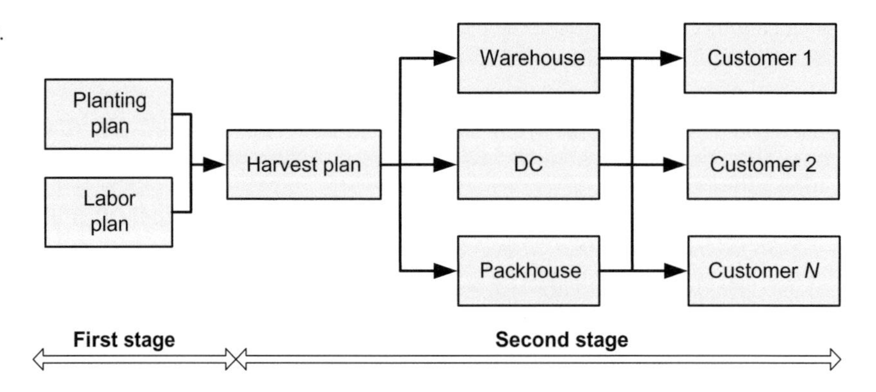

FIG. 6 Production and distribution decisions.

As mentioned before, the benefit of using stochastic programs (SPs) is that, unlike the deterministic solutions, which are based on overall expectations, the stochastic approach can be used to consider specific scenarios that occur depending on the realizations of the different random variables explicitly considered in the model program (Birge and Louveaux, 1997). Moreover, with the information provided by the solution of each scenario, it is possible to implement more meaningful risk metrics that are both relevant and better understood by the growers.

3.2 Stochastic tactical models

This section presents the formulation of the two-stage SP for the tactical model related to the formulation of a tactical model (Ahumada and Villalobos, 2011a). The first-stage decisions for the stochastic model are formed by the planting decisions, which assumes, in accordance with the two-stage approach, that costs and resources in the first stage are deterministic, thus leaving the second stage or recourse variables as the only random functions. In the second stage of the problem the stochastic parameters are the crop yield ($y_{tjt'}$) and the market prices (p_{tki}) for product k at shipping period t, and customer i, which are represented by the scenarios S. Then the model for the first-stage problem is expressed in the following way:

$$\max \left[\begin{array}{l} \sum_{s} pr^s \sum_{ts} Q_t^s - \lambda \sum_{ts} pr^s \left(T - Q_t^s\right)_+ - \sum_{tjl} Plant_{tjl} \cdot Cplant \\ - \sum_{tl} Opl_{tl} \cdot Clabor - \sum_{tl} Hire_{tl} \cdot Chire - \sum_{tl} Opt_{tl} \cdot Ctemp \end{array} \right] \tag{1}$$

s.t.

$$\sum_{j} \sum_{p} Plant_{pjl} \le LA_l \qquad \text{all } l \text{ where } l \in L, p \in TP(j,l), \text{ and } j \in J \tag{2}$$

$$\sum_{j} \sum_{p} \sum_{l} Plant_{pjl} \cdot Cplant_{jl} \le Inv \tag{3}$$

$$\sum_{j} \sum_{p} \sum_{l} Plant_{pjl} \cdot Water_j \le W \tag{4}$$

$$\text{Min}_j \cdot Y_{jp} \le Plant_{pjl} \le \text{Max}_j \cdot Y_{jp} \qquad \text{all } j, p, \text{ and } l \tag{5}$$

$$\sum_{tjl} Plant_{tjl} \cdot c_{t',j} \le M \quad \text{where } t' \in TP(j) \tag{6}$$

$$Opl_{tl} + Opt_{tl} \ge \sum_{pj} Plant_{pjl} \cdot LabP_{ptj} + \sum_{pjl} Harvest_{phjl} \cdot LabH_{phj} \qquad \text{all } t, l \text{ where } h = t \tag{7}$$

$$Hire_{tl} + Fire_{tl} = Opl_{tl} - Opl_{t-1l} \qquad \text{all } l \text{ and } t < t_m \tag{8}$$

$$Hire_{t_m l} + Opl_{t_m - 1l} \ge Opl_{tl} \qquad \text{all } l \text{ and } t_m \le t \tag{9}$$

$$Hire_{tl} \le Mfix, Opt_{tl} \le Mtemp \qquad \text{all } t, l \tag{10}$$

For the first-stage decisions, constraint (1) states that the farmer cannot plant more than the land he has, constraint (2) makes sure that the farmer does not spend more money than that available for investments; constraint (3) verifies that the results satisfy water restrictions, and the fourth constraint provides a lower bound and an upper bound on the amount of hectares plant to plant in a single block. The first-stage decisions include the timing and crops to plant (Plant) and the labor to hire for the whole season (Hire). The objective is to maximize the overall revenue for the producers, which is obtained from the second-stage expected revenues Q_t^sminus the cost generated by the first stage. In the second stage of the problem, given a solution from the first stage of the LP, we have that for every period t and scenario s we obtain a second-stage LP as follows:

$$Q_t^s(x) = \max \left[\begin{array}{l} \sum_{tki} \left(\sum_{fr} SC_{tkfir}^s + \sum_{hw} SW_{htkwir}^s + \sum_{\le h} \sum_{d} SD_{htkdir}^s \right) \cdot price_{tki}^s + \sum_{hj} K_{hj}^s Psalv_j \\ - \sum_{tkqfir} SC_{tkqfir}^s CT_{fir} - \sum_{htkqwir} SW_{htkqwir}^s CTW_{wir} - \sum_{htkqdir} SD_{htkqdir}^s CTD_{dir} \\ - \sum_{htkqfdr} SPD_{htkqfdr}^s CTPD_{fdr} - \sum_{htkqwdr} SWD_{htkqwdr}^s CTWD_{wdr} \sum_{htkqfwr} SPW_{htkqfwr}^s \end{array} \right] \tag{11}$$

s.t.

$$Harvest_{phjl}^s = plant_{pjl} \cdot Yield_{phjl} \cdot Total_{pjl}^s \qquad \text{all } p, h, j, l \text{ where } h \in TH(j,l) \tag{12}$$

$$\sum_{f} SP_{phjlf}^s = Harvest_{phjl}^s \qquad \text{all } p,\ h,\ j,\ l, \text{ where } f \in PF \tag{13}$$

$$Opl_{tl} + Opt_{tl} \ge \sum_{pj} Plant_{pjl} \cdot LabP_{ptj} + \sum_{pjl} Harvest_{phjl} \cdot LabH_{phj} \qquad \text{all } t, l \text{ where } h = t \tag{14}$$

$$Pack_{hkqf} = Col_{hkq} \sum_{phjk} SP_{phjlf} \left(1 - Salv_{phjl}\right) \cdot Pod_{phjk} / Weight_k \qquad \text{all } h,\ f,\ k \text{ where } k \in K(j) \tag{15}$$

The stochastic parameters in the second stage of the model include the price of the products ($price^s_{tki}$) and the total expected harvest from the crops ($Total^s_{pjl}$), and the rest of the parameters retain the same values as in the deterministic model. In (14) the planting decisions in the first stage are fixed for the second stage (plant), and the same for the hired labor (Opl_{tl} and Opt_{tl}).

The SP is required to run the second-stage model for each one of the scenarios to have the best overall solution to the stochastic model. With such a model, it is possible to calculate the best solution for all the potential scenarios, thus finding a planting plan that will maximize the expected income of farmers regardless of the outcome of the random variables. However, the size of the problem, and of the second-stage subproblems, might make the problem hard to solve. In the next section we present the solution methods developed for solving the current problem.

3.3 Scenario development

As mentioned in the previous sections, one of the most important features of SP is the ability to optimize first-stage decisions over the expected outcomes in the second stage. To estimate the realizations of the uncertain second-stage parameters, the stochastic distributions can be replaced by a discrete approximation to the true stochastic variables. These discrete approximations, or scenarios, as they are called in the SP literature, are used in the SP instead of the original distributions. There are some rules and methodological steps needed to assure that the scenarios developed are adequate. In the remainder of this section we present the methodology developed to generate the scenarios for the SP.

The first objective of the scenario generation procedure (see Fig. 7) is to develop scenarios that can adequately represent the uncertainty of the growers' environment. The second objective is to represent that uncertainty with a limited number of scenarios that keeps the SP from growing very large (Kaut and Wallace, 2003) and at the same time are representative of the planning conditions.

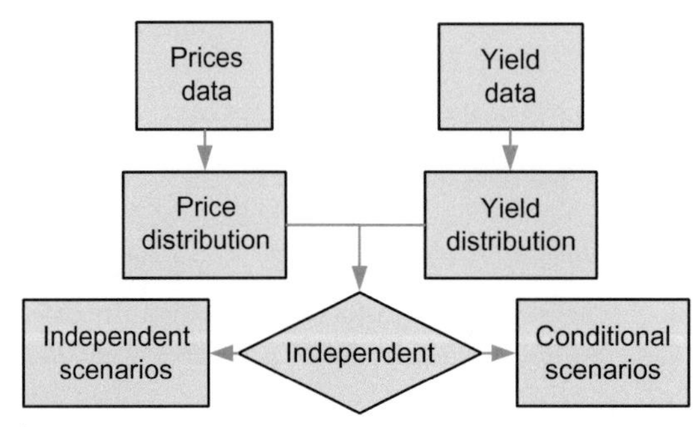

FIG. 7 Methodology for generating scenarios.

The two factors considered for developing the scenarios are crop yields and the crop prices. If there are multiple crops (e.g., tomatoes, peppers, cucumbers, etc.), each product is treated as a stochastic variable. We assume the formation of the scenarios depends on the combination of each one of the stochastic variables.

3.3.1 Estimation of price distributions

The methodology followed to generate the discrete approximations is presented in Fig. 8, and each one of the steps is demonstrated with the case of tomatoes and peppers.

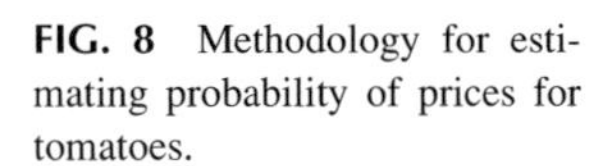

FIG. 8 Methodology for estimating probability of prices for tomatoes.

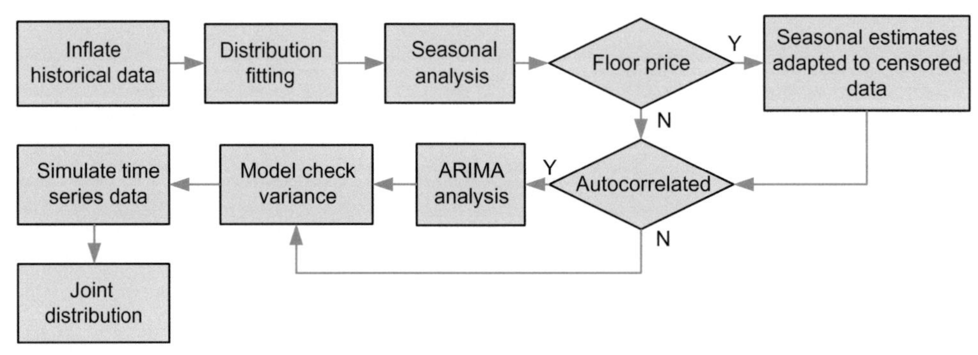

The data used in this example are the weekly historical prices of peppers and tomatoes for the 1999–2006 seasons. The data is publicly available and stored and maintained by the USDA (2007). The first activity in the methodology is to bring the historical prices of the crops to the present-day buying power by using inflation indexes provided by the US Department of Labor.

We used historical monthly data from USDA (2006), which presents monthly prices for the case of tomatoes in the month of January. The data suggest the use of the lognormal distribution, which was verified with statistical tests. Using the fact that the lognormal distribution can be transformed to a normal distribution by applying the natural logarithm function, we used this transformation on the weekly data (1999–2006).

An example of the results obtained by this procedure can be observed in Figs. 9 and 10. These graphs show the average 2.5th percentile and the 97.5th percentiles of the prices for tomatoes and peppers, respectively. The main difference between the two graphs is that the peppers do not have a floor price as in the case of tomatoes. From these same graphs the effect of the lognormal distribution is evident, since the resulting distribution is asymmetric and is skewed to the right (higher prices).

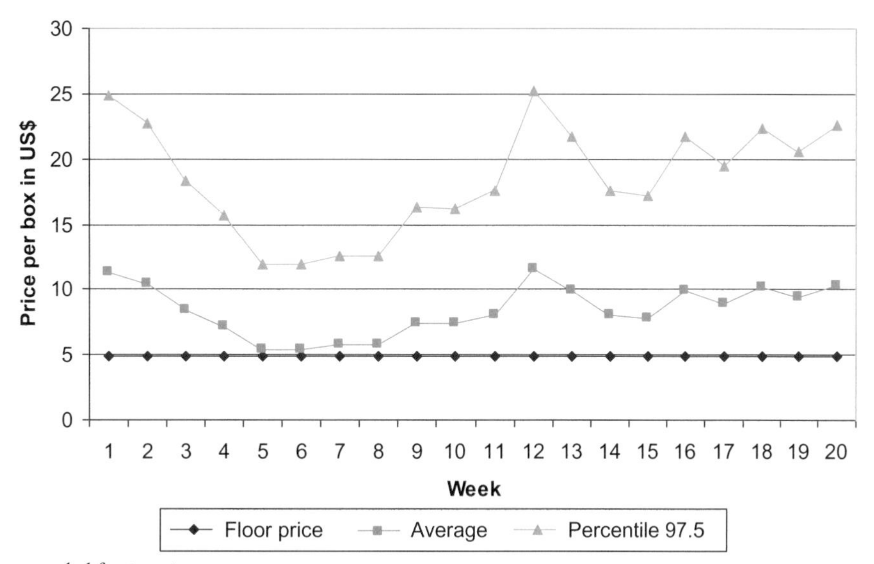

FIG. 9 Range of prices sampled for tomatoes.

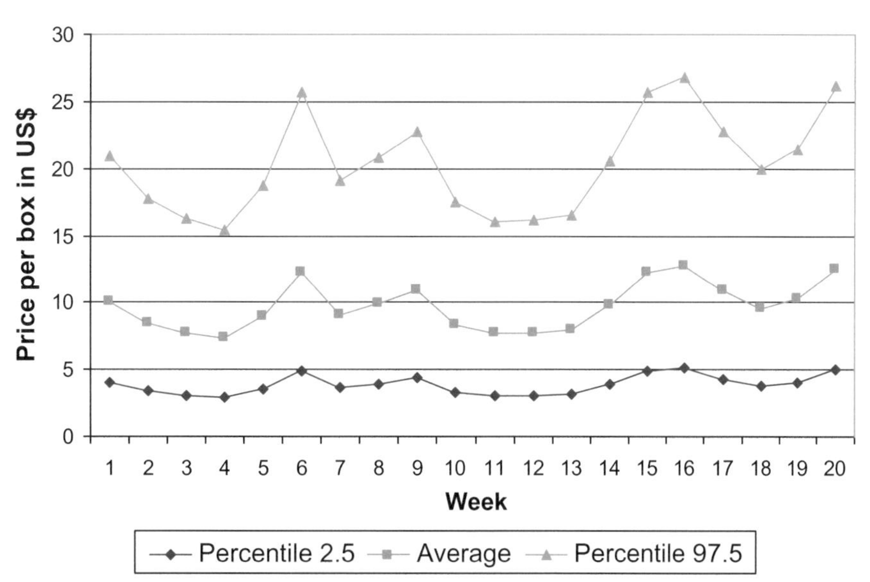

FIG. 10 Range of prices sampled for peppers.

3.3.2 Estimation of distribution of yields

The second part of the methodology consisted of estimating the crop yields with two components. One determines the total amount of yield obtained for a certain year or season, which corresponds in the model to the total quantity to harvest ($Total_{pjl}$). The second component determines the distribution of the harvest in each week ($yield_{phjl}$). From the perspective of parametric distributions, the most common alternatives have been the normal and the beta distributions (Norwood et al., 2004). The best fit was the beta distribution, using historical data for production at the county level.

3.3.3 Generation of scenarios

The development of the scenarios requires a combination of the joint distribution of prices and the crop yields. These scenarios are a discretized representation of the sampling from the distribution of prices and yields for the different crops in the grower's portfolio (Table 2).

TABLE 2 Range of prices sampled for peppers.

	Yield Tom	Yield Pep	Price Tom	Price Pep
Yield Tom	1			
Yield Pep	0.731	1		
Price Tom	−0.085	0.382	1	
Price Pep	0.379	0.503	0.498	1

3.3.4 Results from stochastic model

The main results from the stochastic model (Ahumada et al., 2012) are that, when compared to the deterministic model (Ahumada and Villalobos, 2011a), the recommendations tend to favor planting more hectares of other products, not just tomatoes (in this example only peppers are considered); the stochastic model significantly changes the proportion between peppers and tomatoes, still recommending a larger crop of tomatoes but with more peppers. These results signal crop diversification, since tomato is on average very profitable, but is also subject to large swings in prices from season to season and even within the same year.

The models could then be adjusted to different risk profiles, and the growers could use the recommendations provided to better inform their planting and production decisions; these models could help them to estimate the benefits of contracting versus operating in the open market and also to estimate the proportion of their crops that is worthwhile to contract.

Given these results, one potential extension of this work could consider other crops and even some perennials, as part of a portfolio to be considered by those growers who want to hedge their risk.

3.4 Operational models for revenue maximization

Operational models aid in making production and distribution decisions during the harvest with the objective of maximizing the revenue obtained by producers. Usually, the profitability of producers is highly dependent on the handling of short-term planning in the harvest season (Soto-Silva et al., 2016). Among the main issues that need to be considered in short-term planning are the management of labor, the preservation of the value of perishable crops, and the use of efficient transportation modes that provide the best trade-off between time (quality of products) and cost. These different issues are interrelated, and their management is fundamental for achieving the goals of producers.

Ahumada and Villalobos (2011b) proposed a model to integrate the management of labor, quality, and transportation, while considering the trade-offs of preserving the value of crops and the costs incurred to preserve that value. For example, produce is usually consumed fresh, and consumers demand certain quality attributes from their fresh food, which implies that products must retain their attributes at an acceptable level from the time of harvest up to the time they are purchased by the consumer. This implies that if those quality attributes are conserved, then the value of the crops is preserved. But to preserve those attributes the growers might incur higher costs in terms of labor and transportation (e.g., air transportation).

3.4.1 Problem description

The operational model proposed determines the quantity and timing of harvest, which are dependent on labor costs, the availability of controlled storage facilities, market demand, and distance from the production areas to the destination markets (Csizinszky, 2005). Other decisions include the selection of target markets, the selection of the best transportation mode, storage of the products at different facilities, and labor requirements at the fields and at the packinghouses (Fig. 11).

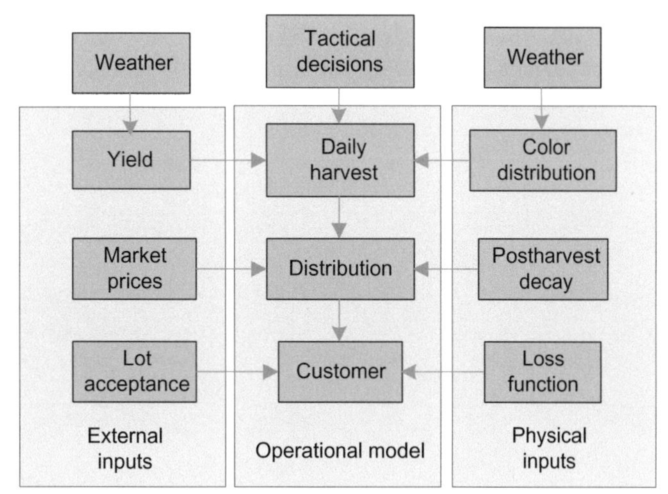

FIG. 11 Planning problem diagram.

3.4.2 Assumptions and planning horizon

The operational model assumed that most tactical decisions, such as production planning and logistics design, have already been made in the strategic and tactical planning phases. With respect to the planning horizon, the look-ahead horizon is related to the shelf life of the crops, since it involves harvest, packing, and distribution decisions. For the case to be discussed, the planning horizon was set to 2 weeks (see Fig. 12).

Mo	Tu	We	Th	Fr	Sa	Su	Mo	Tu	We	Fr	Sa	Su	Mo
1	2	3	4	5	6	7	8	9	10	11	12	13	14

Harvesting
Packing
Storage/transportation
Shelf life

FIG. 12 Planning horizon for the operational model.

3.4.3 Market for the fresh produce industry

Fresh produce has a short shelf life, which prevents the prolonged storage of crops. This limitation forces producers to supply demand with their current production. Those supply restrictions cause the industry to under- and overproduce while aiming to satisfy demand.

One example is the case of the market for fresh tomatoes, where prices increase when supply is low (winter months) and decrease when supply is ample (summer months). These variations in the markets and their effects on the profitability of the growers is what makes short-term decision making increasingly important.

3.4.4 Estimation of fruit quantity and maturity

There are several models that have been used to estimate the maturity and development of tomato and other crops. According to the literature on these models, the most important measures used to sample the quality of fresh produce are grade, shelf life, color, texture, and external appearance (Gary and Tchamitchian, 2001). For the case of tomatoes, two quality attributes are relevant, color and firmness (Schouten et al., 2007b). In particular, color is used as an external index for firmness, shelf life, and overall quality, and it is usually the preferred measure used by produce distribution companies (Tijskens and Evelo, 1994).

Harvesting decisions affect the observed quality and the freshness of the harvested crops. One way to estimate these effects is by predicting the changes in the color of the harvested crops (Hertog et al., 2004; Schouten et al., 2007a), given that the observed colors of the harvested crops provide an estimate of freshness and shelf life.

For a detailed explanation of Hertog's methodology and the data obtained, the reader is advised to consult the cited papers. The results obtained by Hertog et al. (2004) are given in hue angle, which is a color metric obtained from a spectrometer. In the North American produce industry, tomatoes have traditionally been classified into six ripeness stages (Saltveit, 2005): mature green (color 1), breaker (color 2), turning (color 3), pink (color 4), light-red (color 5), and red-ripe (color 6). The compatibility between hue color and the color scheme used in industry is presented in Table 3, Color classification scheme for tomatoes.

TABLE 3 Color classification scheme for tomatoes

Stage of development	Hue	USDA
Mature green	115.00	1
Breaker	83.90	2
Turning	72.85	3
Pink	61.80	4
Light-red	48.00	5
Red-ripe	41.30	6
Over-ripe	37.00	6+

For instance, given a harvest frequency of every 2 days (Fig. 13), it is expected that the fruits collected would have a distribution of colors based on the time spent on the plant, since the last harvest (−1, −2) and the speed of growth induced by the environmental temperature. If the frequency decreases to harvest every 3 days, then the range of color distribution is expected to increase, given that fruits have remained exposed to the temperature for a longer time, as can be observed in the left side of Fig. 13. This implies that the expected distribution of colors at the time of harvest is dependent on the frequency of harvest and the environmental conditions of the days before harvest occurs. One potential effect of this is that the frequency of harvest needs to increase at higher temperatures if a homogeneous coloring of tomatoes and lower levels of color are required.

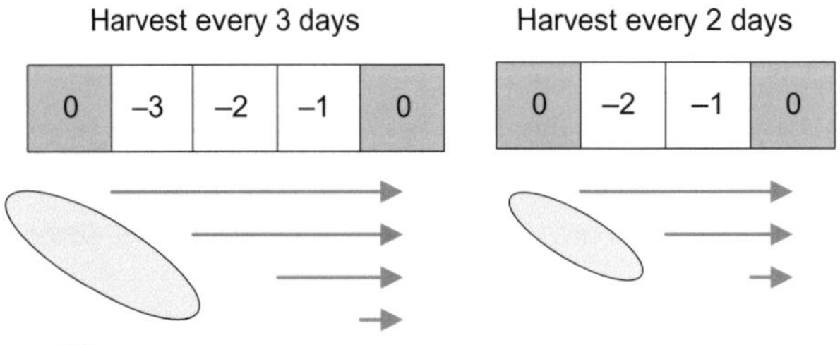

FIG. 13 Maturity of crops based on harvest frequency.

3.4.4.1 Normal distribution applied to maturity distribution

To determine the distribution of colors given a particular harvesting policy when there is limited data available, the authors developed the methodology (Ahumada, 2008) described in the following paragraphs.

The first element of the methodology is finding a function that can represent the effects of weather on development of crops, such as the one presented in the previous section. The second element is the distribution of colors exhibited by the fruits on the plant at the time of harvest. The third element required to estimate the color distribution of fruits is to determine the limits for the different color classes. For determining the class that each one of the crops belong to, it was necessary to set up the ranges used to classify the populations.

Combining these elements, a random function that estimates the distribution of colors for tomatoes was developed. Function (16) estimates the probability that the fruits belong to a certain class of color, since it is assumed the color of the fruits follows a normal distribution (Schouten et al., 2004) and (Hertog, 2002). The technique used to determine the function is the cumulative density function (Bain and Engelhardt, 1987).

$$P(H_b) = \Theta\left[\mu - \frac{\left(\dfrac{H_b H\max - H_b e^{kt(H\min - H\max)} H\min + e^{kt(H\min - H\max)} H\min H\max - H\min H\max}{H_b - He^{kt(H\min - H\max)} + e^{kt(H\min - H\max)} H\max - H\min}\right)}{\sigma}\right] - \Theta\left[\mu - \frac{\left(\dfrac{H_a H\max - H_a e^{kt(H\min - H\max)} H\min + e^{kt(H\min - H\max)} H\min H\max - H\min H\max}{H_a - He^{kt(H\min - H\max)} + e^{kt(H\min - H\max)} H\max - H\min}\right)}{\sigma}\right] \quad (16)$$

where H_b, selected limit for class b; H_a, selected limit for the contiguous class a; μ, mean value of the initial color of fruits at the plant (Hue = 91); σ, standard deviation of the initial color of fruits (17).

Using the parameters from the sample data and a modified version of the normal standard cumulative function, it is possible to determine the probability that the fruits harvested belong to a certain color class by using Eq. (16). This function estimates the density of a normal distribution for each one of the color classes.

The estimated probabilities for the different color classes given by the temperature and the frequency of harvest are presented in Table 4.

TABLE 4 Distribution of colors for tomatoes harvested (normal approximation)

	1st week of January (64°F)				3rd week of March (69°F)		
Days	Crop	Color	%	Days	Crop	Color	%
1	TA	2	0.45	1	TA	2	0.37
1	TA	3	0.34	1	TA	3	0.35
1	TA	4	0.17	1	TA	4	0.22
1	TA	5	0.04	1	TA	5	0.05
2	TA	2	0.34	2	TA	2	0.20
2	TA	3	0.36	2	TA	3	0.32
2	TA	4	0.24	2	TA	4	0.35
2	TA	5	0.06	2	TA	5	0.13
3	TA	2	0.24	3	TA	2	0.10
3	TA	3	0.33	3	TA	3	0.21
3	TA	4	0.32	3	TA	4	0.40
3	TA	5	0.10	3	TA	5	0.29
4	TA	2	0.16	4	TA	2	0.05
4	TA	3	0.28	4	TA	3	0.11
4	TA	4	0.38	4	TA	4	0.33
4	TA	5	0.18	4	TA	5	0.51

3.4.5 Postharvest modeling of perishable crops

The third factor to consider at the operational level is the shelf life of the product. To estimate the decay (or loss of value) of products during storage and distribution, adequate metrics are required for measuring the product quality. A related issue is the desired quality demanded by consumers and their acceptance behavior. Both can be combined in the concept of keeping quality, which has been defined as the time until a commodity becomes unacceptable because certain quality attributes of the crops have deteriorated beyond the acceptance limit of customers (Tijskens and Polderdijk, 1996).

The keeping quality of tomatoes is mainly dependent on the color development of the fruit when it reaches the customers. However, the development of the fruit color, under regular handling and storage circumstances, depends on their initial maturity and the time and temperature during storage (Tijskens and Evelo, 1994). Usually tomatoes are harvested at the mature green or breaker stage (or higher), depending of their market target and the labor costs of the growers. Given these market conditions, growers must handle the trade-offs between what the market demands and the labor costs from a higher harvesting frequency.

To balance the market preferences and the higher cost of labor, the color development of the crops is used. It was assumed that after harvest the fruit is conserved at a constant temperature (e.g., temperature of 59°F) throughout the supply chain. Similar versions of this function have also been used to determine the acceptability of product batches (Hertog et al., 2004; Schouten et al., 2007b), based in the color development of the fruits. Furthermore, the estimation of color development can provide a reasonable approximation of the behavior of tomato fruits through time, which can be used to determine the quality of tomatoes across the distribution chain.

An example of the estimation of color development is presented in Fig. 14, which shows the ripening process for a fruit harvested at the breaker stage and stored at a constant temperature of 59°F. According to this model, it will take around 15 days for the fruit to reach the light-red stage, which is the common ripeness found at the retail market.

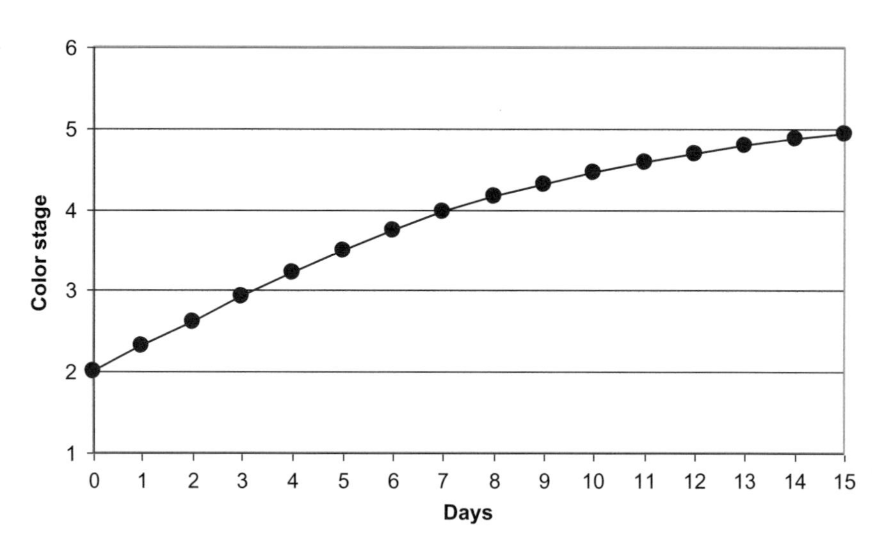

FIG. 14 Postharvest ripening of tomatoes (temperature 59°F).

Based on the ripeness model and the initial maturity of the fruits, it is possible to determine the postharvest ripeness curve for each batch of fruits based on the color selection at packing. The probability that some of the fruits have reached a certain color level after some days of storage is:

$$F(H)=\frac{\left(-\dfrac{H\cdot H\max-H\cdot e^{kt(H\min-H\max)}H\min+e^{kt(H\min-H\max)}H\min\cdot H\max-H\min\cdot H\max}{\mathrm{H}-\mathrm{H}\cdot \mathrm{e}^{\mathrm{kt(Hmin-Hmax)}}+\mathrm{e}^{\mathrm{kt(Hmin-Hmax)}}\cdot H\max-H\min}-a\right)^2}{(b-a)(c-a)} \tag{17}$$

$$F(H)=1-\frac{\left(b+\dfrac{H\cdot H\max-H\cdot e^{kt(H\min-H\max)}H\min+e^{kt(H\min-H\max)}\cdot H\min\cdot H\max-H\min\cdot H\max}{H-H\cdot e^{kt(H\min-H\max)}+e^{kt(H\min-H\max)}\cdot H\max-H\min}\right)^2}{(b-a)(b-c)}$$

where a, lower limit of the triangular distribution at the time of harvest; b, upper limit of the triangular distribution at the time of harvest; c, mode of the triangular distribution at the time of harvest; H, the value at which we want to estimate the distribution $F(H) = P(x \leq H)$.

The values for parameters a, b, and c are given by the ranges from the classes of tomato color (Fig. 14); in the case of tomatoes picked at the breaker stage, the value of $a = 90$, $b = 78$ and $c = 84$. With (17) and the values for the population (a, b, and c), the probability that the lot of tomatoes at the breaker stage (color 2) reaches color H after t days in storage at a temperature T can be estimated; for further details the reader is advised to review (Ahumada, 2008).

3.4.6 Operational planning model

This section provides a detailed description of the operational model designed for the harvesting and distribution of fresh horticultural products. As mentioned before, one of the main decisions of this model include which products to harvest, their frequency (times per week), and the days they should be harvested. These decisions should consider time and labor restrictions, and the effects of harvesting decisions on the quality of the products. In this model we propose to address these decisions by using predetermined patterns for harvesting a plot; for example, a plot could be harvested daily, three times per week, two times per week, etc. Even when there could be many different patterns, only a finite number of these patterns could be used for commercial purposes for the proposed planning horizon in harvest decisions. Moreover, once a pattern is selected, it could be used for several weeks, depending on the market and environmental factors that influence the harvesting decisions.

The potential patterns to select along the harvest horizon are presented in Table 5.

Table 5 shows the days selected for harvest with a changing frequency of harvest between 1 and 4 days. For example, pattern II has a frequency of three times per week (harvest every 2 days) and pattern III has a frequency of two times per week. There are two patterns of type II, three of type III, and four of type IV, which cover all of the 12 harvesting periods in the planning horizon.

TABLE 5 Patterns for harvesting a crop

	Day											
	1	2	3	4	5	6	7	8	9	10	11	12
Pattern	Mo	Tu	We	Th	Fr	Sa	Mo	Tu	We	Th	Fr	Sa
I	1	1	1	1	1	1	1	1	1	1	1	1
II	1	0	1	0	1	0	1	0	1	0	1	0
III	1	0	0	1	0	0	1	0	0	1	0	0
IV	1	0	0	0	1	0	0	0	1	0	0	0

One assumption made for the use of patterns is that more than one pattern can be used for harvesting a plot for a given planning horizon. For example, half the plot can be harvested with pattern II starting at day 1 and the other half can be harvested with the same frequency as pattern I, but starting at day 2. Because of this formulation, one part of a plot can be harvested in day 1 and the other part in day 2. Through this discussion it is evident that the pattern followed to harvest crops is relevant for the optimization of the operational model. The developed operational model, together with the details of the mathematical formulation for the model, are explained in more detail next.

Indices and sets

$t \in T$	Planning periods (days)
$h \in TH(j, l) \subseteq T$	Set of feasible harvesting days for crop j in location l
$p \in P$	Plot formed by the area in location l planted with crop j
$j \in J$	Different crops

$k \in K(j)$	Products of crop j (package, grade)
$C(k)$	Compatible products for storage and transportation in the same container
$w \in W$	Warehouses available for storage
$i \in I$	Customers
$d \in D$	Distribution centers
$f \in PF$	Packaging facilities
$r \in TM$	Transportation mode
$q \in Q$	Quality of products (color)
$v \in V(j)$	Harvesting patterns for crop j

Parameters

AP_p	Total area planted in plot p (hectares)
EH_{hpv}	Expected harvest at period h of plot p harvested by pattern v (boxes)
SH_{hv}	If pattern v is required to harvest in period h
VQ_{vjq}	Percentage of crop j with quality q in by pattern v
VG_{hpk}	Percentage of product k from plot p at period h
VS_{hj}	Percentage of the crop j that is salvaged at period h (fraction)
LAH_h	Labor available for harvesting in period h (in hours)
LRH_j	Boxes of crop j harvested per hour
LBH_j	Labor hours required to cover 1 ha. of crop j
SL_{kq}	Shelf life of product k with quality q
KTC	Capacity of container in pallets
KTW	Capacity of container in weight
KP_f	Capacity of plant f (boxes per period)
DW_{tki}	Expected demand from customer i of product k in time t
DM_{tk}	Expected demand (open market) of product k in period t
Ti_{fir}	Time from packing facility f to customer i by transportation mode r
TiW_{wir}	Time from warehouse w to customer i by transportation mode r
TiD_{dir}	Time from DC d to customer i by transportation mode r
$TiPW_{fwr}$	Time from packing facility f to warehouse w by transportation mode r
$TiPD_{fdr}$	Time from packing facility f to DC d by transportation mode r
$TiWD_{wdr}$	Time from warehouse w to DC d by transportation mode r
RW_k	Amount of crop weight per box of product k
RC_k	Number of boxes of product k in a pallet
COL_{t-hkq}	Expected color product k with initial color q after n days of harvest ($n = t - h$)
MOP	Extra man-hours available from day laborers
$PROB_{t-hkq}$	Estimate of the probability that the product with color q is not accepted by customers based on the time elapsed ($n = t - h$)

Cost parameters

PN_{tk}	Price per product k on period t in the open market
PC_{tki}	Price per product k on period t sold to customer i
PS_j	Salvage price of crop j
CF_j	Fixed cost per box of crop j
CH_j	Cost of harvesting a box of crop j
CK_k	Cost of packing a box of product k
CI_w	Cost of inventory at warehouse w per pallet of product (pallet/day)
CID_d	Cost of inventory at DC d per pallet of product (pallet/day)
CT_{fir}	Cost of transportation from facility f to customer i by mode r
CTW_{wir}	Cost of transportation from warehouse w to customer i by mode r
CTD_{dir}	Cost of transportation from DC d to customer i by mode r
$CTPW_{fwr}$	Cost from packing facility f to warehouse w by mode r
$CTPD_{fdr}$	Cost of transportation from facility f to DC d by mode r
$CTWD_{wdr}$	Cost of transportation from warehouse w to DC d by mode r
$Clabor$	Cost of an hour of labor at the field

Decision variables

X_{pv}	Area of plot p harvested using pattern v
QH_{hpq}	Harvest (boxes) of quality q from plot p in period h
SP_{hpqf}	Quantity of crop with quality q to ship from plot p to facility f in period h
QS_{hj}	Quantity salvaged of crop j in harvesting period h
QP_{hkqf}	Quantity of product k with quality q packed at facility f in period h
SC_{tkqfir}	Product k of quality q shipped from facility f to customer i in period t by mode r
$SPD_{htkqfdr}$	Product k of quality q harvested at h shipped from facility f to DC d in period t by mode r
$SPW_{htkqfwr}$	Product k of quality q harvested at h shipped from facility f to warehouse w in period t by mode r
$SD_{htkqdir}$	Product k of quality q harvested at h shipped from DC d to customer i in period t by mode r
$SWD_{htkqwdr}$	Product k of quality q harvested at h shipped from warehouse w to DC d in period t by mode r
$SW_{htkqwir}$	Product k of quality q harvested at h shipped from warehouse w to customer i in period t by mode r
SWO_{htkqw}	Product k with quality q harvested at h sold from warehouse w in period t
$Invw_{htkqw}$	Inventory of product k at period t with quality q in warehouse w harvested at h
$Invd_{htkqd}$	Inventory of product k at period t with quality q in DC d harvested at h
Z_{tkw}	Quantity to purchase of product k, in period t for warehouse w
Opl_h	Operator hours hired in the field at time h
NTI_{tfir}	Number of containers sent to customer i from facility f in period t by mode r
NTD_{tdir}	Number of containers sent to customer i from DC d in period t by mode r
NTW_{twir}	Number of containers to customer i from warehouse w in period t by mode r

NTP_{tfwr}	Number of trucks sent from facility f to warehouse w in period t by mode r
NTK_{tfdr}	Number of containers sent to DC d from facility f in time t by mode r
NTC_{twdr}	Number of containers sent to DC d from warehouse w in time t by mode r

where X_{pv}, QH_{pq}, SP_{hpqf}, QP_{hkqf}, SC_{tkqfir}, $SPD_{htkqfdr}$, $SW_{htkqwir}$, $SPW_{htkqfwr}$, $SWD_{htkqwdr}$, $SD_{htkqdir}$, $SWO_{htqkwir}$, $Invw_{htkqw}$, $Invd_{htkqd}$, Z_{tkw}, $Opl_h \in R^+$ and NTI_{frir}, NTD_{tdir}, NTW_{twir}, NTP_{tfwr}, NTK_{tfdr}, $NTC_{twdr} \in Z^+$.

$$\begin{aligned}
\max \sum_{tki} PC_{tki}\cdot\left(\sum_{qfr} SC_{tkqfir} + \sum_{hqwr} SW_{htkqwir} + \sum_{hqdr} SD_{htkqdir}\right) + \sum_{tk} PN_{tk}\cdot\left(\sum_{hqw} SWO_{htkqw}\right) \\
-\sum_{hpqf} CK_k\cdot QP_{hpqf} - \sum_{hqp\in P(j)} \left(CH_j + CF_j\right)\cdot QH_{hpq} - \sum_{tfir} NTI_{tfir} CT_{fir} - \sum_{twir} NTW_{twir} CTW_{wir} \\
-\sum_{tdir} NTD_{tdir} CTD_{dir} - \sum_{tfwr} NTP_{tfwr} CTPW_{fwr} - \sum_{tfdr} NTK_{twir} CTPD_{fdr} - \sum_{twdr} NTC_{twdr} CTWD_{wdr} \\
-\sum_{tkw} Invw_{tkw} CI_w - \sum_{tkd} Invd_{tkd} CID_d - \sum_{tkw} Z_{tkw}\cdot PN_{tk} - SP_{hpf} CSP - \sum X_{pv} LBH_j Clabor \\
-\sum_{tkqfir} SC_{tkqfir} PC_{tki} PROB_{tkq} - \sum_{htkqwir} SW_{htkqwir} PC_{tki} PROB_{t-hkq} - \sum_{htkqdir} SD_{htkqdir} PC_{tki} PROB_{t-hkq} \\
-\frac{1}{8}\sum_{tkqfir} SC_{tkqfir} PC_{tki} COL_{tkq} - \frac{1}{8}\sum_{htkqwir} SW_{htkqwir} PC_{tki} COL_{t-hkq} - \frac{1}{8}\sum_{htkqdir} SD_{htkqdir} PC_{tki} COL_{t-hkq} \\
-\frac{1}{8}\sum_{htkqwr} SWO_{htkqwr} PN_{tk} COL_{t-hkq} - \frac{1}{8}\sum_{htkqwr} SWO_{htkqwr} PN_{tk} COL_{t-hkq} + \sum_{hj} PS_j QS_{hj}
\end{aligned} \tag{18}$$

The objective function (18) aims at maximizing the revenue obtained from the products by selling to the most profitable customer up to their maximum demand (DW) and selling the remainder of products in the open market (SWO) when it is profitable to do so. The main costs include the cost of growing the crop (CFj), transportation (e.g., CTfir), harvesting (CHj), and packing costs (CKk) for delivering the product to the customers. The benefits obtained include the price paid by the customers and the revenue from the open market. If it is profitable to sell, then the harvesting, packing, and delivery decisions will be put into effect; otherwise the model will not advise to sell in the open market. Also included in the objective function is the expected cost from rejected shipments. This expected cost is determined by the probability (PROB) that the crop's color passes a determined threshold predetermined by the customer. An alternative is penalizing the deterioration of the products along the chain by estimating the change in color (COL) from the time of harvest to the time it reaches the customer. The function shown in the formulation penalizes with half the value of the products if the color reaches the higher level of maturity ($(6-2)/8=0.5$). However, this function can be modified according to the preferred level of freshness required in the supply chain.

Constraints

$$QH_{hpq} = \sum_v X_{pv} EH_{hpv} VQ_{vjq} \qquad \text{For every } h,p,q \tag{19}$$

$$QS_{hj} = \sum_v X_{pv} EH_{hpv} VS_{hj} \qquad \text{For every } h,\ j \text{ where } j\in P(j) \tag{20}$$

$$\sum p\left(LRH_j\cdot QH_{hpq} + X_{pv} SH_{hv} LBH_j\right) \le LAH_h + OPl_h \qquad \text{For every } h \text{ where } j\in P(j) \tag{21}$$

$$\sum f\left(SP_{hpqf}\right) = QH_{hpq} \qquad \text{For every } h,p,q \tag{22}$$

$$QP_{hkqf} = \sum_{p} VG_{hpk} \cdot SP_{hpqf} \qquad \text{where } k \in K(p) \text{ for each } h,p,q,f \tag{23}$$

$$\sum_{k} QP_{hkqf} \leq KP_f \qquad \text{For every } h,f \tag{24}$$

$$\sum_{v} X_{pv} = AP_p \qquad \text{For every } p \tag{25}$$

$$OPl_h \leq MOP \qquad \text{For every } h \tag{26}$$

The amount harvested (*QH*) in (19) is dependent on the plots harvested according to pattern *v* (*X*), the expected production (*EH*), and the expected color distribution of the crops (*VQ*). The amount harvested in (21) is also restricted by the available labor and the required labor for each box of crop *j*. The amount shipped to each packing facility is limited by the amount harvested (22). The production (*QP*) at the packaging plant in (23) is dependent on the pounds per crop required by each package and their estimated grade (*VG*), since products are formed by the combination of the type of crop, grade, and color. This production is limited by the capacity of the production line (24). In (25) the quantity of crops to harvest by pattern selected is limited to the number of hectares in that plot (*AP*). There is also a maximum number of day laborers available in (26).

$$\sum_{fir} SC_{t_1kqfir} + \sum_{tfwr} SPW_{ht_2kqfwr} + \sum_{tfdr} SPD_{ht_3kqfdr} = QP_{hkqf} \quad \text{for every } k,\ h,\ q,\ f \text{ where}$$

$$t_1 = h + Ti_{fir},\ t_2 = h + TiPW_{fwr},\ t_3 = h + TiPD_{fdr} \tag{27}$$

$$Invw_{htkqw} = Invw_{ht-1kqwr} + \sum_{f} SPW_{htkqfwr} - \sum_{i} SW_{ht_4kqwir} - \sum_{d} SWD_{ht_5kqwdr} + \sum_{f} SWO_{htkqw} + Z_{tkw}$$

$$\text{For all } t \geq h,k,w,q \qquad \text{where } t_4 = t + TiW_{wir},\ t_5 = t + TiWD_{wdr} \tag{28}$$

$$Invd_{htkqd} = Invd_{ht-1kqdr} + \sum_{f} SPD_{htkqfdr} - \sum_{i} SD_{ht_6kqdir} + \sum_{w} SWD_{htkqwdr}$$

$$\text{For all } t \geq h,\ k,\ d,\ q \qquad \text{where } t_6 = t + TiD_{dir} \tag{29}$$

$$\sum_{fr} SC_{tkqfir} + \sum_{hwr} SW_{htkqwir} + \sum_{hdr} SD_{htkqdir} = DW_{tki} \quad \text{all } t,\ k,\ i, \text{where } t - SL_{kq} \leq h \leq t \tag{30}$$

$$\sum_{hw} SWO_{htkqw} \leq DM_{tk} \quad \text{all } t,\ k, \text{where } t - SL_{kq} \leq h \leq t \tag{31}$$

In (27) the quantity produced is transported from the packinghouse directly to customers (*SC*), to warehouses (*SPW*), or to DCs (*SPD*). The products available at the warehouses depend on the sales of the past period and the arrivals (28); as explained before there could be contracted sales (*SW*) and open market sales (*SWO*) at the warehouses. The amount in hand at the DC at (2) also depends on arrivals from the packinghouse (*SPD*) and the warehouse *SWD* and the sales (*SD*). The shelf life of the product is being considered by restricting the shipping decisions to those products that can withstand the time required for transportation (30). The contracted sales (*SW*, *SD*, and *SC*) are satisfied with products that are within the allowed shelf life based on the time of harvest (2). The open market sales, on the other hand, are not restricted to have a minimum volume, just the maximum allowed by the market (*DM*), but still have to satisfy shelf life constraints (31). Finally, constraints (32)–(37) assure that the shipments sent between the facilities of the company and to the customers do not exceed the allowed capacity of the trucks in terms of weight and pallets.

$$\sum_{k} \text{Max}\left\{ \left\lceil \frac{RW_k \cdot SC_{tkfir}}{KTW} \right\rceil, \left\lceil \frac{RC_k \cdot SC_{tkfir}}{KTC} \right\rceil \right\} = NTI_{tfir} \quad \text{for every } t,f,i,r \text{ where } k \in C(k) \tag{32}$$

$$\sum_{k} \text{Max}\left\{ \left\lceil \frac{RW_k \cdot SD_{htkqdir}}{KTW} \right\rceil, \left\lceil \frac{RC_k \cdot SD_{htkqdir}}{KTC} \right\rceil \right\} = NTD_{tdir} \quad \text{for every } t,\ d,\ i,\ r \tag{33}$$

where $k \in C(k)$.

$$\sum_{k} \text{Max}\left\{ \left\lceil \frac{RW_k \cdot SW_{htkqwir}}{KTW} \right\rceil, \left\lceil \frac{RC_k \cdot SW_{htkqwir}}{KTC} \right\rceil \right\} = NTW_{twir} \quad \text{for every } t,\ w,\ i,\ r \tag{34}$$

where $k \in C(k)$.

$$\sum_k \text{Max}\left\{\left\lceil\frac{RW_k \cdot SPW_{htkqfwr}}{KTW}\right\rceil, \left\lceil\frac{RC_k \cdot SPW_{htkqfwr}}{KTC}\right\rceil\right\} = NTP_{tfwr} \quad \text{for every } t, f, w, r \tag{35}$$

where $k \in C(k)$.

$$\sum_k \text{Max}\left\{\left\lceil\frac{RW_k \cdot SPD_{htkqfdr}}{KTW}\right\rceil, \left\lceil\frac{RC_k \cdot SPD_{htkqfdr}}{KTC}\right\rceil\right\} = NTK_{tfdr} \quad \text{for every } t, f, d, r \tag{36}$$

where $k \in C(k)$.

$$\sum_k \text{Max}\left\{\left\lceil\frac{RW_k \cdot SWD_{htkqwdr}}{KTW}\right\rceil, \left\lceil\frac{RC_k \cdot SWD_{htkqwdr}}{KTC}\right\rceil\right\} = NTC_{twdr} \quad \text{for every } t, w, d, r \tag{37}$$

where $k \in C(k)$.

The resulting model formulation is a mixed integer problem (MIP), where most of the variables are continuous and only the variables that represent the number of shipments are integers. For a detailed presentation of the case study and the results, the reader is advised to consult (Ahumada and Villalobos, 2011b).

4 Conclusions and future research

In this chapter we have presented some examples of tactical and operational planning models for perishable agricultural products, in particular for annual fresh fruits and vegetables. These models are applicable to the existing fresh supply chain. However, the US and global markets for fresh fruits and vegetables are going through a transformation that will affect the underlying supply chains of this industry. This transformation will undoubtedly result in the redefinition of the roles of the current players of the supply chain and the level of control they maintain in their particular echelons.

In particular, sensors, real-time market information, on-line demand, and the automated replenishment food systems being developed will undoubtedly revolutionize the current food supply chains. Switching to these supply chains using new information to efficiently match supply and demand will especially benefit the market for fresh fruits and vegetables, which are most likely to perish under current push-style production systems. Thus, it is urgent to identify, develop, and deploy the new fresh supply chains and the underlying planning models that will emerge under the current technological and market conditions. A challenge that the community of researchers will have to address in the near future.

References

Ahumada, Omar. 2008. "Models for planning the supply chain of agricultural perishable products." Ph. D. Dissertation, Arizona State University.

Ahumada, O., Villalobos, J.R., 2011a. A tactical model for planning the production and distribution of fresh produce. Ann. Oper. Res. 190 (1), 339–358.

Ahumada, O., Villalobos, J.R., 2011b. Operational model for planning the harvest and distribution of perishable agricultural products. Int. J. Prod. Econ. 133 (2), 677–687.

Ahumada, O., Rene Villalobos, J., Nicholas Mason, A., 2012. Tactical planning of the production and distribution of fresh agricultural products under uncertainty. Agr. Syst. 112, 17–26.

Bain, L.J., Engelhardt, M., 1987. Introduction to Probability and Mathematical Statistics. Brooks/Cole.

Birge, John, and François Louveaux. 1997. "Introduction to Stochastic Programming." Series in Operations Research, Vol. 421. Springer. https://doi.org/10.1057/palgrave.jors.2600031.

Csizinszky, A.A., 2005. Production in the Open Field. Crop Production Science in Horticulture 13, p. 237. Cab International.

Gary, C., Tchamitchian, M., 2001. Modelling and management of fruit production: the case of tomatoes. In: Food Process Modelling. Elsevier, pp. 201–229.

Gustavsson, J, C Cederberg, U Sonesson, R van Otterdijk, and A Meybeck. 2011. Global food losses and food waste: extent, causes and prevention. International Congress: Save Food!. https://doi.org/10.1098/rstb.2010.0126.

Hertog, M.L.A.T.M., 2002. The impact of biological variation on postharvest population dynamics. Postharvest Biol. Technol. 26 (3), 253–263. https://doi.org/10.1016/S0925-5214(02)00044-3.

Hertog, M.L.A.T.M., Lammertyn, J., Desmet, M., Scheerlinck, N., Nicolaï, B.M., 2004. The impact of biological variation on postharvest behaviour of tomato fruit. Postharvest Biol. Technol. 34 (3), Elsevier: 271–284.

Huang, Sophia. 2004. Global Trade Patterns in Fruits and Vegetables. USDA-ERS Agriculture and Trade Report No. WRS-04-06.

IMechE, 2013. Global food waste not, want not. Institute of Mechanical Engineers, p. 31. http://www.imeche.org/docs/default-source/reports/Global_Food_Report.pdf?sfvrsn=0.

Kaut, Michal, and Stein W Wallace. 2003. Evaluation of Scenario-Generation Methods for Stochastic Programming. Humboldt—Universität zu Berlin, Mathematisch-Naturwissenschaftliche Fakultät II, Institut für Mathematik.

Lowe, T.J., Preckel, P.V., 2004. Decision technologies for agribusiness problems: a brief review of selected literature and a call for research. Manuf. Serv. Operat. Manag. 6 (3), 201–208.

Norwood, B., Roberts, M.C., Lusk, J.L., 2004. Ranking crop yield models using out-of-sample likelihood functions. Am. J. Agric. Econ. 86 (4), 1032–1043. https://doi.org/10.1111/j.0002-9092.2004.00651.x.

Rae, A.N., 1977. Crop Management Economics. Crosby Lockwood Staples.

Saltveit, M.E., 2005. Fruit Ripening and Fruit Quality. Crop Production Science in Horticulture 13, Cab International, p. 145.

Schouten, R.E., Jongbloed, G., Tijskens, L.M.M., Van Kooten, O., 2004. Batch variability and cultivar keeping quality of cucumber. Postharvest Biol. Technol. 32 (3), 299–310. https://doi.org/10.1016/j.postharvbio.2003.12.005.

Schouten, Rob E, Tanja P M Huijben, L M M Tijskens, and Olaf van Kooten. 2007a. "Modelling quality attributes of truss tomatoes: linking colour and firmness maturity." Postharvest Biol. Technol. 45 (3). Elsevier:298–306.

Schouten, R.E., Huijben, T.P.M., Tijskens, L.M.M., van Kooten, O., 2007b. Modelling the acceptance period of truss tomato batches. Postharvest Biol. Technol. 45 (3), Elsevier: 307–316.

Soto-Silva, W.E., Nadal-Roig, E., González-Araya, M.C., Pla-Aragones, L.M., 2016. Operational research models applied to the fresh fruit supply chain. Eur. J. Oper. Res. 251 (2), 345–355. https://doi.org/10.1016/j.ejor.2015.08.046.

Tijskens, L M M, and R G Evelo. 1994. "Modelling colour of tomatoes during postharvest storage." Postharvest Biol. Technol. 4 (1–2). Elsevier:85–98.

Tijskens, L.M.M., Polderdijk, J.J., 1996. A generic model for keeping quality of vegetable produce during storage and distribution. Agr. Syst. 51 (4), 431–452. https://doi.org/10.1016/0308-521X(95)00058-D.

USDA, 2006. Fresh Tomatoes: U.S. Monthly grower price 1960, 2003. U.S. Tomato Statistics, Economic Research Service.

USDA, 2007. Custom Report. U.S. Department of Agriculture. Market News. Available from: http://marketnews.usda.gov/portal/fv (Retrieved October, 15, 2007).

van Wart, J., Christian Kersebaum, K., Peng, S., Milner, M., Cassman, K.G., 2013. Estimating crop yield potential at regional to national scales. Field Crop Res 143, 34–43. https://doi.org/10.1016/j.fcr.2012.11.018.

Chapter 23

Modeling by-products and waste management in the meat industry

Riccardo Accorsi, Marco Bortolini and Andrea Gallo
Department of Industrial Engineering, Alma Mater Studiorum—University of Bologna, Bologna, Italy

Abstract

As currently designed, food supply chains (FSCs) still provide room for developing new business around the production, distribution, and delivery of primary food. Specifically, beyond primary production some FSCs provide an opportunity to design, plan, and optimize new chains based on the valorization of by-products and waste. A concrete example of such opportunities lies in the exploitation and distribution of by-products from the meat industry.

Traditionally, different high-value meat cuts are distributed to retailers and consumers, but less attention is paid to by-products resulting from the slaughtering processes that generate waste and cost instead of being valorized as raw materials for other transformation industries. In order to exploit such business and to reduce the amount of waste associated with the meat industry, new paths can be followed according to the circular economy paradigm. Particularly, new storage, distribution, collection, that is, logistics operations and processes, must be designed and optimized.

This chapter provides a high-level analytic model to design optimal networks between the primary meat chain and other supply chains using by-products. Some potential valorization chains are identified and tested through a sensitivity analysis, which highlights the impact of the logistic distance between chains in the pursuit of a circular economy model.

1 Introduction and background

While in Western countries meat demand has not increased over the last 50 years (e.g., per capita consumption of 75 kg per year in the United States and 45 kg in European countries), it has doubled in China (up to 40 kg per capita per year), following a trend expected to further accelerate in the near future (Allievi et al., *2015*). The current global meat demand is about 106 million tons of poultry, 115 million tons of pork, and 70 million tons of beef (FAO, *2015*). Among these, the production of beef is undoubtedly the highest environmental stressor. Different concerns contribute to the environmental impact associated with this industry (Elferink et al., *2008*). In addition to the energy, water, and land consumption (Virtanen et al., *2011*), beef processing results in large flows of waste and by-products (Accorsi et al., *2017a,b*).

The current beef supply chains are designed and optimized to provide high quality, safe, and affordable food to primary consumers along the forward logistic flows (Soysal et al., *2014*). However, they provide room for developing new businesses around the primary production and forward distribution chains (Jayathilakan et al., *2012*). Specifically, beyond primary production, these supply chains give the opportunity to design, plan, and optimize new chains based on the valorization of by-products and waste.

This chapter aims at developing methods to investigate the feasibility of by-product and waste valorization chains resulting from the primary production of meat products. To this purpose, a strategic optimization model for the design of such networks is formulated. The environment of application of both the method and the model is the beef industry.

One of the main factors influencing decisions in the meat industry is the so-called *slaughtering yield*, defined as the ratio between the weight of the carcass and the cattle weight (Bhaskar et al., *2007*). The carcass is deprived of its head, limbs,

Sustainable Food Supply Chains. https://doi.org/10.1016/B978-0-12-813411-5.00023-5

intestines, liver and spleen, heart sack, and organs. The slaughtering yield results on average 65% for females and 68% for males. During production the bones, fats, and muscles are removed, so that just 225 kg of meat are sold out of the 450 kg of the average animal's weight (Fig. 1).

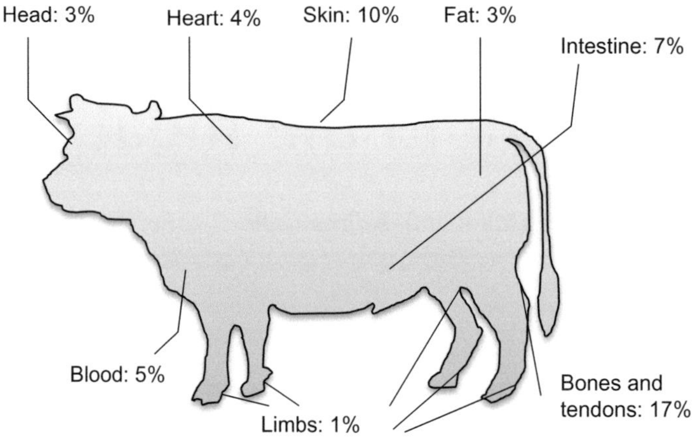

FIG. 1 For cattle, fractions after slaughtering.

As the annual per-capita consumption of beef is between 20 and 30 kg, roughly the same amount of organic material is wasted as a result of the slaughtering process, so that recent statistics account for 1.2 million tons of organic waste produced by the Italian meat industry with significant economic and environmental impacts (Bustillo-Lecompte and Mehrvar, *2017*).

For these reasons, the design of new processes and operations intended to close the loop of waste and valorize secondary materials as by-products becomes of interest for both practitioners and scholars. Meat industry by-products have many potential applications: from biomedical, to healthcare, from cosmetics to fertilizer, or fashion and clothing. To exploit such opportunities and to connect with these supply chains, this chapter presents a linear programming (LP) model for the management of the collection and distribution operations between a slaughterhouse and a set of facilities involved in the transformation and processing of meat by-products.

According to the chapter goals, the remainder of this chapter is organized as follows: Section 2 provides a conceptual framework of the logistics entities involved in the problem; Section 3 formulates the model, while Section 4 presents a numerical application and discusses the results; and Section 5 concludes the chapter with suggestions for future research applications and developments.

2 Problem framework

For the design of collection and distribution networks for meat by-products, the following logistics nodes are identified:

- *Meat production facilities*: these are the *slaughterhouses* (Set P) where cattle from the farms are killed and slaughtered. The whole production of meat cuts is carried out at these nodes.
- *Treatment facilities*: these are the facilities (Set T) involved in the transformation of secondary materials received from the slaughterhouses.
- *Distribution centers*: these nodes (Set S) represent the customers for the by-products market.
- *Disposal facilities*: these are the nodes (Set D) where the waste from the secondary transformation is collected because it is no longer revenue-generating.

The flows of primary meat cuts from the slaughterhouse P to the primary distribution chain (e.g., retailers) are not considered, because they are not differential in our problem, whose boundaries are illustrated in Fig. 2. Some other assumptions are considered and explained in Table 1, categorized by actor.

On the other side, the integration of new supply chains able to transform waste into secondary materials complicates the network and generates new logistics operations: the transport of waste to the treatment facilities, the transport of residual waste from the treatment facilities to the disposal centers, the transport of by-products from the treatment facilities to the distribution centers, as well as the production and transformations at the treatment facilities. Each of these operations generates new costs and the overall profit of the network should be carefully considered and optimized to sustain the development of the circular economy. The following LP formulation is intended to identify the trade-offs between by-product production and waste disposal that maximize the network profit.

TABLE 1 General model's assumptions

Entity	Assumption Description
Meat production facility	– Flows of primary meat products not considered. – Supply of waste to by-products calculated per type and facility as the percentage of waste of that type multiplied by the weight of the cattle slaughtered by day. – As a consequence the waste to be managed will vary linearly with the number of slaughtered cattle. – Per each production facility, one disposal center (i.e., the closest) is considered. Hence the parameter $dist_{pd}$, which indicates the distance between the nodes *p* and *d*, is greater than 0 just for one combination of *p* and *d*.
Treatment facility	– The treatment facility receives waste flows from the production facility as secondary materials. The more organic waste shipped to the treatment facilities, the less the waste to dispose will be. – Treatment facilities are a business opportunity for the meat producers, which can ship waste instead of paying for its disposal. – Per each treatment facility, one disposal center (i.e., the closest) is considered. Hence, the parameter $dist_{td}$, which indicates the distance between the nodes *t* and *d*, is greater than 0 just for one combination of *t* and *d*.

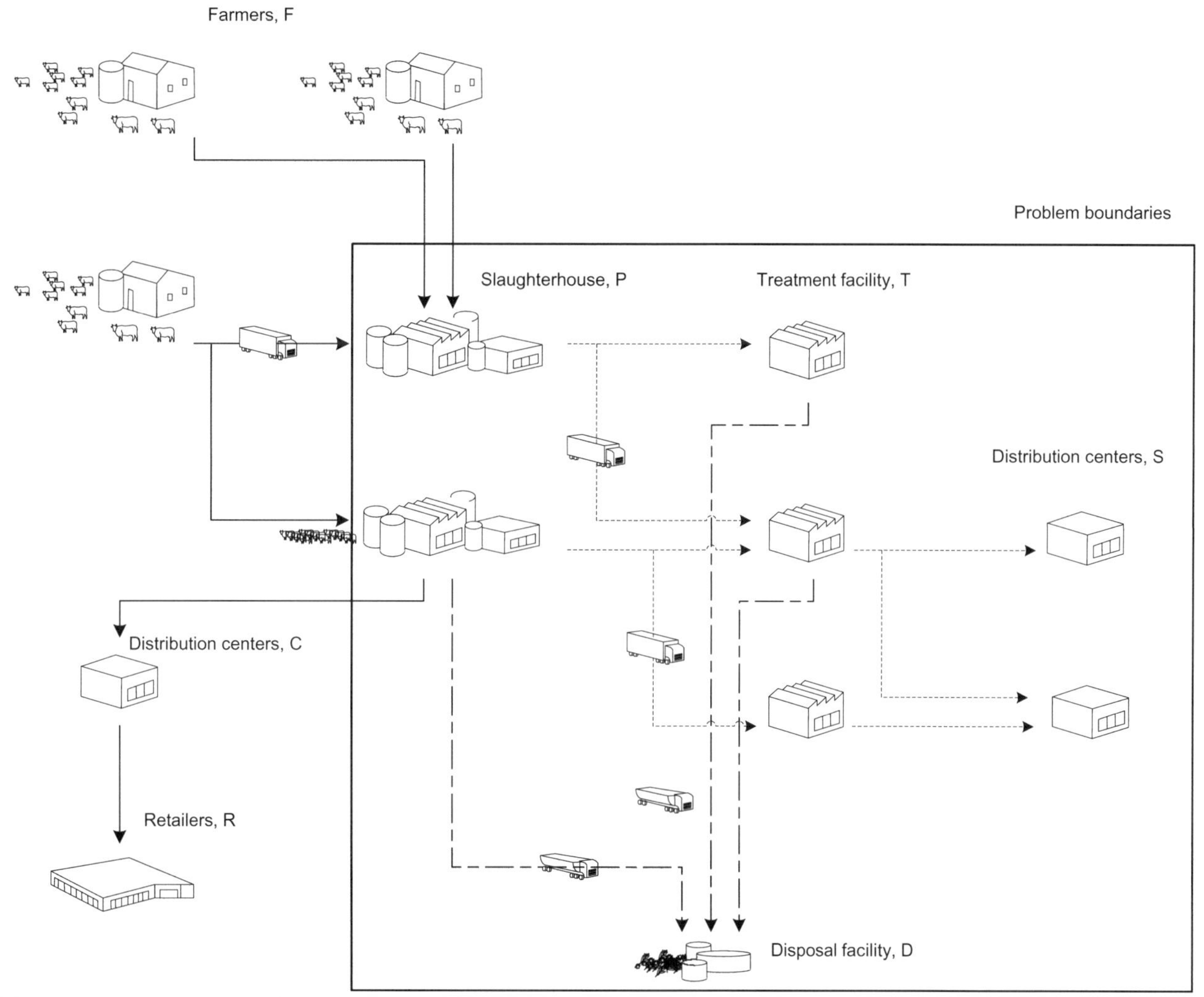

FIG. 2 Considered network, entities, and boundaries.

3 Model

Given the set of entities and network flows illustrated in Fig. 2, the LP model is formulated through sets, parameters, decision variables, objective function, and linear constraints as follows:

Sets and indices

$b \in B$	Set of treated products (i.e., by-products)
$c_b \in C_b$	Set of secondary materials to be treated to produce the by-product b
$p \in P$	Set of slaughterhouses or production facilities
$t \in T$	Set of treatment facilities
$s \in S$	Set of distribution centers
$d \in D$	Set of disposal facilities
$\tau \in \Psi$	Set of periods (e.g., weeks)

Parameters

cf_t	Fixed costs for treatment facility t (€/period)
cd_{bt}	Variable cost for disposal of by-product b at facility t (€/kg)
$cd_{c_b p}$	Variable cost for disposal of secondary material c_b at facility t (€/kg)
cv_{tb}	Variable treatment cost for each by-product b at facility t (€/kg)
pr_b	Price of by-product b (€)
$tcap_{tb}$	Treatment capacity of by-product b by facility t
$dem_{sb\tau}$	Demand of by-product b by distribution center s during period τ (kg)
$cattle_{p\tau}$	Slaughtered cattle at plant p during period τ [no. of animals]
w_{cattle}	Average weight of cattle
w_b	Weight of one unit of by-product b (kg/unit)
w_{c_b}	Weight of one unit of secondary material c_b (kg/unit)
cut_{c_b}	Fraction of cattle generating secondary material c_b (%)
wst_b	Waste fraction for by-product b (%)
uf_{bc}	Usage factor of component c_b on by-product b (kg/unit)
ctr	Transportation cost (€/kg km)
α_{tr}	Incremental factor for transportation costs
$dist_{ab}$	Distance between the generic nodes "a" and "b"

Decision variables

$x_{c_b pt\tau}$	Product flow of secondary material c from producer p to treatment facility t during period τ [kilogram/period]
$x_{bts\tau}$	Product flow of by-product b from facility t to distributor s during period τ [unit/period]

$$\max \sum_{b\in B}\sum_{t\in T}\sum_{s\in S}\sum_{\tau\in\Psi} x_{bts\tau}\cdot pr_b$$
$$-\sum_{t\in T} cf_t\cdot|\Psi|$$
$$-\sum_{b\in B}\sum_{t\in T}\sum_{s\in S}\sum_{\tau\in\Psi} x_{bts\tau}\cdot\left(w_b\cdot dist_{ts}\cdot ctr\cdot\alpha_{tr}+cv_{tb}+w_b\cdot wst_b\cdot cd_{bt}\right)$$
$$-\sum_{b\in B}\sum_{t\in T}\sum_{s\in S}\sum_{\tau\in\Psi}\sum_{d\in D} x_{bts\tau}\cdot\left(w_b\cdot dist_{td}\cdot ctr\cdot\alpha_{tr}\cdot wst_b\right)$$
$$-\sum_{b\in B}\sum_{p\in P}\sum_{t\in T}\sum_{\tau\in\Psi}\sum_{c_b\in C_b} x_{c_bpt\tau}\cdot\left(w_{c_b}\cdot dist_{pt}\cdot ctr\cdot\alpha_{tr}-cd_{c_bp}\right)$$
$$+\sum_{c_b\in C_b}\sum_{p\in P}\sum_{b\in B}\sum_{\tau\in\Psi}\sum_{d\in D}\sum_{t\in T} x_{c_bpt\tau}\cdot\left(w_{c_b}\cdot dist_{pd}\cdot ctr\right) \tag{1}$$

$$\text{s.t.}\sum_{t\in T} x_{bts\tau}\le dem_{sb\tau}\quad\forall s,b,\tau \tag{2}$$

$$\sum_{s\in S} x_{bts\tau}\le tcap_{tb}\quad\forall t,b,\tau \tag{3}$$

$$\sum_{t\in T} x_{c_bpt\tau}\le cattle_{p\tau}\cdot cut_{c_b}\cdot w_{cattle}\quad\forall p,c_b,b,\tau \tag{4}$$

$$\sum_{c_b\in C_b}\sum_{p\in P} x_{c_bpt\tau}\ge\sum_{s\in S} x_{bts\tau}\cdot\left(1+wst_b\right)\cdot uf_{bc_b}\quad\forall\tau,t,b:uf_{bc_b>0} \tag{5}$$

$$x_{c_bpt\tau},x_{bts\tau}\ge 0\quad\forall p,c_b,b,\tau,s \tag{6}$$

The objective function (1) accounts for the profit of the meat producers intended as the price of the by-products multiplied by the quantity delivered to fulfill the secondary demand, minus the costs of management (i.e., treatment and disposal) and transportation of such flows. Fixed costs paid by the treatment plants are considered in the function, although they are not differential. Moreover, the avoided costs of waste disposal and waste transportation at/from the slaughterhouses are considered in the function with a positive value. Constraints (2) bound the flow of by-products to the demand by the distribution centers. Constraints (3) and (4) impose the consideration of the transformation capacity at the treatment facilities, and limit the outgoing flow of secondary materials from the slaughterhouses to the weight and number of slaughtered animals, respectively. Constraints (5) guarantee the balancing of flows throughout the nodes, while Constraints (6) denote the continuous and positive nature of the decision variables.

This formulation can be quickly implemented on high-level mathematical programming languages (e.g., AMPL, GAMS, MatLab) and used to solve real-world instances and to support decision makers on the management of by-product valorization networks.

4 An application from the meat industry

The proposed model is intended for the management of the collection of secondary materials from the slaughterhouses and their transformation into by-products for other supply chains. According to the scheme of Fig. 2, the entities involved in this problem are the meat producers (i.e., the slaughterhouses), the transformation facilities, and their customers (i.e., distribution centers). We applied this model to a (significant) case study provided by the main Italian meat producer. In this problem, five secondary materials and waste from the primary production are considered to produce five different by-products. These secondary materials are valve tissues, skin, fats, bones, and blood, while the resulting by-products are heart valves, bags and clothes, soaps, toys, and organic fertilizers intended for the biomedical, clothing, and agricultural industries.

The remainder of this section summarizes the main inputs resulting from the data collection and shows the results from the application of the model.

5 Data collection

The data collection is conducted for 6months with the collaboration of the company managers (i.e., production managers, logistics managers) and through the company's enterprise resource planning (ERP). A tailored and comprehensive database is fed with the collected records (Accorsi et al., *2018a*). The main tables are summarized in the following paragraphs.

TABLE 2 Meat producers and by-products treatment facilities: locations and parameters

Slaughterhouse (p)	Address	City	$Dist_{pd}$ (km)	
$p=1$	Via Spilamberto, 30/c	Castelvetro	19.5	
$p=2$	Viale Europa, 10	Ospedaletto	22.5	
$p=3$	Contrada Tierzi, 8	Flumeri	47.0	
$p=4$	Viale delle Scienze, 1	Rieti	3.8	
Treatment Facility (t)	**Address**	**City**	**$Dist_{td}$ (km)**	**cf_t (€/month)**
$t=1$	Via Provinciale Est, 145	Nonantola	9.6	75,558
$t=2$	Mezzana Casati	San Rocco al Porto	41.0	98,281
$t=3$	Contrada Lagoni	Sant Angelo dei Lombardi	49.2	133,574
$t=4$	Cascina in Garone	San Giorgio di Lomellina	34.0	112,734
$t=5$	Frazione Santa Margherita	Belgioioso	15.4	166,555
$t=6$	Via Convertite, 8	Faenza	8.1	98,215
$t=7$	Via della Vecchia Fornace	Perugia	9.6	62,146

Table 2 reports the address and location of each slaughterhouse p and treatment facility t involved in the network and the parameter $Dist_{xd}$ which is the traveling distance between each plant/facility (p/t) and the closest disposal center d. Table 2 also reports the monthly fixed costs cf_t for each treatment facility t.

In Table 3, the records report the parameters regarding the nature of the secondary materials and the associated by-products. Specifically, it provides the values of the selling prices for each unit of by-product supplied to the distribution centers, the weight of such unit, and the waste percentage generated during the transformation process at the treatment facilities. Furthermore, this table describes the unit weight of each secondary material and the usage factor uf_{bc}, which represents the weight of secondary material necessary to generate one unit of by-product. It is worth noting how, in such a numerical example, one secondary material is associated with just one by-product, although the proposed formulation allows defining a set of secondary materials involved in the production of each by-product.

Figure 3 illustrates on the Italian map the locations of the 30 distribution centers s considered as secondary markets for the meat by-products. Finally, Fig. 4 shows the average number of cattle slaughtered by the four meat plants along the observed time horizon. The parameters of transportation and disposal costs are set to 0.015 (€/kg km) and 0.126 (€/kg), respectively.

TABLE 3 By-product and secondary material parameters

Code	By-product	Price pr_b (€/unit)	Weight w_b (kg/unit)	Treatment waste wst_b (%)
$b=1$	Heart valves	78	0.6	0.21
$b=2$	Leather bags	23	3.2	0.16
$b=3$	Soaps	0.72	0.2	0.09
$b=4$	Pet toys	5.9	0.7	0.13
$b=5$	Fertilizers	2.1	1	0.08
Code	**Secondary material**		**Weight w_{c_b} (kg/unit)**	**Usage factor uf_{bc} (kg/unit)**
$c_1=1$	Valve tissues		4.5	$Uf_{11}=0.5$
$c_2=1$	Skin		45	$Uf_{21}=2.9$
$c_3=1$	Fats		13.5	$Uf_{31}=0.18$
$c_4=1$	Bones		76.5	$Uf_{41}=0.6$
$c_5=1$	Blood		22.5	$Uf_{51}=0.85$

FIG. 3 The distribution centers (s) served by the treatment facilities (t).

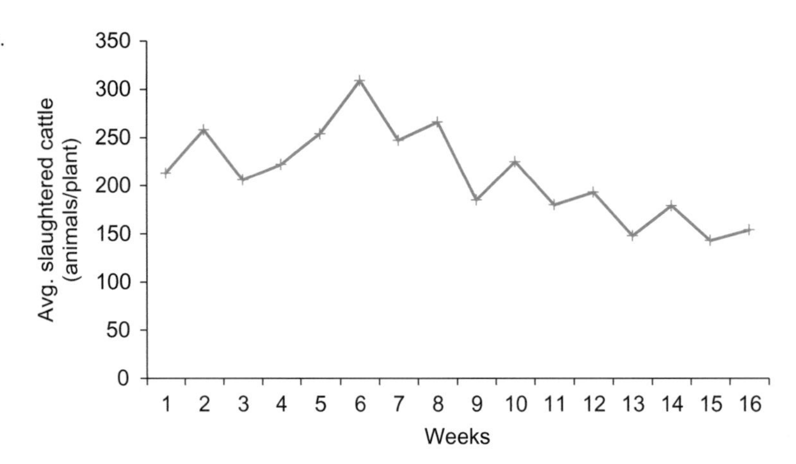

FIG. 4 Slaughtered animals over weeks.

6 Results

The LP model is written in the AMPL language and the instance solved in a few seconds through the commercial linear solver Gurobi run on a workstation configured with Intel Quad Core 2.4-GHz processors and 8 GB of RAM.

The first finding deals with the flows of secondary materials supplied from the slaughterhouses to the treatment facilities, instead of being disposed as waste at the disposal centers. The results for the first 4 weeks of the planning horizon are reported in Fig. 5.

By observing Fig. 5, it can be seen how the traveling distance and the configuration of the logistics network play a crucial role in the management of such flows. In particular, the distance from the disposal centers discourages a plant from disposing of the waste and to pay for its transportation without any revenue. Furthermore, the location of the plant with respect to the treatment facilities is a lever to use to enhance the valorization of secondary materials into by-products. These drivers lead to the different behaviors of the slaughterhouses, as illustrated by the histograms.

Similar findings are obtained from Fig. 6, which illustrates the flow of by-products resulting from the treatment process and delivered to the distribution centers. Throughout the supply chains of secondary materials and by-products, further drivers contribute to the optimal flow allocation, such as the price of the by-products, their weight and waste fraction, as well as the distance between the treatment facilities and the distribution centers.

To further extend the analysis and to investigate the impact of logistics and transportation activities on the valorization of by-products, a sensitivity analysis is proposed. The parameter α_{tr} is used to scale linearly the transportation costs between the network nodes and to measure the impacts of higher costs on the profit generated by the by-product valorization. Specifically, 11 scenarios of the network are generated by varying α_{tr} from 1 to 6 with a step of 0.5. This variation is intended to discern how the overall profit decreases by increasing the sprawl of the network and the traveling distances among the nodes.

Compared to the as-is scenario (i.e., $\alpha_{tr} = 1$), Fig. 7 shows the percentage decrease of flows of by-products delivered to the distribution centers per each scenario.

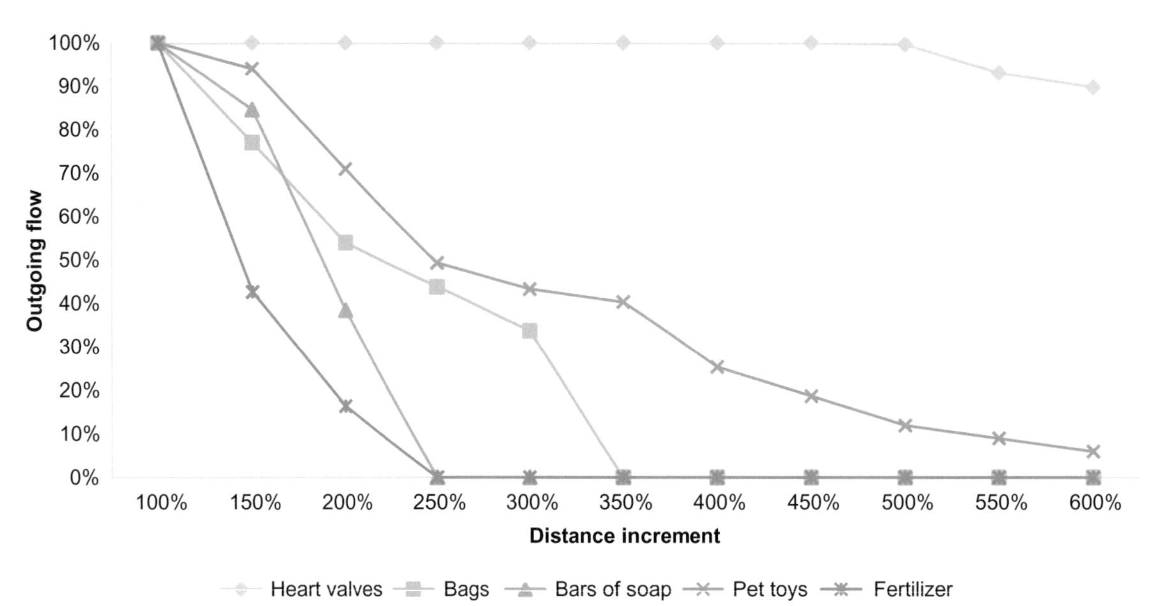

FIG. 7 Optimal by-products flows managed by valorization chains/networks of increasing distance.

It turns out that different by-products show different behaviors in view of their weight and value, the waste they generate across the network, and the geographic distribution of the demand around the treatment facilities. For example, when the network sprawl is two and a half times the as-is configuration (i.e., distances are multiplied by 2.5) the treatment of bones into pet toys and their distribution to the customers is no longer convenient. Conversely, the high value of biomedical by-products such as heart valves makes these flows less sensitive to the logistics distance between the network's nodes.

7 Conclusions

This chapter focuses on the management of the flows of secondary materials, waste, and by-products resulting from the meat industry and provides a decision-support model for the optimal allocation of such flows across a valorization network. The actors involved are slaughterhouses, treatment facilities, distribution centers, and disposal centers, so that the model is intended for meat production systems but can be quickly adapted to other primary food supply chains (e.g., fish, agricultural commodities). The decision-support model is based on a linear programming (LP) formulation operating at the tactical level, to aid the managers in the decisions about allocating the weekly flows of secondary materials to valorization chains rather than to disposal as waste.

The lessons learned from the application of this model to a real Italian case study are as follows. As already stated for other closed-loop supply chains (Accorsi et al., *2015*):

- The traveling distances among the network nodes for primary and secondary products are a critical lever to consider when assessing the profitability of by-product valorization;
- The unit weights, prices, and usage factors of the by-products contribute significantly to the profit of the actors involved in the valorization processes;
- The optimal trade-off between the waste disposal and the by-product production is hard to identify without the adoption of an adequate decision-support approach or planning tools.

Future research development may investigate the significance of each problem parameter (i.e., driver) to the profit secondary chain's actors, through other sensitivity analyses built by using simulation techniques and tools (Accorsi et al., *2017b*). Furthermore, methods and models for cold-chain design and planning (Gallo et al., *2017*) might be integrated to this problem, particularly for those by-products that are sensitive to temperature variations. Lastly, some developments might be devoted to the implementation of new models working at an operational level and involving other problems like vehicle routing (VR) in the presence of cooperation or competition among the network players (Accorsi et al., *2018b*).

Acknowledgments

This research is part of the S.O.FI.A. Project that has received funding from the M.I.U.R under the Italian CLUSTER Funding Programme and Grant agreement code CL.A.N Agrifood CTN01_00230. This research project gained the support of the company Inalca S.p.A., which the Authors would like to thank in the name of Dr. Giovanni Sorlini. Two engineering students joined the research group at University of Bologna and deserve mention. Indeed, Eng. Sara Maccagnani and Eng. Francesco Maria Marchi actively participated in this project and supported the data collection, manipulation, and analysis.

References

Accorsi, R., Manzini, R., Pini, C., Penazzi, S., 2015. On the design of closed-loop networks for product life cycle management: economic, environmental and geography considerations. J. Transp. Geogr. 48 (2015), 121–134.

Accorsi, R., Manzini, R., Baruffaldi, G., Bortolini, M., 2017a. On reconciling sustainable plants and networks design for by-products management in the meat industry. In: Smart Innovation, Systems and Technologies. Vol. 68. Springer, pp. 682–690.

Accorsi, R., Bortolini, M., Baruffaldi, G., Pilati, F., Ferrari, E., 2017b. Internet-of-things paradigm in food supply chains control and management. Proc. Manuf. 11, 889–895.

Accorsi, R., Cholette, S., Manzini, R., Tufano, A., 2018a. A hierarchical data architecture for sustainable food supply chain management and planning. J. Clean. Prod. 203, 1039–1054.

Accorsi, R., Baruffaldi, G., Manzini, R., Tufano, A., 2018b. On the design of cooperative vendors' networks in retail food supply chains: a logistics-driven approach. Int. J. Logist. Res. Applications 21 (1), 35–52.

Allievi, F., Vinnari, M., Luukkanen, J., 2015. Meat consumption and production—analysis of efficiency, sufficiency and consistency of global trends. J. Clean. Prod. 92, 142–151.

Bhaskar, N., Modi, V.K., Govindaraju, K., Radha, C., Lalitha, R.G., 2007. Utilization of meat industry by products: protein hydrolysate from sheep visceral mass. Bioresour. Technol. 98, 388–394.

Bustillo-Lecompte, C.F., Mehrvar, M., 2017. Treatment of actual slaughterhouse wastewater by combined anaerobic–aerobic processes for biogas generation and removal of organics and nutrients: an optimization study towards a cleaner production in the meat processing industry. J. Clean. Prod. 141, 278–289.

Elferink, E.V., Nonhebel, S., Moll, H.C., 2008. Feeding livestock food residue and the consequences for the environmental impact of meat. J. Clean. Prod. 16 (12), 1227–1233.

Food And Agriculture Organization Of The United Nations (FAO) (2015). FAOSTAT, Food Balance Sheets (Available online at *http://faostat3.fao.org/home/index.html*).

Gallo, A., Accorsi, R., Baruffaldi, G., Manzini, R., 2017. Designing sustainable cold chains for long-range food distribution: energy-effective corridors on the silk road belt. Sustainability 9 (11), 2044.

Jayathilakan, K., Sultana, K., Radhakrishna, K., Bawa, A.S., 2012. Utilization of byproducts and waste materials from meat, poultry and fish processing industries: a review. J. Food Sci. Technol. 49 (3), 278–293.

Soysal, M., Bloemhof, J.M., Van der Vorst, J.G.A.J., 2014. Modelling food logistics networks with emission considerations: the case of an international beef supply chain. Int. J. Prod. Econ. 152, 57–70.

Virtanen, Y., Kurppa, S., Saarinen, M., Katajajuuri, J.-M., Usva, K., Mäenpää, I., Mäkelä, J., Grönroos, J., Nissinen, A., 2011. Carbon footprint of food—approaches from national input-output statistics and a LCA of a food portion. J. Clean. Prod. 19 (16), 1849–1856.

Chapter 24

Recipe-driven methods for the design and management of food catering production systems

Riccardo Accorsi*, Federica Garbellini*,†, Francesca Giavolucci*, Riccardo Manzini* and Alessandro Tufano*

*Department of Industrial Engineering, Alma Mater Studiorum—University of Bologna, Bologna, Italy †CAMST—La Ristorazione Italiana Soc. Coop. A r.l., Villanova di Castenaso, Italy

Abstract

The food catering industry provides ready-to-eat meals to consumers who eat away from their homes at many heterogeneous points of demand, for example, schools, hospitals, and company canteens. Given the number of consumers served, many diets, tastes, and expectations must be satisfied, and this fact enlarges the variability and complexity of the menus, made from many hundreds of recipes.

Such complexity also affects the production process and results in a specific organization of the production system and the associated logistic activities. First, the whole production needs to be organized in agreement with the sequence of processing tasks of the recipes, in a typical job-shop system. Second, the details regarding the raw ingredients, work in progress (WIP), and finished products of each recipe can be used to set the hierarchy of processing tasks and to support production scheduling accordingly. However, the lack of knowledge and data relating to the task sequence beyond the recipes limits the implementation of adequate food catering production processes and control systems.

This chapter deals with the organization of production activities in a food catering facility and introduces a methodology for the design of accurate task sequences for each recipe starting from the bill-of-materials of the recipe's ingredients. The deliverable of the proposed method is the hierarchy of the processing/handling/packing task necessary for each recipe, which could be quickly implemented in a spreadsheet or management software for the scheduling of the production activities. Specifically, an algorithm is defined to provide a graph of the task priority according to precedence constraints, and a data architecture illustrated to store the resulting working cycle.

Finally, this chapter discusses the potential application of such methods in the food catering industry in agreement with the expected growth of food service demand and the standardization of production.

1 Introduction

The food service industry emerged in Europe in the 1970s, when companies started to serve meals to their employees with a nutritional profile suitable for their individual jobs. Early on, the service was organized by each company independently with on-site kitchens and canteens. Then the service was progressively outsourced and assigned to third-party service companies, called *mass catering* providers (Europe Food Service, 2017).

Nowadays, mass catering is a niche of the wider food service industry, intended to serve people who have lunch or dinner outside their homes for different reasons, for example, work, study, or illness (Fusi et al., 2015). With greater numbers of target consumers, the complexity of the production mix grows. The resulting production process, which generates chilled or warm ready-to-eat products from raw ingredients, involves a large number of entities, constraints, and decisions to be made simultaneously. Although such complexity is poorly perceived by the final consumers, many efforts are made by catering companies to maintain the expected service level. Most of the management focus is devoted to the procurement of the raw ingredients, which have to comply with the standards and contracts imposed by important public organizations like schools and hospitals. However, less attention is paid to the production of the meals and the organization of the daily processing operations, which are typically manual, meaning time and cost intensive.

Food catering providers operate according to one of these two service models (Ciappellano, 2009):

1. *Cook-serve service*: The client maintains its assets (i.e., kitchen, equipment, and tools) and hosts personnel from the provider who cook food and directly serve to final consumers.

Sustainable Food Supply Chains. https://doi.org/10.1016/B978-0-12-813411-5.00024-7

2. *Deferred service*: The client outsources the entire service and the catering company delivers a certain number of meals according to the daily orders of a portfolio of clients.

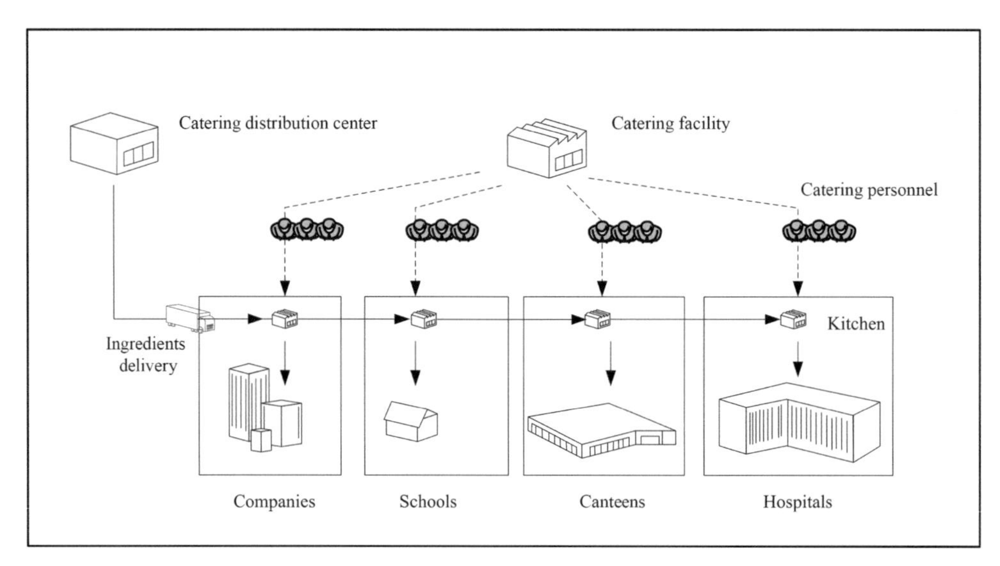

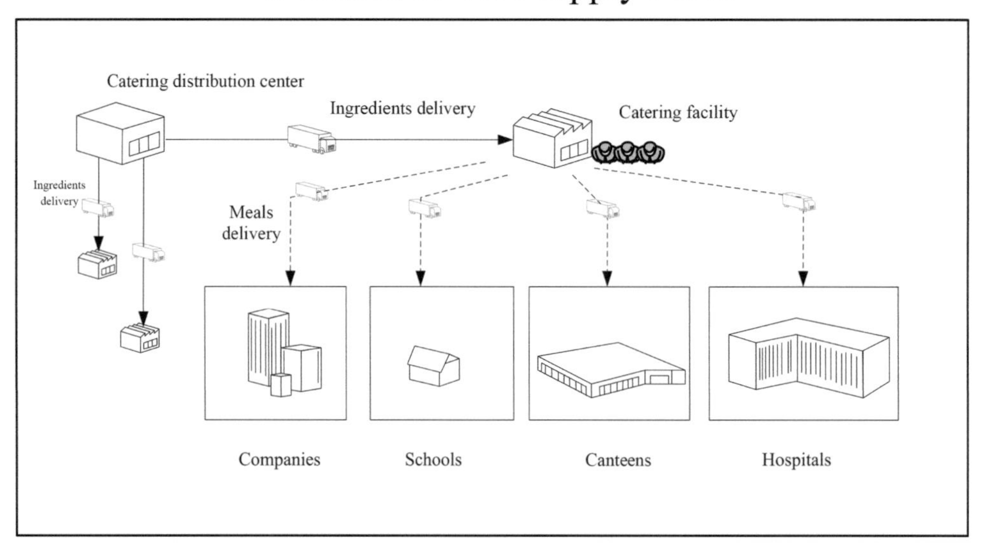

FIG. 1 Cook-served supply chain vs. deferred service supply chain.

Fig. 1 compares these service models. In a cook-serve model the mass catering company provides raw materials, personnel, and cooking equipment directly to the kitchen of the client. Meals are prepared and served on-site, and the cooking area must comply with the hazard analysis critical control point (HACCP) protocol. Nevertheless, this service is expensive, since economies of scale are difficult to obtain unless the number of consumers at the single point of service is large.

According to the deferred service model, a network of centralized kitchens (CEKI) collects orders from many customers and prepares meals that are delivered to the final points of consumption, for example, schools, hospitals, and companies. This supply chain configuration allows economies of scale since a single CEKI may serve tens of clients, that is, customers, and thousands of final consumers. However, the complexity of such a service increases with the centralization of the recipe-processing activities. Special attention must be paid not only to HACCPs within each CEKI, but also to logistic processes and operations, including storage activities, material handling, the facility and processing resources layout, the packaging process, and many other operational issues, such as the scheduling of processing tasks on production resources and the delivery activities. Along such operations, both in the CEKI and during transportation, the temperature of the meals must be carefully monitored and maintained within a safety range (Manzini et al., 2014, 2017), and this condition further affects the cost (i.e., of energy) and the risk behind the deferred service model. As the handling and storage of WIPs and meals are extremely critical, two levers can be tried by the managers to enhance the service level, guarantee the safety and quality of the products, and contain the service costs: the design of the CEKI and the processing resources layout (see Chapter 7), and the organization and scheduling of the processing tasks necessary to complete the recipes.

The scope of this chapter is to investigate the integration of facility layout and recipe scheduling levers on the design and planning of catering production systems according to the deferred service model. After a discussion of the main factors influencing the layout and recipe scheduling in catering facilities, the chapter provides a method to build a priority graph of the processing tasks required to finalize a recipe, starting from available information regarding the recipe's ingredients and the processing resources. Such a graph, typical of assembling systems, can be used to fuel production scheduling tools and aid recipe-driven production planning in catering systems.

1.1 Layout and production technologies

A CEKI is a facility owned and managed by a food caterer able to perform deferred services to hundreds of customers and thousands of final consumers in agreement with some service contract specifications and terms (e.g., delivery times, variability of quantities, order cut-off, etc.). In the following, a nonexhaustive list of issues affecting the design, management, and control of a CEKI is proposed:

1. *Safety compliancy*, for example, the control of ingredients and product contamination before cooking, and the bacteria proliferation after cooking.
2. *Layout and production system configuration*, that is, the determination of the type, number, and location of the production resources and equipment within the facility. This also includes the control of the level of utilization of manufacturing resources, as well as the load of auxiliary plants supplying steam, water, heat, electricity, and collecting waste.
3. *Material handling (MH) issues*, for example, the management of the flow of materials throughout the production system, including MH devices and automation, and the storage of ingredients, WIPs, and finished meals.
4. Operational issues, which involve the scheduling and sequencing of processing tasks and the control and supervision of transformation, handling, packing, storage, and shipping phases. One of the key performance levers in the control of the production system is reducing bottlenecks and delays.

Safety issues are crucial since they are deeply connected with the quality and safety of a finished product and served meals (Williams, 1996). The safety of products is periodically checked by the hygiene authority and it must comply with standards and rules. Thus in order to avoid legal actions and economic sanctions, it is a prime responsibility of the caterer to ensure that food is prepared and supplied in agreement with such standards and rules.

The compliance with such rules depends first on the configuration of the CEKI and the layout of the production areas. The generic CEKI typically adopts a job-shop layout configuration where the processing resources (i.e., stations) are organized into homogeneous departments. Each recipe determines the sequence of visit of the departments in agreement with the tasks of the working cycle. Therefore the layout of the departments is set according to the majority of the recipe's working cycles with the purpose of avoiding reverse flows. An effective method for organizing the layout of the processing departments according to the recipe's working cycle was provided by Hollier (1983), intended for generic manufacturing systems.

Assume that the recipe i is made by a sequence of task $t \in T_i$, which are assigned, each to a processing resource or a working station j. Let D_j be the department of homogeneous stations of type j. According to Hollier (1983), the location of D_j within the CEKI department can be determined through the calculation of the ratio between the outgoing and the incoming flows of material from/to each department (Table 1). Assume the following notations:

TABLE 1 Parameters for the application of the Hollier's method in CEKI layout design

Sets/ parameters	Description
i	Recipe
j	Working station, processing resource (i.e., machine)
D_j	Department of homogeneous machines of type j
$t \in T_i$	Processing tasks of recipe i
$f^i_{jj'}$	1 if working station j' is visited just after j in the working cycle of recipe i; 0 otherwise
q_i	Average daily production batch quantity of recipe i
$cap^{tool}_{jj'}$	Handling tool capacity from station j to j'

Then, the material from/to chart, a square matrix $|J| \times |J|$, can be obtained through Eq. (1), which calculates the value of each cell $\alpha_{jj'}$:

$$\alpha_{jj'} = \sum_i f^i_{jj'} \cdot \left\lceil \frac{q_i}{cap^{tool}_{jj'}} \right\rceil \forall j, j' \in J \tag{1}$$

For each department j, a flow exchange metric can be calculated using Eq. (2) as:

$$ef^{From/To}_j = \frac{\sum_{j'} \alpha_{jj'}}{\sum_{j'} \alpha_{j'j}} \forall j \in J \tag{2}$$

Finally, the department D_j will be located along a straight line from the inbound side to the outbound side according to the decreasing value of the metric $ef^{From/To}_j$ as exemplified in Fig. 2. An example of the application of the Hollier model to design a CEKI layout is provided in Tables 2 and 3, which respectively report the recipes' working cycles and the $f^i_{jj'}$ square matrix.

TABLE 2 Examples of recipe's working cycles

Recipe i = 1: "Ratatouille" q_i = 10	*Recipe i = 2: "Ragout Sauce" q_i = 8*	*Recipe i = 3: "Roast beef" q_i = 6*
Ingredients picking	Ingredients picking	Ingredients picking
Vegetable preparation	Vegetable preparation	Meat preparation
Vegetable cooking	Meat preparation	Meat cooking
Packing	Sauce cooking	Meat cutting
	Pasta cooking	Packing
	Packing	

TABLE 3 Task sequence flows (*Recipe i = 2: "Ragout Pasta"*)

$f^2_{jj'}$	Storage chambers	Raw prep. area	Braiser	Boiler	Packing machine
Storage chambers		1		1	
Raw prep. area			1		
Braiser					1
Boiler					1
Packing machine					

The n tasks of recipe $i = 2$ correspond to $n - 1$ cells with a value equal to 1 in the $f^i_{jj'}$ matrix of Table 2. The ingredients are retrieved from the storage chambers and sent to preprocessing areas (made by working tables) or to the boiling station (i.e., pasta), while the cuts of vegetables and meats are mixed into the braiser and cooked for 2 h. When the ragout sauce is ready, the past is boiled for 8 min and, the meal mixed and packed in the portioning and packaging areas. Table 4 proposes an example of Table $\alpha_{jj'}$ built upon the standard daily production mix, while the calculation of the flow exchange metric is exemplified in Table 5 and results in the layout of Fig. 2. It is worth noting how the capacity of the handling system, represented by the parameter $cap^{tool}_{jj'}$, plays a relevant role in the determination of the proper sequence of the departments and the facility layout.

TABLE 4 From-to material flow chart

$\alpha_{jj'} = \sum_i f^i_{jj'} \cdot \left\lceil \frac{q_i}{cap^{tool}_{jj'}} \right\rceil$	Storage chambers	Raw prep. area	Braiser	Boiler	Packing machine
Storage chambers		18		12	
Raw prep. area			16	6	6
Braiser					12
Boiler		4	8		12
Packing machine					

TABLE 5 Exchange flow chart, $ef^{From}/_{To}$

Department D_j	$\sum_{j'} \alpha_{jj'}$	$\sum_{j'} \alpha_{j'j}$	$ef^{From}/_{To}$
Storage chambers	30	0	∞
Raw prep. area	28	22	1.27
Braiser	12	24	0.5
Boiler	12	18	0.67
Packing machine	0	30	0

In addition, within a CEKI, the location of the processing departments should contribute to avoiding the so-called cross-contamination that results from the transfer of pathogens and bacteria between raw and cooked ingredients, WIPs, meals, and between and food and the personnel or surfaces of equipment and other tools (Ministero della Salute, 2012). Such constraint results in departments typically being located along a straight line from the preprocessing areas (i.e., peeling, washing, cutting), through the cooking areas and lastly the packing areas.

The typical CEKI layout assumes hence the shape of a nonreturnable straight production system (see Fig. 2), where the raw ingredients are received and stored at the inbound side (on the left), and pass through the facility to complete their working cycle through to the packing, storage, and shipping areas at the outbound side (on the right), minimizing, as much as possible, the return flows of WIPs, handling tools, and workers.

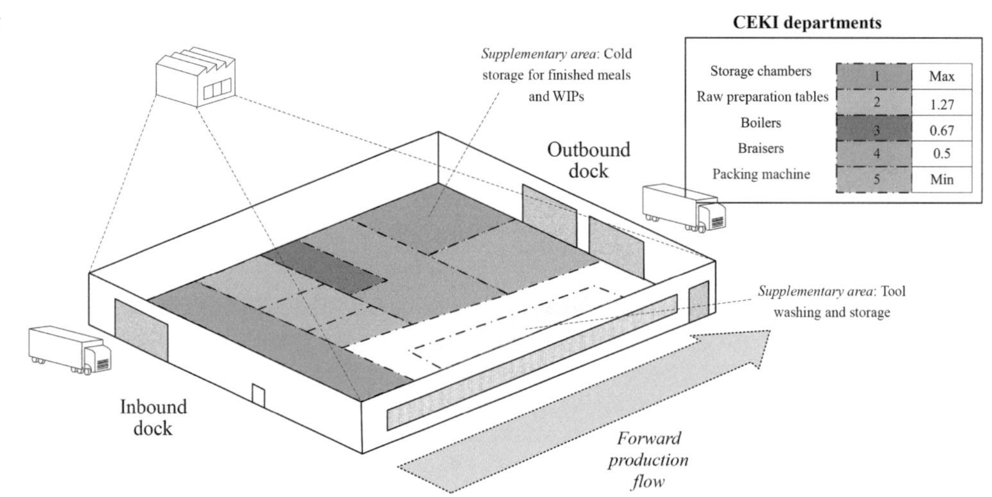

FIG. 2 Example of layout of a CEKI.

To address the safety issues, the layout and location of the processing resource are not the only levers that can be adopted by the manager. Other opportunities lie in the selection of adequate cooking technology and packaging solutions. These should be intended to guarantee meals out of the so-called "dangerous temperature range" (i.e., 10–65°C) where bacteria proliferate (European Parliament, 2004). Concerning the cooking technologies, three main opportunities exist with different temperature profiles to respect in order to deliver safe products to the consumers. These are as follows:

- *Cook-warm*: Meals are cooked and maintained hot (above 65°C) after production through oven and adiabatic packages, until they are delivered to the final consumer.
- *Cook-chill*: Meals are cooked and swiftly blast-chilled, dropping them to a temperature lower than 10°C (within a maximum of 3h). Meals are kept chilled during delivery and then rewarmed just before being served.
- *Cook-freeze*: Meals are treated as in the case of cook-chill, but they are frozen (at a temperature lower than −18°C) instead of being blast-chilled.

These three temperature profiles imply different production resources and processing tasks. Fig. 3 illustrates the alternative production cycles for each temperature profile.

The decision on the proper and most convenient production technology (i.e., temperature profile) for each recipe depends on a wide range of factors, such as the cost and the number of the processing resources, the due times for meal delivery, the distance between the CEKI and the points of consumption, the number of parallel recipes processed by the facility, and the resulting utilization of the working stations.

Such complexity requires advanced tools and methodologies able to embrace the multidisciplinary nature of this strategic decision. An example of these tools is from Tufano et al. (2018). Others (Penazzi et al., 2017) use simulation to study the impacts of different production resources on the throughput and the energy consumption of a CEKI, and explore the interactions between the following entities:

- *Products*: Each product is defined by a recipe, that is, a list of ingredients organized into a bill-of-material (BOM), and a list of tasks organized through a working cycle.
- *Clients*: Each client belongs to a category, for example, school, hospital, and industrial company. The contract between the client and the caterer usually defines the products and temperature profiles, the raw materials (e.g., origin, organic, nutritional values), and the packaging. In addition, the delivery time window is determined for each point of consumption.
- *Assets*: that is, machinery and production equipment and tools to handle and store products during the production cycle and the catering service. Each machine, tool, or handling vehicle is defined by capacity and throughput parameters as well as by the energy consumption rate.

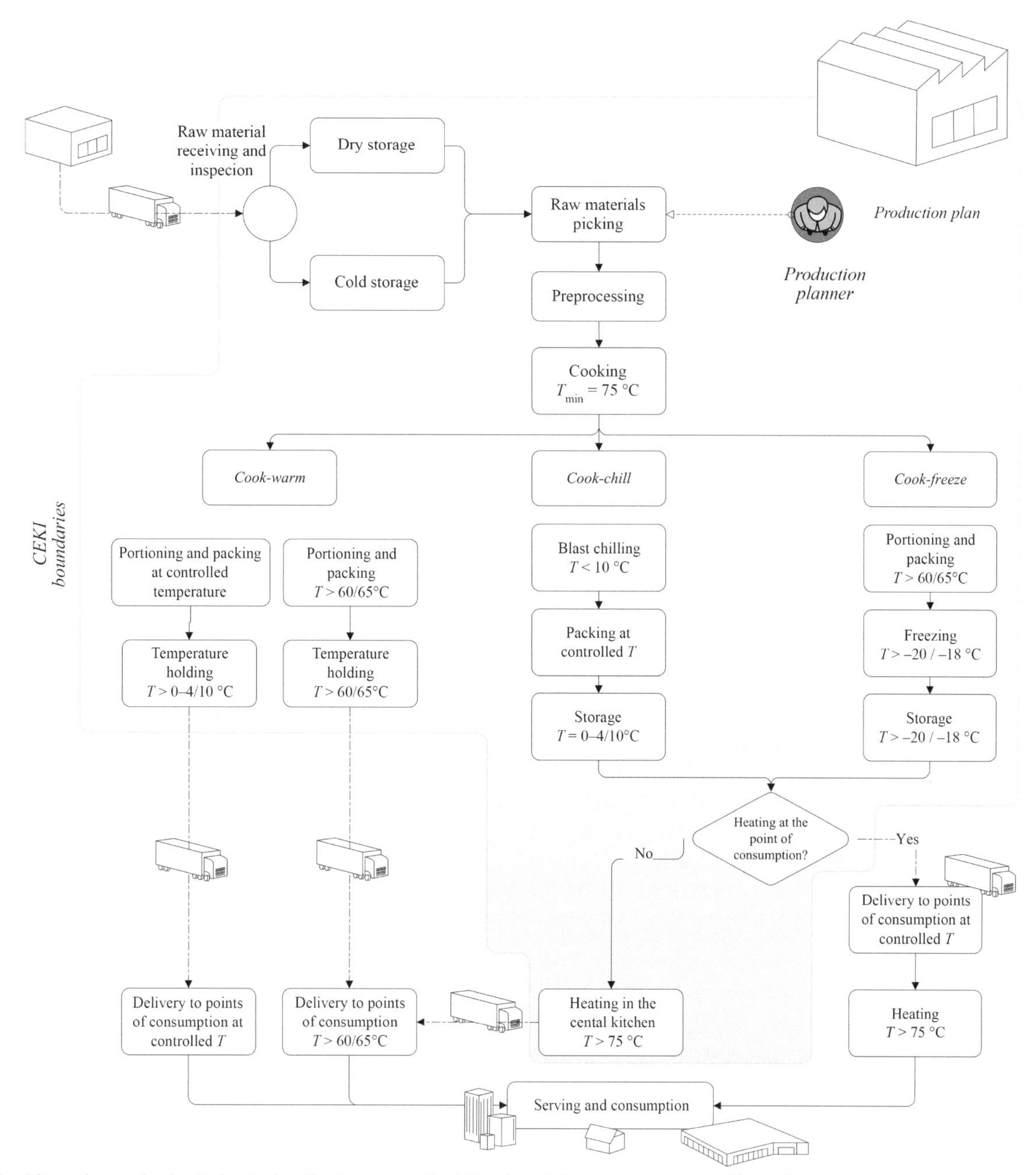

FIG. 3 Alternative production technologies: *Cook-warm*, *cook-chill* and *cook-freeze* temperature profiles and processes.

- *Packages*: Primary and secondary packages are selected in agreement with physical and chemical characteristics of the product, the portioning phase, and the temperature profile chosen for that recipe.
- *Storage systems*: Used for raw materials, semifinished, and finished products, these are areas or queuing stations where climate conditions (temperature and relative humidity) are properly maintained.

In an average-size CEKI serving 6000 complete meals per day to 50–100 different points of consumption, about 50–100 working stations and 5–10 storage rooms can be counted. Nevertheless, the expectations of consumers and the customization of food habits impose high flexibility on the caterer, who is called to offer a wide production mix (i.e., menu) made of hundreds of recipes delivered in time, with a high level of quality and safety (Adler-Nissen et al., 2013). For this reason, a CEKI has an annual menu containing about 400–500 different recipes, which means thousands of different tasks and hundreds of processes performed each day within a few hours (e.g., before lunch) in a typical nonautomated production system like a CEKI.

To deal with such complexity in the type and number of entities and processes to manage, the design of catering production systems must be preceded by preliminary and exhaustive data gathering and manipulation. Data collection is concerned with the datasheets of the assets (i.e., working stations), the technical parameters of the contracts with the clients, and also with the BOMs of the recipes and their working cycle. The resulting knowledge supports the *ex-ante* analysis of the technical and economic feasibility of the production activities and system. Once data is gathered and properly manipulated, the layout design should follow the systematic approach rigorously introduced by Drira et al. (2007) for a generic production facility. It encompasses seven steps:

1. Data collection and analysis of the production mix (i.e., mix of recipes) and the order profiles;
2. Setting up the recipe BOMs and building the working cycle hierarchy (i.e., task priority graph) underlying the relationships (hierarchies and precedencies) between tasks and resources;
3. Setting up a strategic and draft scheduling of the recipes (i.e., priority list);
4. Analysis of the material flows exchanged between resources or working station in agreement with the schedule of *Step 3*;
5. Determination of the proper number of working resources (see Chapter 7);
6. Determination of the area required by the set of working stations and the departments;
7. Design of the plant layout.

The remainder of this chapter will discuss the main factors influencing the production scheduling in catering systems, and then will focus on the preliminary data manipulation (*Steps 1* and *2*) aimed at building the task-priority graph for each recipe, which represents a crucial input of any operations scheduling and layout design activity.

1.2 Production scheduling

Operations in the food industry must comply with tailored parameters and aspects that are distinctive for each product and recipe. Nevertheless, most of the decisions are made without the support of information systems able to track, collect, and store the necessary data for data-driven analysis. Even the scheduling of production is typically managed through experience or best-practice (i.e., rules of thumb), as the information used by the scheduling tools is hard to collect (Houba et al., 2000) in nonautomated production systems such as CEKIs.

The scheduling of catering production activities should build upon the concept of a *recipe* which can be defined as a set of ingredients required to be processed by certain resources in a given sequence to obtain a unit of finished product. The recipe encompasses the overall information required for the proper planning of the production operations. This includes:

1. The set of tasks to be performed and their duration (i.e., machine working time, which often depends on the lot size);
2. The sequence of tasks in the working cycle of the task priority graph;
3. The quantities of raw materials (i.e., ingredients) required to produce a unit (or a lot) of finished product.

Although all this information exists, it is not usually tracked, collected, or stored adequately within catering production systems. As a consequence, the production control and scheduling are poorly aided by advance data-driven decision-support tools.

The design of a CEKI is deeply interconnected with the schedule of the working cycles and the processing tasks. For example, different schedules affect the number of production resources (i.e., machines) necessary to fulfill the production mix differently, as treated and discussed in Chapter 7.

Therefore, many efforts have been made in the literature to address the scheduling of food production and catering operations. Minimization of the operator idle time is a typical objective of the proposed methodologies for the scheduling (Guley and Stinson, 1984). The variability of the cooking time (i.e., task duration), which is extremely affected by the recipe and the quantity of processed product, increases the imbalance between the theoretical and real time to completion of batches and among the completion times of the recipes within the production mix. To address this unavoidable issue, which results in quality decay and losses, the role of buffering areas throughout the production stages has been investigated by scholars (Akkerman and Van Donk, 2007) together with the hierarchy of the working cycles (Akkerman and Van Donk, 2009). Others focus on the alignment between production and shipping activities through integrated scheduling and routing optimization model scheduling (Meinecke and Reiter, 2014; Devapriya et al., 2017; Farahani et al., 2009).

Instead of providing advanced scheduling models for catering production systems, this chapter faces the preliminary phase of building information from the product BOMs issue for the design of task-priority graphs and working cycles able to fuel existing scheduling models and tools. The aim of the proposed methodology is to support managers in the analysis of the precedences among the tasks of the working cycles, thereby enforcing an *at-the-latest* scheduling policy. The at-the-latest production scheduling is intended to reduce the waiting time of finished products before shipping, as well as the associated decay of quality and safety.

The next section introduces the methodology for the design of the task-priority graph of each recipe, while Section 3 provides a numerical proof of concept and Sections 4 and 5 discuss the main remarks.

2 Methodology

This section introduces the methodology used to build knowledge of the recipes (1) through the definition of a tailored data structure supporting the organization of useful information and the modeling of the hierarchy of technological precedences associated with the recipe's working cycle (2).

The recipe brings and contains the knowledge required by any scheduling system. This is the list of ingredients required and the quantities of each ingredient (i.e., BOM), but is also the priority of the tasks to be executed to complete it. While the first data is typically stored in the company's ERP system (i.e., with a list of suppliers for purchasing purposes and cost accounting), the sequence of tasks and their hierarchy are often just kept in the operator's mind and are not tracked or collected through effective information-support systems.

At an information level, a recipe can be described through a set of *nodes* representing a state of the product (i.e., an ingredient, a WIP, or the finished product itself), and a set of *arcs* representing the transformation of a state of the product into the following throughout an assembling/transformation hierarchy (i.e., from the ingredient to WIP), as shown in Fig. 4.

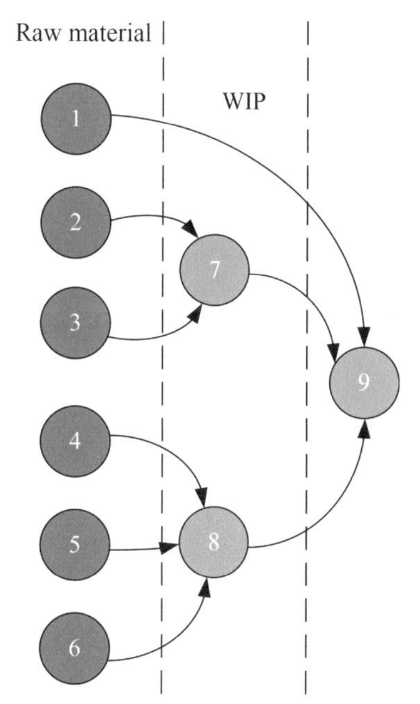

FIG. 4 Recipe hierarchy: graph of states of a product (nodes) and transformation steps (arcs).

The combination of some nodes and arcs (e.g., *Node 2* and 3 ➔ *Node* 7 in Fig. 4) corresponds, at a physical level, to a given processing task of the work cycle of the observed recipe. As a consequence, data collection should follow a structure able to identify and track the link between the recipe hierarchy, illustrated in Fig. 4, and the recipe's working cycle, which is the sequence of processing tasks.

In order to address to this scope, the typical data structure of the recipe's BOM needs to be revised and integrated by further data field as shown in Table 6.

TABLE 6 Revision of the BOM data structure

Data field	Description	Information to track
Progressive Ingredient ID	Identification code	Traditional
Recipe Code	Code of the recipe	Traditional
Primary Ingredient Code	Code of the primary ingredient	Traditional
Primary Ingredient Descr.	Description of the primary ingredient	Traditional
Primary Ingredient Type	Type of the ingredient (Raw material: *RM*; WIP: *SF*; Finished product: *FP*)	New
Secondary Ingredient Code	Code of the secondary ingredient. It is equal to *Primary Ingredient Code* for RM, SF, FP. When different, it indicates an assembly between RM/SF and SF/FP	New
Secondary Ingredient Descr.	Description of the secondary ingredient	New
Secondary Ingredient Qty	Quantity of secondary ingredient per unit of finished product (kg)	Traditional

With such upgraded distinctions between primary and secondary ingredients, the new BOM supports the identification of the assembling steps, which means that a raw ingredient turns into a WIP, or a WIP turns into a finished product. Specifically, when a record of the new BOM presents the same code in the *primary* and *secondary* ingredient fields, a node of the hierarchy in Fig. 4 is identified. Conversely, when a record has different *primary* and *secondary* ingredient codes, an arc of Fig. 4 is set.

Through the proposed data structure, the connection between the BOM (i.e., list of ingredients) and the task sequence of each recipe can be identified, stored, and manipulated by scheduling software and tools. This methodology results in the proper description of a working cycle, which can be gathered and stored through the data structure proposed in Table 7.

TABLE 7 Working cycle data structure

Data field	Description
Recipe Code	Recipe identification code
Task ID	Task identification code
Task Description	Task description
Ingredient Code	Code of the ingredient to be processed
Machine	Machine performing the task
Working Time Type	(0) Picking process (1) Fixed-time process or (2) process with time proportional to the batch quantity
Working Time	Task duration [min] according to its working time type
Handling Tool	Handling tool used to perform the task
Department	Department where the task is to be performed

Finally, an algorithm is introduced and hereby illustrated to automatically design the task-priority graph (like in Fig. 4) for each recipe, starting from the new data structure.

2.1 Graph design algorithm

The assumption behind this algorithm is that the tasks listed in the recipe's working cycle are not necessarily in sequence, but some of them could be performed at the same time. Thus, the output of this algorithm is the identification of a list of predecessors (i.e., the tasks that have to be completed before a given task may start) for each task.

This algorithm derives the task hierarchy from the recipe hierarchy retrieved from the new BOM structure. Indeed, each node of the recipe hierarchy corresponds to a state of the recipe (i.e., raw material, WIP, or finished product) which waits for (and requires) some processing tasks. Consequently, a state of the product can be replaced by a sequence of nodes, each representing a task.

The algorithm is here illustrated with respect to the aforementioned data structure. It goes through the list of ingredients of a recipe starting from the last ingredient to the first. For each row, if the secondary ingredient equals the primary ingredient, it creates a list of tasks that process that ingredient in the working cycle (ordered by increasing values of the Task ID) and sets these tasks as predecessors accordingly (see for reference the example of Fig. 5). Otherwise, it sets the last phase that processes the secondary ingredient as a predecessor of the first phase that processes the primary ingredient (see the arcs connecting the rectangles in Fig. 5).

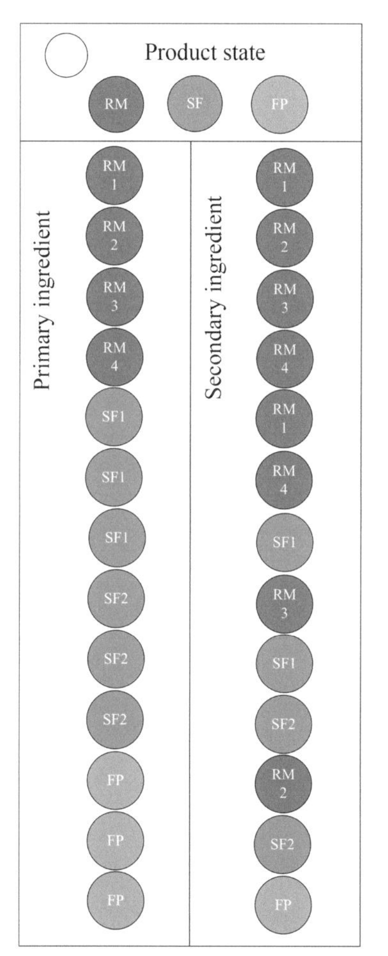

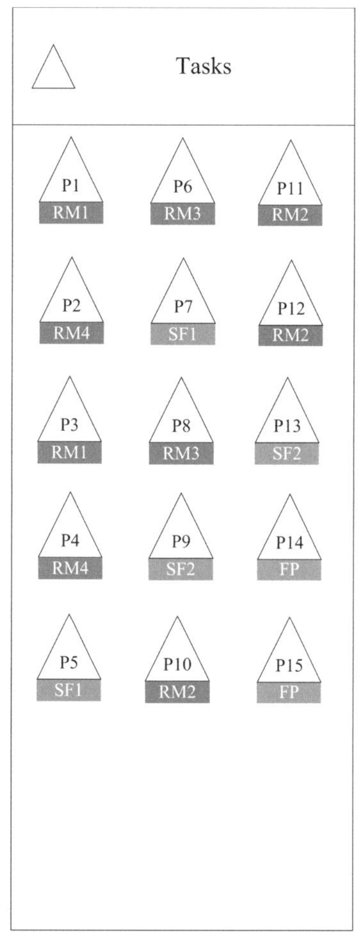

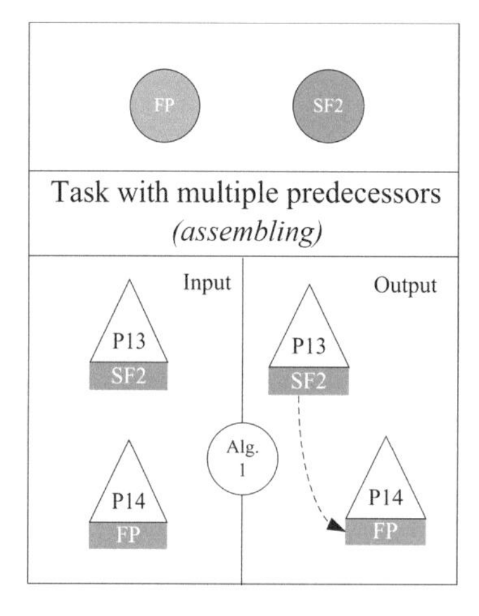

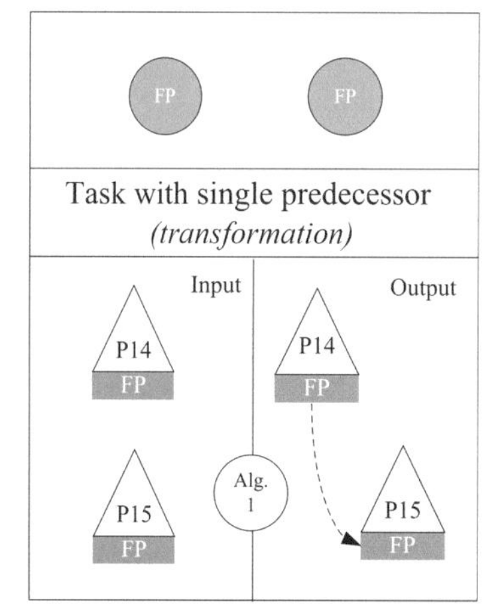

FIG. 5 Entities involved in the proposed algorithm.

In the following, the pseudocode of the algorithm is reported.

```
Input :
g_j, j = 1,...,n ingredients of a recipe ∈ G
p_i, i = 1,...,m tasks of a recipe ∈ P
a_ij = 1 if j processes i as a primary ingredient
b_ij = 1 if j processes i as a secondary ingredient
Output: R list of predecessors for each task
for j ← n to 1 do
    R_i=empty set
    P=j|a_ij = 1
    S=empty set
    for i ← n to 1 do
        if b_ij == 1 then
            Add j to S
        end
    end
    foreach jj ∈ S do
        if jj = j then
            if j ≥ 2 then
                for i ← m to 1 do
                    R_i = i − 1
                end
            end
        end
    end
end
```

ALGORITHM 1 Definition of predecessors

The output of Algorithm 1 is an oriented priority graph of tasks representing the precedences of tasks in the working cycle. Traditional project management or scheduling tools may be implemented/used upon this graph to estimate processing time and resource utilization, and to schedule tasks on resources properly according to the *at-the-latest* scheduling policy.

3 Proof of concept

This section introduces an application of the proposed methodology to show the potential of this technique and discusses its implications for the design and management of the CEKI's production operations.

The recipe for baked potatoes has been chosen as a proof of concept. Table 8 reports the records of the BOM according to its upgraded structure. This represents the starting point and the input of the algorithms. The table lists the ingredients and their main characteristics and properties and maps the assemblies of WIPs and finished products through the reference of primary and secondary ingredients code.

TABLE 8 BOM records of the recipe: baked potatoes

Recipe Code	Progressive Ingredient ID	Primary Ingredient Code	Primary Ingredient Descr.	Primary Ingredient Type	Secondary Ingredient Code	Secondary Ingredient Descr.	Secondary Ingredient Qty
100	1000	1000	Potatoes	RM	1000	Potatoes	0.2
100	1001	1001	Oil	RM	1001	Oil	0.01
100	1002	1002	Sage	RM	1002	Sage	0.0005
100	1003	1003	Rosemary	RM	1003	Rosemary	0.005
100	1004	1004	Salt	RM	1004	Salt	0.0005
100	1005	1005	Pepper	RM	1005	Pepper	0.0005
100	1006	1011	Seasoning	SF	1001	Oil	0.01
100	1007	1011	Seasoning	SF	1002	Sage	0.0005
100	1008	1011	Seasoning	SF	1003	Rosemary	0.005

TABLE 8 BOM records of the recipe: baked potatoes—cont'd

Recipe Code	Progressive Ingredient ID	Primary Ingredient Code	Primary Ingredient Descr.	Primary Ingredient Type	Secondary Ingredient Code	Secondary Ingredient Descr.	Secondary Ingredient Qty
100	1009	1011	Seasoning	SF	1004	Salt	0.0005
100	1010	1011	Seasoning	SF	1005	Pepper	0.0005
100	1011	1011	Seasoning	SF	1011	Seasoning	0.0165
100	1012	1012	Baked potatoes	FP	1000	Potatoes	0.2
100	1013	1012	Baked potatoes	FP	1011	Seasoning	0.0165
100	1014	1012	Baked potatoes	FP	1012	Baked potatoes	0.2165

The second input required by the algorithm is the table of the working cycle, that is, the list of tasks to perform to realize a batch of baked potatoes. Table 9 reports the sequence of tasks (with no definition of priority or predecessors) with their main characteristics, for example, the description, the ingredient involved, the working time, and the description of the handling tool eventually used to perform the task.

TABLE 9 Working cycle of baked potatoes

Recipe Code	Task ID	Task Desc.	Ingr. Code	Machine Code	Working Time Type	Working Time	Handling Tool	Dep.
100	2000	Picking	1000	Operator	0	–	Service Trolley	Cold Storage Cell
100	2003	Picking	1003	Operator	0	–	Service Trolley	Cold Storage Cell
100	2002	Picking	1002	Operator	0	–	Service Trolley	Cold Storage Cell
100	2001	Picking	1001	Operator	0	–	Service Trolley	Foodstuffs Storage
100	2004	Picking	1004	Operator	0	–	Service Trolley	Foodstuffs Storage
100	2005	Picking	1005	Operator	0	–	Service Trolley	Foodstuffs Storage
100	2006	Spilling	1002	Teflon Table	2	0.005	Service Trolley	Vegetable Processing
100	2007	Spilling	1003	Teflon Table	2	0.005	Service Trolley	Vegetable Processing
100	2008	Washing	1002	Vegetable Washer	1	1	Basket	Vegetable Processing
100	2009	Washing	1003	Vegetable Washer	1	1	Basket	Vegetable Processing
100	2010	Mixing	1001	Stainless Steel Table	2	0.005	Service Trolley	Vegetable Processing

Continued

TABLE 9 Working cycle of baked potatoes—cont'd

Recipe Code	Task ID	Task Desc.	Ingr. Code	Machine Code	Working Time Type	Working Time	Handling Tool	Dep.
100	2011	Mixing	1002	Stainless Steel Table	2	0.005	Service Trolley	Vegetable Processing
100	2012	Mixing	1003	Stainless Steel Table	2	0.005	Service Trolley	Vegetable Processing
100	2013	Mixing	1004	Stainless Steel Table	2	0.005	Service Trolley	Vegetable Processing
100	2014	Mixing	1005	Stainless Steel Table	2	0.005	Service Trolley	Vegetable Processing
100	2015	Sack Opening	1000	Stainless Steel Table	2	0.0005	Service Trolley	Cooking Area
100	2016	Seasoning Addition	1011	Stainless Steel Table	2	0.005	Service Trolley	Cooking Area
100	2017	Seasoning Addition	1000	Stainless Steel Table	2	0.005	Service Trolley	Cooking Area
100	2018	Backing	1012	Oven	1	25	Tray Trolley	Cooking Area

Finally, Fig. 6 presents the output of the algorithm as an oriented graph in which tasks are aggregated into blocks of tasks that may be scheduled at the same time.

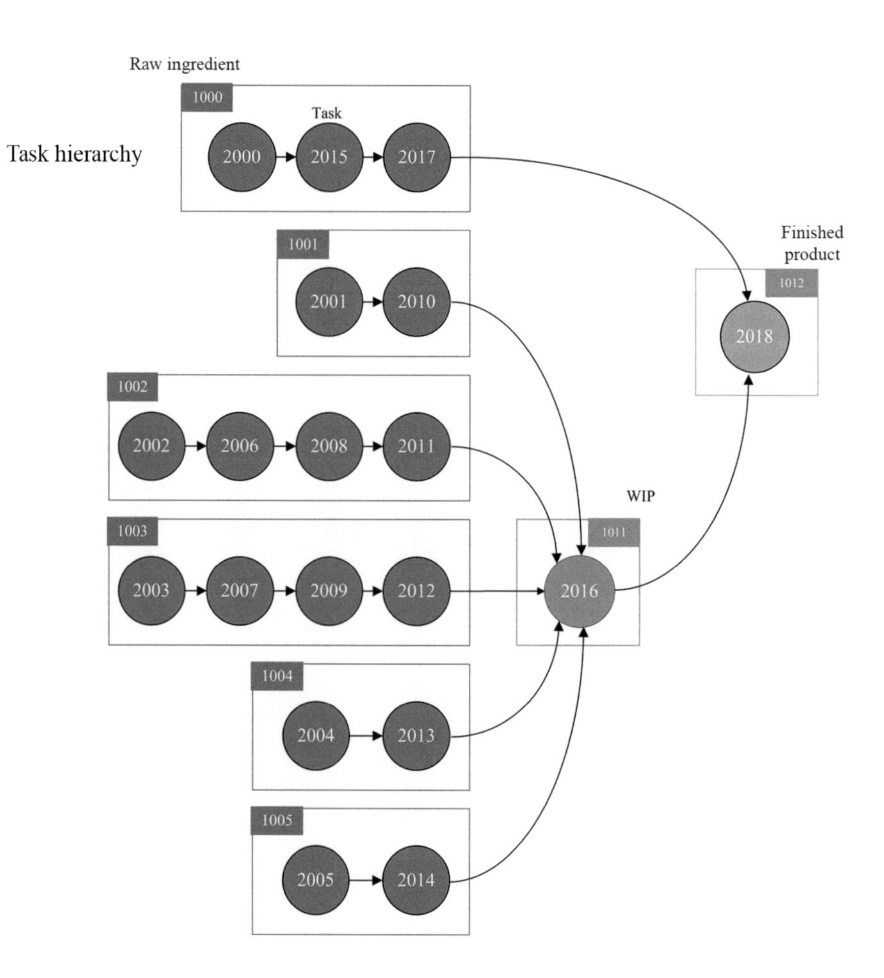

FIG. 6 Tasks priority graph with group predecessors per each task.

4 Discussion

The previous section shows an application of the proposed methodology in the case of recipe coding inspired by a catering company. In such an application, we obtain the typical input of any commercial scheduling tool (e.g., MS Project) starting from the recipe BOMs. This required an upgrading of the data structure, with some new data fields recorded to allow better data manipulation and knowledge building. The proposed case emphasizes the importance of solid datasets and data organization in the company's ERP and information system (Liu et al., 2014), and recognizes data storage and maintenance as the preliminary but necessary activity for any data-driven decision in the catering industry.

An in-depth revision of the caterer information systems, and particularly the data structure used to store the production details (i.e., Entity-Relationship Diagram of the production database), could aid not only in the application of scheduling tools and algorithms, but also the ingredients clusterization and the maintenance of the suppliers register to obtain economies of scale in purchasing activities. Each ingredient is, in fact, a generic raw material that can be purchased from many suppliers. By defining generic types/categories of ingredients (e.g., *potato from vendor A* and *potato from vendor B* belong to the same category of the generic ingredient *potato*), the buyer might choose the supplier each time, to exploit economies of scale without affecting how the recipe is recorded at the information level by the company's ERP.

Tailored data structure could also support industrial engineers in the analysis and rationalization of the working cycle (e.g., avoid redundant phases or tasks), as well as in the accounting of production costs. It would indeed be possible to identify which type of machine is involved in a given task and which is called to process a given ingredient, and to calculate the fractions of the recipe full cost accordingly. Finally, a traditional job-shop/assembling scheduler can be applied to the task-priority graph in order to avoid production delays and balance the utilization of the production resources.

The main contribution from the industrial application of the proposed methodology lies on the identification, through the task-priority graph, of the set of tasks of a recipe that can be performed at the same time instead of being done in sequence. Early, the task-priority graph was not formalized or tracked by any information system and was just kept in mind by the production operators. Thus any optimization or rationalization was limited. The new data structure and the building of the task-priority graphs for all the recipes of the production mix have provided the following benefits:

- Better understanding of the hierarchy of a recipe, such as when the state of the product changes (e.g., from raw ingredient to WIP);
- Opportunities for splitting the working cycle at a certain phase or step, identifying the most convenient breakpoint, and setting the layout (e.g., buffers) accordingly;
- Decoupling the working cycle (e.g., preprocessing of raw materials and cooking when the order is confirmed) (Van Kampen and Van Donk, 2014);
- Analyzing when (e.g., for which recipe) the production of WIPs, stored into refrigerated chambers until an order occurs, is convenient;
- Selecting which recipes should be devoted to *cook-warm* or *cook-chill* temperature profiles;
- Studying the variability of the customer orders and predicting their impact on congestions and resource utilization throughout the production process in agreement with the logistic and distribution capabilities (Accorsi et al., 2018).

With regard to these benefits, the proposed methodology paves the way for an increasing modularity of the food catering production systems.

5 Conclusions

This chapter explores the role of recipes in the design and control of catering production systems, with a particular focus on centralized kitchens (CEKI). The complexity of such production lies in the wide production mix, which includes hundreds of recipes, and on the order profile, which counts thousands of meals daily delivered to schools, hospitals, and private companies in a very short time segment. Many issues affect this type of production, including the compliance with food safety and quality standards, the reduction of delays, and the organization and control of production and delivery operations.

This chapter first provides a brief discussion of main factors affecting the management of a CEKI and the description of a renowned recipe-driven layout support-design method.

Then, it introduces new data structures for the collection and organization of a production key details study, which supports knowledge building upon recipes and BOMs and the implementation of production scheduling tools and software. An algorithm is proposed and used to build a task-priority graph and a task hierarchy from the records of the recipe's BOM. The algorithm provides the list of tasks that can be performed at the same time, thus supporting the optimization of the recipe's working cycle. Indeed, the output of this algorithm could directly fuel commercial scheduling tools (e.g., MS Project, Lekin) intended for the optimization of production activities, the balancing of resource utilization, and the reduction of delays and queues.

Acknowledgments

The authors would like to heartily thank the company CAMST Soc. Coop. a r.l.—La Ristorazione Italiana, which was deeply involved in this applied research. Special thanks are due to Dr. Andrea Lenzi, Dr. Enrico Stoppa, Mauro Fabbri, Dr. Simone Gozzi, and Eng. Daniele Capelli for their valuable inputs, support, and willingness to cooperate in this research project.

References

Accorsi, R., Baruffaldi, G., Manzini, R., Tufano, A., 2018. On the design of cooperative vendors' networks in retail food supply chains: a logistics-driven approach. Int. J. Logist. Res. Appl. 21 (1), 35–52. https://doi.org/10.1080/13675567.2017.1354978.

Adler-Nissen, J., Akkerman, R., Frosch, S., Grunow, M., Løje, H., Risum, J., Wang, Y., Ørnholt-Johansson, G., 2013. Improving the supply chain and food quality of professionally prepared meals. Trends Food Sci. Technol. https://doi.org/10.1016/j.tifs.2012.08.007.

Akkerman, R., Van Donk, D.P., 2009. Analyzing scheduling in the food-processing industry: structure and tasks. Cognit. Technol. Work 11 (3), 215–226. https://doi.org/10.1007/s10111-007-0107-7.

Akkerman, R., Van Donk, D.P., 2007. Product prioritization in a two-stage food production system with intermediate storage. Int. J. Prod. Econ. 108, 43–53. https://doi.org/10.1016/j.ijpe.2006.12.018.

Ciappellano, Salvatore. 2009. Manuale Della Ristorazione.

Devapriya, P., Ferrell, W., Geismar, N., 2017. Integrated production and distribution scheduling with a perishable product. Eur. J. Oper. Res. 259 (3), 906–916. https://doi.org/10.1016/j.ejor.2016.09.019.

Drira, A., Pierreval, H., Hajri-Gabouj, S., 2007. Facility layout problems: a survey. Annu. Rev. Control. 31 (2), 255–267. https://doi.org/10.1016/j.arcontrol.2007.04.001.

Europe Food Service, 2017. European industry overview. http://www.ferco-catering.org/en/european-industry-overview.

European Parliament. 2004. Regolamento (CE) N. 852/2004 del parlamento europeo sull'igiene dei prodotti alimentari.

Farahani, P., M. Grunow, and H. O. Günther. 2009. "A heuristic approach for short-term operations planning in a catering company." In IEEE International Conference on Industrial Engineering and Engineering Management, 1131–35. https://doi.org/10.1109/IEEM.2009.5372966.

Fusi, A., Guidetti, R., Azapagic, A., 2015. Evaluation of environmental impacts in the catering sector: the case of pasta. J. Clean. Prod. 132, 146–160. https://doi.org/10.1016/j.jclepro.2015.07.074.

Guley, H.M., Stinson, J.P., 1984. Scheduling and resource allocation in a food service system. J. Oper. Manag. 4 (2), 129–144. https://doi.org/10.1016/0272-6963(84)90028-7.

Hollier, R.H., 1983. Automated Material Handling. Springer, p. 1983.

Houba, I.H.G., Hartog, R.J.M., Top, J.L., Beulens, A.J.M., Van Berkel, L.N., 2000. Using recipe classes for supporting detailed planning in food industry: a case study. Eur. J. Oper. Res. 122 (2), 367–373. https://doi.org/10.1016/S0377-2217(99)00239-8.

Van Kampen, T., Van Donk, D.P., 2014. Coping with product variety in the food processing industry: The effect of form postponement. Int. J. Prod. Res. 52, 353–367. https://doi.org/10.1080/00207543.2013.825741.

Liu, Y., Wang, D., Len, X., Xiong, Y., 2014. Study on frozen food Industry's integrated information management strategy based on ERP. Adv. J. Food Sci. Technol. 6 (11), 1249–1254. https://doi.org/10.19026/ajfst.6.191.

Manzini, R., Accorsi, R., Ayyad, Z., Bendini, A., Bortolini, M., Gamberi, M., Valli, E., Gallina Toschi, T., 2014. Sustainability and quality in the food supply chain. A case study of shipment of edible oils. Br. Food J. 116 (12), 2069–2090. https://doi.org/10.1108/BFJ-11-2013-0338.

Manzini, R., Accorsi, R., Piana, F., Regattieri, A., 2017. Accelerated life testing for packaging decisions in the edible oils distribution. Food Pack. Shelf Life 12, 114–127. https://doi.org/10.1016/j.fpsl.2017.03.002.

Meinecke, C., Reiter, B.S., 2014. A representation scheme for integrated production and outbound distribution models. Int. J. Logist. Syst. Manag. 18. https://doi.org/10.1504/IJLSM.2014.062817.

Ministero della Salute, 2012. Manuale Di Corretta Prassi Operativa per La Ristorazione Collettiva.

Penazzi, S., Accorsi, R., Ferrari, E., Manzini, R., Dunstall, S., 2017. On the design and control of food job-shop processing systems. A simulation analysis in catering industry. Int. J. Logist. Manag. 28 (3), 782–797.

Tufano, A., Accorsi, R., Garbellini, F., Manzini, R., 2018. Plant design and control in food service industry. A multidisciplinary decision-support system. Comput. Ind. 103, 72–85.

Williams, P.G., 1996. Vitamin retention in cook/chill and cook/hot-hold hospital foodservices. J. Am. Diet. Assoc. 96 (5), 490–498. https://doi.org/10.1016/S0002-8223(96)00135-6.

Index

Note: Page numbers followed by *f* indicate figures, *t* indicate tables, and *b* indicate boxes.

Printed in France by Amazon
Brétigny-sur-Orge, FR